The Repair of Vehicle Bodies

This revised and updated seventh edition of the best-selling reference manual on vehicle body repair brings the book up to date for the current body repair trade. It serves as a comprehensive guide covering the vocationally related qualification (VRQ) required by the modern student and apprentice, as well as providing the CPD essential for all working professionals.

The entire book is overhauled to reflect current industry trends with regards to materials, processes and procedures. New additions include:

- An entirely new section on the work of the MET technician (mechanical, electrical and trim)
- New developments in body repair methodology such as repair pods and the greater use of alignment equipment
- Greater emphasis on the environment with new sections on hybrid vehicles and the hazards of starting current vehicles with high levels of technology
- Details on both the historic and the current joining methods for the vintage and modern markets
- Full coverage on the legalities surrounding insurance work for bodyshop staff
- Updated tables and illustrations

This book not only provides the knowledge and skills for body repair, it helps to develop a real understanding of the how and why behind this information. It will be essential for anyone studying Levels 1–3 Vehicle Body Repair, Vehicle Refinishing and MET courses, including the new apprenticeships and technical certificates from the IMI, Pearson-BTEC and C&G. HNC and degree Automotive Engineering students will find the text valuable to develop skills and knowledge for practical project work. Industry professionals, vehicle restorers and car DIY enthusiasts will continue to find it an essential and comprehensive source of information.

Andrew Livesey served a full apprenticeship in the motor industry before going to the University of Manchester as a mature student: he is a Fellow of the Institute of the Motor Industry and Member of the Institution of Mechanical Engineers. He also runs an independent Engineering Consultancy in Kent and teaches part time. He is former Head of Motorsport at Oxford Brookes University, has worked for McLaren and has run a company building race and rally cars.

The late **Alan Robinson** was formerly Section Leader for Vehicle Bodywork and Vehicle Body Repair, as well as Course Coordinator for CGLI 398 Vehicle Body Competencies and CGLI 385 Vehicle Bodywork at Gateshead College.

The Repair of Vehicle Bodies

Seventh Edition

ANDREW LIVESEY

Routledge
Taylor & Francis Group

LONDON AND NEW YORK

Seventh edition published 2019
by Routledge
2 Park Square, Milton Park, Abingdon, Oxon, OX14 4RN

and by Routledge
711 Third Avenue, New York, NY 10017

Routledge is an imprint of the Taylor & Francis Group, an informa business

First edition published by Heinemann Educational Books Ltd 1973
Sixth edition published by Routledge 2013

NOTICE

No responsibility is assumed by the publisher for any injury and/or damage to persons or property as a matter of products liability, negligence or otherwise, or from any use or operation of any methods, products, instructions or ideas contained in the material herein. Because of rapid advances in the medical sciences, in particular, independent verification of diagnoses and drug dosages should be made.

British Library Cataloguing-in-Publication Data
A catalogue record for this book is available from the British Library

Library of Congress Cataloging-in-Publication Data
Names: Livesey, Andrew, author.
Title: Repair of vehicle bodies / Andrew Livesey.
Description: 7th edition. | Boca Raton : Routledge, 2018. | Earlier editions by Alan Robinson. | Includes bibliographical references and index.
Identifiers: LCCN 2018007662 (print) | LCCN 2018012006 (ebook) | ISBN 9781351230643 (Adobe PDF) | ISBN 9781351230636 (ePub) | ISBN 9781351230629 (Mobipocket) | ISBN 9780815378709 (hardback : alk. paper) | ISBN 9780815378693 (pbk. : alk. paper) | ISBN 9781351230650 (ebook : alk. paper)
Subjects: LCSH: Automobiles—Bodies—Maintenance and repair.
Classification: LCC TL255 (ebook) | LCC TL255 .R56 2018 (print) | DDC 629.2/60288—dc23
LC record available at https://lccn.loc.gov/2018007662

ISBN: 978-0-815-37870-9 (hbk)
ISBN: 978-0-815-37869-3 (pbk)
ISBN: 978-1-351-23065-0 (ebk)

Typeset in Times New Roman
by Apex CoVantage, LLC

MIX
Paper from
responsible sources
FSC
www.fsc.org   FSC® C013056

Printed and bound in Great Britain by
TJ International Ltd, Padstow, Cornwall

To my three favourite teachers at Barden Boys School in Burnley. Stuart Burrows – he introduced me to the work of Robert Thompson; Alan Towers – he showed me how to modify my bike; Alfred Lawson – he helped me write my first book. These great teachers started my quest to learn practical skills and write about them.

Contents

Preface

This book has been revised to reflect the on-going changes in the vehicle body repair industry and the developments in technical qualifications and apprenticeships including the new T Level qualification of the Department for Education and Trailblazer apprenticeships being developed by the Institute for Apprenticeships. The author has responded to changes suggested by readers including more information related to MET and detailing. The author remains conscious that this text is also used by mature readers and learners in countries other than the UK and has ensured that the classical craft skills sections remain thorough and detailed.

I hope that you get pleasure reading it and that it helps you develop both the skills and knowledge needed by both industry professionals and vehicle restorers alike. I am always open to new suggestions for the next edition.

Andrew Livesey
MA CEng MIMechE FIMI

Car body construction

Starting out

It's worth starting out by asking the question why do we have car bodies in the first place? Then moving on to why are they made like they are? This will help you to understand more about your job. So, let's have a detailed look at vehicle bodies in a general sense.

We now have more variations in designs of car bodywork than we have ever had; we'll come back to that later in this chapter. When the early car designers, probably better known as inventers and engineers back in the day, wanted to get their cars on the road they needed to do it quickly and cheaply so that they could sell their products and thereby make a living – like us in the body repair industry now. So, about 120 years ago, with no online auction sites, in fact not even a telephone, they sourced their materials locally. All towns and villages had blacksmiths, and bigger towns had horse-drawn cart and carriage makers; these were the source of parts. The early cars were made by little more than attaching a petrol engine to what would have been a horse-drawn carriage.

The early cars had the engine underneath the carriage, meaning that the body and its passengers were high off the road, giving a high centre of gravity (CG). This meant that the cars were unstable and liable to roll over if the driver tried to take a corner too quickly.

As the number of cars started to grow, other engineers started to research and design separate parts of the car. These engineers made money by patenting their designs and selling the right to use their designs to car manufacturers or making the parts for manufactures to use. You probably have heard of some of them, such as Ackermann, Lucas, Borg Warner, Bosch, Girling and Dunlop.

Obviously, a petrol-engined carriage would not suit everybody. What's more, the cost of them was horrendous. An early car cost about £2,000 when the wage of a skilled worker was about £1 per week. Yes, 40 years to work to pay for it. So lots of variations started to appear: engines and gearboxes and transmissions in different locations and, more important to us, lots of different body types.

Vehicle construction techniques tend to follow a number of similar patterns from the variety of manufacturers. In this chapter we will discuss some of the general principles involved in how cars and light commercial vehicles are made. You will realize that many vehicles are sold with different names, as though made by different firms, which are almost identical apart from their badges and levels of trim or options. This is referred to as badge engineering; it has been used since the dawn of motoring. The reason for this is that it offers economy of scale in manufacturing and greater customer choice – it is win–win for us all. For the body repairer or vehicle restorer, it means that some parts, for example a sill panel, may fit many different makes and models of vehicles. These parts are also frequently available from independent parts dealers at lower prices than through dealerships

Vehicle safety testing

EuroNCAP – one of the main reasons why cars are designed and made as they are is to make them as safe as possible. With smaller cars, this must be done within a tight budget, as cars must be priced to fit within what are referred to as price points. That is, if you look at the motoring magazines and websites, you will see that each of the vehicle manufacturers makes cars which are very similar in performance and specification at very similar prices. They tend to be fairly equal when it comes to safety of the driver, the passengers and other road users. EuroNCAP is an abbreviation for **European New Car Assessment Programme** and assesses the vehicle safety using a number of standard tests.

- *Frontal impact test* – the test for frontal impact is carried out at 64 kph (40 mph). The car strikes a deformable barrier which is offset from the centre. This is similar to many road accidents in which one car strikes another during an overtaking accident or strikes another object like a road bollard.
- *Side impact test* – this is similar to many accidents in which one car is hit in the side by another – often referred to as 'T-boning' after the shape of a piece of steak. The deformable trolley is towed

Figure 1.1a Frontal impact test

Figure 1.1b Protection rating colour coding

Figure 1.1c Side impact test

Figure 1.1d Pole test

Figure 1.1e   Pedestrian impact tests

Figure 1.1f   Typical drivers

on rails into the driver's side of the car at 50 kph (30 mph), typical of accidents at road junctions.

- *Pole test* – about a quarter of all serious or fatal accidents happen because of side impacts. The car being tested is propelled sideways on a trolley at 29 kph (18 mph) into a rigid pole. The pole is similar in size to a telegraph pole or lamppost. The main reason for this test is to assess head injury to the driver. The roof strength and operation of side airbags are important to prevent head injury.
- *Pedestrian impact tests* – a series of tests are carried out to replicate accidents involving both child and adult pedestrian impact at 40 kph (25 mph). This speed is now being adopted as a maximum in high-density build-up areas to reduce pedestrian deaths. This series of tests has led to vehicles having smooth rounded fronts with the absence of bonnet badges or other protruding features.

- *Star ratings* – each vehicle which is tested is given a star rating for its protection of adult occupant, child occupant and pedestrians. The star ratings are made up from points scoring against each test. The ratings of current vehicles can be found on the EuroNCAP website www.euroncap.com.

Elements of vehicle construction

Front end construction – the main purpose of the front end is to locate the engine, the gearbox and the front suspension. The type of construction depends on the drive layout of the vehicle. If the vehicle is of conventional layout – that is, front-mounted engine driving the rear wheels – then it is usually of **integral (or mono) construction** with the passenger compartment. With this type of construction, the chassis rails will be similar at the rear of the vehicle to locate the rear axle and final drive. When being constructed in the factory, the engine and the gearbox are usually lowered into the body assembly. With the sub-frame type of construction, the engine and the gearbox are mounted onto the sub-frame and offered up to the body during manufacture. It is important that the front end of the vehicle is correctly **aligned** to ensure that the vehicle runs straight and true.

Floor pan – this is the main strength of the passenger compartment. The ribs and pressed-out shapes are there to make the floor pan stiff and strong. This type of

Figure 1.2   Front end construction (a) integral or mono (b) composite and (c) front and rear subframes, these are bolted separately to the body assembly

construction was invented by Porsche for the first VW in the 1930s to make it light and simple to construct. The floor pan must be strong enough to locate the seats and seat belts as well as connect the front of the car to the rear. The floor pan is usually made of thicker metal than other body panels. Various cross members and the sills are attached to the floor pan to give the passenger compartment extra strength. The floor pan must be resistant to twisting.

Body side and roof – this is assembled onto the floor pan to complete the passenger compartment. Each of these panels is pressed out and welded together to form a tough cell-like structure for the driver and passengers.

Vehicle type approval – Many industrial sectors are subject to some form of approval or certification system, but road vehicles are a special case because of their importance to and impact upon society and have

Figure 1.3 Main floor assemblies and boot floor assemblies *(Proton)*

1 Reinforced parking brake lever
2 Cross member assembly backbone
3 Reinforced assembly backbone
4 Bracket A-Frame LH
5 Bracket A-Frame RH
6 Pan front floor
7 Cross member assembly front floor front RH
8 Cross member assembly front floor rear RH
9 Cross member assembly front floor rear LH
10 Cross member assembly front floor front LH
11 Reinforced seat belt side LH/RH
12 Bracket anti zipper
13 Sill front floor side inner LH/RH
14 Reinforced side member front floor LH/RH
15 Side member front floor LH/RH

NOTE:
A large **reinforcement** has been added to the front floor backbone (reinforcement assembly backbone), and this is coupled with the side sill, side member and cross member to provide increased rigidity to the total floor.

been subject to specific technical standards almost from their first invention. Within Europe, two systems of type approval have been in existence for many years. One is based around EC Directives and provides for the approval of whole vehicles, vehicle systems and separate components. The other is based around UNECE (United Nations Economic Commission) Regulations and provides for approval of vehicle systems and separate components but not whole vehicles.

Figure 1.4   Body side assemblies, roof, BC-post, front and rear door of a hatchback *(Proton)*

1	Panel outer rear door	18	Reinforced pillar front inner lower
2	Panel inner rear door	19	Pillar front inner lower
3	Panel outer front door	20	Side member front
4	Panel inner rear door	21	Panel front fender
5	Panel hood	22	Pan front floor
6	Panel cowl top inner	23	Reinforced front pillar lower
7	Panel cowl top outer	24	Extension upper frame outer
8	Panel assembly dash	25	Side structure
9	Reinforced radiator side RH	26	Reinforced front pillar centre
10	Stay hood lock	27	Pillar front inner upper
11	Cross member front end	28	Rail roof side inner
12	Reinforced radiator side LH	29	Rail roof front
13	Bar front end upper	30	Pillar centre inner
14	Bulkhead front pillar lower	31	Pillar centre outer
15	Bulkhead front pillar side sill	32	Pillar rear inner
16	Bracket cross member front	33	Rail roof rear
17	Bracket cross member	34	Panel roof

Type approval is the confirmation that production samples of a design will meet specified performance standards. The specification of the product is recorded, and only that specification is approved.

Automotive EC Directives and UNECE Regulations require third-party approval – testing, certification and production conformity assessment by an independent body. Each member state is required to appoint an Approval Authority to issue the approvals and a Technical Service to carry out the testing to the Directives and Regulations. An approval issued by one Authority will be accepted in all the member states. Vehicle Certification Agency (VCA) is the designated UK Approval Authority and Technical Service for all type of approvals to automotive EC Directives and ECE Regulations. The VCA also has offices in Europe, North America and the Far East.

Approved parts carry the E-mark. That is a letter 'e' or 'CE' followed by a number that indicates the country of approval.

Vehicle information code plate

Location – the vehicle information code plate is usually riveted on the scuttle inside the engine compartment. It has important information about the vehicle which is often needed for repair processes.

The plate shows model code, engine model, transmission model and colour code.

Figure 1.5 Location of vehicle information code plate

Figure 1.6 Code plate descriptions

Position	Description
A	VEHICLE IDENTIFICATION NUMBER (VIN)
B	ENGINE
C	TRANS AXLE
D	EXTERIOR CODE
E	INTERIOR CODE
F	OPTION CODE
G	GROSS VEHICLE WEIGHT
H	GROSS COMBINATION WEIGHT
I	FRONT AXLE WEIGHT
J	REAR AXLE WEIGHT

Brief history of the motor vehicle

The development of the motor vehicle fits into a timeline stretching over 300 years. The vehicle developed alongside other products and technology. Just as you might go to a trade show and buy a new tool, then discover that it'll work for another purpose, car designers do the same. A recent example is LED lighting used in many cars; a few years ago, it came to our notice by being used for decorative lights in the house. Somebody had the idea to make them more powerful – we now have lots of use for LED lights – and they have the advantage of being very efficient compared to tungsten bulbs. This timeline lists some of the important events which have brought us to have the current range of vehicles.

1688   Ferdinand Verbiest, a missionary in China, makes a model steam vehicle using the steam turbine principle.

1740   Jacques de Vaucansen makes a clockwork carriage in Paris.

1765   Scottish scientist and engineer James Watt develops the steam engine.

1769   Louis XV of France funds Nicholas Cugnot to build a steam-powered tractor for military use. It carries four people at just over 2 mph for ten minutes.

1771   Cugnot crashes the steam tractor into a wall – the first motoring accident. Cugnot was put into prison for this.

1784   James Watt patents a steam carriage but does not build it. William Murdock builds a model steam vehicle in Redruth, Cornwall – it works.

1801   Lots of steam engines are being installed in tin mines in Cornwall – some of these are featured in the TV film *Poldark*. Another Cornish man, Dick Trevithick, builds a steam carriage and drives it up Cambourne Beacon. Two years later this is in use in London for transporting passengers.

1817   The bicycle is invented – but it hasn't pedals, it is like a full-size version of current children's toy the balance bike.

1824   Lots of steam carriages are built and used in London and other parts of England.

1845   Pneumatic tyre is invented.

1859   Oil is discovered in the United States.

1860   Gas engine invented which is used in many factories for power instead of steam – it is quicker to start up.

1861   Michaux fits pedalling mechanism to the bicycle – a name still associated with cycle parts.

1876   Otto makes 4-stroke petrol engine.

1885   Karl Benz produces his first car – this is the first car as we know it.

As you can see it took 200 years from concept to production of the first useable car. As the year 1900 approached, truck and van bodies started to appear. Prior to this, the of cost manufacture was so high that only very rich people could afford them – and they were very much curiosities or toys, like hyper-sports-cars are now. A rolling chassis was made to which a car, a van or a truck body could be fitted. The bodies were made by body-building companies using a mixture of wood, metal and fabric as had been horse-drawn carriage bodies – these were in fact the same companies responding to the new market. Customers could order bodies built to order to fit onto a rolling chassis. Larger cars had their chassis extended to accept limousine bodies – rather like the extended limos used by rock stars and others for fun evenings out.

The First World War led to car chassis being adapted to carry big machine guns and the design of battle tanks and other armoured vehicles. After WWI there was a great increase in car manufacturers – more than 2,000 individual car manufacturers were in business in the UK in 1922. This was responding to the increased desire to own a car – manufacturers looked at ways to make them cheaper – which usually also meant lighter. Cars like Ford, Austin and Morris became available for £100 brand new – but as now, you could spend the same amount of money on accessories – accessories in 1930 included lights and bumpers.

This led to two very significant advances in body design and manufacture. Standard was the first make of car to have 'standardized' parts – parts could be interchanged between models – so vehicle parts departments – then called stores – could stock easy-to-fit parts. The other change was the building of cars with unitary construction – where the body formed the load-bearing member, so no need for a separate chassis – the Morris Minor produced from the 1940s to the 1960s is a classic example of this technique.

At the end of the 1950s was the introduction of what we now call the classic Mini by Austin and Morris – the same car with different badges. It had a mixed reception until a member of the British Royal Family used one for fun. The body innovation significance of the Mini was in two parts. One significance was that it used two sub-frames – the front sub-frame located the engine and steering, the rear sub-frame just the rear wheels and suspension. This construction allowed the use of cheaper mass-production components and the interchangeability of different engines and gearboxes. The other significance was to have the minimum of interior trim, giving the maximum interior space for such a small exterior size. In other words, this is an example of excellent design, giving minimum weight and maximum carrying capacity. The Mini was also the cheapest car available – its target market was the motorcyclist commuter, a market that it took by storm.

The next big change in body design was the introduction of the hatchback, now almost universal in current small cars. The secret of the design was to use the car roof as a structural load-bearing member so that the equivalent of a door could be cut at the rear. So we use the terms "three-door' and 'five-door cars'. The first car to do this was the 1958 Austin A40 Mk2, but the rear door was split horizontally into two parts. The 1974 VW Golf, quickly followed by the Ford Fiesta, had a one-piece top-hinged door – as is almost universal on current hatchbacks.

The classic Mini morphed into the current BMW Mini in 2000. The change with this model was the level of specification of trim and electronics in such a small outer shell. Where the classic Mini was a showcase for minimalism and utility, the new Mini is a showcase for comfort and functionality. Since its introduction, it has grown in size and model range. In this time period many vehicle models have also grown but kept the same badge and model name range.

In 1970, the Range Rover was introduced. This combined 4 × 4 off-road capability with a luxury interior. This was a bodywork concept to be followed by many other manufacturers leading to the term 'sport utility vehicle' – SUV. The Range Rover was followed by others of different sizes, such as intermediate sports utility vehicles – ISUV – and smaller ones designated SSUV. Now the concept is applied to two-wheel-drive versions with the same bodywork.

To some extent the design of vehicles has gone full circle. The rolling chassis has become a floor pan – that is, a series of structural members with engine, transmission and suspension may be fitted with a full range of bodies. The current BMW 3 Series is a good example. It is available in convertible, coupe, saloon, estate and SUV forms. All use the same platform, the name used for the rolling chassis or floor pan assembly. In addition, different platforms are available for different sizes of engines within the same range. This concept is not new; it was first used by Ferdinand Porsche in the design of the 1930s VW beetle, then applied to make the first Porsche sportscars and a whole series of one-off vehicles. The use of this concept is largely due to the developments in design and manufacturing. The introduction of computer-aided (CAD) and associated computer-aided manufacturing (CAM) means that new designs can be made in minutes, and production lines are not linear but are based on cells that can be reconfigured quickly to move model production to suit design requirements.

Customers may now order their new car online choosing platform, body, colour and trim. This translates into a manufacturing specification and can all be carried out without speaking to a person.

Skills and questions

As well as knowledge about vehicle construction, you need to develop a range of skills, which will depend on the type of vehicle that you are working on and your previous skill level. The following tasks and questions will help you to develop your skills and understanding of vehicle construction:

1 Identify the construction of two different vehicles, making notes and sketches of the vehicle layout and how the major components are attached to the passenger compartment.

2 Obtain a body parts manual for one of the vehicles you work on and make sketches to show how the panels are attached to each other.

3 Look at two different variants of the same vehicle, for example saloon and convertible or saloon and light commercial, and find how many panels are interchangeable.

4 Remove and refit any major components you get the opportunity to do.

5 Describe how the body code and the trim code are identified from the code plate on a vehicle which you work on – write down the details.

2

Health and safety and the environment

Health and safety and the environment are controlled by a number of Acts of Parliament and subsequent regulations and statutory instruments. These may have regional variations or specifics relating to the various parts of the vehicle repair trade. These topics are also impacted by other laws, for instance those relating to the country side. Perhaps the most important aspect of the health and safety laws is that they are vicarious. That is, if you are a manager or supervisor of a bodyshop, you may be prosecuted for any injury or death caused by one of the technicians or other staff actions, as well as the staff involved being prosecuted. Breaking health and safety or environmental laws may result in custodial sentences as well as fines and damages to the injured parties. Meeting the requirements of the health and safety and environmental laws and regulations is the responsibility of everybody. Ignorance of the law is not an excuse, so you need to take care

Personal health and safety procedures

Skin care (personal hygiene) systems All employees should be aware of the importance of personal hygiene and should follow correct procedures to clean and protect their skin in order to avoid irritants causing skin infections and dermatitis. All personnel should use a suitable barrier cream before starting work and again when recommencing

work after a break. There are waterless hand cleaners available which will remove heavy dirt on skin prior to thorough washing. When the skin has been washed, after-work creams will help to restore its natural moisture.

Many paints, refinishing chemicals and bodyshop materials will cause irritation on contact with the skin and must be removed promptly with a suitable cleansing material. Paint solvents may cause dermatitis, particularly where skin has been in contact with peroxide hardeners or acid catalysts: these have a drying effect which removes the natural oils in the skin. There are specialist products available for the bodyshop which will remove these types of materials from the skin quickly, safely and effectively.

Hand protection Body technicians and painters are constantly handling substances which are harmful to health. The harmful effect of liquids, chemicals and materials on the hands can be prevented, in many cases, by wearing the correct type of gloves. To comply with COSHH Regulations, vinyl disposable gloves must be used by painters to give skin protection against toxic substances. Other specialist gloves available are: rubber and PVC gloves for protection against solvents, oil and acids; leather gloves for hard wearing and general repair work in the bodyshop; and welding gauntlets, which are made from specially treated leather and are longer than normal gloves to give adequate protection to the welder's forearms.

The Repair of Vehicle Bodies, Andrew Livesey, 9780815378709, 2019, Taylor & Francis

Protective clothing is worn to protect the worker and his clothes from coming into contact with dirt, extremes of temperature, falling objects and chemical substances. The most common form of protective clothing for the body repairer is the overall, a one-piece boiler suit made from good quality cotton, preferably flame-proof. Worn and torn materials should be avoided, as they can catch in moving machinery. Where it is necessary to protect the skin, closely fitted sleeves should be worn down to the wrist with the cuffs fastened. All overall buttons must be kept fastened, and any loose items such as ties and scarves should not be worn. Protective clothing worn in the paint shop by the spray painters should be either good-quality washable nylon garments, anti-static, and complete with hood, elasticated wrists and ankles; or low-lint disposable coveralls, which offer a liquid barrier protection from splashes, airborne dusts and paint overspray. The coveralls must withstand continuous exposure to a variety of chemicals and must be suitable for protection when using iso-cyanate based two-pack paints; they also prevent the environment being contaminated by particles from the operator's clothing and hair. They can be of the one-piece variety or can have separate disposable hoods.

Head protection is very important to the bodyworker when working underneath a vehicle or under its bonnet while it is being repaired. A light safety helmet, normally made from aluminium, fibreglass or plastic, should be worn if there is any danger from falling objects and will protect the head from damage when working below vehicles. Hats and other forms of fabric headwear keep out dust, dirt and overspray and also prevent long hair (tied back) becoming entangled in moving equipment.

Eye protection is required when there is a possibility of eye injury from flying particles when using a grinder, disc sander, power drill or pneumatic chisel, or when removing glass windscreens or working underneath vehicles. Many employers are now requiring all employees to wear some form of safety glasses when they are in either the repair or the paint areas of the bodyshop, because in any bodyshop location, there is always the possibility of flying objects, dust particles, or splashing liquids entering the eyes. Not only is this painful, but it can, in extreme cases, cause loss of sight. Eyes are irreplaceable: therefore, it is advisable to wear safety goggles, glasses or face shields in all working areas. The following types of eye protection are available:

- *Lightweight safety spectacles* with adjustable arms and with side shields for extra protection. There is a choice of impact grades for the lenses.
- *General-purpose safety goggles* with a moulded PVC frame which is resistant to oils, chemicals and water. These have either a clear acetate or a polycarbonate lens with BS impact grades 1 and 2.
- *Face shields* with an adjustable head harness and deep polycarbonate brow guard with replaceable swivel-up clear or anti-glare polycarbonate visor BS grade 1, which gives protection against sparks, molten metal and chemicals.
- *Welding helmet or welding goggles* with appropriate shaded lens to BS regulations. These must be worn at all times when welding. They will protect the eyes and face from flying molten particles of steel when gas welding and brazing and from the harmful light rays generated by the arc when MIG/MAG, TIG or MMA welding.

Figure 2.1   Lightweight safety glasses

Figure 2.2 Face protection mask

Figure 2.3 Automatic welding mask

Foot protection – Safety footwear is essential in the bodyshop environment. Boots or shoes with steel toecaps will protect the toes from falling objects. Rubber boots will give protection from acids or in wet conditions. Never wear defective footwear, as this becomes a hazard in any workshop environment.

Respiratory protection – lungs – one of the most important hazards faced by the bodyshop worker is that of potential damage to the lungs. Respirators are usually needed in body repair shops even though adequate ventilation is provided for the working areas. During welding, metal or paint preparation or spraying, some form of protection is necessary. Under the COSHH Regulations, respiratory protection is essential and therefore must be used. Respirators give protection against abrasive dusts, gases, vapours from caustic solutions and solvents, and spray mist from undercoats and finishing paint by filtering the contaminated atmosphere before it is inhaled by the wearer. They may be either simple filtering devices, where the operator's lungs are used to draw air through the filter, or powered devices incorporating a battery-driven fan to draw contaminated air through the filters and deliver a flow of clean air to the wearer's face. There are four primary types of respirator available to protect the bodyshop technicians: dust respirators, cartridge filter respirators, powered respirators and constant-flow air line breathing apparatus.

Dust respirators: (masks), the most basic form of respiratory protection is the disposable filtering half-mask, typically used when preparing or finishing bodywork such as by rubbing down or buffing, and where dust, mist and fumes are a problem. This face mask provides an excellent face seal while at the same time allowing the wearer to speak freely without breaking the seal. Breathing resistance is minimal, offering cool and comfortable use. Various types of mask are available for use in a variety of environments where contaminants vary from nuisance dust particles to fine dusts and toxic mists. These masks can only be used in atmospheres containing less than the occupational exposure limit of the contaminant.

Cartridge filter mask: The cartridge filter or organic vapour type of respirator, which covers the nose and mouth, is equipped with a replacement cartridge that removes the organic vapours by chemical absorption. Some of these are also designed with a pre-filter to remove solid particles from the air before the air passes through the chemical cartridge. They are used in finishing operations with non-toxic paints but not with isocyanate paints. For the vapour/particle respirator to function correctly, it is essential that it fits properly against the face. Follow the manufacturer's instructions for changing the cartridges when spraying over a continuous period.

Powered respirators using canister filters offer protection against toxic dusts and gases. The respirator draws contaminated air through filters with a motor fan powered by a rechargeable battery and supplies clean air to the wearer's face. This avoids discomfort and fatigue caused by the effort of having to inhale the air through filters, permitting longer working periods. These devices find great use both in the spray shop and in the repair shop when carrying out welding.

Constant-flow compressed air line breathing apparatus is designed to operate from an industrial compressed air system in conjunction with the spray gun. Using a waist-belt-mounted miniature fixed-pressure regulator and a pre-filter, the equipment supplies breathing-quality air through a small-bore hose to a variety of face masks and visors to provide respiratory protection for paint spraying (such as with isocyanates), cleaning and grinding. The COSHH Regulations have made it mandatory for all respiratory protection equipment to be both approved and suitable for the purpose, for the operatives to be correctly trained in the equipment's use and maintenance, and for proper records to be kept.

Ear protection – the **Noise at Work Regulations** define three action levels for exposure to noise at work:

- A daily personal exposure of up to 80 dB. Where exposure exceeds this level, suitable hearing protection must be provided on request
- A daily personal exposure of up to 85 dB. Above this second level of provision, hearing protection is mandatory.
- A peak sound pressure of 87 dB.

Tech note

Noise is measured in decibels (dB) – this is measured on a scale based on logarithms. That is to say that increases do not follow the normal arithmetic scale in terms of increase in noise. An increase of 3 dB, from say 84 dB to 87 dB, will give a doubling of the noise heard. So cutting the reading by 3 dB will reduce the noise heard by half.

Where the second or third levels are reached, employers must designate ear-protection zones and require all who enter these zones to wear ear protection. Where the third level is exceeded, steps must be taken to reduce noise levels as far as is reasonably practicable. In every case where there is a risk of significant exposure to noise, assessment must be carried out and action taken to minimize hearing damage.

The first two noise action levels relate to exposure over a period (one day) and are intended to cater for the risks of prolonged work in noisy surroundings. The third level is related to sudden-impact noises like those occurring in metalworking procedures.

Fire precautions

The Fire Precautions (workplace) Regulations 1992/1993/1997 replaced and extended the old Fire Precautions Act 1971. These regulations are aligned with standard practice in EC Directives in placing the responsibility for compliance on the employers. They require employers not only to assess risks from fire but to include the preparation of an evacuation plan, to train staff in fire precautions and to keep records. Workplaces with fewer than 20 employees may require emergency lighting points and fire warning systems. The self-employed who do not employ anyone but whose premises are regularly open to the public may only require fire extinguishers and warning signs; they will, however, need to be able to demonstrate that there is a means of escape in case of fire. Where five or more persons work on the premises as employees, all assessments need to be recorded in writing.

Most of these requirements were already covered by existing legislation. There is a detailed requirement for the recording of assessments, the provision of training and the requirement that means of fighting fire, detecting fire and giving warning in case of fire be maintained in good working order.

What is fire? Fire is a chemical reaction called combustion (usually oxidation resulting in the release of heat and light). To initiate and maintain this chemical reaction, or in other words for an outbreak of fire to occur and continue, the following elements are essential:

- *Fuel*: A combination substance, solid, liquid or gas.
- *Oxygen*: Usually air, which contains 21 per cent oxygen.
- *Heat*: The attainment of a certain temperature – once a fire has started it normally maintains its own heat supply.

Methods of extinction of fire. Because three ingredients are necessary for fire to occur, it follows logically that if one or more of these ingredients are removed, fire will be extinguished. Basically three methods are employed to extinguish a fire: removal of heat (cooling); removal of fuel (starving); and removal or limitation of oxygen (blanketing or smothering).

Removal of heat – If the rate of heat generation is less than the rate of dissipation, combustion cannot continue. For example, if cooling water can absorb heat to a point where more heat is being absorbed than generated, the fire will go out.

Removal of fuel – This is not a method that can be applied to fire extinguishers. The subdividing of risks can starve a fire, prevent large losses and enable portable extinguishers to retain control; for example, part of a building may be demolished to provide a fire stop. The following advice can contribute to a company's fire-protection programme:

- What can cause fire in this location, and how can it be prevented?
- If fire starts, regardless of cause, can it spread?
- If so, where to?
- Can anything be divided or moved to prevent such spread?

Removal or limitation of oxygen – It is not necessary to prevent the contact of oxygen with the heated fuel to achieve extinguishment. It will be found that where most flammable liquids are concerned, reducing the oxygen in the air from 21 to 15 per cent or less will extinguish the fire. Combustion becomes impossible even though a considerable proportion of oxygen remains in the atmosphere. This rule applies to most solid fuels, although the degree to which oxygen content must be reduced may vary. Where solid materials are involved, they may continue to burn or smoulder until the oxygen in the air is reduced to 6 per cent. There are also substances which carry within their own structures sufficient oxygen to sustain combustion.

Fire risks in the workshop

Fire risks in the vehicle body repair shop cover all classes of fire: class A is paper, wood and cloth; class B is flammable liquids such as oils, spirits, alcohols, solvents and grease; class C is flammable gases such

Put all three together
to result in

Fuel Heat

Oxygen

Fuel Heat

Oxygen

Fire

Figure 2.4   The fire triangle

as acetylene, propane, butane; and also electrical risks. It is essential that fire is detected and extinguished in the early stages. Workshop staff must know the risks involved and should be aware of the procedures necessary to combat fire. Bodyshop personnel should be aware of the various classes of fire and how they relate to common workshop practice.

Class A fires: wood, paper and cloth

Today wood is not used in cars, although there are exceptions. Cloth materials are used for some main trim items and are therefore a potential fire hazard. The paper used for masking purposes is a prime area of concern. Once it has done its job and is covered in overspray, it is important that it is correctly disposed of, ideally in a metal container with a lid, and not scrunched up and thrown on the floor to form the potential start of a deep-seated fire.

Class B fires: flammable liquids

Flammable liquids are the stock materials used in the trade for all body refinishing processes: gun cleaner to clear finish coats and cellulose for the more modern finishes can all burn and produce acrid smoke.

Class C fires: gases

An increasing percentage of cars run on liquid gas (LPG/LNG), also welding gases or propane space heaters not only burn but can be the source of ignition for A or B fires.

Electrical hazards

Electricity is not of itself a class of fire. It is, however, a potential source of ignition for all of the fire classes mentioned above. The Electricity at Work Regulations cover the care of cables, plugs and wiring. In addition, in the bodyshop, the use of welding and cutting equipment produces sparks, which can,

in the absence of good housekeeping, start a big fire. Training in how to use firefighting equipment can stop a fire in its early stages. Another hazard is the electrical energy present in all car batteries. A short-circuit across the terminals of a battery can produce sufficient energy to form a weld and in turn heating, a prime source of ignition. When tackling a car fire, a fireman will always try to disconnect the battery, as otherwise any attempt to extinguish a fire can result in the re-ignition of flammable vapours.

Body filler is a further possible source of ignition to be aware of as the result of mixing two materials to use as a body filler material. The result of mixing in the wrong proportions can give rise to an exothermic (heat-releasing) reaction; in extreme cases the mix can ignite.

General precautions to reduce fire risk

- Good housekeeping means putting rubbish away rather than letting it accumulate.
- Read the manufacturer's material safety data sheets so that the dangers of flammable liquids are known.
- Only take from the stores sufficient flammable material for the job in hand.
- Materials left over from a specific job should be put back into a labelled container so that not only you but anyone (and this may be a fireman) can tell what the potential risk may be.
- Take care when welding that sparks or burning underbody coating does not cause a problem, especially when working in confined areas of vehicles.
- Be extremely careful when working close to plastic fuel lines.
- Petrol tanks are a potential hazard: supposedly empty tanks may be full of vapour. To give some idea of the potential problem, consider 5 litres (1 gallon) of petrol: it will evaporate into 1 m^3 (35 $feet^3$) of neat vapour, which will mix with air to form 14 m^3 (500 $feet^3$)

of flammable vapour. Thus the average petrol tank needs only a small amount of petrol to give a tank full of vapour.

The keys to fire safety are:

- Take care.
- Think.
- Train staff in the correct procedures before things go wrong.
- Ensure that these procedures are written down, understood and followed by all personnel within the workshop.

Tech note

Any carbon-based material will burn in air if at the temperature needed for combustion – be aware of this with dust in the body or paint shop.

Types of portable fire extinguishers

Water is the most widely used extinguisher agent. With portable extinguishers, a limited quantity of water can be expelled under pressure and its direction controlled by a nozzle.

There are basically two types of water extinguishers. The gas (CO_2) cartridge operated extinguisher, when pierced by a plunger, pressurizes the body of the extinguisher, thus expelling the water and producing a powerful jet capable of rapidly extinguishing class A fires. In stored-pressure extinguishers the main body is constantly under pressure from dry air or nitrogen, and the extinguisher is operated by opening the squeeze grip discharge valve. These extinguishers are available with 6-litre or 9-litre-capacity bodies and thus provide alternatives of weight and accessibility.

Foam is an agent most suitable for dealing with flammable-liquid fires. Foam is produced when a solution of foam liquid and water is expelled under pressure through a foam-making branch pipe, at which point air is entrained, converting the solution into foam. Foam extinguishers can be pressurized either by a CO_2 gas cartridge or by stored pressure. The standard capacities are 6 and 9 litres.

Spray foam

Unlike conventional foams, aqueous film forming foam (AFFF) does not have to be fully aspirated in order to extinguish fires. Spray foam extinguishers expel an AFFF solution in an atomized form which is suitable for use on class A and class B fires. AFFF is a fast and powerful means of tackling a fire and seals the surfaces of the material, preventing re-ignition. The capacity can be 6 or 9 litres, and operation can be by CO_2 cartridge or stored pressure.

Carbon dioxide

Designed specifically to deal with class B, class C and electrical fire risks, these extinguishers deliver a powerful concentration of carbon dioxide gas under great pressure. This not only smothers the fire very rapidly but is also non-toxic and is harmless to most delicate mechanisms and materials.

Dry powder

This type of extinguisher is highly effective against flammable gases, open or running fires involving flammable liquids such as oils, spirits, alcohols, solvents and waxes, and electrical risks. The powder is contained

Figure 2.5 Types of portable fire extinguishers

in the metal body of the extinguisher, from which it is supplied either by a sealed gas cartridge or by dry air or nitrogen stored under pressure in the body of the extinguisher in contact with the powder.

Dry powder extinguishers are usually made in sizes containing 1 to 9 kg of either standard powder or (preferably and more generally) all-purpose powder, which is suitable for mixed-risk areas.

Choosing and siting portable extinguishers

Because there is such a variety of fire risks in bodyshops, it is important to analyse these risks separately and (with the help of experts such as fire officers) to choose the correct firefighting medium to deal with each possible fire situation. It should be noted that portable fire extinguishers are

Class of fire	Water	Foam (AFFF)	CO_2 gas	Powder
A Paper Wood Textile Fabric	✔	✔		✔
B Flammable liquids		✔	✔	✔
C Flammable gases			✔	✔
⚡ Electrical hazards			✔	✔
🚗 Vehicle protection		✔		✔

Figure 2.6   Choice of fire extinguishers

classified as first-aid firefighting and are designed for ease of operation in an emergency. It is important to realize that because they are portable they have only a limited discharge. Therefore their siting, together with an appreciation of their individual characteristics, is fundamental to their success in fighting fire.

Safety signs in the workshop

It is a legal requirement that all safety signs used in a bodyshop comply with BS 5378: Part 1. Each of these signs is a combination of colour and design, within which the symbol is inserted. If additional information is required, supplementary text may be used in conjunction with the relevant symbol, provided that it does not interfere with the symbol. The text can be in an oblong or square box of the same colour as the sign, with the text in the relevant contrasting colour, or the box can be white and the text black. BS 5378 divides signs into four categories:

- *Prohibition* signs have a red circular outline and crossbar running from top left to bottom right on a white background. The symbol displayed on the sign must be black and placed centrally on the background, without obliterating the crossbar. The colour red is associated with 'stop' or 'do not'.
- *Warning* signs have a yellow triangle with a black outline. The symbol or text used on the sign must be black and placed centrally on the background. This combination of black and yellow identifies caution.
- *Mandatory* signs have a blue circular back ground. The symbol or text used must be white and placed centrally on the background. Mandatory signs indicate that a specific course of action is to be taken.
- *Safe condition* signs provide information for a particular facility and have a green square or rectangular background to

Key to British and European Standard safety signs

| Prohibition | Warning | Safe condition | Mandatory |
| Don't do | Risk of danger | The safe way | Must do |

Figure 2.7 Standard safety signs

accommodate the symbol or text, which must be in white. The safety colour green indicates 'access' or 'permission'.

Fire safety signs are specified by BS 5499, which gives the characteristics of signs for fire equipment, precautions and means of escape in case of fire. It uses the basic framework concerning safety colours and design adopted by BS 5378.

General safety precautions in the workshop

The Health and Safety Act is designed to ensure that:

- Employers provide a safe working environment with safety equipment and appropriate training
- Employees work in a safe manner using the equipment provided and follow the guidance and training which is provided
- Customers and others entering any business premises are safe and protected

The following section looks at some of the details related to health and safety. In all cases you should ask yourself the following questions:

- Are there any regulations relating to this, what are they, and what do I need to do to follow them?
- What is the risk involved, and how can I remove or reduce the risk?
- Is any documentation needed?

> **Remember**
> It is all about keeping yourself, your colleagues and your customers safe, as you would want them to keep you safe too.

Particular hazards may be encountered in the bodyshop, and safety precautions associated with them are as follows:

1 Do wash before eating, drinking or using toilet facilities to avoid transferring the residues of sealers, pigments, solvents, filings of steel, lead and other metals from the hands to the inner parts and other sensitive areas of the body.
2 Do not use kerosene, thinners or solvents to wash the skin. They remove the skin's natural protective oils and can cause dryness and irritation or have serious toxic effects.
3 Do not overuse waterless hand cleaners, soaps or detergents, as they can remove the skin's protective barrier oils.
4 Always use barrier cream to protect the hands, especially against fuels, oils, greases, hydrocarbon solvents and solvent-based sealers.
5 Do follow work practices that minimize the contact of exposed skin and the length of time liquids or substances stay on the skin.
6 Do thoroughly wash contaminants such as used engine oil from the skin as soon as possible with soap and water. A

waterless hand cleaner can be used when soap and water are not available. Always apply skin cream after using waterless hand cleaner.

7 Do not put contaminated or oily rags in pockets or tuck them under a belt, as this can cause continuous skin contact.

8 Do not dispose of dangerous fluids by pouring them on the ground or down drains or sewers.

9 Do not continue to wear overalls which have become badly soiled or which have acid, oil, grease, fuel or toxic solvents spilt over them. The effect of prolonged contact from heavily soiled overalls with the skin can be cumulative and life threatening. If the soilants are or become flammable from the effect of body temperature, a spark from welding or grinding could envelop the wearer in flames, with disastrous consequences.

10 Do not clean dusty overalls with an air line: it is more likely to blow the dust into the skin, with possible serious or even fatal results.

11 Do wash contaminated or oily clothing before wearing it again.

12 Do discard contaminated shoes.

13 Wear only shoes which afford adequate protection to the feet from the effect of dropping tools and sharp and/or heavy objects on them, and also from red-hot and burning materials. Sharp or hot objects could easily penetrate unsuitable footwear such as canvas plimsolls or trainers. The soles of the shoes should also be maintained in good condition to guard against upward penetration by sharp or hot pieces of metal.

14 Ensure gloves are free from holes and are clean on the inside. Always wear them when handling materials of a hazardous or toxic nature.

15 Keep goggles clean and in good condition. The front of the glasses or eyepieces can become obscured by welding spatter adhering to them. Renew the glass or goggles as necessary. Never use goggles with cracked glasses.

16 Always wear goggles when using a bench grindstone or portable grinders, disc sanders, power saws and chisels.

17 When welding, always wear adequate eye protection for the process being used. MIG/MAG welding is particularly high in ultraviolet radiation, which can seriously affect the eyes.

18 Glasses, when worn, should have 'safety' or 'splinter-proof' glass or plastic lenses.

19 Always keep a suitable mask for use when dry flatting or working in dusty environments and when spraying adhesive, sealers, solvent-carried waxes and paints.

20 In particularly hostile environments such as when using volatile solvents or isocyanate materials, respirators or fresh air–fed masks must be worn.

21 Electric shock can result from the use of faulty and poorly maintained electrical equipment or misuse of equipment. All electrical equipment must be frequently checked and maintained in good condition. Flexes, cables and plugs must not be frayed, cracked, cut or damaged in any way. Equipment must be protected by the correctly rated fuse.

22 Use low-voltage equipment wherever possible (110 volts).

23 In case of electric shock:
   (a) Avoid physical contact with the victim.
   (b) Switch off the electricity.
   (c) If this is not possible, drag or push the victim away from the source of the electricity using non-conductive material.
   (d) Commence resuscitation if trained to do so.
   (e) Summon medical assistance as soon as possible.

Electrical hazards

The Electricity at Work Act 1989 fully covers the responsibilities of both the employee and the employer. As a body repairer, you are obliged to follow these regulations for the protection of yourself and your colleagues. Some of the important points to be aware of are given below.

Voltages – the normal mains electricity voltage via a three-pin socket outlet is 240 volts; heavy-duty equipment such as vehicle hoists uses 415 volts in the form of a three-phase supply. Both 240-volt and 415-volt supplies are likely to kill anybody who touches them. Supplies of 415 volts must be used through a professionally installed system. If 240 volts is used for power tools, then a safety circuit breaker should be used. A safer supply for power tools is 110 volts; this may be wired into the workshop as a separate circuit or provided through a safety transformer. Inspection hand-lamps are safest with a 12-volt supply; but for reduced current flow, 50-volt hand-lamp systems are frequently used.

Checklist

Before using electrical equipment, the body repairer is advised to check the following:

1 Cable condition – check for fraying, cuts or bare wires.
2 Fuse rating – the fuse rating should be correct for the purpose as recommended by the equipment manufacturer.
3 Earth connection – all power tools must have sound earth connections.
4 Plugs and sockets – do not overload plugs and sockets; ensure that only one plug is used in one socket.
5 Water – do not use any electrical equipment in any wet conditions.
6 PAT testing – it is a requirement of the Electricity at Work Regulations that all portable electrical appliances are tested regularly. They should be marked with approved stickers and the inspection recorded in a log.

COSHH

The **Control of Substances Hazardous to Health** regulations require that assessments are made of all substances used in the body repair shop, for instance paint and body filler. This assessment must state the hazards of using the materials and how to deal with accidents arising from misuse. Your wholesale supplier will provide you with this information as set out by the manufacturer in the form of either single sheets on individual substances or a small booklet covering all the products in a range.

RIDDOR

The Reporting of Injuries, Diseases and Dangerous Occurrences Regulations 1995 require that certain information is reported to the Health and Safety Executive (HSE). This includes the following:

1 Death or major injury – if an employee or member of the public is killed or suffers major injury the HSE must be notified immediately by telephone.
2 Over-three-day injury – if as the result of an accident connected with work an employee is absent for more than three days an accident form must be sent to the HSE.
3 Disease – if a doctor notifies an employer that an employee suffers from a reportable work-related disease then this must be reported to the HSE.
4 Dangerous occurrence – if an explosion or other dangerous occurrence happens, this must be reported to the HSE. It does not need to involve a personal injury.

Maintain the health, safety and security of the work environment

It is the duty of every employee and employer in the motor industry to comply with the statutory regulations relating to health and safety and the associated guidelines which are issued by the various government offices. That means you must work in a safe and sensible manner. A body repairer is expected to follow the health and safety recommendations of his/her employer; employers are expected to provide a safe working environment and advise on suitable safe working methods.

Health and safety law states that organizations must

- *provide a written health and safety policy (if they employ five or more people);*
- *assess risks to employees, customers, partners and any other people who could be affected by their activities;*
- *arrange for the effective planning, organization, control, monitoring and review of preventive and protective measures;*
- *ensure they have access to competent health and safety advice;*
- *consult employees about their risks at work and current preventive and protective measures.*

Tech note
Everybody in an organization has a duty of care related to health and safety. The HSE may bring about prosecutions or lesser prohibitions subject to timed actions – for instance, being given a short period of time to rectify a machine fault. However, the final consequences can be devastating for a firm and its employees. Possible outcomes are:

- unlimited fine
- imprisonment
- closing down of the business
- disqualification from working in that job or type of business

---

You must work in a safe manner or you are breaking the HSWA and are liable to a fine and possible imprisonment and maybe disqualification from working in that job. This means following the safe working practices which are normally used within the industry. The guidelines published by the Health and Safety Executive (HSE) and motor vehicle textbooks usually identify industry-accepted

safe working practices. Examples of important procedures are:

1   Always use axle stands when a vehicle is jacked up.
2   Always use an exhaust extractor when running an engine in the garage.
3   Always wear overalls, safety boots and any other personal protective equipment (PPE) when it is needed, for example safety goggles when grinding or drilling and a breathing mask when working in dusty conditions.
4   Always use the correct tools for the job.

Even if you are working in a safe and careful manner, you are still likely to spill the odd small amount of fluid or snag the air line, which will then create a hazard. The procedure here is always to remove the hazard, no matter how it was created, immediately. If you spill petrol or oil when you disconnect a pipe from the engine, you should clean it up immediately, or else you or a colleague may slip and fall. Absorbent granules should be used for this job, as they will soak up the liquid without causing a fire hazard or making the floor more slippery. Brake fluid is a special hazard because if it is spilled on the vehicle's paint work it will soften the paint and may cause it to peel off, just like paint stripper. Therefore any spilled brake fluid should be wiped off immediately and the paint surface washed and polished if needed. Antifreeze spilled on paintwork will soften the paint surface and cause discoloration, so it too must be wiped off immediately with absorbent paper roll or towel and washed down if needed.

When working on any system which contains fluids, it is good practice to use a drip-tray to catch any possible spillages. This saves having to clean the floor as well as ensuring that all the used oils and fluids are disposed of safely, that is, you can pour them from your drip-tray into your disposal container. The Environment Protection Act requires that you dispose of used oils and other fluids in a way which will not cause pollution. In practice, many vehicle manufacturers now collect used oil and brake fluid for recycling or safe disposal. Smaller garages without franchises will dispose of these liquids either through a private waste-collection company or through a scheme in conjunction with the local authority. No waste oil, petrol, brake fluid or similar chemicals must be allowed to enter the drainage system.

Exhaust fumes are very dangerous. They can kill you. Small intakes of exhaust fumes will give you bad headaches and over time can cause lung and/or brain diseases, so ensure that you do not run a vehicle in a workshop without an exhaust extractor. Also ensure that the extractor pipe is correctly connected and is not leaking.

The air line used in most garages operates at between 100 and 150 psi (7 and 10 bar). This is a very high pressure, so it must be handled with great care. When you are using an air line, always wear safety goggles to prevent dust entering your eyes. You must not use an air line for dusting off components, especially brake and clutch parts, as the very fine dust can cause damage to the throat and lungs. Before you use an air line, ensure that the coupling is fitted firmly into the socket and that the pipe is not leaking anywhere along its length. Any damage or leaks should be immediately reported to your supervisor or manager so that they can be repaired. The high pressure of the air can quickly turn a small leak in an air line into a large gash, which in turn may make the air line whip around and cause damage to colleagues or customers' vehicles. Another area of potential danger is when using electrical equipment such as an electric drill, hand-lamp or grinder. Most mains-operated equipment runs at 240 volts; an electric shock from such a voltage is most likely to kill you straight away. Some companies use 110-volt equipment which is operated through a transformer. This is much safer, especially if the transformer is fitted with an overload cut-out. Hand-lamps should operate at 50 volts, or preferably 12 volts to give

the highest level of safety. Plugs should only be fitted to electrical equipment by skilled persons. At the same time, a fuse of the correct amperage rating should be fitted and the equipment tested and logged in accordance with the Portable Appliance Testing Regulations (PAT testing). PAT tested equipment should be numbered and carry a test date label. Before you use any electrical equipment, visually check it for signs of damage and check that the cable is not frayed or split. Then ensure that you plug it into the correct voltage outlet. Do not attempt to use any electrical equipment which you suspect may be faulty; report the fault immediately to your supervisor.

Where identified hazards cannot be removed, appropriate action is taken immediately to minimize risk to own and others' health and safety

This section is about those situations in which the hazards cannot be readily removed – that is, how do you behave in accident situations or when equipment malfunctions and you can see an accident about to happen? Most mechanics are only likely to encounter such problems every few years, but the professional is the person who can save the day. The following is an example of when a service manager colleague came to the rescue. The central locking and the car alarm on a vehicle malfunctioned. A small child was trapped in the vehicle; it was a very hot day at a local car boot sale. The mother and child were hysterical, and the father had gone for the fire brigade; other members of the public just watched. The mother shouted for help. My quick-witted friend grabbed a screwdriver off one of the stalls, inserted it behind the rear quarter-light rubber and levered out the glass, put his hand inside the car, opened the door and released the child. The panic was over.

There were other ways in which this situation could have been dealt with, but this one was acceptable because it provided a very quick solution and caused the minimum amount of damage to the vehicle. The important point is that people come first and property second, although the amount of damage to the property should be kept to the minimum.

Working by the roadside is always hazardous, but you can minimize the risks by following a few simple rules:

1   Always wear a high-visibility safety vest.
2   Use the warning triangle and hazard lights.
3   Use the flashing lamp on your recovery vehicle.
4   Only work on hard and level ground.
5   If you are on the hard shoulder, get as far over to the left as possible.
6   Always use props when vehicle cabs or other panels are raised.
7   Think through the possible hazards; never take risks.
8   Have the vehicle towed to the workshop if necessary.

Often it is better to do nothing than to cause damage; this is referred to as preserving the situation. Many times things look different after a cup of tea or you have had time to check it out with a colleague.

Dangerous situations are reported immediately and accurately to authorized persons

As a trainee in the motor industry your company will require you to report any dangerous situations to your supervisor; this will be a person that you know as the charge-hand, foreperson or service manager. Any internal matter should in the first instance be reported to one of these people – you will know who this is from your induction training. However, if you are working alone or the matter is not a company one, then you must inform the relevant authority. The four emergency

services in the UK are Police, Fire, Ambulance and Coast Guard. To call them, use any telephone and dial 999.

Suppliers' and manufacturers' instructions relating to safety and safe use of all equipment are followed

Many pieces of garage equipment are marked 'only to be used by authorized personnel'. This is mainly because incorrect use can cause damage to the equipment, the workpiece or the operator. Do not operate equipment which you have not been properly trained to use and have not been given specific permission to use.

The suppliers of garage equipment issue operating instructions, and as part of your training, you must read these instruction booklets so that you will understand the job better. You will also find that certain safety instructions are marked on the equipment. The vehicle hoist (ramp), hydraulic jack and other lifting equipment are marked with the Safe Working Load (SWL) in either tonnes or kilogrammes. You must ensure that you do not exceed these maximum load figures. Some items of equipment have two-handed controls or dead-man grips – do not attempt to operate these items incorrectly.

Approved/safe methods and techniques are used when lifting and handling

Do not attempt to manually carry a load which you cannot easily lift and which you cannot see above and around. The maximum weight of load that you should lift is 20 kilogrammes, but as a trainee this may still be too heavy for you.

When you are lifting items from the floor, always keep your back straight and bend your knees. Bending your back whilst lifting can cause back injury. If you keep your feet slightly apart, this will improve your balance. It is always a good idea to wear safety gloves when manually lifting.

Hoists and jacks are available for lifting vehicles; hydraulic or chain-operated equipment is available for lifting engines; hydraulic devices are available to lift gearboxes. For moving equipment and heavy components you should have either a trolley or a sack-truck.

You are advised to seek the assistance of a colleague when moving a heavy load even when you are using lifting equipment.

Required personal protective clothing and equipment are worn for designated activities and in designated areas

The following table lists typical items of personal protective equipment (PPE) and states when they must be worn.

You will often see safety notices requiring you to wear certain PPE in some areas at all times. This is because other people are working in the area, and you may be at risk. Hard hats are sometimes required when working underneath vehicles on a hoist.

Table 2.1 PPE Usage

No.	PPE	Usage
1	Cotton overalls (boiler suit)	All the time
2	Safety footwear	All the time
3	Disposable gloves	Dealing with dirty or oily items
4	'Rubber gloves'	Operating the parts cleaning bath
5	Reinforced safety gloves	Handling heavy, sharp-edged items
6	Dust mask	Rubbing down
7	Breathing apparatus and paper coveralls	Spray painting
8	Goggles	Using grinder or drill
9	Waterproof overalls and boots	Steam or pressure washing

Injuries involving individuals are reported immediately to competent first aiders and/or appropriate authorized persons, and appropriate interim support is organized to minimize further injury

Should there be an accident, the first thing to do is call for help. Either contact your supervisor or a known first-aid person. Should any of these not be available, and it is felt appropriate, call for your local doctor or an ambulance.

You are not expected to be a first-aid expert, nor are you advised to attempt to give first aid unless you are properly qualified. However, as a professional in the motor industry, you should be able to preserve the scene, that is, prevent further injury and make the injured person comfortable. The following points are suggested as ones worth remembering:

1 Switch off any vehicle or power source.
2 Do not move the person if injury to the back or neck is suspected.
3 In the case of electric shock, turn off the electricity supply.
4 In the case of a gas leak, turn off the gas supply.
5 Do not give the person any drink or food, especially alcohol, in case surgery is needed.
6 Keep the person warm with a blanket or coat.
7 If a wound is bleeding heavily, apply pressure to the wound with a clean bandage to reduce the loss of blood.
8 If a limb has been trapped, use a safe jack to free the limb.

Visitors are alerted to potential hazards

The best policy is not to let customers into the workshop – many garages have a notice to this effect on the workshop door. For MOT purposes, garages must have a customer viewing area. However, it is not always possible to keep people out of the workshop. Insurance company assessors and RAC/AA engineers will probably also require entry to the workshop, as will some customers who are concerned about their vehicles. So, before allowing them into the workshop you should warn them of potential hazards, for instance the dangers of oil and grease and the requirement to wear a hard hat.

It is always a good idea to accompany customers when they are in the workshop. This way you can advise them in the event that they may do something potentially dangerous or if there is a hazard of which they may not be readily aware.

Injuries resulting from accidents or emergencies are reported immediately to a competent first aider or appropriate authority

If a person is injured, the first action must be to ensure that first aid is given by a competent first aider or other suitable person. Most companies have a designated first aider who is trained to deal with accidents and emergencies. If your company has no such person on the staff, then you will have a designated person who you must contact in the event of a colleague being injured. That person may be your supervisor or another senior member of the staff. If no manager or other senior person is available, you should either dial 999 for an ambulance or telephone your company doctor, then inform the garage manager.

Incidents and accidents are reported in an accident book

By law all companies are required to maintain records of accidents which take place at work. These records are usually kept in an accident book. Accident books may be inspected by HSE inspectors. They must be kept for a period of at least three years from the date of the last entry.

The information which is required to be recorded in the accident book is:

- Name and address of injured person
- Date, time and place of accident/dangerous occurrence
- Name of person making the report and date of entry
- Brief account of accident and details of any equipment/substances which were involved.

It is always a good idea to keep a notepad to help remind you which way round things go when working on unfamiliar vehicles, this would also be useful for making any other notes, such as those about an accident.

Where there is a conflict over limitation of damage, priority is always given to the person's safety

You can always buy a new wing for a car, but you cannot buy a new arm for a mechanic. In the event of an accident, people come first. For instance, if a building is on fire, do not re-enter to retrieve your belongings; wait until the fire is out and there is no risk before going back into the building. If a car is about to fall off a jack, get out of the way. Do not try to catch the car with your hands or some such other dangerous action.

Professional emergency services are summoned immediately by authorized persons in the event of a fire/disaster

An authorized person is somebody who has the task of carrying out a specific job. Anybody may call the emergency services if they are needed.

The four emergency services are:

1 **Police**
2 **Fire**
3 **Ambulance**
4 **Coastguard**

All are called by dialling 999 on an outside-line telephone. The emergency services operator will ask you which service you require. In certain cases, the police will automatically be called, for instance in the case of a severe fire.

All emergency telephone calls are recorded on tape at the telephone exchange. You will be asked for your name, the place where the emergency is and where you are calling from. With the introduction of electronic telephone exchanges, the number which you are calling from is automatically recorded, and you will be asked for the number to help confirm that your call is not a hoax.

Many companies have a direct telephone line to the fire station, and these automatically call the fire service if a fire is detected by sensors or if someone breaks the glass of a fire alarm. In such cases, if the fire service is called out and there is not a fire, they may charge the company a large fee. So do not tamper with such a device unless you are authorized to do so or there is a dangerous fire.

Alarm/alert/evacuation systems are activated immediately by authorized persons

Generally only senior staff (managers) are allowed to operate alarms and other forms of alert/evacuation systems. This is because of the costs which may be involved if the fire service is called out wrongly and the damage which may occur if staff and/or customers panic.

In the case of a fire, the normal alarm is a form of siren or bell. For other emergencies, say a serious injury, an audible warning from a speaker announcement system may be used. In a garage, these are usually operated from the service department office.

In the event of warnings, procedures for isolating machines and evacuating premises are followed

If you hear a fire/emergency warning, you must follow the company's evacuation procedure – this is usually stated on the workshop wall. If you hear a fire alarm, a typical evacuation procedure is:

1 Shut off the electricity by pressing the emergency stop button.
2 Leave the premises by the nearest route and go to the assembly point or muster point, which is in the customers' car park.
3 Do NOT re-enter the building until your supervisor tells you that it is safe to do so.

In the event of discovering a fire, raise the alarm by breaking the glass of the alarm button.

Reports/records are available to authorized persons and are complete and accurate

The Social Security (Claims and Payments) Regulations 1979 require employers to maintain an accident book as well as regulation 7 of the HSWA. This book requires brief details of any accident or dangerous occurrence to be recorded. An approved book BI 510 is available from the HSE direct or through most good bookstores. For more detailed information, HSE Form 2508 should be completed. HSE staff have a statutory right to see a completed accident book or Form 2508, and they may also ask for further information. If you are personally involved in an accident, you are advised to keep a copy of the book entry and any completed forms as well as your own notes on the event. These may be useful in the event of legal proceedings.

Machines and equipment are isolated, where appropriate, from the mains prior to cleaning and routine maintenance operations

You must always isolate an electrical machine from the mains supply before either cleaning it or carrying out any maintenance or repairs. There are two reasons for this: first, if you touch an electrically live part, you may get an electric shock; second, the machine may be accidentally started, which could cause injury or damage.

With portable electrical appliances this simply means switching off and taking the plug out of the socket.

With fixed machinery, for instance a pillar drill, you will need to switch off the power supply at the isolator switch. This is usually found on the wall near the machine. Isolating this way is fine while cleaning the machine, but for carrying out maintenance or repair work it is advisable to remove the supply fuse from the isolator box. With the fuse removed, the machine cannot be restarted if the isolator is accidentally turned on by a colleague who confuses the isolator for the one on an adjacent machine.

Safe and approved methods for cleaning machines/equipment are used

There are three main items of cleaning equipment used in the garage: the cleaning bath (or tank), the pressure washer and the steam cleaner.

The cleaning bath uses a chemical solvent. This is usually used for cleaning dirty/oily components. The components are submerged in the solvent and dirt is loosened with a stiff-bristled brush.

The pressure washer is used for cleaning the mud off the underside of vehicles; water at very high pressure will clean off mud. For hard-to-remove dirt, detergent can be added to the pressure washer. The steam cleaner, often referred to as a steam jenny (jenny = generator), produces hot pressurized water with the option of detergent. This is used for removing very stubborn grease and dirt, like that found on the underside of high-mileage goods vehicles.

If you are cleaning an engine or electrical equipment, it is important not to get water inside. Before cleaning inside an engine bay with a pressure cleaner the electrical components and engine inlets should be covered over with polythene.

When cleaning portable electrical appliances be careful not to get water on the plug. This could cause a short circuit.

The mechanical parts of fixed machines may be cleaned with solvents, then dried with absorbent paper towel.

Appropriate cleaning and sanitizing agents are used according to manufacturer's instructions

Before using any solvent, detergent or sanitizing agent such as bleach you must read both the label on the container and the COSHH sheet which the manufacturer or your company has prepared.

Solvent should only be used in the cleaning bath for which it is designed.

The pressure washer or steam jenny should only be used with the recommended detergent.

Electrical items can be cleaned with one of the many aerosol sprays which are available for this purpose, but the volatile fumes which are given off must not be breathed in.

You should remember that all cleaning agents should be kept away from your mouth and eyes, and contact with your skin may cause irritation or a more serious skin disease. Always wash your hands and any other exposed areas of skin with toilet soap after carrying out a cleaning task.

Used agents are safely disposed of according to local and statutory regulations

The Environmental Protection Act (EPA) and local by-laws in most areas require that used cleaning solvents must be disposed of safely. This means that they must be put into drums and either collected by a refuse disposal firm or taken to a local authority amenity site, where they are put into a large tank for bulk incineration. Several local authorities, for instance Surrey and Hampshire, are now looking at ways of using the energy produced by burning waste material to produce electricity. Emptying used solvents into the drain can lead to a heavy fine or even imprisonment.

Detergents are by their nature biodegradable, that is, they break down, do not build up sludge and will not explode, unlike solvents. However, if you use large quantities of detergents, wash bays which are fitted with the correct type of drainage system should be used.

Machinery, equipment and work areas are cleaned according to locally agreed schedules

In your company's Health and Safety Policy document, there will be reference to the cleaning of the floors and equipment in the garage and general amenities such as toilets and rest areas. Also there will be maintenance and repair records for the workshop equipment, which will include a regular schedule of cleaning and inspection.

Most companies work on the basis of sweeping down fixed machinery and floors at the end of each day, unless the generation of dirt requires more frequent attention.

On a weekly basis there will be a more thorough cleaning programme which may include window cleaning and wet cleaning certain areas.

Workshop equipment is usually cleaned and inspected on a monthly basis unless there is reason, such as a fault, for a more regular treatment.

Appropriate safety clothing and equipment is used when working with hazardous cleansing agents and equipment

To protect yourself from the cleaning agents which you are using, you must, where appropriate, wear personal protective equipment (PPE). Most cleaning agents are poisonous and cause irritation or more serious complaints if allowed to come into contact with your eyes or skin.

Whenever you are working on a motor vehicle, it is expected that you wear cotton overalls and safety footwear. In addition, the HSWA requires that employers provide and employees wear the appropriate PPE for hazardous jobs such as using cleaning equipment. The general requirements are as follows:

1 Cleaning bath – rubber protective gloves which extend over the user's wrists, plus goggles and plastic apron. Avoid getting solvent on your overalls as this can lead to skin irritation, be especially careful not to put solvent-soaked or oily rags in your overall pockets.
2 Pressure washer – rubber protective gloves and goggles, waterproof (plastic) over-trousers and jacket and finally rubber boots (wellingtons). The idea is to be able to take the waterproof gear off and be dry underneath.
3 Steam cleaning plant – The hazard here is that as well as being wet, the water is scolding hot. So the waterproof clothes must be of such a manufacture that they will protect the wearer from the high-temperature, high-pressure steam. This means thick and strong over-trousers, coat, boots, gloves and a hat. A full-face mask is used to give complete protection.

Notes
1 Health and safety issues are further discussed in the appropriate chapters.
2 Examples of a Risk Assessment and a COSHH Data Sheet can be found in Appendix 1.

Skills and questions

Care for health and safety and the environment is something which MUST be part of EVERY job that you do EVERY day. However, these skills need developing and do not always come easily. Keep practising until health and safety are second nature for you. Meanwhile, try these questions:

1 State five basic rules concerning dress and behaviour which demonstrate personal safety in the workshop environment.
2 List five necessary precautions for safety in the workshop and describe each one briefly.
3 What is meant by a skin-care system as used in the workshop?
4 Explain the importance of eye and face protection in the workshop environment.
5 Explain the importance of protective clothing for a body repairer and a paint sprayer.
6 Explain the significance of headwear and footwear while working in the workshop.
7 State the minimum noise level at which ear protection must be used.
8 With the aid of a diagram, explain the fire triangle.
9 Identify the correct colour code for the following fire extinguishers: water, foam, CO_2, and dry powder,
10 Explain the precautions which must be taken when handling toxic substances in a workshop environment.

3

Hand and power tools

Hand tools used in body repair work

Vehicles are designed for **volume production,** not for ease of repair. Most of the large manufacturers produce cars at the rate of about one per minute, taking about a day to produce each car. Every second of production counts in the vehicle's price, so production methods are very finely tuned. They are not designed to be undone. That means that to replace panels and carry out repairs requires special techniques, skills, and tools. Specialist tools have been designed to suit the varying contours and shapes of the present all-steel bodies and panels. These tools are made of high-carbon tool steel, which is forged and then heat treated to give long service in the hands of a skilled body repair worker. In a body repair toolkit the basic tools are the **hammer** and **dolly.** All other tools have been developed around these, giving us the body repair specialist tools which are now currently available. A repair job cannot be successfully carried out before one has completely mastered the skill of using the **planishing hammer and dolly** in coordination with each other, as this skill is the basis of all body repair work involving the use of hand tools.

Hammers

The **planishing hammer,** or panel hammer, is used more than any other tool in the body repair trade, and for this reason the best hammer available should be obtained. The principal purpose of the panel hammer is for smoothing and finalizing of a panel surface after it has been roughed out to the required shape. The planishing hammer should have a true and unmarked face, and it must be kept polished and free from road tar, underseal and paint, which readily adhere to its working faces during use. This tool is designed solely to be used in conjunction with a dolly block; it must never be used for chiselling or any other work which might mark or impair the faces of the hammer, for if the face of the hammer became marked, the marks would be transferred to the surface of the panel.

These particular hammers are generally made with one face square. This face is usually flat, while the other, round end is slightly domed or crowned. However, hammers are available with the square end domed and the round end flat, or alternatively with both faces flat or both faces crowned. In practice the square end, which is usually flat, is used for planishing on a curved surface of a panel, in corners or against swaged recessed sections and the crowned end for reasonably flat panels. The difference in the faces stops the panel surface being marked with the edge of the hammer when used in conjunction with the dolly block. The weight of the planishing hammer for general and new work ranges from 12 oz. (340 g) to 16 oz. (450 g), and the handles, which are usually very thin at the neck of the shaft for balance purposes,

The Repair of Vehicle Bodies, Andrew Livesey, 9780815378709, 2019, Taylor & Francis

Figure 3.1 Body repair hammer and dolly set

are made of hickory or ash to give the hammer a good rebound action when used with a dolly block.

Standard bumping hammer. This hammer is used for initial roughing out of work on damaged panels. It is also used for finalizing and finishing. The round face is 1 3/8 inch (35 mm) in diameter and the square face is 1½ inch (38 mm) square, and the total weight is 14 oz. (395 g). The hammer is made with either flat or crowned faces.

Light bumping hammer. This tool is ideal for work on light-gauge materials. It is used in the same manner as the standard bumping hammer. The squared face is 1 in. (25 mm) square and the round face is 1¼ in (32 mm) in diameter, and it has a weight of 12 oz. balanced and gives a very good finish when used correctly. Hammers can be obtained for light or heavy work with weights from 12 oz. (340 g) to 18 oz. (510 g).

Shrinking hammers are similar in design to a normal planishing hammer but have faces which, instead of being smooth, are serrated, giving a cross-milled effect like a file. The purpose of these serrations is to achieve a shrinking effect when the hammer is used in conjunction with a dolly block. This is caused by the fact that the contact area between hammer and metal is greatly

reduced by the serrations on the face. This tool is used largely when beating the surface on over-stretched panel areas which have to be hot-shrunk in order to return them to their normal contours. Hammers are available for light or heavy shrinking, according to the depth of the serrations.

Dinging hammer. This hammer is a long-reach planishing hammer and is designed for careful, controlled finishing work. It is extremely well balanced and gives a very good finish when used correctly. Hammers can be obtained for light or heavy work with weights from 12 oz. (340 g) to 18 oz. (510 g).

Pick and finishing hammer. This tool is used in place of or in conjunction with the planishing hammer. Its main use is to pick up small, low areas on the surface of a panel which is in the process of being repaired by planishing. On panels that are reasonably flat, such as door panels, parts of roof panels and bonnets, this method of raising low areas is quick and, if carried out correctly, does not unduly stretch the metal. To lift a low area with a pick hammer, one or two taps with the pick end of the hammer are directed from underneath the panel under repair to the centre of the low area. The blows stretch the metal sufficiently to raise the surface surrounding the point of the low spots where the blows were struck. This slightly raised area is next tapped down lightly with a planishing hammer or the finishing end of the pick hammer onto a suitably shaped dolly block, and the panel is finished off by filing with a panel file. When one becomes proficient in using this tool, it is possible to raise the surface with light blows and finish off by filing only. However, without sufficient experience there is a danger of over-stretching the metal owing to the inability to direct the blow accurately onto the low area under repair. The pick and finishing hammer has a pointed end which is suitable for removing low spots and is also a useful finishing hammer, having a crowned surface on a round face of 1½ inch (38 mm) diameter; it is well balanced and weighs 14 oz. (395 g).

Straight pein and finishing hammers. These hammers are used in a similar manner to the pick and finishing hammer but are designed with either a straight or a curved peined end, which acts like a chisel, and a domed round end which is used for planishing. They are suitable for roughing out prior to planishing, or in the finishing stages of planishing for stretching small low areas. These hammers can be used to dress out sections which are difficult to work on owing to their awkward shape or position, such as around lamp openings and in recessed and moulded sections on panels.

Curved pein and finishing hammer. This tool is identical in use to the straight pein and finishing hammer except that its curved pein end allows for greater flexibility in dressing out sections which are difficult to work on owing to their awkward shape or position.

Fender bumping hammer. This tool has a long curved head with one face which is circular to reduce the effect of stretching the metal when in use. This hammer is used for roughing out and dressing out damaged sections on panels to restore them to their correct shape and curvature before planishing begins. The heavy weight of this hammer, together with the curve, makes it very effective for hammering out difficult and inaccessible sections.

File hammer or beating file. This tool is designed to be used like a hammer in conjunction with a dolly block, although it is actually a file with a serrated face and is suitably shaped for holding in the hand. The milling on the file blade tends to shrink the panel as well as leaving a regular rough patterned surface ideal for locating low spots on the panel under repair and for finishing with a body file. The tool is used in conjunction with a hand dolly and with a glancing blow. It is most effective on large, flat sections, where it will be found ideal for smoothing and levelling out wavy panels. Two types of beating files are available; one is flat for use on low- and high-crowned surfaces, while the other is half-round in

Figure 3.2   Body mallet in use

shape and is used on convex or reverse-curved panel sections.

Mallets can be of the round or pear-shaped type, made from boxwood or lignum vitae, or can be rubber, aluminium or plastic faced. Some mallets have interchangeable heads so that the correct head can be used for the material being worked. A mallet is greatly used in the initial stages of smoothing and roughing out of a panel prior to planishing. When hot shrinking, the mallet is the tool to achieve a most successful shrink because a normal planishing hammer would tend to stretch the metal rather than shrink it. Without a mallet, aluminium work would be most difficult, as this metal is so easily marked and stretched. The working faces of the mallet must be kept in first-class order, or marks on the surface of the metal will result.

Hand dollies

These are either cast or drop forged steel blocks, heat treated to provide the correct degree of hardness. The shapes of the dolly blocks have been designed to provide a working surface that is highly polished and suitable for use on the many contours found on motor vehicle bodies. They are used in conjunction with the planishing hammer or beating file and act as a support or anvil to smooth out the surface area of panels that have been damaged.

These dollies, together with the planishing hammers, are the most essential tools for the panel beating trade. Obviously one dolly block will not be suitable for all shapes requiring planishing; therefore it is advisable to have a set of these dollies which would be suitable for a wide range of the shapes and contours encountered on the ever-changing body styles of the modern motor vehicle. When selecting a dolly block for a particular job, it should always be remembered that as flat a dolly block should be used as possible for the job in hand; then the dolly will not only cover the panel area more quickly because of its bigger face area but will smooth out the metal without excessive stretching. When working on panel contours, it must be borne in mind that a dolly block having a high-crowned surface will tend to stretch the metal much quicker than one having a low-crowned surface; hence the choice of block depends on whether the particular section under repair needs to be stretched quickly or just smoothed and planished without stretching.

It is very common for the faces of the dolly blocks to become coated with paint, road tar or anti-drum compounds, which it picks up from the underside of panels under repair. This coating must be removed from the surface of the dolly block so that when it is used with the hammer there is a metallic contact (metal to metal) between dolly block and work and hammer. This contact should be heard as a ringing noise if successful planishing is to be achieved.

Some details are as follows.

Double-ended hand dolly is conveniently designed for a good grip with two useful surfaces, one high crowned and one low crowned. It is ideal for general planishing and roughing out. Its weight is 3 lb 4 oz. (1473 g).

Utility dolly forms the basis of every body kit. It offers a variety of useful faces, and is found ideal for working in confined spaces.

General-purpose dolly is very well shaped and easy to hold and offers wide, low-crowned faces which are essential when working on the new body styles. Its weight is 2 lb 12 oz. (1247 g).

Heel dolly has a low flat face with radiused corners. It is suitable for corner and

angle work and is easy to handle because of its fairly small weight of 2 lb 14 oz. (1020 g). It is used in places where other dollies cannot be used. Its weight is 2 lb 4 oz. (1020 g).

Toe dolly combines a large flat face with a low crown on other faces and is long and thin for easy handling in narrow sections. The bottom face is ground flat which adds to the versatility of this block. Its weight is 3 lb (1360 g).

Thin toe dolly has similar surfaces to the toe dolly but is thinner and is ideal for working in awkward places where other dollies cannot be used. Its weight is 2 lb 4 oz. (1020 g).

Shrinking dolly block is designed for shrinking welded seams and reducing stretched areas prior to filling. It has a groove down the centre of the block into which the stretched or welded seams can be dressed. Its weight is 3 lb (1360 g).

Grid dolly is similar in shape to the toe dolly but has a large crowned grid face on the upper surface. The base has a plain flat face for normal finishing work. It has been designed to act as a shrinking dolly when used in combination with a shrinking hammer and the application of heat. The weight is 2 lb 4 oz. (1130 g).

Curved dolly or comma block has a long curved face, combined with the high- and low-crowned and the tapering face; it is extremely well suited to the modern body design and is comfortable to hold in difficult areas and narrow corners. The weight is 3 lb 6 oz. (1530 g).

Round dolly is light and small and very easy to hold. It offers high- and low-crowned faces for work on small areas of damage. The weight is 1 lb 10 oz. (736 g).

Body spoons

These tools are made from a high-grade steel which has been drop forged and heat treated. They are sometimes called prying spoons because the spoon end is used in the same manner as a dolly in conjunction with a hammer. The body spoon really does the same job as a dolly block but is designed

Figure 3.3   Body spoon set

for use in confined spaces where a normal dolly block cannot be held in the hand, e.g. between door frames and outer door panels. The spoon end, which acts as the dolly, must be kept in good condition and free from anti-drum compound so that it gives a metal-to-metal contact when used in conjunction with a hammer. The spoon can also be used for roughing or easing out by wedging the spoon in between panels.

Surface spoon has an extra-large working area which is very slightly crowned, and it is ideal for working in between and round struts and brackets without much dismantling.

General-purpose spoon is a double-ended body spoon and is also a basic item in most panel beaters' toolkits. The applications for a spoon offering such different curves are many and varied and meet the needs of modern body shapes.

High-crowned spoon has been designed to offer a broad working blade with a high crown to work in positions where other dollies and spoons cannot be used.

Drip moulding spoon has a special lip which can be hooked under the drip moulding of the roof section and hence simplifies the repair. The curved surface is fully finished to enable the tool to be used as a short standard spoon.

Spring hammering spoon is a light pressed-steel spoon which is designed for spring hammering on panels which have minor blemishes in their finishing stages.

The broad blade spreads the blow evenly over a larger contact area, thereby reducing the possibility of sinking the metal. This spoon is not made for prying or levering.

Heavy-duty pry spoon is ideal for heavy prying and roughing-out work. It has a long, low, curved blade which can be inserted into very thin sections.

Pry and surfacing spoon has a short handle which gives easy access in limited spaces and can be used efficiently for prying behind brackets and panel edges.

Long-reach dolly spoon is a special long-handled body spoon designed to be used in restricted spaces between double-skinned panels such as doors and quarter panels.

Thin spoon is a very special thin-bladed spoon, the blade having a very slow taper which permits entry into restricted spaces between double-skinned panels which would not be possible by using a normal spoon.

Body files

Flexible panel file. This tool is designed with a two-position handle and has a 14 in. (30 cm) spring steel backing plate to give adequate support over the whole blade. Positioned between the two hand grips is a turnscrew threaded left and right hand for adjusting the blade to concave or convex positions to suit the user's requirements. The main use of this tool is to assist in the final planishing of the work. First and most important, it locates areas which are low on the surface of the panel under repair; second, it files out small marks or defects on the panel surface. It can be adapted to file the surface of almost any shaped panel by setting

Figure 3.4   Body file

the blade either straight, concave or convex. The file blades are specially designed so that they do not remove too much metal, and the milled serrations are wide apart and curved to prevent clogging when filing metal which has been painted, soldered or plastic filled. It is important to release the tension on the file blade after use in order to reduce the risk of breakage if it is dropped or struck by accident.

Flat file. This file has a solid wood holder designed to take a standard 14 in. (30 cm) blade, which must be used in a rigid position.

Half-round body file. This tool is similar in design to the flat file, but the blade is half-round. It is very useful for curved surfaces.

Abrasive file. This is not a normal file but an abrasive holder with a wood handle having spring clips at either end onto which can be attached an abrasive grit paper. This tool is of greatest use when rubbing down plastic body fillers to their final finish.

Body file blades. These flexible, double-sided blades are produced from a special alloy steel and heat treated. The milled teeth allow a smooth filing operation and are specially shaped to reduce clogging. The blades are either flat or radiused according to the surface to be filed.

Supercut blades are 8 tpi (teeth per inch) general-purpose standard blades.

Standard cut blades are 9 tpi and used for soft metals such as lead and solder fillers.

Fine cut blades are 13 tpi and used for aluminium, copper and brass.

Extra fine cut blades are 17 tpi and suitable for cast iron, steel or any narrow section of metal.

Plasticut blades are 6.5 tpi and used for plastic body fillers only.

Hand snips

The offset combination of universal snips is preferred by the panel beater when cutting thin-gauge metal. Universal snips are suitable for cutting straight lines, outside and inside curves. A right- and left-hand pair of combination snips will be suitable for

Figure 3.5 Aircraft snips

most of the sheet metal cutting that will be encountered by the panel beater, and there is no need for any curved-blade snips. When the more popular right-hand snips are used, the waste metal forms a coil to the left of the cutting blades, thus causing little distortion to the surface of the sheet or panel being cut; similarly, with the left-hand pair of snips, the waste metal passes to the right of the cutting blades, leaving the undistorted sheet or panel on the left. Snips can be obtained in varying sizes with either straight or crank handles. A straight pair of snips is often necessary for long straight cuts.

Specialist panel beating tools

Panel puller. This tool comprises a long steel rod with a cross T-piece at the top which acts as a hand grip. At the other end a fixed hexagonal nut holds a strong self-tapping screw, while a heavy cylindrical weight can slide up and down the shaft against a stop which is near the handle. This tool has been designed to pull out dents and creases from the face side of the panel without the necessity of removing the trim and lining materials to gain access behind the dents. It can also be used on double-skinned panels where access is impossible with conventional tools.

The panel puller is used by drilling a 1/8th-inch (3 mm) -diameter hole at the deepest part of the dent, then inserting the self-tapping screw in the hole and screwing it until a firm grip is obtained. One hand holds the T-bar while the other pulls the sliding weight towards the stop; there the weight rebounds, forcing the dented panel out under the impact of the blow.

The Zipcut spot-weld remover is used with an electric or air drill, and is ideal for

Figure 3.6 Zipcut spot-weld remover *(Sykes-Pickavant Ltd)*

removing spot welds on all areas of body-work and subframes. The cutter blade A is reversible with two cutting edges. Adjustment B provides for varying depth of cut so that only the upper panel is released (C), leaving the original spot weld behind on the lower panel.

Cone drills or variable hole cutters are special alloy steel drill bits which have been hardened to give a fine cutting edge. They are shaped like a cone and fit in the chuck of a standard power drill. The hole diameter can vary from 6 to 40 mm depending on the hole cutter in use. These cutters are used for fitting wing mirrors, aerials and rubber grommets into body structures.

Cobalt drills or spot drills are special alloy steel drills to which cobalt has been added to give a very hard sharp cutting edge. The design of the cutting edge allows spot welds to be drilled out of panel assemblies without creating any panel distortion.

The impact driver will loosen or tighten the most stubborn screws, nuts and bolts. The tool is supplied with a complete range of interchangeable bits.

Body trim tool is a universal spring steel tool for removing body trims and clips and also weatherstrips, door trims, headlamp fittings, windscreen clips, plastic mouldings, motifs and badges.

Door handle spring clip removers are specially designed pliers for fitting or removing the spring clips used on many modern

door handles. One jaw sets in the neck of the spring whilst the other grips the other end, and the whole spring is then levered out.

Trim panel remover. This tool is designed to remove vehicle trim panels, upholstery and roof liners without damage. It also fits a wide variety of button-type fasteners used on vehicle trim.

Door hinge pin remover and replacer is designed for the removal and replacement of hollow hinge pins used on vehicle door hinges.

Corrosion assessment tool. This tool has been designed for vehicle inspection. It acts as a combined tapping hammer, blunt scraper and short lever. It has been introduced to standardize the methods of corrosion assessment employed by MOT testers and is ideal for bodyshops conducting pre-MOT checks and assessment for corrosion.

Bendable file. This tool is a standard round file, but it can be bent to the desired shape, which allows it to be used in difficult locations where a normal straight file would not have easy access.

Clamps

Sheet metal clamp is a general-purpose clamp which can be locked onto sheet metal with a powerful quick-release grip, making panel assembly and tack welding procedures much easier operations.

Welding clamp has a powerful grip which holds parts in alignment while leaving both hands free for the welding operation. The deep-throated jaws and centre opening provide the operator with maximum visibility and full access to the welding area.

C-clamp has a wide jaw opening with a relatively small gripping area, which produces pressure without permanent jaw damage to panel surfaces. It allows accurate close-up working in restricted areas and on awkward shapes, for example sill panels.

Figure 3.7   Door skinner tool

Punches

Hole punches have interchangeable heads to punch holes of either 5 mm (3/16 inch) or 6 mm (1/4 inch) diameter, and enable joining panels to be accurately aligned for welding. The 5 mm diameter hole is for gas welding or brazing, and the 6 mm hole is for MIG welding.

Wing punch is a hole punch with a specially designed head which allows it to be used on wing panels and channel sections and also fit over roof gutters and wheel arches.

Edge setters or jogglers

The **edge setter** is a portable, hand-operated tool designed to provide a 'joggled' joint or stepped edge on a repaired or new panel, thus creating a flush-fitting lap joint. The joggled panel edge provides a stiff joint and helps prevent distortion when welding. It can be used on mild steel up to 1.2 mm (18 gauge) and applied in many situations which require this type of joint. The **Rolastep edge setter** is a portable tool used on the vehicle or a panel assembly. Its rollers produce a smooth, uniform stepped panel edge to create a flush-fitting lap joint, allowing a panel replacement to be inserted.

Door skinner. This is a special tool for crimping flanges tightly on replacement door skins. It is used by first bending the flange to an acute angle and then tightening

or crimping the flange to the door frame. During this operation, interchangeable tough nylon pads prevent damage to the surface of the door skin.

Inter-grip welding clamp. The inter-grip clamp is approximately 25 mm cube shaped and uses a wing nut and toggle bar to firmly clamp pieces of metal together edge to edge and perfectly level, leaving only a small gap which will allow full penetration of the weld and filler rod. Once the two sections are tack welded together the wing nut is slackened, the toggle bar removed and the clamp lifted from the job prior to final welding. The end result is a first-class butt weld requiring very little dressing. The clamps are ideal when welding patch repairs in floors, wings and door skinning or when assembling fabricated or new panels.

Temporary sheet metal fastener system. This sheet metal holding system is a set of pins, installed with special pliers, which quickly and easily clamps panels together before final welding, bonding or riveting. The clamps will securely hold together body panels and sections where no other clamp will reach because there is no rear access. The 3 mm blind fasteners have an expanding pin which, when inserted through the panel and released, exerts over 8 kg (18 lb) holding pressure. The edge grips have 12 mm deep jaws to clamp flanges and other edge work. To install or remove either, the pliers are needed to overcome the powerful spring pressure.

Figure 3.8 (a) Engineer's ball-pein hammer (b) stretching and flanging hammer (c) wiring hammer (d) hollowing or blocking hammer

Sheet metal bench stakes

Quite often when making panels and parts of panels in sheet metal, it is necessary to perform some operation such as bending, flanging, seaming, shaping and planishing which cannot be done by machine. In such cases these operations will have to be carried out by hand using bench stakes. There are many different types of stake in use; the smaller types are usually made from cast steel, the heavier are of cast iron and for special work stainless steel stakes may be used.

All these stakes must have highly polished working surfaces and should be kept in a dry atmosphere to reduce the possibility of corrosion on the working surface. Certain stakes are used only occasionally, and to maintain a bright, smooth surface, they must be given a coating of light oil from time to time. The condition of the stakes has much to do with the workmanship of the finished job, for if a stake has been roughened by hammer or chisel marks the completed job will look rough and lack finish. For this reason stakes should not be used to 'back up' a sheet of metal or a panel directly when cutting it with a chisel; hammer blows should be aimed with care; and a mallet should be used whenever possible when forming sheet metal over the stakes.

Special reinforced square holes of various sizes are made at intervals in the bench for insertion of the bases of the stakes, which are usually square in cross-section and taper in shape to facilitate a good grip in the bench. There are many different types of stake available, and those most commonly used are as follows:

- *Pipe stake* consists of one or two cylindrical arms having different diameters. This stake is used for forming pipes and cylindrical work pieces.
- *Grooving stake* has a grooved slot cut along the top face. The slots are used for wiring and swaging.
- *Funnel stake* has a thick tapered head and is used in forming, riveting and seaming tapered articles such as funnels.
- *Anvil stake* has a flat, square-shaped head with a short shank and is used for general working operations.
- *Side stake* consists of one long elliptical shaped head. It is used for bending, seaming and forming curved radii.
- *Beck iron stake* has a square tapering flat head on one side and a round tapered head on the other side. It serves as a general-purpose anvil for riveting and shaping round and flat surfaces, straight bending and corner seams.

Figure 3.9 Sheet metal bench stakes

1 Beck iron stake
2 Horse stake with heads
3 Round-bottom stake
4 Half-moon stake
5 Tinman's mandrel
6 Hatchet stake
7 Pipe stake
8 Anvil stake
9 Side stake
10 Funnel stake
11 Grooving stake

- *Horse stake with heads* has holders (the horse part of the stake) which can take various-shaped small heads. This stake is useful in many operations for which other stakes are not suitable.
- *Round-bottom stake* consists of a single vertical piece with a flat round head on top. It is used for flanging circular and curved work.
- *Tinman's mandrel* is a single horizontal metal bar, one section of which has a flat surface and the other section of which is rounded. The flat length has a slot cut in it which permits the stake to be fastened directly to the bench. The stake is used for riveting, seaming and forming.
- *Half-moon stake* consists of a single vertical shank with a half-rounded head on top. The stake is used for forming or shaping curved flanges.
- *Hatchet stake* consists of a horizontal, sharp, straight edge and is used for making straight sharp bends.
- *Panel head stake* is shaped like a mushroom and is made in various sizes. It is used for raising, tucking and planishing.
- *Ball stake* is shaped like a ball and is used for the planishing of hollowed or raised articles. It is also used for raising domes or hemispherical shapes by mallet or raising hammer.

Hand grooving tools

The **hand groover** is used when making a seam joint by hand between two pieces of sheet metal. The end of the tool is recessed to fit over the joint formed when the edges of the two pieces of metal are pre-bent into a U-shape and interlocked. The grooving tool is then placed over the interlocked edges and hammered. One edge of the tool acts as a guide while the other forces the metal down into a crease, thus forming a groove. This crease or set in the metal should be made gradually, the tool being hammered while tilted very slightly in the direction in which the crease is to be made. This will avoid marking the metal on the opposite side of the joint to the crease. A tool a size larger than

that of the grooved joint should be chosen to give clearance in use. The tools are manufactured in various sizes from 1/8th in. to ¾ in (3–19 mm). These grooved joints can be used to advantage when fastening large flat panels together as used on commercial vehicles.

Rivet sets

A rivet set is made of tool steel. A deep hole in the bottom of the set is used to draw the rivet through the metal sheet before riveting, and a cup-shaped hole is used to form the finished head on the rivet. The drawing-up hole and the cup hole are offset slightly to the direction of the vertical hammer blows, and therefore great care should be taken to hold the set in an upright position when riveting. If the tool is allowed to tilt, it is liable to make indentations on the area of the metal which has been riveted. Hand rivet sets are made in various sizes ranging from 3 mm to 19 mm (1/8 inch to ¾ inch).

Cutting tools

Hand snips These are used to a great extent in the fabrication or repair side of the panel beating trade, for cutting thin-gauge mild steels and aluminium alloys. The limits to which these snips can cut satisfactorily, leaving a good edge, is a thickness of 1 mm (20 SWG) in mild steel and slightly greater thicknesses in aluminium. Snips are available in a variety of types and sizes and are made from the finest tool steel forgings. The blades are ground and set for the accurate cutting of all sheet material.

The most popular hand snips are the **universal combination snips**, which are made in sizes 10 in. to 13 in. (254–280 mm) with narrow or broad blades, with straight or cranked handles, and left or right handed. The handles are offset on the cranked variety so that straight and curved cuts can be made in any direction in one operation with the sheet kept flat and the operator's hand clear of the metal, the cut being followed from above. The word 'universal' means that this single pair of snips can be used for cutting straight lines and both external and internal curved edges. **Standard pattern straight snips** are also very popular, although limited in their use to straight-line cutting or large external curves. Sizes vary from 6 in. to 12 in. (152–305 mm). **Standard pattern curved or bent snips** are designed for cutting circular outlines and inside circles or curved shapes, as the actual cutting blades themselves are curved.

The cutting tools known as **aviation snips** were originally developed for the aircraft industry. In contrast to the single pivot of ordinary tin snips, aviation snips feature compound leverage for greater shearing power, being able to cut up to 1.2 mm (18 gauge) steel and 1.6 mm (16 gauge) aluminium. The blades are made from heat-treated forged chrome steel and are radiused in order to maintain a constant cutting angle from the back of the jaw to the front, allowing good cutting pressure for the user.

The terms 'left hand' and 'right hand' used for snips in this country have the continental meaning. *Left hand* means that the lower blade, which should be on the underside of the sheet of metal being cut, is on the left-hand side. *Right hand* means that the lower blade is on the right-hand side of the metal being cut. The US reference is exactly opposite to that used in this country.

Monodex cutters. This tool is designed for cutting thin sheet metal. Unlike snips, it has a flat face with one central serrated blade which in use cuts a slot equal in width to the thickness of the blade, 3 mm (1/8 inch). As the material is cut, it coils up in front of the cutters. The advantage of this tool is that it can cut sheet metal internally, and work along either a straight or curved line by drilling a small access hole in the metal to allow the blade to penetrate through the metal to commence cutting. Allowance must be made for the wastage of 3 mm strip (1/8 inch) when cutting to a final size.

Cold chisels. These are often used by the body repair worker for cutting out components which have been damaged, chipping off rusted bolts which cannot be removed by normal methods and chiselling sections which cannot be cut by snips or hacksaws. Chisels are made from high-quality steel. The cutting edge is specially hardened and tempered, while the remainder of the chisel is left soft to enable it to withstand hammer blows.

The chisel is subdivided into cutting edge, shank and head. The cutting end has a wide, flat taper and is ground on both sides to form the cutting edge. Metalworking chisels should have sufficient length to allow the shank to be gripped properly, as short chisels are difficult to handle and may lead to accidents through hammer blows if the head projects insufficiently beyond the gripping hand. On the other hand, excessively long chisels cannot be guided so well and consequently may vibrate and break easily. Only the cutting edge of the chisel is hardened. When the chisel has been in use for some time, a ridge forms on its head which is called a burr; this needs to be removed because it may lead to accidents during use. Chisel joint angles vary between about 30° and 60°, the sharper angles being used for cutting softer metals. The metalworker's chisel is flat and is used for working on flat or curved surfaces of thin sheet. It is also used for cutting solid bolts and rivets.

Hacksaws. The hacksaw is used by the body repair worker to advantage where a clean, neat cut is required, usually where two pieces have to be cut to form a welded joint, and is ideal for cutting irregular-shaped panels which could not be cut with snips or chisel. There are many different types of hacksaws, but in general they consist of a fixed or adjustable frame fitted with a renewable blade of hardened and tempered high-carbon or alloy steel. The hacksaw blades are classified by length and the number of teeth to the inch. Coarse blades have 14 to 16 teeth to 1 in. (25.4 mm) and are used to cut soft metals; finer blades have 18 to 24 teeth to 1 in. (25.4 mm) and are used to cut harder metals. The blades must be fitted with the teeth pointing away from the operator to the front of the frame, and the blade should be slightly tensioned in the frame.

Mini sheet metal cutter. This is a light-weight metal shear which will produce curved and straight cuts in mild steel up to 1.6 mm (16 gauge). A lever-and-ratchet action operates the roller cutters, which are adjustable for metal thickness, and the cutting action can be reversed for cutting internal angles. The cutter can be bench mounted or used in a vice. This tool is ideally suited for the small bodyshop where limited cutting is required.

Drill cutting attachment. This is a drill attachment for cutting sheet metal up to 1.2 mm (18-gauge) mild steel and 1.6 mm (16-gauge) aluminium). This tool provides the user with clear visibility of the cutting line, allowing an accurate cut. When cutting, it removes a continuous 3 mm strip from a panel without distortion. It will fit any drill chuck, air or electric.

Bending and swaging tools

Sheet metal folder This is a small, compact folder constructed and designed for mounting in a vice. It can be used to fold mild steel sheet up to 1.2 mm (18 gauge) with a bend length of 600 mm (24 inch). The tool is capable of producing bends of up to 100° and will produce both channel and box sections. It is a useful tool for the fabrication of small metal sections needed in vehicle repair.

Swaging tool or edge setter This tool is designed for bench mounting or to be used in a vice. The fixed rollers produce a smooth, uniform stepped panel edge on a straight or curved panel. The bottom roller is adjustable to accept varying gauges of material up to 1.2 mm (18 gauge) mild steel. The swaged or joggled edge can be used to strengthen panel edges or to create a flush-fitted lap joint when working on panel replacement.

Shrinker and stretcher machine This is a versatile sheet metalworking tool, and will quickly and easily pull curved shapes into steel up to 1.2 mm (18 gauge) or aluminium up to 2.0 mm (14 gauge). It shrinks or stretches the metal into smooth even curvature with a radius as small as 75 mm (3 inch). It is used to fabricate and repair panels where curves are needed in angle sections, especially around head lights, edges of window frames, door edges and wheel arches.

General-purpose assembly and dismantling tools

Having the right tool for the job is a sign of a prepared body repair worker. Knowing how to use the tool is a mark of an experienced repair worker. Knowledge and experience come with study and time, but without the right tools, even the best bodyworker cannot carry out quality repair work.

Spanners or wrenches

A complete collection of spanners or wrenches is indispensable for the body repair worker. A variety of vehicle body parts, accessories and related parts all utilize common bolt and nut fasteners as well as a range of special fasteners. The word 'wrench' means 'twist'. A wrench is a tool for twisting or holding bolt heads and nuts. The width of the jaw opening determines the size of the spanner, so that it fits around a nut and a bolt head of equal size. The larger the spanner size, the longer the spanner: the extra length provides the user with more leverage to turn the larger nut or bolt. Spanners are drop forged and made from chrome-vanadium steels.

Open-ended spanner. This is the most commonly used spanner. However, it has the disadvantage that it fits the nut or bolt on two sides only; consequently there is a greater tendency for the open-ended spanner to slip off the bolt or nut, resulting in rounded nuts and injured hands. Open-ended spanners can be single-ended or double-ended to fit nuts of consecutive sizes. The head of the spanner is usually set at an angle for use in constricted spaces: when the spanner is turned over, the nut can be approached from a different angle. The open-ended spanner fits both square head (four-cornered) or hexagonal (hex) head (six-cornered) nuts. These spanners come in sets ranging from 3 mm to 36 mm.

Ring spanner. The ring spanner is made in a variety of sizes, points and offsets and may be flat or cranked. It fits the nut all round by means of teeth cut in the inner surface of the ring. This is a very popular type of spanner, as it enables the nut to be tightened quickly and can be used where movement is restricted. The ring spanner is much safer to use, as more force can be applied without slipping and rounding the corners of the nut. The handle of the ring spanner can be straight or offset to provide hand clearance. Each end is usually a different size. The ring spanner does have limitations, for there must be sufficient clearance for the jaws to fit over and around the head of the nut or bolt. Ring spanners are available in 6, 8 or 12 points. The advantage of the 12-point ring spanner is that it can hold the nut in 12 different positions, which is ideal for working in confined spaces. Ring spanner sets are available in sizes from 3 mm to 50 mm.

Combination spanner. The combination spanner has an open-ended jaw at one end and a ring on the other end. Both ends are the same size. The combination spanner is probably the best choice of second set, for it complements either open-ended spanner or ring spanner sets. Combination spanner sets are available sizes 3 mm to 36 mm.

Adjustable wrench. An adjustable wrench has one fixed jaw and one movable jaw. The movable jaw is adjusted by means of a worm wheel fitted in the handle, which is meshed with teeth in the jaw. These wrenches are useful because they are manufactured in a range of sizes, and each one can be adjusted to fit many sizes of nuts.

Socket sets. In many situations a socket ratchet is much faster and easier to use than an open-ended or ring spanner because of its versatility in use, and in some body repair applications it is absolutely essential. A basic socket ratchet set consists of a ratchet, bar extensions, a universal joint and a set of sockets. The barrel-shaped socket fits over and around a given size of nut; inside it is shaped like the ring spanner, having a set of teeth to grip the points of the nut. Sockets are available with 6 pointed teeth (hexagonal) and 12 pointed teeth (bi-hexagonal). A 6-point socket gives a tight hold on a hex nut, minimizing slippage and rounding of the nut's points. The 12-point socket does not have the holding power of the 6-point socket, but its numerous positions maximize the possible turning radius.

The closed end of the socket has a square hole into which the drive shaft of the ratchet fits Socket sets can be purchased in ¼ inch, ⅜ inch, ½ inch and ¾ inch drive sizes. The smaller sizes are used for turning small fasteners, badges and trim fasteners, where little torque is required. The larger drive sizes with the corresponding longer handles are used where greater torque is needed. The size of the individual sockets in a set depends on the drive size of the set as well as the number of sockets in the set. The socket size is the same as the bolt head. For reaching nuts or bolts in limited-access areas long sockets are available. Socket set accessories multiply the usefulness of the socket set. A good socket set has a variety of accessories such as:

- Ratchet
- Ratchet attachment
- Extension bars
- Speed brace
- Long-hinged handle with universal joint
- Sliding T-bar
- Universal joint
- Flexible spinner
- Flexible extension
- Coupler

Screwdriver attachments are also available for use with socket ratchets. These attachments are very useful when a fastener cannot be loosened with a regular screwdriver owing to lack of access and space. The leverage that the ratchet handle provides is often all it takes to move a stubborn screw.

Screwdrivers

A variety of threaded fasteners used in the automotive industries are assembled by the use of a screwdriver. Each fastener requires a specific kind of screwdriver, and a body repairer should have several sizes of each type.

All screwdrivers, regardless of type, are designed to have several things in common. The size of the screwdriver is determined by the width or diameter of the blade and the length of the shank. It is important to select the blade tip accurately by matching it to the size of the slot or the type of fastener being used. The larger the handle, the better the grip and the more torque it will generate when turned.

Standard screwdriver (for slotted heads). A slotted screw accepts a screwdriver with a standard tip. The standard-tip screwdriver is probably the most common type used. The blade should match the slot: an oversize screwdriver will not reach the bottom of the slot, and an undersized screwdriver will float in the slot, and either condition could result in damage to the tool and the fastener.

Phillips screwdriver (for cross-heads). Phillips screws have a four-pronged funnel-shaped depression in the screw head and are known as cross-heads. The tip of the Phillips screwdriver has four prongs that fit the four slots in the screw head. The four surfaces enclose the screwdriver tip so that there is less likelihood that the screwdriver will slip off the fastener. The most useful set of Phillips screwdrivers ranges from number 0 to number 4.

Pozidriv screwdriver. This screwdriver is like a Phillips in that the head is in the form of a cross, but the tip is flatter and blunter. The square tip grips the screw head and slips less than a Phillips screwdriver. Although Pozidriv and Phillips look alike, they are not interchangeable. The most useful set of Pozidriv screwdrivers ranges from number 1 to number 4.

Torx screwdriver. The Torx fastener is used quite a lot on vehicle bodies. The

Figure 3.10 Tool chest and tools

screwdriver has a six-pronged star profile, which provides greater turning power and less slippage and also ensures optimum pressure to prevent the tool from slipping out of the fastener's head.

Speciality screwdrivers. These are normally in sets of screwdriver tips and assorted drivers which are all interchangeable and prove most useful in vehicle body repair work.

Tool chests

A **cabinet-type tool chest** as shown is standard equipment in most bodyshops. A portable toolbox on top of the chest holds large hand and power tools. A chest of drawers holds spanners, sockets and all the necessary assembly tools. The tool chest is on castors so that it can be conveniently located in any working area of the bodyshop.

Power tools used in body repair work

Choosing between pneumatic or electric power tools has always been difficult because, while they are designed to perform basically the same function, the particular advantages and disadvantages of each are so different.

Electric tools are simple to set up. They require only a source of electric current and so are suitable for immediate use anywhere in the workshop. The disadvantage is that because they have their own power supply (the motor) incorporated, they tend to be heavy and awkward for some jobs. Not only that, but over-heavy use can wear the motor, causing it to burn out.

Pneumatic tools, by contrast, are more complex. Among the advantages they offer are the regulation of air flow to suit the job and the fact that they are usually lighter, smaller and easier to handle. They also require less maintenance than a comparable electric tool. These factors can only be achieved, however, if a high-performance air compressor system is available. Moreover, it has long been a disadvantage that, despite filters and regulators, pneumatic tools can still blow oil, rust, dust and other impurities into the atmosphere and onto the work being treated.

Power tools play an important part in the modern techniques used by the panel beater in the repairing of crash damage to motor vehicles. Their use reduces man hours to a minimum, and they also enable the operators to achieve better results than those possible using hand tools. A wide range of equipment is available to suit all types of crash damage.

Air power chisel. This tool is designed for the removal of damaged panels so that essential stripping can be carried out easily and quickly. It consists of a chisel bit which is retained by a spring through one end of a barrel and a piston which moves to and fro within the barrel, striking the end of the chisel bit as it does so. The barrel must be kept dirt-free to eliminate the possibility of undue wear by the reciprocating movement of the chisel. A feather-action trigger in the handle allows for variation in the strength of the blow according to the amount of pressure applied. To provide additional safety, a beehive or volute spring is fitted over the

end of the barrel to prevent the chisel leaving the tool accidentally.

A wide range of special chisel bits is produced for the power tool. Most of these are cutting chisels of various sizes, used for general dismantling, cutting rivets, nuts, bolts and the removal of spot-welded sections. There is also a special thin-gauge metal cutting chisel for use on panel work, and a recessing tool which can be fitted on the end of the chisel. The recessing tool can be used on the edges of panels to form a lip or joggled section so that one panel passes over the face of the other, forming a flush finish which is ideal for spot-weld connections and joints in panels.

Metal shears. They are usually electrically operated but in some cases can be driven by compressed air. They are designed to be used in the hand with an on/off trigger control built into the handle. The blade-holding shoe is designed in a spiral so that it parts the metal as it cuts it. The machine is based on the principle of shearing metal, and this is done by a pair of very narrow blades, one of which is usually fixed while the other moves to and fro from the fixed blade at high speed. The blades have a very pronounced angle or set to permit the blade to pierce the sheet for internal cutting. The tool can be easily turned when cutting owing to the narrow width of the blades and so can cut curved as well as straight shapes. The shears

Figure 3.11 Air chisel set

are capable of cutting sheet metal up to a maximum of 2 mm (14 SWG) or 3 mm (10 SWG) aluminium, which more than covers the automobile requirements. The shear is used extensively in the cutting of light materials for the construction of new vehicles but also plays an important role in the removal of damaged sections in crash repairs, as it leaves a nice, neat cut, and there is no risk of flame damage as with the more traditional methods of oxy-acetylene cutting.

Pneumatic metal shear. This tool uses the same action as snips, having two fixed blades and one blade moving vertically between them. This central movable blade cuts out a thin ribbon of metal which coils and is discarded. Cutting can be started from the edge of the panel or from any point on the panel surface using a predrilled hole large enough to insert the central cutting blade. Radiuses up to 200 mm (8 inch) can be cut on material up to 1.2 mm (18 gauge). Care must be taken not to damage the panel surface by causing buckling during the cutting action.

Metal cutting nibbler. This is a portable nibbler which uses a punch and die. The actual cutting is done by the edge of the punch which is reciprocated at high speed, leaving a slot or channel in the metal. Cutting can be commenced from the panel edge, or from a hole predrilled in the panel surface to allow entry of the punch. Allowances must be made for the width of the cut, and when accurate positioning of the cut is essential, templates can be used as a guide.

Power saw. This is a high-speed pneumatic tool using interchangeable, reciprocating hacksaw blades. It is able to cut from straight lines to tight curves. Cutting can be started anywhere on a panel without the need for pilot holes. Single- or double-skinned panel assemblies can be easily cut owing to the short blade stroke. The tool can be used to cut a wide variety of materials including light-gauge steel, aluminium, reinforced glass fibre, wood and hardboard. Its fine cut is ideally suited for panel replacement where joints have to be cut and welded. Another power saw is the rotary or circular saw.

Power drills

Standard drills. There are two types of power drill used in the bodyshop: the majority are usually electrically driven (**corded**) and require a 13 A electrical power point, while the others are **air powered** and require a compressor and air points. The most popular sizes in common use are the 10 mm (3/8 inch) pistol drill and 19 mm (3/4 inch) heavy-duty machines. Drills are used mostly for drilling out spot welds along damaged panels which need removing, and also for drilling out broken rivets and bolts. As well as the standard size drills, there is a special drill called an angle drill, which is extremely useful in bodywork, as it can be used in confined spaces and enables many otherwise impossible drilling tasks to be carried out.

Cordless drill. This means that it does not have a wire or cord. Instead it uses a rechargeable battery to supply the power, with a fast-charge facility for recharging from a mains supply. It can be single speed or two speed, and can drill up to 10 mm-diameter holes in

Figure 3.12 Cordless drill-driver (*Bosch*)

steel. When fitted with a reversible adjustable clutch this tool can be used as a power screwdriver, thus making it a tremendously versatile piece of equipment, especially in the body-building industry.

Specialized air drill (spot-weld remover). This is a hand-held clamp-type drilling tool designed for cutting out spot welds. It is powered by compressed air with controllable speeds and has a removable cutting drill supplied in two sizes, 6 mm and 8 mm; the depth of cut is also adjustable. When there is a flange at the rear of the spot weld, it is supported by the clamping action of the C-shaped arm while the drill is cutting out the spot weld to a pre-set depth, allowing the panels to separate. Owing to the shape and clamping action of the tool, it is restricted to removing spot welds from an open panel edge only. Welds on panel edges which have no access from the rear would entail the cutting of the panel to be replaced to obtain access for the tool clamp. This tool could be a worthy addition to any bodyshop because of its time saving over conventional methods and its neatness of finish.

Sanding machines

Whether they are disc, belt, orbital, flat or finishing, sanders are the most frequently used power tools in the body repair shop.

Disc sander is the most popular sander. It is a powerful machine with a hard rubber flexible pad onto which the actual sanding disc is fastened by means of a centre nut. This nut is threaded in such a way that it tightens as the machine rotates, thus avoiding the risk of a disc coming loose. Disc sanders are driven either electrically or by compressed air.

The main uses of the disc sander in the repair shop are sanding sections on panels which are either rusted or have been repaired by planishing and require sanding to a smooth finish in readiness for painting; cleaning sections that need to be filled with solder prior to tinning; stripping paint prior to planishing or dressing out; dressing off

surplus weld metal after welding new sections; feather edging minor paint blemishes; and blending in body solder or plastic filler in sections that have been filled.

Belt sander can be used for repair to bodywork and is extremely good on convex or concave surfaces where access with a disc sander would be difficult. This type of machine uses an endless sanding belt, which is 3 in. (16 mm) wide, instead of the normal round disc. Some models are fitted with a vacuum unit so that the dust is kept to an absolute minimum in the workshop.

Pneumatic sander/filer. This is a dual-purpose tool for either sanding or filing. For sanding, an abrasive paper sheet is clipped to the tool. It has a reciprocating straight-line action which speeds up feather edging or filing materials and results in a smooth surface finish. A standard file blade can be fitted to the tool's base to convert it into a power-operated body file with a speed of up to 3000 strokes per minute.

Dual-action sander. The dual action (**DA**) on this sander makes the pad oscillate counter-clockwise while the head revolves clockwise, resulting in a non-repeating surface pattern that eliminates scratches on the surface of the panel.

Geared orbital sander/polisher. This is a multi-orbital action machine which

Figure 3.13 Dual-action sander

eliminates swirl marks and allows coarse-grit paper to be used for high stock removal rates. On paint and fibreglass, it leaves a perfect key for primers. The multi-orbital action reduces work surface heat generation, which is particularly important when sanding fillers and paint. The tool can also be used for polishing because of its rotary gearing. It can also be connected to a centralized or free-standing dust collection system.

Small angle grinder, the small angle grinder is sometimes known as a mini-grinder or a **grinderette**. It is a versatile one-handed machine and will cut away damaged metal, remove rust, grind down spot, MIG and gas welds, smooth in-fills and generally grind down prepared surfaces. Owing to its small size and light weight, it is ideal for grinding in confined spaces and tight corners. These machines normally use a hard disc for grinding.

Other power tools

Impact air screwdriver. The impact air screwdriver has a quick-change bit holder for inserting different screwdriver bit heads: slot, Phillips and Pozidriv. It is ideal for removal of rusted-in screws and other fasteners in locations such as door locks and door hinges.

Bonded windscreen electric cutter. This cutter greatly simplifies what was formerly a time-consuming task for two people. With this tool, windows bonded in with polyurethane adhesive can be taken out without difficulty. There is no environmental hazard and no fumes which might constitute a health risk. Residual adhesive can be removed without damaging the body.

The special cutter uses an AC/DC motor. An electric circuit permits the blade oscillating frequency to be varied to suit the required cutting speed. A wide selection of special-purpose steel cutting blades is available for all current car models. To simplify operation at various points on the body, the cutter blade has a twelve-sided mount to vary its position in relation to the tool. This precision

Figure 3.14   Bonded screen seal cutter

tool is used to cut through seals, joints and hard-bonded connections with exceptional speed and ease. Bonded-in car windscreens, for example, can be removed without damaging either the glass or the frame.

Dust extraction for power tools

Dust extraction is built into many power tools and other equipment, for it is essential that airborne dust levels in bodyshops are reduced to a minimum for the safety of the bodyshop employees. Sanding, grinding and cutting operations in the bodyshop usually generate quantities of air-borne dust; this is a major hazard not only to health but also for refinishers wanting a quality paint finish. Systems therefore have been specifically designed to overcome problems of dust, noise and oil pollution. Typically, the equipment consists of a high-performance dust-extractor unit and a range of plug-in air tools which includes random orbital sanders, orbital sanders, saws and grinders. Dust created during the sanding or sawing process is immediately extracted into a handy, easily emptied vacuum pump. In the case of the sanders, a series of holes in the pad itself is used to remove the dust and debris, which would otherwise form a cloud and obscure the surface to be worked on as well as necessitating breathing apparatus for safety's sake. A single quick-fix hose connects the hand tool to the power source. The hose has separate compartments for compressed air supply, exhaust and dust extraction. Most sanding and flatting tools have adjustable speed control. They have housings made of high-impact GRP, which makes them light in weight and suitable for prolonged usage without operator fatigue.

There is detailed legislation on the control of dust and its disposal. Most bodyshops run one of the many proprietary systems which are available. These systems usually include service contracts and insurance on the extractor systems. Depending on the system usage this will include regular inspection at pre-set intervals. Disposal of bodyshop waste must be carried out by licenced companies and records kept to show that it has been disposed of in an environmentally friendly manner.

Skills and questions

To become a skilled body repairer, you need to be able to use the hand tools, power tools and equipment which are found in the bodyshop. Bodyshops use different tools and equipment depending on the type of work they do. What is important is that you can use the tools and equipment particular to your company; but for your tests and examinations, you will need to know about the full range of tools and equipment. The following tasks will help you to develop your skills. Take your time, as skills take time to learn and, when learnt, need practice and application.

1 Use at least four hand tools, or power tools, or other pieces of equipment, when you have done this sketch the items and label them – noting any particular safety points to be aware of.
2 List the hand tools which you would expect to find in a body repairer's toolkit.
3 Describe the dust-extraction and disposal system used by the company where you work – make sure that you state the part which you play in this environmental control system.

4 Explain how you would use a dolly and hammer to straighten a bend piece of metal – use sketches if appropriate.
5 Describe how to change the blade on a hacksaw or a power saw.
6 Sketch and describe each of the following hand tools, giving an example of the type of work it would be used for: (a) cross-pein and finishing hammer, (b) body spoon, (c) planishing hammer.
7 List the hand tools which you would expect to find in a body repairer's toolkit.
8 Describe a suitable application for the use of a flexible body file.
9 Describe a suitable application for (a) a toe dolly, (b) an anvil dolly.
10 Explain a repair situation in which a body spoon may be used instead of a panel dolly.
11 Sketch and describe four essential hand tools used by the body repairer.
12 Name and explain the use of a hand tool which would have the effect of spreading a blow, thus reducing the stretching of the metal.
13 For what purpose would a sandbag and a wooden hollowing block be used?
14 State a typical use for a pear-shaped mallet.
15 Describe a typical repair application in which a pick hammer would be used.
16 Describe a repair situation in which each of the following special hand tools would be used: (a) impact driver, (b) panel puller, (c) edge setting tool, (d) door skinner.
17 Describe how the following four power tools could be used by the body repair worker: (a) power saw, (b) spot-weld remover, (c) bonded windscreen cutter, (d) miniature belt sander.
18 Describe the movement of a dual-action (DA) sander.
19 In what circumstances would a grid dolly be used?
20 Compare the advantages of a beating file with those of a planishing hammer.
21 When referring to sheet metal snips, what do the terms 'right hand' and 'left hand' mean?
22 Describe three types of sheet metal bench stakes, and sketch one of them.

23 Explain two safety precautions which must be observed when using a cold chisel.

24 What safety precautions must be observed before using an electric drilling machine?

25 Describe the safety measures which would be necessary when using hand and small power tools in a workshop.

26 What action should be taken to render a panel hammer safe for further use after its head has become loose?

27 Why are the teeth on a body file milled in a curved formation?

28 Some beating files and panel hammers have serrated faces. Explain the reasons for this.

29 Name a pneumatic tool which can be used for the removal of a damaged panel section.

30 Name three types of spanner that could be used by a body repair worker.

31 Describe the advantages of socket sets in removing and replacing damaged panels.

32 Sketch a Torx screwdriver bit that could be used with an impact screwdriver.

33 State the function of an air transformer and its importance when used with power tools.

34 Explain the two different types of air hose couplings used with pneumatic tools.

35 Explain the reason why the face of a planishing hammer should be maintained in perfect condition.

36 State the dangers of carrying sharp or pointed tools in pockets or protective clothing.

37 When a fault develops on an electrical power tool, what action must be taken?

38 Explain the reasons why some tools have a built-in dust-extraction system while others are connected to a centralized dust-extraction system.

39 Identify and sketch two types of mole grips used in body repair applications.

40 With the aid of a sketch, identify the BS symbol used on power tools to indicate double insulation.

4

Materials used in vehicle bodies

Vehicles are built from a very wide range of materials. This range includes metals such as steel and aluminium; natural materials such as wood and leather; and synthetic (man-made) materials such as PVC and GRP. The choice and the range of materials are constantly changing. Some materials like leather and wood are only found on premium-quality cars because of the costs involved; other cars only use the lightest synthetic materials to help them go as fast as possible.

Skills

The skills that you need to develop in relation to the materials used in vehicles are:

- Being able to identify the different materials correctly so that you can select both the correct material and the appropriate process to work on it
- Having the ability to cut and join the different materials to give a professional finish to the job
- Identifying the ways in which to safely handle and store the different materials to prevent damage to the material as well as the personnel involved
- Being able to explain about the materials used in the construction of the vehicle to the customer. This is often needed to explain why some processes are time consuming.

Steel-making process

In the manufacture of **steel coils**, the material needed for vehicle bodies, the raw material **iron ore** is fed into a **blast furnace**, together with limestone and coke; the coke is used as a source of heat, while the limestone acts as a flux and separates impurities from the ore. The ore is quickly reduced to molten iron, known as pig iron, which contains approximately 3 to 4 per cent carbon. In the next stage of manufacture, the iron is changed into steel by reheating it in a steel-making furnace and blowing oxygen either into the surface of the iron or through the liquid iron, which causes oxidation of the molten metal. This process burns out impurities and reduces the carbon content from 4 per cent to between 0.08 and 0.20 per cent.

Because of the constantly shrinking supply of raw materials and the increased awareness of the need to protect the environment, more and more **scrap** metal is being used in the production of steel. That is, old cars are crushed into blocks at the breaker's yard, and then melted in the furnace to make fresh steel. From the furnace, the steel typically undergoes a series of processes which are covered in the next paragraphs.

Casting – The steel is cast into **ingots**; these are either heated in a furnace and rolled down to a slab or, more commonly, continuously cast into a slab. Slabs by either casting process are typically 200 to 250 mm

(8–10 inches) thick, ready for further rolling. These slabs are reheated prior to rolling in a computer-controlled continuous hot strip mill to a strip around twice the thickness required for body panels. The strip is closely wound into coils ready for further processing.

Pickling – Before **cold rolling**, the surfaces of the coils must be cleaned of **oxide** or black scale formed during the hot rolling process, which would otherwise ruin the surface texture. This is done by pickling the coils in either dilute **hydrochloric acid** or dilute **sulphuric acid** and then washing them in hot water to remove the acid. The acid removes both the oxide scale and any dirt or grit which might also be sticking to the surface of the coil.

Cold rolling – In the cold rolling process, the coil is rolled either in a single-stand reversing mill (narrow mills using either narrow hot mill product or slit wide mill product) or in a multiple-stand tandem mill to the required thickness. Most mills are computer controlled to ensure that the thickness is exactly maintained and employ specially prepared work rolls to ensure that the right surface standard is achieved on the rolled strip. The cold rolling process hardens the metal, because mild steel quickly hardens. The cold rolled coils are suitable for applications such as panelling where no bending or very little deformation is needed. At this stage the coil is still not suitable for the manufacture of the all-steel body shell, and it must undergo a further process to soften it; this is known as annealing.

Annealing of the coils of sheet steel used for the manufacture of a car body must not only have a bright, smooth surface but must also be soft enough for bending, rolling, shaping and pressing operations, and so the hardness of cold rolled coils to be used for car bodies must be reduced by annealing. If annealing were carried out in an open furnace this would destroy the bright surface of the coil, and therefore oxygen must be excluded or prevented from attacking the metal during the period of heating the coils.

The normal method of annealing the coils is box annealing. The coils are stacked on a furnace base, covered by an inner hood and sealed. The atmosphere is purged with nitrogen and hydrogen to eliminate oxygen. A furnace is then placed over the stack and fired to heat the steel coils to a temperature of about 650°C for around 24 hours, depending upon charge size and steel grade.

Temper rolling – During the process of annealing the heat causes a certain amount of buckling and distortion, and a further operation is necessary to produce flat coils of body steel. The annealed coil is de-coiled and passed through a single-stand temper mill using a specially textured work roll surface, where it is given a light skin pass, typically of 0.75 to 1.25 per cent extension. This is necessary to remove buckles formed during annealing, to impart the appropriate surface texture to the strip and to control the metallurgical properties of the strip. The strip is then rewound, ready for despatch or finishing as appropriate.

Finishing – The temper rolled coils can be slit to narrow coil, cut to sheet, re-inspected for surface critical applications or flattened for flatness-critical applications as appropriate. Material can be supplied with a protective coating of oil and packed to prevent damage or rusting during transit and storage.

Specifications of steels used in the automobile industry

The motor body industry uses many different types of steel. Low-carbon steel is used for general constructional members. High-tensile steels are used for bolts and nuts which will be subjected to a heavy load. Specially produced deep-drawn steel including micro-alloyed steel is used for large body panels which require complex forming. Zinc-coated steel sheets are increasingly being specified for automobile production, both for body and chassis parts, as improved corrosion protection is sought. **Stainless steel is used for its non-rusting, hard-wearing and decorative qualities.**

The many different types of springs used in the various body fittings are produced from spring steel, while specially hardened steels make the tools of production. Drills, chisels, saws, hacksaws and guillotine blades are all produced from special alloy steels, which are made from an appropriate mixture of metals and elements. Steel varies from iron chiefly in carbon content; **iron contains 3 to 4 per cent carbon**, while **carbon steels may contain from 0.08 per cent to 1.00 per cent carbon**. The chemical composition and mechanical properties of these carbon steels, especially when alloyed with other elements such as nickel, chromium and tungsten, have been gradually standardized over the years, and now the different types of steels used are produced to specifications laid down by the British Standards Institution. A British Standard specification defines the chemical composition and mechanical properties of the steel and also the method and apparatus to be used when testing samples to prove that the mechanical properties are correct. The tensile strength, and in the case of sheet and strip steel the bend test, are the properties of most interest, but the British Standard specification also defines the elongation, the yield point and the hardness of the steel.

The steels used in the motor trade may be grouped as follows:

1   Cold forming steels
2   Carbon steels
3   Alloy steels
4   Free cutting steels
5   Spring steels
6   Rust-resisting and stainless steels.

Carbon steel

Carbon steels can be classified as **low-carbon steel, carbon-manganese steel, micro-alloyed steel, medium-carbon steel**, and **high-carbon steel.**

The properties of plain carbon steel are determined principally by carbon content and microstructure, but it may be modified by residual elements other than carbon, silicon,

manganese, sulphur and phosphorus, which are already present. **As the carbon content of the steel increase, so do the strength and hardness, but this is at the expense of the ductility and the malleability.**

Low-carbon steel – for many years low-carbon steel (sometimes referred to as mild steel) has been the predominant auto-body material. Low-carbon coil, strip and sheet steel have been used in the manufacture of car bodies and chassis members. This material has proved to be an excellent general-purpose steel offering an acceptable combination of strength with good forming and welding properties. It is ideally suited for cold pressings of thin steel sheet and is used for wire drawing and tube manufacture because of its ductile properties. Low-carbon steel is soft and ductile and cannot be hardened by heating and quenching but can be case hardened and work hardened. It is used extensively for body panels, where its high ductility and malleability allows easy forming without the danger of cracking. In general, low-carbon steel is used for all parts not requiring great strength or resistance to wear and not subject to high temperature or exposed to corrosion. However, new factors such as a worldwide requirement for fuel conservation for lighter-weight body structures and safety legislation requiring greater protection of occupants through improved impact resistance are bringing about a change in materials and production technology. This has resulted in the range of micro-alloyed steels known as high-strength steels (HSSs) or high-strength low-alloy steels (HSLAs).

Micro-alloyed steel is basically a carbon-manganese steel having a low carbon content, but with the addition of micro-alloying elements such as niobium and titanium. Therefore it is classed as a low-alloy, high-strength steel within the carbon range. As a result of its strength, toughness, formability and weldability, the car body manufacturers are using this material to produce stronger, lighter-weight body structures. A typical composition utilized for a micro-alloyed high-strength steel (HSS) is as follows:

Table 4.1 HSS steel

No.	Alloying Material	%
1	Carbon (C)	0.05–0.08
2	Manganese (Mn)	0.80–1.00
3	Niobium (Nb)	0.015–0.065

The percentage of niobium used depends on the minimum strength required. Formable HSSs were developed to allow the automotive industry to design weight out of the car in support of fuel-economy targets. A range of high-strength formable steels with good welding and painting characteristics has been developed. The steels are **hot rolled for chassis and structural components and cold reduced for body panels**. Through carefully controlled composition and processing conditions, these steels achieve high strength in combination with good ductility to allow thinner gauges to be used: a reduction from 0.90 mm down to 0.70 mm is a general requirement.

There was a danger that the new HSS, thinner but just as strong, would lack the ductility which allows it to be press formed into shape. This problem was overcome by the use of micro-alloying. In a crystalline microstructure of atoms the atoms are in layers; when the crystals are stretched as they are in the forming process, one layer of atoms slides over another. The layers of atoms slide like playing cards in a pack, and in doing so are changing shape in a ductile way. The sliding can be controlled by adding additional elements to the steel such as niobium or titanium. The element reacts with the carbon to produce fine particles which spread through the steel. The element controls the ductility and strengthens the steel, thereby improving the properties of the material.

These steels are low carbon, employing solution strengthening, cold reduced rephosphorized, or precipitation hardening that is hot rolled and cold reduced micro-alloyed

elements to produce fine-grained steels which are suitable for welding by spot and MIG processes only.

Tech note
If possible, get your lecturer to show you a variety of steels under a high resolution microscope. This way you can see the structure at first hand.

Medium-carbon steel can be hardened by quenching, and the amount it can be hardened increases with its carbon content. This type of steel can be used for moving parts such as connecting rods, gear shafts and transmission shafts, which require a combination of toughness and strength, but it is being replaced in the car industry by high-alloy steels. Medium-carbon steel can be found on many commercial vehicles – it is used for truck chassis and split-rim wheels. It is also occasionally used for load-bearing cross members on larger cars.

High-carbon steel can be hardened to give a very fine cutting edge but with some loss of its ductile and malleable properties. It is used for metal- and wood-cutting tools, turning tools, taps and dies, and forging and press dies because of its hardness and toughness but is seldom used now for motor vehicle parts because of the introduction of high-alloy steels.

Zinc-coated steels are used in the automotive industry in the quest to provide extended warranties. A six-year anti-corrosion warranty is the norm for a passenger car. The use of zinc-coated steels is one way of achieving this. Modern automobiles must be not only of high quality but also durable and economical, as perceived by their purchasers. These vehicles are expected to exceed an average of seven years without structural or cosmetic deterioration due to corrosion. Increasingly aggressive environmental influences tend to shorten the life of the car, whereas ever more specialized steel

Figure 4.1   Body shell panels showing galvanized protection (*Motor Insurance Repair Research Centre*)

1	Front grille panel	B	25	Front door frame stiffener RH and LH	A
2	Reinforcement	B	26	Front door skin RH and LH	A
3	Front bumper mounting reinforcement RH and LH	B	27	Rear door skin RH and LH	A
4	Bonnet lock reinforcement RH and LH	B	28	Rear door frame RH and LH	B
5	Front wing RH and LH	B	29	Rear door frame stiffener RH and LH	A
6	Bonnet lock panel	B	30	Rear door frame gusset RH and LH	B
7	Front suspension turret stiffener RH and LH	A	31	Rear door skin stiffener RH and LH	A
8a/8b	Front wheel arch gussets RH and LH	B(a)/C(b)	32	Rear door lock stiffener RH and LH	A
9	Front bulkhead stiffener RH and LH	A	33	Front window frame stiffener RH and LH	A
			34	Front door frame RH and LH	B
10	Front inner wing RH and LH	B	35	Front door skin stiffener RH and LH	A
11	Front suspension turret RH and LH	A	36	Front door lock stiffener RH and LH	A
12	A-post reinforcement RH and LH	B	37	Sill rear gusset RH and LH	A
13	Body side gussets RH and LH	B	38	Rear chassis leg RH and LH	A
14	Body side RH and LH	A	39	Floor/heelboard gussets	A
15	Inner sill RH and LH	A	40	Rear chassis leg gusset RH and LH	B
16	Inner sill reinforcement RH and LH	A	41	Boot floor cross members	A
17	B-post gusset RH and LH	A	42	RH boot floor brace closing panel	A
18	Upper dash panel	B	43	Boot floor brace RH and LH	A
19	Bonnet skin	B	44	Boot floor bracket RH and LH	A
20	Bonnet frame	B	45	Rear bumper mounting gusset RH and LH	B
21	Dash stiffener RH and LH	B	46	Boot lid frame	B
22	Front chassis leg closing panel RH and LH	A	47	Exhaust rear hanger stiffener	B
			48	Boot lid skin	A
23	Front chassis leg gusset RH and LH	B	49	Boot lid lock mounting stiffener	B
24	Front chassis leg RH and LH	A	50	Rear wing RH and LH	A
			51	Rear suspension turret RH and LH	A
			52	Suspension turret capping RH and LH	B
			53	Boot floor side extension RH and LH	A
			54	Outer rear wheel arch RH and LH	A

A = Galvanized one side only
B = Galvanized both sides
C = Galvanized layer applied individually

sheets are being incorporated into vehicle construction in the battle against corrosion.

The use of zinc-coated steels has dramatically been increased to meet these challenges. Different areas of a vehicle require different zinc coatings and coating weights to meet appearance and performance criteria. Typically these coatings are available in both hot-dipped and electrolytically deposited versions, in a range of coating weights or thicknesses. Both types of coatings offer both barrier protection and sacrificial corrosion protection and the choice of product depends on the particular application and requirements. The hot-dip product, which is available as either plain zinc or iron-zinc alloy, is generally used for underbody parts. The electrolytically deposited protection product is used for exposed body panels, where a full-finish surface quality is available to ensure that a showroom paint finish is achieved. The electrolytically deposited product can be applied to one side only or to both sides of a body panel.

Single-sided zinc-coated steel – Free zinc is applied to one side of a steel sheet by either the hot-dip or the electrolytic process for this material. Its uncoated side provides a good surface for paint appearance, so it is used mainly for outer body panels. Since free zinc is towards the inside of the car, it protects against perforation corrosion.

One-and-a-half-sided zinc-coated steel – in this case, one side of the sheet is coated with free zinc, and a thin layer of zinc-iron alloy is formed on the other side. This product is produced mainly by the hot-dip process. It is primarily used for exposed panels, where the zinc-iron layer is on the outside for cosmetic protection and the free zinc side provides perforation protection.

Double-sided zinc-coated steel is manufactured by applying free zinc to both sides of the sheet with equal or different coating weights. All types are readily paintable and weldable. However, care should be taken to ensure that welding conditions are comparable with the material used: for example, it may be necessary to use higher weld current ratings on the thicker zinc coatings.

Alloy steel is a general name for steels that owe their distinctive properties to elements other than carbon. They are generally classified into two major categories, these are:

- *Low-alloy steels*, which possess similar microstructures to, and requires similar heat treatments to, plain carbon steels
- *High-alloy steels*, which are defined as a steel having enhanced properties owing to the presence of one or more special elements or a larger proportion of element than is normally present in carbon

Table 4.2 Alloying element properties used with steel

No.	Alloying Element	Properties
1	Chromium	Increased hardness and resistance to corrosion.
2	Cobalt	Increased hardness, especially at high temperatures.
3	Manganese	High tensile strength, toughness and resistance to wear.
4	Molybdenum	Increased hardness and strength at high temperatures.
5	Nickel	Increased tensile strength, toughness, hardness and resistance to fatigue.
6	Niobium	Strong carbide forming effect; increases tensile strength and improves ductility.
7	Silicone	Used as a deoxidizing agent, and has the slight effect of improving hardness.
8	Titanium	Strong carbide forming element.
9	Tungsten	Greater hardness, especially at high temperatures; improved tensile strength and resistance to wear.
10	Vanadium	Increased toughness and resistance to fatigue.

steel. This section is concerned primarily with high-alloy steels.

Alloy steels usually take the name of the element or elements, in varying percentages, having the greatest influence on the characteristics of the alloy. Those typically found in the automotive industry are:

Tech note

The correct heat treatment is essential to develop the properties provided by alloying elements. You must be careful when welding not to damage the heat treatment, as this could cause the part to fail at a later date.

There are many alloy steels containing different combinations and percentages of alloying elements, of which some of the most popular are as follows:

- *High-tensile steel* is used whenever there is an essential need for an exceptionally strong and tough steel capable of withstanding high stresses. The main alloying metals used in its manufacture are nickel, chromium and molybdenum, and such steels are often referred to as nickel-chrome steels. The exact percentage of these metals used varies according to the hardening processes to be used and the properties desired. Such steels are used for gear shafts, engine parts and all other parts subject to high stress.
- *High-speed steels* are mostly used for cutting tools because they will withstand intense heat generated by friction and still retain their hardness at high temperatures. It has been found that adding tungsten to carbon steel forms alloy steel, which will retain a hard cutting edge at high temperatures. High-speed steels are based on tungsten or molybdenum or both as the

primary heat-resisting alloying element; chromium gives deep hardening and strength, and vanadium adds hardness and improves the cutting edge.

- *Manganese steel.* The addition of a small percentage of manganese to steel will produce an alloy steel which is both extremely tough and very resistant to wear. It is used extensively in the manufacture of chains, couplings and hooks.
- *Chrome-vanadium steel* contains a small amount of vanadium, which has the effect of intensifying the action of the chromium and the manganese in the steel. It also aids in the formation of carbides, hardening the alloy and increasing the ductility of the steel. These steels are valuable where a combination of strength and ductility are desired. Most spanners are made from chrome-vanadium. This alloy is also used for axle shafts, connecting rods, springs and torsion bars.
- *Silicon-manganese steel* is spring steel using the two elements of manganese and silicon. These steels have a high strength and impact resistance and are used for road springs and valve springs.
- *High-strength steels* have been introduced into automotive production slowly only because of the need for very expensive specialized press tools to form the body panels from this stronger material. The die tools need to be much harder than for normal low-carbon (LC) steel, and the presses need to be stronger and more accurate. HSS came about because of the need to make vehicles lighter following the 1970 fuel crisis. The lighter the car is the better fuel economy will be. This concept has led to the American car makers forming the **Ultra-Light Steel Auto Body (ULSB)** group and the **Ultra-Light Steel Body – Advanced Vehicle Concepts (ULSB – ACV)** group. This further research has led to the concept of advanced high-strength steel (AHSS) as the materials have been developed and understood.

Tech note
You can identify cars which are made out of HSS by their shape. HSS cars usually have longer flowing lines than other cars; their panels have long swage lines to give the panels extra strength. These styling clues are most noticeable on very stylish German cars. HSS panels, when tapped by the finger ends, produce a higher-pitched note than those of more conventional construction, providing that no sound deadening material is in place.

- *Cost* – Steel costs about one-fifth of the price of aluminium when bought in the quantities needed by a car maker. Also the iron and steel industry has hundreds of years of practical experience in shaping and forming steel compared to the other materials which could be used to make vehicle bodies.
- *Properties of HSS* – High-strength steel has a yield strength ranging from 300 to 1200 MPa compared to LC steel, which has a range of 140 to 180 MPa. However, although the metal is stronger, it is not necessarily stiffer. That is, the body parts can not necessarily be made of thinner metal, as they are likely to sag. If you look at the swage lines on the latest vehicles, you will see that many panels are stiffened by the use of swaging. The current modern shapes are to allow the usage of thinner sheet steel, which is lighter and of course cheaper. Oddly, however, the new vehicles are not lighter in weight; this is because of the addition of electrical body controls such as electric windows and seats. HSS is not as easy to form as LC steel; also some types of HSS can be drawn better than others. Generally the extra strength of HSS is brought about by changes in the steel microstructure during the steel processing. The following paragraphs discuss the different types of HSS and AHSS steels.

- *HSS* are also known as re-phosphorized – added phosphorous; isotropic – added silicone and bake hardened – strain age hardened. The two most common types used in vehicle body construction are MSLA and HSLA.
- *Medium-strength low-alloy steel* has a yield strength of between 180 and 300 MPa. This steel is made by dissolving more phosphorous or manganese alloy into the molten steel during manufacture.
- *High-strength low-alloy steel* has a yield strength of between 250 and 500 MPa. This is made by adding small amounts of titanium or niobium to the molten steel which produces a fine dispersion of carbide particles.
- Advanced high-strength steel types are aimed at producing steel with suitable mechanical properties for the forming of vehicle body parts, usually through the hydro-forming process – using water pressure to mould the metal over the die.
- *Dual-phase (DP) steel* has a yield strength of between 500 and 1000 MPa. It is made by adding carbon to enable the formation of (hard) martensite in a more ductile ferrite matrix. Manganese, chromium, vanadium or nickel may also be added. The DP steel may have its strength triggered by either bake hardening or work hardening when it is stressed under the stamping or other forming process.
- *Transformation-induced plasticity (TRIP) steel* has a yield strength of between 500 and 800 MPa, with greater figures attainable in some cases. TRIP steels may be alloyed with higher quantities of carbon and silicone and aluminium. The strength is triggered by work hardening by the stress induced during the stamping or forming process. That is, the retained austenite is transformed into martensite by the increasing strain during the stamping or other forming process.
- *Complex-phase (CP) steel* has a yield strength of 800–1200 MPa. CP steel has a very fine microstructure using the same

alloying elements as in DP or TRIP steel with the possible additions of niobium, titanium and/or vanadium. Again the high strength is triggered by applied strain.

Applications of AHSS

TRIP and CP steel is ideal for use in crash zones. It is excellent for absorbing energy during impact. CP steel is often used for 'A' and 'B' posts and bumper attachments. Increasingly, AHSS steel is used for strengthening members to which other steel panels are welded, in other words a steel composite structure.

Repair of HSS and AHSS panels

One of the problems is that it is not possible to recognize HSS and AHSS panels by sight. Therefore it is essential to follow the guidelines offered by the vehicle manufacturer on recommended repair methods. As a general guide, look out for parts such as 'A', 'B' and 'C' posts, screen pillars, cant rails and strengthening cross members on cars made after about 1990. On new cars look out for panels with large swage lines. Remember that the pressing process triggers the hardening of the metal and so makes the panel stiff both in shape and in microstructure. Damaged AHSS and HSS panels are not readily repairable, as the impact changes the microstructure, making the metal harder. So panel beating is not an option, and replacement panels must be fitted in most cases.

The normal method of joining HSS and AHSS panels is by spot welding. The heat and pressure involved in the welding process changes the microstructure of the metal, so great care is needed in this process. Again, follow manufacturer's instructions on these repairs. It is sensible to do tests before spot welding the new panels. You can use the undamaged sections of the panels which you have removed for test welds, changing weld time and current, then cutting through the welded area to check for penetration and adhesion.

Any form of applied heat to HSS and AHSS panels should be avoided. This includes trying to anneal or soften the panel for the purposes of straightening, heat shrinking or oxyacetylene cutting. Neither MIG plugging nor cold working is recommended. Remember that the nature of these processes will effect a change to the properties of the steel and that the energy of the impact will have had the same effect on the panel.

Guide to spotting AHSS and HSS panels

* Look for reinforced areas such as door pillars and cross-bracing areas and where two or more panel parts overlap each other.
* Look for body panels with pronounced sharp swage lines.
* Feel for very thin panels.
* Listen for panels which when tapped gently give a crisp, metallic ring.

Boron steel

A particular type of HSS is referred to as boron steel, containing a very small amount of boron, typically 0.001 to 0.004 percent. This steel is used by many of the major manufacturers such as Vauxhall and requires special repair techniques. To join boron steel panels the method used is often MIG brazing.

Stainless steel

The discovery of stainless steel was made in 1913 by Harry Brearley of Sheffield while he was experimenting with alloy steels. Amongst the samples which he threw aside as unsuitable, there was one containing about 14 per cent chromium. Some months later, he saw the pile of scrap test pieces and noticed that most of the steels had rusted, but the chromium steel was still bright. This led to the development of stainless steel. The classic Rolls-Royce radiator was one of the first examples of the use of stainless steel.

The designer, engineer or fabricator of a particular component may think that stainless steel is going to be both difficult to work and expensive. This is quite wrong and perhaps stems from the fact that many people tend to fall into the trap of the generic term 'stainless steel'. In fact, this is the title for a wide range of alloys. Therefore if such materials are to be used effectively and maximum advantage is to be taken of the many benefits they have to offer, there should be very close collaboration and consultation over which grade of stainless steel is best for the particular job in hand. There are more than 25 standard grades of stainless steel specified by British Standards. Each provides a particular combination of properties, some being designed for corrosion resistance, some for heat resistance and others for high-temperature creep resistance. Many, of course, are multipurpose alloys and can be considered for more than one of these functions. In terms of composition, there is one element common to all the different grades of stainless steel. This is chromium, which is present to at least 10 per cent. It is this element which provides the basis of the resistance to corrosion by forming what is known as a 'passive film' on the surface of the metal. This film is thin, tenacious and invisible and is essentially a layer of chromium oxide formed by the chromium in the steel combining with the oxygen in the atmosphere. The strength of the passive film, in terms of resistance to corrosion, increases within limits with the chromium content and with the addition of other elements such as nitrogen and molybdenum. The formation of the passive film, therefore, is a natural characteristic of this family of steels and requires no artificial aid. Consequently, if stainless steels are scratched or cut or drilled, the passive film is automatically and instantaneously repaired by the oxygen in the atmosphere. Stainless steels can be conveniently divided into the following four main groups:

- *Austenitic* – generally containing 16.5 to 26 per cent chromium and 4 to 22 per cent nickel.
- *Ferritic* – usually containing 12 to 18 per cent chromium.
- *Austenitic/ferritic duplex* – usually containing 22 per cent chromium, 5.5 per cent nickel, 3 per cent molybdenum and 0.15 per cent nitrogen.
- *Martensitic* – based on a chromium content of 11 to14 per cent, although some grades may have a small amount of nickel too.

Of the above groups, the austenitic steels are by far the most widely used because of the excellent combination of forming, welding and corrosion-resisting properties that they offer. Providing that the correct grade is selected as appropriate to the service environment and that the design and production engineering aspects are understood and intelligently applied, long lives with low maintenance costs can be achieved with these steels.

HyResist 22/5 duplex is a highly alloyed austenitic/ferritic stainless steel. It has more than twice the proof strength of normal austenitic stainless steels whilst providing improved resistance to stress corrosion cracking and to pitting attacks. It possesses good weldability and can be welded by conventional methods for stainless steel. The high joint integrity achievable combined with good strength and toughness permit fabrications to be made to a high standard. It is being increasingly used in offshore and energy applications.

Aluminium

Figure 4.2 Hydro-electric power generation

Tech note

You must keep aluminium completely separate from steel and other metals to prevent **cross-contamination**. This means completely separate storage facilities and keeping all the tools separate too.

In the present-day search for greater economy in the running of motor vehicles, whether in the private sector, the PCV sector or the commercial sectors, the tendency is for vehicle manufacturers to produce vehicle bodies which, whilst still maintaining their size and strength, are lighter in weight than those of the previous design. Aluminium is approximately one-third of the weight of an equivalent-strength piece of steel, and aluminium alloys can be produced which have an ultimate tensile strength of 340 to 620 MN/m². In the early 1920s the pioneers of aluminium construction were developing its use for both private and commercial bodies; indeed, the 1922 40-hp Lanchester limousine body had an aluminium alloy construction for the bulkhead and bottom frame, and aluminium was used for all the body panels. Before the Second World War aluminium was used mainly for body panels, but since the war aluminium alloys have been and are now being used for body structures except for a limited number of specialized vehicles in the very-high-end price bracket. Although aluminium is more expensive than steel, it is easy to work and manipulate and cleaner to handle. It also has the advantage of not rusting, and, provided that the right treatment is adopted for welding, corrosion is almost non-existent. In recent years the use of aluminium and aluminium alloys for motor bodies in the commercial field has developed enormously, because this allows an increase in vehicle pay-load.

In the modern motor body the saving of weight is its most important advantage, and although on average the panel thickness used is approximately double that of steel, a considerable weight saving can be achieved. One square metre of 1.6 mm-thick aluminium weighs 4.35 kg, while one square metre of 1.00 mm-thick steel weighs 7.35 kg; the use of aluminium results in a saving in weight of just under 40 per cent.

The non-rusting qualities of the aluminium group are well known and are another reason for their use in bodywork. An extremely thin film of oxide forms on all surfaces exposed to the atmosphere, and even if this film is broken by a scratch or chip, it will reform, providing complete protection for the metal. The oxide film, which is only 0.0002 cm thick, is transparent, but certain impurities in the atmosphere will turn it to various shades of grey.

Production of Aluminium

Aluminium has been known of for about for 130 years. This may seem a long time; but in terms of metalworking means it is relatively new. The industrial history of aluminium did not begin until 1886, when Paul Heroult in France discovered the basis of the present-day method of producing aluminium. Aluminium is now produced in such quantities that in terms of volume it ranks second to steel among the industrial metals. Aluminium of commercial purity contains at least 99 per cent aluminium, while higher grades contain 99.5 to 99.8 per cent of pure metal.

In the production of aluminium, the ore bauxite is crushed and screened, then washed and pumped under pressure into tanks and filtered into rotating drums, which are then heated. This separates the aluminium oxide from the ore. In the next stage the aluminium oxide is reduced to the metal aluminium by means of an electrolytic reduction cell. This cell uses powdered cryolite and a very heavy current of electricity to reduce the aluminium oxide to liquid metal, which passes to the bottom of the cell and is tapped off into pigs of aluminium of about 225 kg each. In the UK this process is mainly carried out in the North of Scotland, as hydro-electric power is readily available relatively cheaply.

Figure 4.3 Aluminium extruded strips

Types of aluminium sheet

Sheet, strip and circle blanks are sold in hard and soft tempers possessing different degrees of ductility and tensile strength. Sheet is supplied in gauges down to 0.3 mm, but it is generally more economical to order strip for gauges less than 1.6 mm.

Manufacturing process

Sheet products are first cast by the semi-continuous casting process, then scalped to remove surface roughness and preheated in readiness for hot rolling. They are first reduced to the thickness of plate and then to sheet if this is required. Hot rolling is followed by cold rolling, which imparts finish and temper in bringing the metal to the gauge required. Material is supplied in the annealed, that is soft, condition and in at least three degrees of hardness, H1, H2 and H3 (in ascending order of hardness).

Aluminium alloys

From the reduction centre the pigs of aluminium are re-melted and cast into ingots of commercial purity. Aluminium alloys are made by adding specified amounts of alloying elements to molten aluminium. Some alloys, such as magnesium and zinc, can be added directly to the melt, but higher-melting-point elements such as copper and manganese have to be introduced in stages. Aluminium and aluminium alloys are produced for industry in two broad groups:

1 Materials suitable for casting
2 Materials for the further mechanical production of plate sheet and strip, bars, tubes and extruded sections.

In addition both cast and wrought materials can be subdivided according to the method by which their mechanical properties are improved:

*   *Non-heat-treatable alloys* – Wrought alloys, including pure aluminium, gain in strength by cold working such as rolling, pressing, beating and any similar type of process.
*   *Heat-treatable alloys* – These are strengthened by controlled heating and cooling followed by ageing at either room temperature or at 100 to 200°C.

The most commonly used elements in aluminium alloys are copper, manganese, silicon, magnesium and zinc. The manufacturers can supply these materials in a variety of conditions. The non-heat-treatable alloys can be supplied either as fabricated (F), annealed (O) or strain hardened (H1, H2, H3). The heat-treatable alloys can be supplied as fabricated (F) or annealed (O) or, depending on the alloy, in variations of the heat treatment processes (T3, T4, T5, T6 and T8).

Abbreviations for basic temper

If you are buying aluminium sheet, you will come across a range of abbreviation used in it description. Some of the common ones are:

*   *As fabricated: F* The temper designation F applies to the products of shaping processes in which no special control

over thermal conditions or strain hardening is employed. For wrought products
there are no specified requirements for
mechanical properties.

- *Annealed: O* The temper designation O
  applies to wrought products which are
  annealed to obtain the lowest strength
  condition.

Abbreviations for strain-hardened materials

The temper designation H for strain-hard-
ened products (wrought products only)
applies to products subjected to the application of cold work after annealing (hot forming) and partial annealing or stabilizing in
order to achieve the specified mechanical
properties. The H is always followed by two
or more digits, indicating the final degree of
strain hardening. The first digit (1, 2 or 3)
indicates the following:

- H1 strain hardened only
- H2 strain hardened and partially annealed
- H3 strain hardened and stabilized

The second digit (2, 4, 6 or 8) indicates the
degree of strain hardening, as follows:

- HX2 tensile strength approximately midway between O temper and HX4 temper
- HX4 tensile strength approximately midway between O temper and HX8 temper
- HX6 tensile strength approximately
  midway between HX4 temper and HX8
  temper
- HX8 full hard temper

Abbreviations for heat-treated materials

The temper designation T applies to products
which are thermally treated, with or without
supplementary strain hardening, to produce
stable tempers. The T is always followed by

one or more digits, indicating the specific
sequence of treatments as follows:

- T3 solution heat treated, cold worked,
  and naturally aged to a substantially stable condition
- T4 solution heat treated and naturally
  aged to a substantially stable condition
- T5 cooled from an elevated temperature
  shaping process and then artificially aged
- T6 solution heat treated and then artificially aged
- T8 solution heat treated, cold worked
  and then artificially aged

The additional digits for the T tempers TX51
indicate that the products have been stress
relieved by controlled stretching.

Aluminium alloys used in bodywork

The choice of material and the condition
in which it is required must depend largely
upon design requirements and the manufacturing processes within the factory. Table 4.3
gives the properties and applications of popular aluminium alloys as used in the automotive industry.

> **Tech note**
> A truss panel is a panel which carries a
> heavy load, rather like a roof truss, and
> is responsible for holding major parts
> of the vehicle together. A gusset is a
> filling panel which holds two major
> panels together as they pull in differ-
> ent directions, like you sometimes get
> in shirt tails and underwear.

Rubber

The value of rubber lies in the fact that it
can be readily moulded or extruded to any
desired shape, and its elastic quality makes it

Table 4.3 Aluminium Sheet Applications

ID No	% Mg	% Si	% Mn	Density g/cm³	Application
1200	–	–	–	2.71	Vehicle panels where panel beating is needed, mouldings, trim and cladding
3103	–	–	1.2	2.73	Flat panelling of vehicles, general sheet metalwork, tanker cladding
5251	2.25	–	0.4	2.69	Body panels, head boards, drop sides, truss panels, cab panels, tanker shells and divisions
5154A	3.5	–	0.4	2.67	Welded body construction, tipper bodies and truss panels in buses
5454	2.7	–	0.8	2.65	Tanker shells and divisions
5083	4.4	–	0.5	2.66	Pressurized bulk transport at ambient or low temperature
6082 (Heat treatable)	0.7	1.0	0.5	2.70	All structural sections in riveted vehicle bodies, tipper panes, and highly stressed under-frame gussets and truss panels

capable of filling unavoidable and irregular gaps and clearances. It is an ideal material in door shuts and as the gasket for window glass, and in both instances it provides the means for excluding dust and water, although with windscreens and back-lights additional use has generally to be made of a sealing material. Rubber specifications have been built on the basis of the properties of material which has given satisfaction. One major difficulty has been to ensure and measure resistance to weathering; rubber is subject to oxidation by ozone in the atmosphere. This oxidation causes cracking in the surface of the rubber. In addition to natural rubber, a variety of types of synthetic rubbers are used by the motor industry; these vary in price and characteristics, and all are more expensive than natural rubber. For complete ozone resistance, it is necessary to use either butyl or neoprene rubber; both satisfy atmospheric and ozone ageing tests. Butyl rubber, however, is 'dead' to handle and contains no wax, and so whilst neoprene is costly its use is essential for some parts.

Sponge sealing rubbers can be provided with built-in ozone resistance by giving them a live skin of neoprene, and a further way of providing ozone resistance is to coat the rubber components with Hypalon; more recently, continuously extruded neoprene sponge has been adopted. Apart from weather resistance, the important requirement of door and boot lid seals is that their compression characteristics should ensure that they are capable of accommodating wide variations in clearance without giving undue resistance to door closing.

Various types of foam rubber have been evolved to suit the different parts of the car seating, and the designer's choice of material is governed by cost, comfort, durability, the type of base, the type of car, and whether it is a cushion or a squab, a rear or front seat. When considering the foams available, it is apparent that the number of permutations is large. The types of foam available today fall into seven broad categories. These are:

1 **Moulded latex foam**
2 **Low-grade fabricated polyether**
3 **Fabricated polyether**
4 **Moulded polyether**
5 **Fabricated polyester**
6 **Polyvinyl chloride foam**
7 **Reconstituted polyether**

Latex foam today utilizes a mixture of natural and synthetic latexes to obtain the best qualities of both. After being stabilized with ammonia, natural latex is shipped in liquid form to this country from Malaysia, Indonesia and other rubber-producing countries. Synthetic latex, styrene butadiene rubber, is made as a by-product of the oil cracker plants. Polyether foam can now be made in different grades, and the physical properties of the best grades approach those of latex foam. As a general guide, service life and physical properties improve as the density increases for any given hardness. As the cost is proportional to weight, it follows that the higher-performance foams are more costly.

Flexible polyurethane foam seats are replacing heavy and complicated padded metal spring structures. Moulded seats simplify assembly, reduce weight and give good long-term performance. A major innovation here has been the cold cure systems. These produce foams of superb quality, particularly in terms of strength, comfort and long-term ageing. The systems are particularly suitable for the newer seat technologies such as dual hardness, where the wings are firm to give lateral support, leaving the seat pad softer and more comfortable.

Sealers

The history of sealers is longer than that of the motor car. Mastic, bitumen compounds and putties of various kinds, referred to as caulking, have been used on wooden ships thousands of years, and they found their way onto horse drawn carriages and from there to the modern vehicle in the early 1920s. This was when mass production of the pressed steel body became popular. in 1927 one of the first truly effective sealers was introduced. It is known as Dum Dum and is still in use. It was a modified roof sealer and proved to have many applications in body production. Amongst the first developed was an inter-weld sealing compound, primarily to prevent corrosion. Since then, particularly in the post-war years, there have been remarkable developments, probably accelerated by criticisms from the overseas markets that British cars were susceptible to dust and water entry. Companies specializing in the manufacture of mastic compounds have developed a range of materials which are now used not only for welding and for general putty application but also for floor pans, drip rails, body joints, exterior trim and many other points, leading to car bodies which are very well sealed against dust and water ingress.

Application		Material description	Application equipment
Bolted joints		Performed strip	Hand or palette/putty knife
		Zinc-rich primer	Brush or spray
		Seam sealer light	Applicator gun (hand)
Spot-welded joints		Zinc-rich primer	Brush or spray
		Structural adhesive or seam sealer	
		Seam sealer light	Applicator gun (hand)
Bonded joints		Metal-to-metal adhesive semi-structural	Caulking gun

Figure 4.4 Types of sealed joints used in vehicle bodywork

The term 'sealer' covers a wide variety of materials used in the motor industry for sealing against water and dust, from products which remain virtually mastic (soft, not hardened) throughout their life to others which harden up but still retain some measure of elasticity. They range from mixtures of inert fillers and semi-drying oils to heat curing plastisols which may be applied in a thin paste form as an interweld sealer or as extruded beads. Sealing compounds can be categorized into the following general groups:

• Oil-based compositions,
• Rubber-based compositions
• Synthetic-resin-based compositions

The choice of each of the types will be dependent on the site for application, on the eventual conditions of exposure and often on price. These categories can be subdivided further into the various physical forms in which they can be made available, which include mastic putties for hand application, extruded sections for placing in precise locations, gungrade compositions, which have the advantage of speed and economy of application, and pouring and spraying grades.

The properties of sealers will obviously vary according to their type and to their application. Thus preformed strip or putty sealers must adhere to the surfaces to which they are applied and must not harden or crumble

Table 4.4 Sealants, adhesives and anti-corrosion materials

No.	Item	Properties	Applications
1	Thermosetting sealant	• Hardens when heated • Non-running	Sealant and adhesive
2	Spot sealer	• Electro-conductive and spot-welding can be done after application • Excellent water resistance	Spot-welding locations
3	Body sealant	• Highly solid, no volume shrinkage • Non-running	Sealing of sheet metal seams – drip rails, floor, body sides, panels and trunk
4	Structural adhesive	• Two-agent mixture adhesive • Low viscosity	Bonding (gluing) places that cannot be spot-welded, such as roofs and aluminium-construction vehicles
5	Interior trim adhesive	• Highly solid, non-running • Good heat resistance • No rubber swelling • Non-drying	Sealing for grommets, packing and metals
6	Quick-drying high-strength adhesive	• Quick drying – 10 seconds to 5 minutes • Dries colourless and transparent	Adhesives may be made for different applications
7	Wax injection	• Resists build-up of water	Used inside body sections
8	Rocker panel primer	• Resists corrosion	Priming rocker panels
9	Underbody anti-corrosion agent	• Surface protection against water and dirt	Coating vulnerable body panels
10	Undercoating	• Non-running • Thick application possible • Will not fail when subjected to low temperatures and constant wetness	Undercoating vulnerable body panels

in service. Glazing sealers must be capable of being readily applied from a gun, with the ability to harden off on the surface, but must remain mastic in the assembly so that they are capable of maintaining a leak-tight joint whatever deflection the body undergoes. Heat-gelling sealers must be capable of being readily applied by extrusion or possibly by spraying and then must set up when cured but still retain a degree of flexibility. As a result of soaring energy costs together with the need for car aerodynamic design, direct glazing of windscreens and fixed body glass was introduced, and an adhesive was required to bond glass windscreens to the metal aperture. The material used is polyurethane adhesive sealant. It possesses a combination of adhesion, sealant and gap-filling qualities; it is a one-component adhesive and sealing compound of permanent elasticity. This dual-purpose material is based on special moisture-cured polyurethane with an accelerated setting time. The curing time is dependent upon the humidity levels prevailing, as well as the temperature. For example, at 20°C with a relative humidity level of 65 per cent, a 6 mm-diameter bead will be tack free within 1 hour and fully cured in 24 hours.

Tech note

Structural adhesives are used widely on aluminium bodied vehicles. They are used both for securing panels to die-cast components and between panels which are secured by blind rivets. Blind rivets on aluminium vehicles generally only pierce one panel to prevent water leakage.

Sound-deadening, thermal insulating and undersealing materials

The type of material used for sound damping or deadening depends on whether it is also required to provide undersealing. A material required for sound deadening only will

normally be applied to the interior of a vehicle, whereas one required to provide sound deadening and undersealing properties will be applied to the underside of the vehicle. Thus the former need not be fully water resistant, whereas the latter must be water resistant in addition to many other necessary requirements. The sound-deadening properties of a material are related to its ability to damp out panel vibrations, and this in turn is related to some extent (but not solely) to its weight per unit volume. Thus the cheapest sound-deadening materials are based on mixtures of sand and bitumen, although these tend to be brittle. A better material is bitumen filled with asbestos; although this is probably less effective as a panel damper than sand-filled bitumen, it is nevertheless more suitable owing to its better ductility. In general, those sound deadeners applied to the interior of the vehicle are water-based bitumen emulsions with fillers, whereas sound deadener/sealers applied externally should be solvent-based materials. An effective sound deadener is clay-filled water dispersed polyvinyl acetate (PVA) resin emulsion; this has damping characteristics approximately three times better than the bitumen-based material, but naturally it is more expensive.

Other sound-absorbing materials are now used for insulation in the automotive industry. Needle felts are blends of natural and manmade fibres locked together by needle punching. These are used in die-cut flat sheet forms for attachment to moulded carpets, floor boot mats and as anti-rattle pads. Bonded, fully cured felts are similar blends of fibres bonded together with synthetic resins. All the binder is cured during the felt-making process. As with needled materials, these are mainly used for flat products, especially where low density is required such as in sound-absorption pads and floor mats. In moulded felts and moulded glass wool the binder is only partially cured during the felt-making process. The curing sequence is completed under the action of temperature and pressure in matched die compression moulded tools to produce components which have three-dimensional form and a controllable degree of rigidity. Fully cured and needled products can be given various surface treatments including abrasion-resistant and waterproof coatings such as latex, PVC or rubber, or they can be combined with bitumen to improve the sound insulation properties. Moulded felt can be supplied covered with a range of woven textile covers or with various grades of PVC and heavy-layer bitumen EVA products.

Polyurethane flexible moulded foam can be modified or filled to meet different insulating requirements in the vehicle. The foam can also be moulded directly onto the hard layer, allowing simple tailoring of insulation thickness.

Figure 4.5   Leather seat

Interior furnishings

Carpets and floor coverings – The interior body engineer has a choice of materials, ranging from carpeting of the Wilton type for prestige vehicles through polyamide and polypropylene moulded needle felts to rubber flooring for economy versions. The main requirements of flooring are wear resistance, colour fastness to light and water and adequate strength to enable the customer to remove the flooring from the car without it suffering damage. The method of manufacture of pile carpets varies: in some cases the pile is bonded to hessian backing; in other cases it is woven simultaneously into the backing and then anchored in position with either a rubber coating or a vinyl coating on the back surface. Quality is normally controlled by characteristics such as number of rows of tufts of pile per unit length, height and weight of the free pile, overall weight of the carpet, strength as determined by a tensile test in both the warp and weft directions, together with adhesion of pile if applicable. Rubber flooring generally has a vinyl coating to provide colour.

Leather (hide) – Large numbers of motor vehicle users all over the world continue to specify hide upholstery when the option is available and will gladly pay the extra cost

involved for a material which defies complete simulation. Great advances have been made in the development of suitable substitutes and the best of the plasticized materials, which are quite undistinguishable by eye from leather. The unique character of leather lies in its microstructure, the like of which is not obtained in any manmade material. Under a microscope leather can be seen to consist of the hairy epidermis and under that the corium, or bulk of the hide, this being the basis of the leather as we know it. By virtue of the millions of minute air spaces between the fibres and bundles of fibres, leather is able to 'breathe'. To the motorist this means that leather does not get hot and uncomfortable in warm weather or cold and inflexible in winter, and although permeable to water vapour it offers sufficient resistance if it is exposed to normal liquids. It is also strongly resistant to soiling, and when it does get dirty the dirt can usually be removed fairly easily without special materials. Unlike some plasticized materials, leather does not appear to attract dirt and dust as a result of static electricity. Many people, moreover, regard the distinctive smell of leather as an asset, and this defies imitation by manufacturers of substitute materials. However, natural hide has to go through many complex processes before it attains the form familiar to the upholsterer, trimmer or motorist.

> **Tech note**
> Hides of different animals produce leather with different properties. Ask your tutor to let you look at different leathers under the microscope.

Plastics

The word plastic really means that something can be bent and will stay in that shape. The opposite is elastic, that is something can be bent and will spring back to shape. Think of plasticine modelling material and elastic bands to get the idea.

Thermoplastics and thermosetting plastics

The simplest way of classifying plastics is by their reaction to heat. This gives a ready subdivision into two basic groups: thermoplastics and thermosetting plastics. Thermoplastic materials soften to become plastic when heated, with no chemical change taking place during this process. When cooled they again become hard and will assume any shape into which they were moulded when soft. Thermosetting materials, as the name implies, will soften only once. During heating, a chemical change takes place and the material cures; thereafter the only effect of heating is to char or burn the material. As far as performance is concerned, these plastics can be divided into three groups:

1. General-purpose thermoplastics
   - Polyethylene
   - Polypropylene
   - Polystyrene
   - SAN (styrene/acrylonitrile copolymer)
   - Impact polystyrene
   - ABS (acrylonitrile butadiene styrene)

- Polyvinyl chloride (PVC)
- Poly vinylidene chloride
- Poly methyl methacrylate
- Poly ethylene terephthalate

2. Engineering thermoplastics
   - Polyesters (thermoplastic)
   - Polyamides
   - Polyacetals
   - Polyphenylene sulphide
   - Polycarbonates
   - Polysulphone
   - Modified polyphenylene ether
   - Polyimides
   - Cellulosics
   - RIM/polyurethane
   - Polyurethane foam
3. Thermosetting plastics
   - Phenolic
   - Epoxy resins
   - Unsaturated polyesters
   - Alkyd resins
   - Diallyl phthalate
   - Amino resins

Amorphous and crystalline plastics

An alternative classification of plastics is by their shape. They may be crystalline (with shape) or amorphous (shapeless).

Amorphous plastics

Amorphous plastics basically are of three major types:

- ABS: acrylonitrile butadiene styrene
- ABS/PC blend
- PC: polycarbonate.

Amorphous engineering plastics have the following properties:

- High stiffness
- Good impact strength
- Temperature resistance
- Excellent dimensional stability
- Good surface finish
- Electrical properties

Table 4.5 Typical mechanical properties of representative plastics

Material	Modulus of elasticity E (MN/m²)	Tensile strength (MN/m²)	Compressive strength	Elongation (%)
LD polyethylene	120–240	7–13	9–10	300–700
HD polyethylene	550–1050	20–30	20–25	300–800
Polypropylene 900–140 32–35 35 20–300				
GFR polypropylene	1500+	34–54	40–60	5–20
Flexible PVC	3500–4800	10–25	7–12	200–450
Rigid PVC	2000–2800	40	90	60
Polystyrene	2400–4200	35–62	90–110	1–3
ABS copolymer	1380–3400	17–58	17–85	10–140
Perspex	2700–3500	55–75	80–130	2–3
PTFE	350–620	15–35	10–15	200–400
Nylon 11	1250–1300	52–54	55–56	180–400
GFR nylon 6	7800–800	170–172	200–210	3–4

- Flame retardance (when required)
- Excellent transparency (polycarbonate only).

In the automotive industry use is made of the good mechanical properties (even at low temperatures), the thermal resistance and the surface finish. The applications are:

1 Body embellishment
2 Interior cladding
3 Lighting where, apart from existing applications of back lamp clusters, polycarbonate is expected to replace glass for headlamp lenses.

Semi-crystalline plastics

Semi-crystalline plastics are in two basic types:

- Polyamide 6 and 66 types
- Polybutylene terephthalate (PBT).

Semi-crystalline plastics have the following properties:

- High rigidity
- Hardness
- High heat resistance
- Impact resistance

- Abrasion, chemical and stress crack resistance.

The semi-crystalline products find major application in the automotive sector, where full use is made of the mechanical and thermal properties, together with abrasion and chemical resistance. Examples include:

1 Under-bonnet components
2 Mechanical applications
3 Bumpers, using elastomeric PBT for paint on-line
4 Body embellishment (wheel trims, handles, mirrors)
5 Lighting, headlamp reflectors.

Blended plastics

Blended plastics have been developed to overcome inherent specific disadvantages of individual plastics. For large-area body panels, the automotive industry demands the following properties:

- Temperature resistance
- Low-temperature impact resistance
- Toughness (no splintering)
- Petrol resistance
- Stiffness.

Neither polycarbonate nor polyester could fulfil totally these requirements. This led to the combination of PC and PBT to form Macroblend PC/PBT, which is used for injection-moulded bumpers.

Plastics applications

Plastic products can be decorated by vacuum metallizing and electroplating. They have replaced metals in a lot of automotive applications, such as mirror housings, control knobs and winder handles as well as decorative metallic trim. It is a field which uses their advantages to the full without relying on properties they lack.

Thin parts must be tough and resistant to the occasional impact. They must be impervious to attack by weather, road salts, extremes of temperatures and all the other hazards that reduce older forms of body embellishments to pitted, rusted, dull, crumbling metal. They do not need high tensile strength or flexural strength, as they do not have to carry heavy stresses. They must be cheap and capable of being formed into highly individual and complex shapes. All these requirements are satisfied by thermoplastics and thermosetting resins. They can be pressed, stamped, blow moulded, vacuum formed and injection moulded into any decorative shape required.

Apart from their decorative properties, the mechanical properties of acrylic resins are among the highest of the thermoplastics. Typical values are a tensile strength of 35 to 75 MN/m^2 and a modulus of elasticity of 1550 to 3250 MN/m^2. These properties apply to relatively short-term loadings, and when long-term service is envisaged tensile stresses in acrylics must be limited to 10 MN/m^2 to avoid surface cracking or crazing. Chemical properties are also good, the acrylics being inert to most common chemicals. A particular advantage to the automotive industry is their complete stability against petroleum products and salts.

Acetal resins are mostly used for mechanical parts such as cams, sprockets and small leaf springs but also find application for housings, cover plates, knobs and levers. They have the highest fatigue-endurance limits of any of the commercial thermoplastics, and these properties, coupled with those of reduced friction and noise, admirably qualify the acetal resins for small gearing applications within the vehicle.

Plastics can be self-coloured so that painting costs are eliminated and accidental scratching remains inconspicuous, and they can be given a simulated metal finish. For large-scale assemblies, such as automobile bodies, painting is necessary to obtain uniformity of colour, especially when different types of plastics are used for different components. Plastics can also be chrome plated, either over a special undercoating which helps to protect and fix the finish or by metal spraying or by vacuum deposition, in which the plastic part is made to attract metal particles in a high-vacuum chamber. The use of a plastic instead of a metal base for chrome plating eliminates the possibility of the base corroding and damaging the finish before the chromium plating itself would have deteriorated. The chrome coating can be made much thinner and yet have a longer effective life, with a consequent saving in cost.

Until fairly recently polymer materials were joined only by means of adhesives. Now the thermoplastic types can be welded by using various forms of equipment, in particular by hot gas welding, hot plate machines which include pipe welding plant, ultrasonic and vibration methods, spin or friction welding machines and induction, resistance and microwave processes. Lasers have been used experimentally for cutting and welding. The joining of metals to both thermoplastic and thermosetting materials is possible by some welding operations and by using adhesives.

Over the past years, the natural applications for plastics in automobiles (interior fittings, cushioning and upholstery, trim, tail lights, electrical components) have become saturated. The growth for the future can be expected to come from the use of plastic for bodywork and some mechanical components. Already there is a widespread use of plastics for front and rear bumpers. We can expect to see bonnets, boot lids and front wings in plastics.

All the major volume producers of cars are engaged in long-term development work towards the all-plastic car. Whether such targets can be realized remains to be seen. Factors such as energy costs and availability of resources may play a greater part in the total picture than simple objects like vehicle weight reduction.

Abbreviations for automotive plastics

Plastics repair

A car is made up, by weight, of about 65 per cent steel, 5 per cent non-ferrous metal, 15 per cent plastics material and 15 per cent other non-metallic materials. Plastics materials are very light, so in terms of bulk, the percentage is much larger, but nobody appears to have worked out the figures for this yet! What we do know is that the plastics parts are often damaged in even a minor accident, and the replacement of these parts costs insurance companies and private

owners dearly. That's how we make a profit, you might say; but you could make more profit by repairing the damaged plastics part, which would reduce the cost to the customer and reduce the waste of precious natural resources.

Types of plastics

> **Tech note**
> The word *plastics* is being used here because it is technically correct to describe the range of manmade materials. 'Plastic', without the 's', is used to describe the material state where it can be deformed and it will remain in that state after the force has been removed. In conversation it is normal to say 'plastic' for both cases, as it is unlikely that there will be any confusion.

The two main classifications of plastics are thermoplastics and thermosetting plastics (which are referred to as thermosets). One of the key areas of knowledge needed to repair plastics components is an understanding of these two classifications and being able to identify them in a vehicle component.

A thermoplastic is one which melts when it is heated up. If you get a carrier bag from the supermarket and warm it slightly, it will become soft and pliable. It is a thermoplastic. At this point you must remember that plastics are made from petroleum-based chemicals and are therefore easy to set on fire and burn at very high temperatures, so avoid matches and other naked flames when handling them. Conversely, if you put the same carrier bag in a freezer, it would go stiff and make a crackling noise when you handle it. Vehicle thermoplastic components are made to operate normally over a wide temperature range, so obviously they need to get very hot before they will melt and very cold before they become brittle. Now if you heat up a thermoplastic to a high enough temperature,

Table 4.6 Abbreviations for automotive plastics

No	Abbreviation	Full name
1	ABS*	Acronitrile butadiene styrene*
2	PP*	Polypropylene*
3	PE*	Polyethylene* PC
4	PA*	Polyamide*
5	PBT*	Polybutylene tetraphtalate*
6	PU	Polyurethene (thermoset)
7	UP	Unsaturated polyester (thermoset)
8	CS	Chopped strands
9	SMC	Sheet moulding compound
10	MF	Milled fibres
11	WR	Woven roving

* Can be repaired by welding

it is going to melt, which means that you can repair a thermoplastic component by welding. Before you dash out to try to weld that bumper assembly which is sat on the bench, there are a few more things which you need to know.

First is that you need a special plastics welder and second, if it is not a thermoplastic bumper, you may well damage it beyond repair and set the workshop on fire too. You cannot simply identify thermoplastics just by warming them up; we'll look at ways of identifying plastics later in the chapter.

A thermoset is one which uses or generates heat during its setting stage. The first thermoset was Bakelite, the heavy dark-brown plastics material which was used for distributor caps and ignition coil ends. It does not go soft when you try to heat it up; if you subject it to a flame, it will burn and char. It is a brittle material and easily chips. Thermosets cannot be welded; most can be bonded using a suitable bonding agent or glue.

Reinforcement

On their own plastics materials have only a limited amount of strength – interior trim, dashboard panels and lamp lenses are examples of non-reinforced plastics. Apply a strong force from a mechanic's hand and these components will break. So components which are going to take structural loads within the vehicle or be capable of withstanding impact, such as a bumper assembly, need some form of reinforcement. The most common reinforcement material is glass.

You are probably familiar with glass-reinforced polyester (GRP), in which a piece of glass matting is laid up with a mixture of polyester resin and catalyst (hardener) to affect a body repair or for the manufacture of kit cars and small boats. Incidentally, GRP is also used to cover all glass-reinforced plastics. If you look at a strand of the glass from this matting through a microscope or with a very strong magnifying glass, you will see that the glass is in fact made from very small-diameter round tubes. A round tube

gives very high strength, but the length-to-diameter ratio of these tubes is such that they can bend without collapsing, so that they can be laid up on curved surfaces, then when the resin sets they are firmly held in place like roof beams for maximum strength.

The glass reinforcement used for vehicle components is made to suit the application and can vary between the woven cloth-like material which is sandwiched in layers in a bumper assembly and the finely powdered glass particles which are used to strengthen a lamp cluster.

Increasingly, materials other than glass are being used to reinforce plastics, although often this is in addition to glass. Carbon fibre is used either in the form of a continuous thread which is wound around the component or as a woven matting similar to glass reinforcement. Where two different materials are used, they are referred to as composites. So when it comes to identifying a plastics material, check to see if it is reinforced, and if it is, which type of reinforcement is used. If a glass matting is used, you can usually see the woven layers of glass on the underside of the component. Carbon fibre can be recognized by its graphite-grey colour.

Identification markings

Many vehicle manufacturers now mark their products with a code which will enable the identification of the type of plastics used. The reason for this coded marking is mainly

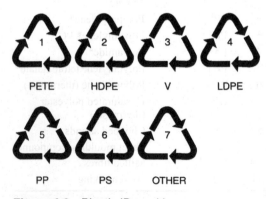

Figure 4.6 Plastic ID markings

Table 4.7 ASPI materials coding system

No.	Material type	Code
1	Polyethylene terephthalate	1PETE
2	High density polyethylene	2HDPE
3	Vinyl	3V
4	Low density polyethylene	4LDPE
5	Polypropylene	5PP
6	Polystyrene	6PS
7	Other	6OTHER

for the identification of genuine parts and recycling purposes. Currently there is no standard system in Europe, nor indeed the UK, for identifying plastics. The British Plastics Federation is trying to encourage all European plastics manufacturers to use the American Society of the Plastics Industry (SPI) material code system. This uses a number and a series of abbreviation letters. The number and abbreviation letters identify a classification of plastics. The code letters used by manufacturers outside the SPI system are often registered trademarks; this creates legal problems as well as identification problems. Again, the SPI classification code is intended mainly for recycling purposes, but it is very useful for general identification information.

Manufacturing processes

There are many different methods of manufacturing plastics components. After a little experience you will be able to work out how components are best manufactured, and this is usually a good guide as to the type of material used. Injection moulding (Inj) is used for items such as grille panels, air vents, dashboard panels, wheel covers and lamp units. The material is likely to be a thermoplastic. These items usually have a smooth surface finish and carry markings which show that they have been in a mould; typically the lettering is raised. They also tend to be fairly flexible. Injection moulding can also be carried out with thermosets.

In this case the material is much stiffer but still shows the mould lines and has a smooth surface finish with raised lettering. The thermoset injection moulding may use a resin (RIM) or a bulk material (BMC). In some cases the BMC may be recycled material filler, but this will have low strength. Bumper assemblies, wheel arch extensions and rear lamp holders are typical applications of injection-moulded thermosets. Thermosets in the form of GRP using woven glass fibre or woven carbon fibre may be hand layed-up (HLU) or compression moulded (com). To affect a good bond with a carbon fibre material, an autoclave is needed to control the finishing process. The texture and colour will allow you to identify GRP and carbon fibre materials.

Safety

Having described the different plastics materials and the manufacturing processes, you should be able to start identifying them. As with all things, you will need to practise until you become skilled, and some mistakes are inevitable. When you try repairing a few items you will get a feel for the job, just like tightening nuts and bolts. You will soon become aware of which parts on which vehicles can be repaired. But before you start to work on plastics materials, you need to look at safety. As well as the normal workshop safety procedures, there are a number of specific hazards relating to plastics materials which you must take extreme care with. Let's have a look at them before discussing some of the repair procedures.

Plastics materials are made from petroleum-based products; this means that they are highly flammable, so you must avoid high levels of heat and naked flames. The most common reinforcing material is glass, but other equally problematic materials may be used. If you start to grind plastics components you will get powdered glass as well as the plastics dust. The powdered glass can cut the blood vessels inside the lungs and stomach. The plastics powder can cause respiratory diseases, and the dust from carbon fibre can cause lung and other internal diseases. So ensure that you are using the correct masks or other breathing apparatus to suit the situation. Heating plastics materials or using solvents or bonding agents can give rise to volatile organic compounds (VOCs); breathing protection is obviously needed in this case. To prevent a build-up of fumes and dust in the workshop the use of an extractor system is advised. Solvents and bonding agents should not come into contact with your skin. Gloves and safety goggles are a first line of defence, and you are reminded to consult the COSHH sheet supplied by the manufacturer with all of these products.

Repair procedures

As a general rule thermoplastics can be welded and thermosets bonded. We'll have a look at a few procedures in detail.

Let us start with something simple. Often in an accident repair a plastics headlamp binnacle is scrapped because one of the lugs is broken off. If you apply a small amount of acetone to both of the broken surfaces you will often find that the lug can be bonded back into place. The acetone (also used as nail varnish remover and not popular amongst mechanics) actually melts the plastics material, and pressing the two parts together causes them to bond and dry.

Dashboards and some flexible bumper assemblies which are made from thermoplastics can be welded. There are two ways of welding thermoplastics. One is to use a hot-air welding gun and the other is to use a soldering iron. The hot-air welding gun blows out a stream of air which will melt the plastic. The temperature is over 100°C. The paint should be cleaned off about 20 mm (3/4 in.) on both sides of the joints. The welded joint is made using the blow gun and a plastics filler rod in the same way as you would oxy-acetylene weld steel. If it is a long joint you should tack weld first. The two parts can be held together with strips of masking tape on the reverse whilst you carry out the welding. If a component has cracked, like a bumper assembly, and internal stress might cause the crack to continue during or after the repair, it is a good idea to drill a small hole at each end of the crack. Usually 4-mm (3/16-in.) holes at each end of the crack will be sufficient to remove the internal stress. These should then be filled after the welding has been completed. If you are using a filler rod it is a good idea to 'vee' the edges of the joint to accommodate the filler. This can be done using a file.

Thin thermoplastic items which will melt without a great deal of heat being applied can be welded using a soldering iron (without the solder). The procedure is as follows. Remove any paint within about 15 mm of the joint (a P40 disc is usually ideal). Drill stress-relieving holes at each end if it is a crack; hold the gap closed with tape on the underside; run the soldering iron over the joint so that the material melts and fuses together. When the repair has cooled, remove any excess or unevenness with the P60 and then finish to feather into the existing paint using P600.

If the component, say a bumper, has been holed, it is possible to weld in a piece from a scrap bumper of the same shape. Cut the damaged section out of the bumper – a round or oval shape will prevent further cracking. Then cut a piece out of the scrap bumper which will just fit into the hole. Remove the paint from all the edges and weld in as in the previous examples (blow gun or soldering iron). Finish with P600 production paper. If the joint or the final contour is not satisfactory this can be corrected using body filler in the normal way.

To repair a thermoset component you need to bond or glue on a patch. If the damage is a crack, such as in a bumper assembly, the procedure is to clean up the damaged area on both sides of the crack, drill stress-relieving holes at each end, then whilst the crack is held closed, bond a patch to the underside of the component. Complete the repair using body filler and P600 paper in the normal way. The patch is preferably the same material as the damaged item; the bonding could be by a number of materials, including superglue. If the bumper has been holed, then cut a patch to fill the hole as you would for a thermoplastic bumper and additionally cut another patch which is larger than the hole. The larger patch is then bonded to the underside so that it attaches to both the original item and the piece which is filling the hole. The job is again completed using body filler and P600 to feather in the paintwork. A small amount of body filler on the underside to blend the patch into the surrounding material will make the job look neat.

Be aware that not all plastics can be repaired. Those which have a waxy finish will not even let superglue stick to their surface.

Painting plastics

Plastics materials require a suitable keying primer and/or undercoat. You should use the one which is recommended by the vehicle manufacturer. A coat of underseal on the rear of any panel which is open to the elements will give added protection.

Safety glass

More and more glass is being used on cars. Pillars are becoming slimmer and glass areas are increasing as manufacturers approach the ideal of almost complete all-round vision and the virtual elimination of blind spots. Windscreens have become deeper and wider. They may be gently curved, semi-wrapped round or fully wrapped. With few exceptions they are of one-piece construction, sometimes swept back as much as 65° from the vertical. Styling trends, together with a growing knowledge of stress design in metal structures, have resulted in a significant increase in the glazed areas of modern car body designs. As a result of this move towards a more open style, the massive increase in the cost of energy has brought growing pressure on vehicle designers to achieve more economic operations, principally in respect to lower fuel consumption

through better power/weight ratios. An outcome of these two lines of development has been a situation in which although the area of glass has increased, the total weight of glass has remained constant or even decreased.

Broadly speaking, motor vehicle regulations specify that windscreens must be of safety glass. To quote one section: 'On passenger vehicles and dual-purpose vehicles first registered on or after 1 January 1959, the glass of all outside windows, including the windscreen, must be of safety glass'. The British Standards Institute defines safety glass indirectly as follows: 'All glass, including windscreen glass, shall be such that, in the event of shattering, the danger of personal injury is reduced to a minimum. The glass shall be sufficiently resistant to conditions to be expected in normal traffic and to atmospheric and heat conditions, chemical action and abrasion. Windscreens shall, in addition, be sufficiently transparent, and not cause any confusion between the signalling colours normally used. In the event of the windscreen shattering, the driver shall still be able to see the road clearly so that he can brake and stop his vehicle safely'.

Two types of windscreen fulfil these requirements – those made from heat-treated (or toughened) glass and those made of laminated glass. In addition there are plastic-coated laminated or annealed safety glasses. Most windscreens and some rear windows fitted in motor vehicles are of ordinary laminated glass. For the main part, toughened glass is confined to door glass, quarter lights and rear windows, where the use of more expensive laminated products has yet to be justified. However, laminated glass is being increasingly used on locations other than windscreens for reasons of vehicle security and also for passenger safety (containment in an accident), especially in estates with seating in the rear. Note that the applicable EEC Directive (see later) has effectively banned the fitment of toughened windscreens from the end of 1992.

Ordinary laminated safety glass is the older of the two types and is the result of a basic process discovered in 1909. Some years earlier a French chemist, Edouard Benedictus, had accidentally knocked down a flask which held a solution of celluloid. Although the flask cracked it did not fall into pieces, and he found that it was held together by a film of celluloid adhering to its inner surface. This accident led to the invention of laminated safety glass, made from two pieces of glass with a celluloid interlayer. An adhesive, usually gelatine, was used to hold them together, and the edges had to be sealed to prevent delaminating. However, despite the edge sealing, the celluloid (cellulose nitrate plastic) discoloured and blistered; hence celluloid was replaced by cellulose acetate plastic, but this, although a more stable product than celluloid, still needed edge sealing.

Nowadays a polyvinyl butyral (PVB) self-bonding plastic interlayer is used; no adhesive is necessary and the edges do not need sealing, making it quite practical to cut to size after laminating. When producing glasses to a particular size, however, the glass and vinyl interlayer are usually cut to size first. In the process the vinyl plastic interlayer is placed between two clean, dry pieces of glass, and the assembly is heated and passed between rubber-covered rollers to obtain preliminary adhesion. The sandwich of glass and interlayer is then heated under pressure for a specified period in an autoclave. This gives the necessary adhesion and clarity to the interlayer, which is not transparent until bonded to the glass. If a piece of laminated glass is broken, the interlayer will hold the splinters of glass in place and prevent them flying.

Plastic-coated laminated safety glass is an ordinary laminated glass which has soft elastic polyurethane films bonded onto the inner surface to provide improved passenger protection if fragmentation occurs. There is some interest in the use of bilayer construction which uses 3-mm or 4-mm annealed glass bonded with a load-bearing surface layer of self-healing polyurethane.

Uniformly toughened glass is produced by a completely different process, involving heating of the glass followed by rapid cooling. Although patents were taken out in 1874 covering a method of increasing the strength of flat glass sheet by heating and cooling it in oil, toughened glass was not in common use until the 1930s. Modern toughened glass is produced by heating the glass in a furnace to just below its softening point. At this temperature it is withdrawn from the furnace and chilled by blasts of cold air. The rapid cooling hardens and shrinks the outside of the glass; the inside cools more slowly. This produces compressional strain on the surfaces with a compensating state of tension inside and has the effect of making the glass far stronger mechanically than ordinary glass. If, however, the glass does fracture in use, it disintegrates into a large number of small and harmless pieces with blunted edges. The size of these particles can be predetermined by an exact temperature control and time cycle in the toughening process, and manufacturers now produce a uniformly toughened safety glass which will, when broken, produce not less than 40 or more than 350 particles within a 50 mm square of glass. This conforms to the British Standard specification.

There are other types of safety glass – mostly crossbreeds of the pure toughened glass screen – which are designed to combine vision with safety. These are modified zone-toughened glasses, having three zones with varying fragmentation characteristics. The inner zone is a rectangular area directly in front of the driver, not more than 200 mm high and 500 mm long. This is surrounded by two other zones, the outer one of which is 70 mm wide all round the edge of the windscreen. This type of windscreen has been fitted to various vehicles since 1962. As a result of ECE Regulations this type of windscreen has been superseded by the fully zebra-zoned windscreen. Many countries, including the United States but with the exception of the UK, legislate against toughened windscreens.

Although sheet and plate glass are manufactured satisfactorily for use in doors, rear lights and windscreens, float glass has now largely superseded their use for reasons of economy and improved flatness. Most laminated windscreens used in the motor vehicle trade are 4.4 mm, 5 mm, 5.8 mm or 6.8 mm in overall thickness, with a 0.76 mm PVB interlayer. However, 4.4 mm is the thinnest laminated glass available, and as this has to be made from two pieces of glass, it needs very careful handling during manufacture and is therefore expensive. Windscreens made from float glass should be a maximum of 6.8 mm thick, whether toughened or laminated. However, some large coaches and lorry windscreens are 7.8 mm thick (4 mm glass + 0.76 mm PVB interlayer + 3 mm glass). This gives immense strength and robustness against stone impact.

Other body glasses, because they can be made from sheet glass and also can be toughened safety glass, are usually between 3 mm and 4 mm thick.

From our brief look at the history of glass manufacture, it is obvious that the curving of glass presents no problems; in fact, the problem has been to produce flat, optically perfect glass. However, to curve safety glass and still retain its optical and safety qualities requires careful control. Glass has no definite melting point, but when it is heated to approximately 600°C it will soften and can be curved. Curved glasses should be specified as 6.8 mm thick, as it is more difficult to control the curving of 4.4 mm glass. Even 6.8-mm-thick glasses will have slight variations of curvature. To accommodate this tolerance, all curved glasses should be glazed in a rubber glazing channel, of which there are many different sections available. Glazed edges of glasses should be finished with a small chamfer known as an arrissed edge, while edges of glasses that are visible or which run in a felt channel should be finished with a polished, rounded edge. Should a glass be required for glazing in a frame, a notch will usually be required to clear the plate used to join the two halves of the frame together. The line of this notch must not have sharp corners because of the possibilities of

cracking. Although laminated safety glass can be cut or ground to size after laminating, toughened safety glass must be cut to size and edge finished before the heat-treating process.

Nearly all fixed glazing is now glazed using adhesive systems. Shapes are becoming more complex, needing very good angles of entry control to meet bonding requirements. The trend is towards aerodynamic designs involving flush glazing and the removal of sudden changes in vehicle shape; therefore corners must be rounded rather than angular as in older vehicle designs. Glass is often supplied with moulded-on finisher (encapsulation). Consequently bending processes are becoming very sophisticated. Adhesive glazing (polyurethane is the adhesive that is normally used) has added considerably to the complexity of vehicle glazing in a scientific sense. It has many advantages. However, if carried out correctly, it will reduce water leaks, it suits modern car construction, it results in a load bearing glazing member, and it lends itself to robotic assembly in mass production. As a consequence of adhesive glazing, all the associated glazing is now printed with a ceramic fired-in black band to protect the polyurethane adhesive from ultraviolet degradation, and also for cosmetic reasons so that the adhesive cannot be seen.

By a Ministry of Transport regulation, safety glass was made compulsory in 1937 for windscreens and other front windows. As already indicated, with effect from 1 January 1959, the Road Traffic and Vehicle Order 359 has demanded that for passenger vehicles and dual-purpose vehicles, all glass shall be safety glass. For goods vehicles, windscreens and all windows in front of or at the side of the driver's seat shall be safety glass.

Skills and questions

To develop your skills related to materials:

1   Using the workshop manuals and the part book in conjunction with a relevant vehicle, identify the different materials used in its

construction so that you can select both the correct material and the appropriate process to work on it – get the help of the other technicians where you work.

2   Cut and join a selection of scrap parts made of as many materials as you can. Make the joint finish as professional as you can.

3   Describe the system used to keep aluminium and steel parts and tools separate in a large bodyshop.

4   Write a paragraph explaining about the materials used in the repair of one part of a vehicle as you would to a customer.

5   Windscreens are usually fitted by specialist firms. Find time and opportunity to observe a screen being replaced. Note the safety precautions taken and the customer care taken in removing the damaged glass.

6   What would the following alloy steels be used for: (a) high-tensile steel, (b) manganese steel, (c) chrome-vanadium steel?

7   List the properties of commercially pure aluminium.

8   Explain why, in the construction of a motor vehicle, commercially pure aluminium has a very limited application.

9   Identify the grades of hardness in aluminium sheet and state how the hardness is achieved.

10   Explain how you would identify the following: (a) low-carbon steel, (b) aluminium alloy, (c) stainless steel.

11   Describe the difference between laminated safety glass and toughened safety glass.

12   Give three requirements of a body sealing compound, and describe one type of sealer used in vehicle repair.

13   Suggest reasons why stainless steel is sometimes used for trim and mouldings.

14   Explain the difference between hide and PVC materials.

15   Explain what is meant by micro-alloyed steel or HSS.

16   Give reasons why the car manufacturers are using zinc-coated steels.

17   Name the three main groups of stainless steel.

18   Explain the following terms in relation to plastic: (a) monomer, (b) polymer, (c) copolymer.

19   Explain the difference between thermoplastics and thermosetting plastics.

20   Which safety glass, used for vehicle windscreens, shatters into small segments on impact?

21   Describe the basic properties required of a body joint sealing compound.

22   Identify the group of plastics that can be softened or remoulded by the application of heat.

23   Steel panels can be strengthened without adding weight. Name and explain the process.

24   Describe three different ways in which the surface of steel can be protected.

25   State the reasons why certain metals need to be protected from the effects of the atmosphere.

26   Describe how some metals can resist attack by the atmosphere.

27   What is the alloying effect when zinc and copper are added to aluminium?

28   Explain the different properties of heat-treatable and non–heat-treatable aluminium alloys.

29   State the reasons, other than weldability, why low-carbon steel is chosen in preference to aluminium as a vehicle body shell material.

30   Define the term 'HSLA steel'.

31   Define what is meant by the term 'non-ferrous metal'.

32   Explain the difference in properties between low-carbon steel and alloy steel.

33   Describe the two processes which can be used to join plastic.

34   Explain where plastic can be used on a vehicle.

35   State the applications where natural rubber has been replaced by synthetic materials in the automobile industry.

36   Describe two ways of attaching a windscreen to a vehicle body.

37   Describe how to replace a glass in an opening quarter-light frame.

38   Why should you never hit a hammer with another hammer?

39   Which material should not be used for axle stand pins?

40   Why is brass often used for drifting bearings?

5

Metal forming processes

Properties of metals

Steel and aluminium both possess certain properties which make them especially suitable for the processes involved in vehicle bodywork, particularly the forming and shaping of vehicle body parts either by press or by hand, and some of the jointing processes. These properties are described in the following sections.

Malleability

A malleable metal may be stretched in all directions without fracture occurring, and this property is essential in the processes of rolling, spinning, wheeling, raising, flanging, stretching and shrinking. In the operation of beating or hammering a metal on a steel block – such as planishing – an action takes place at each blow wherein the metal is squeezed under the blow of the hammer and is forced outwards around the centre point of the blow. The thinner the metal can be rolled or hammered into sheet without fracture, the more malleable is the metal. After cold working, metals tend to lose their malleable properties and are said to be in a work-hardened condition. This condition may be desirable for certain purposes, but if further work is to be carried out, the malleability may be restored by annealing. Annealing, or softening, of

the metal is usually carried out before or during curvature work such as raising and hollowing, provided the metal is not coated with a low-melting-point material. However, the quality of the modern sheet metal is such that many forming operations, such as deep drawing and pressing, may be carried out without the need for an application of heat.

The following are examples in which the properties of malleability are most evident:

*   *Riveting* – Here the metal will be seen to have spread to a marked degree. If splitting occurs, the metal is insufficiently malleable or has been overworked (work hardened).
*   *Shaping* – The blank for a dome consists of a flat disc which has to be formed by stretching and shrinking into a double-curvature shape. The more malleable and ductile the material of the blank is, the more readily it can be formed; the less malleable and ductile, the more quickly does the metalwork hardened thus need more frequent annealing.

The degree of malleability possessed by a metal is measured by the thinness of leaves that can be produced by hammering or rolling. Gold is extremely malleable and may be beaten into very thin leaf. Of the metals used for general work, aluminium and copper are

Table 5.1 Physical properties of metals and alloys

Metal	Melting temperature range (°C)	Density (kg/m³)	Specific heat capacity/ (J/kg/K 10³)	Thermal conductivity (W/m/K)	Electrical conductivity (% IACS)	Coefficient of linear expansion/ (K 10⁻⁶)
Aluminium	660	2.69	0.22	218	63	23
Al-3.5 magnesium	550–620	2.66	0.22	125	25	23
Duralumin type	530–610	2.80	0.21	115–140	20–36	23
Copper	1085	8.92	0.39	393	101	17
70/30 brass	920/950	8.53	0.09	120	17	19
95/5 tin bronze	980/990	8.74	0.09	80	12	17
Lead	327	11.34	0.13	35	8	29
Magnesium	650	1.73	1.04	146	35	30
Nickel	1455	8.90	0.51	83	21	13
Monel	1330/1360	8.80	0.43	26	3	10
Tin	232	7.30	0.22	64	13	20
Titanium	1665	4.50	0.58	17	3	8.5
Zinc	419	7.13	0.39	113	26	37
Iron	1535	7.86	0.46	71	7	12
Mild steel	1400	7.86	0.12	45	31	11

Table 5.2 Typical mechanical properties of metals and alloys

Material	Modulus of elasticity E (kN/mm²)	Tensile and compressive strength (N/mm²)	Elongation (%)	Hardness (HV)
Pure aluminium	68–70	62–102	45–7	15.30
Aluminium alloys	68–72	90–500	20–5	20–80
Magnesium	44	170–310	5–8	30–60
Cast irons (grey)	75–145	150–410	0.5–1.0	160–300
SG cast iron	170–172	370–730	17–2	150–450
Copper	122–132	155–345	60–5	40–100
Copper alloys	125–135	200–950	70–5	70–250
Mild steel	190	420–510	22–24	130
Structural steels	190	480–700	20–24	130
Stainless steel	190	420–950	40–20	300–170
Titanium	100–108	300–750+	5–35	55–90
Zinc	90	200–500	25–30	45–50

outstanding for their properties. The property of malleability is used to advantage in the manufacture of mild steel sheets, which are rolled to a given size and gauge for the motor industry.

It is also evident in the ability of mild steel and aluminium panels to be formed by mechanical presses into complicated contours for body shells. Malleability and ductility are the two essential properties needed in order to mass produce vehicle body shells by pressing.

Tech note

The order of malleability of various metals by hammering is as follows: gold, silver, aluminium, copper, tin, lead, zinc, steel.

Ductility

Ductility depends on tenacity or strength in tension and the ease with which a metal is deformed and is the property which enables a metal to be drawn out along its length, that is drawn into a wire. In wire drawing, metal rods are drawn through a hole in a steel die; the process is carried out with the metal cold, and the metal requires annealing when it becomes work hardened.

Ductile properties are also necessary in metals and alloys used in the following processes:

* *Pressed components* – Special sheets which have extra-deep drawing qualities are manufactured especially for press work such as that used in modern motor vehicle body production. These sheets undergo several deformations during the time they are being formed into components, yet because of their outstanding ductile properties they seldom fracture.
* *Welding electrodes and rods* – Ductility is an essential property in the production of electrodes, rods and wires. The wire-drawing machines operate at exceptionally high speeds, and the finished product conforms to close tolerances of measurement; frequent failure of the material during the various stages of drawing would be very costly.

Tech note
The order of ductility of various metals is: gold, aluminium, steel, copper, zinc, lead.

Tensile strength

A very important property of metals is related to its strength in resistance to deformation; that property is tensile strength, which may be defined as the property by which a metal resists the action of a pulling force. The *ultimate tensile strength* of a metal is a measure of the force which ultimately fractures or breaks the metal under a tensile pull. The ultimate tensile strength (UTS) of a material is normally expressed in tons per in^2 or MN/m^2, and may be calculated as follows:

$$UTS = \frac{Tensile\ Force\ (N)}{Cross\text{-}sectional\ Area\ (mm)}$$

In this case the load is the maximum required to fracture a specimen of the material under test, and the calculation is based on fracture taking place across the original cross-sectional area. In ductile materials a special allowance must be made for wasting or reduction of original cross-sectional area.

High-carbon steels possess a high degree of tenacity, evidence of which can be seen in the steel cables used to lift heavy loads. The mild steels used in general engineering possess a small amount of tenacity, yet a bar of metal of one inch square ($6.5\ cm^2$) cross-section, made from low-carbon steel, is capable of supporting a load in excess of 20 tonnes.

Methods of increasing the tensile strength

It is possible to increase the tensile strength of both sheet steel and pure aluminium sheets by cold rolling, but this has the added effect of reducing their workable qualities. In the manufacture of vessels to contain liquids or gases under pressure it is not always possible to use metals with a high tensile strength; for instance, copper is chosen to make domestic hot water storage cylinders because this metal has a high resistance to corrosion. In this case, the moderate strength of copper is increased by work hardening such as planishing, wheeling and cold rolling. Work hardening has the added effect of decreasing the malleability.

Tech note
The order of tensile strength of various metals in tons per in^2 (MN/m^2) is: steel 32 (494); copper 18 (278); aluminium 8 (124); zinc 3 (46); lead 1.5 (23).

Hardness

When referring to hardness, it should be carefully stated which kind of hardness is meant. For example, it may be correctly said that hardness is that property in a metal which imparts the ability to:

1 Indent, cut or scratch a metal of inferior hardness
2 Resist abrasive wear
3 Resist penetration.

A comparison of hardness can be made with the aid of material testing machines such as those used to carry out the Brinell or Vickers Diamond tests. Hardness may be increased by the following methods:

* *Planishing* – In addition to increasing the tensile strength of a metal, planishing also imparts hardness.
* *Heat treatment* – Medium-carbon and high-carbon steels, such as those used in many body-working tools, can be hardened by heating to a fixed temperature and then quenching.

Tech note
The order of hardness of various metals is as follows: high-carbon steel, white cast iron, cast iron, mild steel, copper, aluminium, zinc, tin, lead.

Toughness

This property imparts to a metal the ability to resist fracture when subjected to impact, twisting or bending. A metal need not necessarily be hard to be tough; many hard metals are extremely brittle, a property which may be regarded as being opposite to toughness.

Toughness is an essential property in rivets. During the forming process the head of the rivet is subject to severe impact, and when in service rivets are frequently required to resist shear, twist and shock loads. Toughness is

also a requisite for steel motor car bodies, which must be capable of withstanding heavy impacts and must often suffer severe denting or buckling without fracture occurring. Further, when repairs are to be made to damaged areas it is often necessary to apply force in the direction opposite to that of the original damaging force; and the metal must possess a high degree of toughness to undergo such treatment.

Compressibility

Compressibility may be defined as the property by which a metal resists the action of a compressing force. The *ultimate compressive strength* of a metal is a measure of the force which ultimately causes the metal to fail or yield under compression. Compressibility is related to malleability in so far as the latter refers to the degree to which a metal yields by spreading under the action of a compressing or pushing force, while the former represents the degree to which a metal opposes that action.

Elasticity

All metals possess some degree of elasticity; that is, a metal regains its original shape after a certain amount of distortion by an external force. The elastic limit of a metal is a measure of the maximum amount by which it may be distorted and yet return to its original form on removal of the force. Common metals vary considerably in elasticity. Lead is very soft yet possesses only a small amount of elasticity. Steel, on the other hand, may reveal a considerable degree of elasticity as, for example, in metal springs. The elasticity of mild steel is very useful in both the manufacture of highly curved articles by press work and in the repair of motor car bodies.

Fatigue

Most metals in service suffer from fatigue. Whether the metal ultimately fails by fracture or by breaking depends on a number of factors associated with the type and conditions of service. When metal structures in service are subjected to vibration over long periods, the rapid alterations of push and pull, that is, compressive and tensile stresses, ultimately cause hardening of the metal with increased liability to fracture. Any weak points in the structure are most affected by the action and become the probable centres of failure, either by fracture or by breaking. Various methods have been devised for testing the capacity of metals to resist fatigue, all of which depend on subjecting the metal specimen to alternating vibratory stresses until failure occurs. The rate and extent of vibration over a specified time form the basis of most fatigue tests.

Weldability

This property is a measure of the ease with which a metal can be welded using one of the orthodox systems of welding. Certain metals are welded very easily by all recognized methods; some metals can only be welded by special welding processes; and other metals and alloys cannot be welded under any circumstances.

Heat treatment of
steel and aluminium

Heat treatment can be defined as a process in which the metal in the solid state is subjected to one or more temperature cycles to confer certain desired properties. The heat treatment of metals is of major importance in motor bodywork. All hand tools used by the bodyworker are made from a type of steel which is heat treatable so that the tools are strong and hard and have lasting qualities. The mechanical parts of a motor vehicle are also subject to some form of heat treatment, and in the constructional field heat-treatable aluminium alloys are being used extensively in commercial bodywork.

Work hardening and annealing

Most of the common metals cannot be hardened by heat treatment, but nearly all metals will harden to some extent as a result of hammering, rolling or bending. **Annealing** is a form of heat treatment for softening a metal which has become work hardened so that further cold working can be carried out. The common metals differ quite a lot in their degree and rate of work hardening. Copper hardens rather quickly under the hammer and, as this also reduces the malleability and ductility of the metal, it needs frequent annealing in order that it may be further processed without risk of fracture. Lead, on the other hand, may be beaten into almost any shape without annealing and without undue risk of fracture. It possesses a degree of softness which allows quite a lot of plastic deformation with very little work hardening. However, copper, though less soft than lead, is more malleable. Aluminium will withstand a fair amount of deformation by beating, rolling and wheeling before it becomes necessary to anneal it. The pure metalwork hardens much less rapidly than copper, though some of the sheet aluminium alloys are too hard or brittle to allow very much cold working. Commercially pure iron may be cold worked to a fair extent before the metal becomes too hard for further working. Impurities in iron or steel impair the cold working properties to the extent that most steels cannot be worked cold (apart from very special low-carbon mild steel sheets used in the car industry), although nearly all steels may be worked at the red-heat condition.

The exact nature of the annealing process used depends to a large extent on the purpose for which the annealed metal is to be used. There is a vast difference in technique between annealing in a steel works, where enormous quantities of sheet steel are produced, and annealing in a workshop, where single articles may require treatment. Briefly, cold working causes deformation by crushing or distorting the grain structure within the metal. In annealing, a metal or alloy is heated to a temperature at which recrystallization occurs and then allowed to cool at a predetermined rate. In other words, crystals or grains within the metal which have been displaced and deformed during cold working are allowed to rearrange themselves into their natural formation during the process of annealing. Iron and low-carbon steels should be heated to about 900°C and allowed to cool very slowly to ensure maximum softness, as far as possible out of contact with air to prevent oxidation of the surface; this can be done by cooling the metal in warm sand. High-carbon steels require similar treatment except that the temperature to which the steel needs to be heated is somewhat lower and is in the region of 800°C. Copper should be heated to a temperature of about 550°C or dull red and either quenched in water or allowed to cool out slowly. The rate of cooling does not affect the resulting softness of this metal. The advantage of quenching is that the surface of the metal is cleaned of dirt and scale. Aluminium may be annealed by heating to a temperature of 350°C. This may be done in a suitable oven or salt bath. In the workshop aluminium is annealed by the use of a blow-pipe, and a stick or splinter of dry wood is rubbed on the heated metal; when the wood leaves a charred black mark the metal is annealed. Sometimes a piece of soap is used instead of the wood; when the soap leaves a brown mark the heating should be stopped. The metal may then be quenched in water or allowed to cool out slowly in air. Zinc becomes malleable between 100 and 150°C and so may be annealed by immersing it in boiling water. Zinc should be worked while still hot, as it loses much of its malleability when cold.

Heat treatment of carbon steel

Steel is an important engineering material because, although cheap, it can be given a wide range of mechanical properties by heat treatment. Heat treatment can both change

the size and shape of the grains and alter the micro-constituents. The shape of the grains can be altered by heating the steel to a temperature above that of recrystallization. The size of the grains can be controlled by the temperature and the duration of the heating and the speed at which the steel is cooled after the heating; the micro-constituents can be altered by heating the steel to a temperature that is sufficiently high to produce the solid solution austenite so that the carbon is dispersed and then cooling it at a rate which will produce the desired structure. The micro-grain structure of carbon steel has the following constituents:

- *Ferrite* – pure iron
- *Cementite* – carbon and iron mixed
- *Pearlite* – a sandwich layer structure of ferrite and cementite

All low-carbon steels of less than 0.83 per cent carbon content consist of a combination of ferrite and pearlite. Carbon steel containing 0.83 per cent carbon is called *eutectoid* and consists of pure pearlite structure. Steels over 0.83 per cent carbon up to 1.2 per cent carbon are a mixture of cementite and pearlite structures. If a piece of carbon steel is heated steadily its temperature will rise at a uniform rate until it reaches 700°C. At this point, even though the heating is continued, the temperature of the steel will first remain constant for a short period and then continue to rise at a slower rate until it reaches 775°C. The pause in the temperature rise and the slowing down of the rate indicate that energy is being absorbed to bring about a chemical and structural change in the steel. The carbon in the steel is changing into a solid solution with the iron and forming what is known as *austenite*. The temperature at which this change in the structure of the steel starts is 700°C, which is known as the **lower critical point**; the temperature at which the change ends is known as the **upper critical point**. The difference between these points is termed the **critical range**. The lower critical point is the same for all steels, but the upper critical point varies with the carbon content. Briefly, steels undergo a chemical and structural change, forming austenite, when heated to a temperature above the upper critical point; if allowed to cool naturally they return to their normal composition.

Steel can be heat treated by normalizing, hardening, tempering and case hardening as well as by annealing, which has already been described.

Normalizing

Normalizing is a process used to refine the grain structure of steel after it has been subjected to prolonged heating above the critical range (as in the case of forging) and to remove internal stresses caused by cold working. The process may appear to differ little from annealing, but as its name suggests the effect of normalizing is to bring the steel back to its normal condition, and no attempt is made to soften the steel for further working. Normalizing is effected by slowly heating the steel to just above its upper critical range for just sufficient time to ensure that it is uniformly heated and then allowing it to cool in still air.

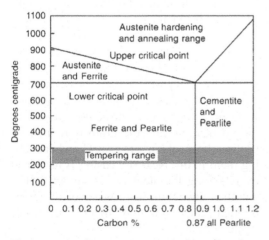

Figure 5.1 Changes in structure of carbon steel with temperature and carbon content

Hardening

It has already been said that if a piece of steel is allowed to cool naturally after heating to above its upper critical point, it will change from austenite back to its original composition. If, however, the temperature of the heated steel is suddenly lowered by quenching it in clean cold water or oil, this change back from austenite does not take place, and instead of pearlite, a new, extremely hard and brittle constituent is formed, called *martensite*. This process makes steels containing 0.3 per cent or more carbon extremely hard, but steels having a carbon content of less than 0.3 per cent cannot be hardened in this way because the small amount of carbon produces too little martensite to have any noticeable hardening effect. The steel to be hardened should be quenched immediately it is uniformly heated to a temperature just above the upper critical point. It is also important not to overheat the steel and to allow it to cool to the quenching temperature. Whether water or oil is used for quenching depends upon the use to which the steel is to be put. Water quenching produces an extremely hard steel but is liable to cause cracks and distortion. Oil quenching is less liable to cause these defects but produces a slightly softer steel. A more rapid and more even rate of cooling can be obtained if the steel is moved about in the cooling liquid, but only that part of the steel which is to be hardened should be moved up and down in the liquid in order to avoid a sharp boundary between the soft and hard portions.

A workshop method of hardening carbon tool steel is to heat the steel, using the forge or oxyacetylene blowtorch, to a dull red colour and then quench it in water or oil. This would harden the article ready for tempering.

Tempering

Hardened steel is too brittle for most purposes, and the process of tempering is carried out to allow the steel to regain some

Table 5.3   Temperature colours for steel

Colour	Temperature (°C)
Black	450–550
Very dark red	600–650
Dark red	700–750
Cherry red	800–850
Full red	850–900
Bright red	950–1000
Dark orange	1050–1100
Light orange	1150–1200
Yellow white	1270–1300
White (welding heat)	1400–1550

Table 5.4   Tempering colours for plain carbon steel

Colour	Temperature (°C)	Type of article
Pale straw	220–230	Metal turning tools, scrapers, scribers
Dark straw	240–245	Taps, dies, reamers, drills
Yellow-brown	250–255	Large drills, wood turning tools
Brown	260–265	Wood working tools, chisels, axes
Purple	270–280	Cold chisels, press tools, small springs, punches, knives
Blue	290–300	Springs, screwdrivers, hand saws

of its normal toughness and ductility. This is done by heating the steel to a temperature below the lower critical point, usually between 200 and 300°C, thereby changing some of the martensite back to pearlite. The exact temperature will depend on the purpose for which the steel is intended; the higher the temperature, the softer and less brittle the tempered steel becomes. Methods used for controlling the tempering process depend upon the size and class of the article to be tempered. One method is to heat the hardened steel in a bath of molten lead and tin, the melting points of various combinations of these two metals being used as an indication of the temperature. Another method of tempering small articles is to polish one face or edge and heat it with a flame. This polished surface will be seen to change colour as the heat is absorbed. The colour changes are caused by the formation at different temperatures of thin films of oxide, called tempering colours (Table 5.4). After tempering, the steel is either quenched or allowed to cool naturally.

Case hardening

Although mild steels having a carbon content of less than 0.3 per cent cannot be hardened, the surface of the mild steel can be changed to a high-carbon steel. Case hardening of mild steel can be divided into three main processes:

1   Carburizing
2   Refining and toughening the core
3   Hardening and tempering the outer case.

A method called pack carburizing is often used in small workshops. After thoroughly cleaning them, the steel parts to be carburized are packed in an iron box so that each part is entirely surrounded by 2.5 cm of carburizing compound. The box is sealed with fireclay, heated in a furnace to 950°C and kept at that temperature for 2 to 12 hours, during which time the carbon in the compound is absorbed by the surface of the steel parts. The steel parts are then allowed to cool slowly in the box. At this stage their cores will have a coarse grain structure due to prolonged heating, and the grain is refined by heating the steel parts to 900°C and quenching them in oil. Next the casing is hardened by reheating the steel to just under 800°C and then quenching it in water or oil. Tempering of the casing may then be carried out in the normal way.

Small mild steel parts can be given a very thin casing by heating them in a forge to a bright red heat and coating them with carburizing powder. The parts are then returned to the forge and kept at a bright red heat for a short period to allow the carbon in the compound to penetrate the surface of the metal. Finally the parts are quenched and ready for use.

Heat treatment of aluminium alloys

An aluminium alloy is heat treated by heating it for a prescribed period at a prescribed temperature and then cooling it rapidly, usually by quenching. The particular form of heat treatment which results in the alloy attaining its full strength is known as **solution treatment**. The alloy is raised to a temperature of 490°C by immersing it in a bath of molten salt. The bath is usually composed of equal parts of sodium nitrate and potassium nitrate contained in an iron tank. This tank is heated by gas burners and, except for its open top, is enclosed with the burners in a firebrick structure which conserves the heat. The temperature of the bath must be carefully regulated, as any deviation either above or below prescribed limits may result in the failure of the metal to reach the required strength. The alloy is soaked at 490°C for 15 minutes and then quenched immediately in cold water.

At the moment of quenching the alloy is reasonably soft, but hardening takes place fairly rapidly over the first few hours. Some alloys, chiefly the wrought materials, harden more rapidly and to a greater extent than others. Their full strength is attained gradually over four or five days (longer in cold weather); this process is known as **natural age hardening**. As age hardening reduces ductility, any appreciable cold working must be done while the metal is still soft. Working of the natural ageing aluminium alloys must be completed within two hours of quenching, or for severe forming within thirty minutes. Age hardening may be delayed by storing solution-treated material at low temperatures. Refrigerated storage, usually at 6–10°C, is used for strip sheet and rivets, and work may be kept for periods up to four days after heat treatment. If refrigerated storage is not used to prevent age hardening it may be necessary to repeat solution treatment of the metal before further work is possible.

Alloys of the hiduminium class may be **artificially age hardened** when the work is finished. Artificial ageing is often called **precipitation treatment**; this refers to the precipitation of the two inter-metallic compounds responsible for the hardening, namely copper and manganese silicon. The process consists of heating the work in an automatically controlled oven to a temperature in the region of 170°C for a period of 10 to 20 hours. Artificial ageing at this temperature does not distort the work. The temperature of the oven must be maintained to within a few degrees, and a careful check on the temperature is kept by a recording instrument. In order to ensure uniform distribution of temperature a fan is fitted inside the oven to keep the air in circulation. At the end of a period of treatment the oven is opened to allow the work to cool down. One of the chief advantages of this process is that work of a complicated character may be made and completed before ageing takes place. Moreover, numerous parts may be assembled or riveted together and will not suffer as a result of the ageing treatment.

How metal is formed to provide strength

It has been established that the strength of a material is governed by its material composition and by the method and direction of loading, that is, tensile, compressive, torsion, shear and bending. Generally the majority of metals are capable of withstanding greater loads in tension than any other type of stress. One of the properties of steel is that, within certain limits, it is elastic: that is, if it is distorted by a load or force it will change shape, but it will return to its original shape when the force is removed. However, above a certain intensity of load (the elastic limit) the metal will remain distorted when the load is removed. Sheet steel, such as is used in the manufacture of car bodies, has reasonable strength in tension but has little resistance to compressive and/or torsional loads. This lack of resistance is due not so much to poor compressive strength as to lack of rigidity. Low-carbon steels are used extensively in the manufacture of vehicle bodies, and the designer has to ensure that the relatively thin sections of material are capable of withstanding the various types of loading. In addition to the permanent stresses present in the material, the vehicle body as a whole is subject to shock stresses due to road conditions, and these must also be taken into consideration by the designer.

With the development of deep drawing steels and better press equipment, large streamlined panels were designed and formed into contours that were more attractive, gave longer life and greater safety, and at the same time reduced the bulky construction previously required to give similar strength. It is known that the shape of any material is held by the stresses set up in the material itself, such as those given by angles, crowns, channels and flanges. The original shape will be maintained until the material is subject to a force sufficient to overcome the initial stresses. Furthermore, it will tend to return to its original shape providing it has not been distorted beyond the point of elasticity.

Crowned surfaces

The building up of stresses at the bend or peak is also an important consideration in the design and manufacture of the modern car body. The most common features of the body are the curved surfaces (Figure 5.2); these are called crowns and may be curved either in one direction or in all directions. A crowned surface is stronger than a flat panel, and whilst it will resist any force

Figure 5.2   Strength in crowned surfaces

Figure 5.3   Strength in angles and flanges

tending to change its shape, it also has the ability to return to its original shape unless distorted beyond its point of elasticity. These are the features of metal sheet which has been formed in a press into a permanent shape, with die-formed stresses throughout its entire area tending to hold the shape. On the bend or crown, one side of the sheet is longer than the other; and the metal at the surface is more dense than at the centre of the sheet. The final action of the press is to squeeze the surface together, thus setting up stresses and greater strength. The greater the crown or curve of the panel, the greater its strength and rigidity to resist change in its shape. This is illustrated by the fact that a low crown, a surface with very little curve, such as a door panel, is springy and is not very resistant to change of shape. On the other hand high crowns, that is surfaces with a lot of curve, like wings, edges of roofs and sill panels, are very resistant to change in shape by an outside force.

Angles and flanges

A further method of giving strength to metal is to form angles or flanges along the edges of sheets. A right-angled bend greatly increases the strength of a sheet, as can be demonstrated by forming a right-angled bend in a thin sheet of metal and then trying to bend the metal across the point of the bend. This method is used on inner door panels and at the edges of wings, edges of bonnets and boot lids and wherever stiffness is required at unsupported edges.

U-channels and box sections

A U-channel, as the name implies, comprises two right-angled bends, one at each of the opposite sides of a piece of metal (Figure 5.4). Much more strength can be gained by making a U-channel instead of a single-angled bend in any reinforcement section. The U-channel is also the most common type of section used in the construction of car frames. Yet another method of increasing strength is known as box construction, which consists of two U-sections welded together to form a square pillar or box. Box sections are used in conjunction with U-channel construction in the manufacture of car chassis frames, underbodies, subframes, cross members and any construction where great strength is necessary.

U-channel section   Box section

Box section using   One-piece box section
two U-channels

Figure 5.4 Strength in U-channel and box sections

Machines for sheet metal fabrication

> **Tech note**
> Illustrations and practical details of machine operation are given in Chapter 13 'Craft techniques and minor accident damage'.

The mass-produced, all-steel bodies of both private cars and light commercial vehicles call for the production of involved panel shapes on a massive scale. This requires heavy power presses employing mating die sets, which in effect press the sheet metal and deform the structure of the metal to a degree where it takes on a natural shape and retains that shape. However, since all pressed panels have some degree of tension in the material structure, it is important that this should be taken into account by the body repairer.

It is often necessary to manufacture sheet steel parts by metal forming machines. The most common of these include wheeling machines, guillotines, fly presses, folders and forming presses. In body repair shops the fly press is often hand operated and is a most useful piece of equipment. Tools for the fly press fit into a central spindle, which is moved down in order to pierce, blank and, to a limited extent, form the metal parts required. Another widely used machine is the folder or bending machine; adjustment of the setting of the blades of this machine enables sharp acute or obtuse angled bends to be produced in long lengths of sheet metal for the making of sections. The swaging machine is used for swaging, wiring, joggling and closing sheet metalwork edges. Finally, the bodyshop is not complete unless it possesses a wheeling machine, which is used to produce double-curvature panels from flat sheets by passing the sheets to and fro between rollers or wheels in order to stretch the sheets and to create a curved shape.

Shearing theory

If a piece of sheet metal is placed on one bottom cutting member or blade, and the top cutting blade is brought to bear on the metal with continuing pressure, after a certain amount of deformation the elastic limit of the metal is exceeded and the top cutting blade penetrates and cuts the surface of the metal. Fractures begin to run into the metal thickness from the points of contact of the top and bottom cutting blades, and if these are positioned correctly relative to each other, the fractures meet and the metal is sheared before the top member penetrates and cuts the whole thickness of the metal. The horizontal distance between the two cutting members is called the clearance, and it is this distance that determines the quality of the cut and the load on the machine or hand tool being used. There is only one correct clearance allowance for any particular sheet, and this depends mainly upon the thickness of the metal being cut and its physical properties. This clearance allowance is usually 5 to 10 per cent of the metal thickness.

1 The notch made by the upper cutting blade
2 Smoothly cut part
3 Torn or fractured section
4 The notch made by the lower cutting blade

Figure 5.5 The cutting section showing a sheared piece of metal

When the clearance is set correctly the metal is cut (sheared) smoothly and cleanly

If the clearance is excessive the material is bent instead of being cut

Figure 5.6 Blade clearance

Cutting machines

Hand lever shears

There are the simplest sheet metalworking machines. For cutting mild steels they are used with blade lengths up to 500 mm for cutting 3 mm material, or with a length of about 200 mm for material up to 6 mm. The cutting members are flat blades, and there is generally no provision for adjustment of blade clearance. The bottom blade is fixed to the machine frame, while the top cutting blade is fixed to a moving member which is pivoted in the main frame and operated by a hand lever via a simple link mechanism. An adjustable hold-down is usually fitted to the top member to hold the sheet down during cutting. The body of the shears may be cranked or offset to permit the cut sheets to be forced beyond the blades so that sheets longer than the blades may be cut. The body of the machine is usually fabricated, and this construction is to be preferred.

Universal shearing machines (nibbling machines)

In these machines the cutting is done by a pair of very narrow blades, one of which is fixed while the other moves up and down from the fixed blade at fairly high speeds. Generally the blades have a very steep rake or angle to permit piercing of the sheet for internal cutting. Since the blades are very narrow the sheet can be easily turned during cutting. The term 'universal' is apt, since the machine will cut straight or curved shapes inside or outside. These machines are powered by compressed air or an electric motor and are held in the hands during the cutting operation. In the machine the top blade is fixed to the moving member or ram and the bottom blade is fitted on an extension which is shaped like a spiral so as to part the material after cutting. The ram and blades are driven at a speed between 1200 and 2100 rev/min, depending on the metal thickness and whether air or electrical power is used.

The maximum thickness of metal to be cut is about 4.5 mm, but the machines find their widest application on lighter gauges of metal 1.00 mm and 0.8 mm. The lighter machines have a maximum cutting radius of about 15 mm and the heavier machines one of about 50 mm.

Tech note

The most important points to remember when using the machine are:

1 See that the blades are set for the correct thickness of metal being cut.
2 Keep the blades sharp (a blunt blade tends to be dangerous).
3 Hold the machine correctly with the blades at right angles to the metal being cut.

Guillotine

A 1.25 m treadle guillotine will meet the requirements of the average panel shop for cutting sheet metal. The guillotine is probably the most widely used of the straight-line cutting machines. The principle employed in the cutting action is very similar to that of the hand bench shears. A concentrated load is applied across the width of the sheet or plate as cutting proceeds. All have a bottom cutting blade which is flat and is fixed horizontally and a top blade which is mounted at a rake to the bottom blade and is fixed to a moving beam. The top beam and blade can be brought to meet the bottom blade by hand, foot, mechanical, air or hydraulic power, depending on the type of machine.

The guillotine consists of a bed, a foot treadle, two cutting blades, and a front, back and two side gauges. The face of the bed sometimes has a graduated scale which permits the cutting of sheets to a specified size. The side gauge, which has steel bars bolted on each side of the bed, can also be graduated for measuring purposes. By placing the edge of the sheet against one of these side gauges, a cut can be made at right angles between the side gauge and the bottom blade. The front gauge consists of a rectangular bar which slides in the bed and can be set at any required distance from the blades. For cutting long sheets, extension arms are fastened to the front of the machine. Slots in the extension arms permit the movement of the front gauge away from the bed to accommodate the larger sheets. The back gauge has an angle-shaped bar which slides on two rods fastened to the rear of the machine. This bar is used to cut a number of pieces of the same length by the adjustment that is allowed on the rods. The lower cutting blade is fastened to the bed of the machine, and the upper cutting blade is attached to a beam which is moved by stepping on the foot treadle. This beam is connected to the treadle by heavy springs which return the treadle to its original position after the metal has been cut. Also attached to this beam and blade is a hold-down device which is located in front of the top blade, and as the treadle is pressed down it clamps and holds the metal in place ready to be cut. A guard is used in front of the upper blade to prevent the operator's fingers from coming into contact with the blade or hold-down device. The standard guillotine blades are usually made from plain carbon steel and are suitable for cutting all types of mild steel and non-ferrous metals. For cutting higher-tensile steels such as stainless steel, alloy steel blades are necessary.

Guillotines should never be used to cut metal which exceeds the capacity of the machine. The capacity range for these models is: a 2 m blade, which will cut up to 1.2 mm; a 2.5 m blade, which will cut up to 1.00 mm; and a 1.25 m blade, which will cut up to 1.6 mm. Most foot-operated machines will cut up to 1.6 mm metal. Under no circumstances should the guillotine be used to cut wire, rod or bar, as it is intended to cut only flat sheet metal.

When using this machine, make sure that the angle finger guard, which is fitted in a position between the operator and the cutting blades, is set in such a way as to protect the hands from both the crushing action of the hold-down device and the cutting action of the blades. Accurate cutting is accomplished by leaning over and sighting the scribed cutting line on the metal at a point perpendicular to the cutting edge of the bottom blade. Check before using the machine that the clearance between the top and bottom blades is set correctly for the thickness of metal to be cut. Do not cut metal of heavier gauge than the machine is designed to cut. Always hold the metal flat and firmly on the base plate of the guillotine; this will prevent drag (the effect of pulling the metal in towards the blades), which often occurs as the cutting edge loses its sharpness.

The guillotine is used more in coach building and the building of new vehicles than in the body repair side of the industry. It is ideal for cutting a number of sheets to a predetermined size because it guarantees greater accuracy than is possible with hand methods.

Safety note

The following safety precautions should be observed when working with guillotines:

1 Keep the blades of the machines sharp. Remember that blunt blades drag the metal and can cause the loss of a finger.
2 Hold the metal firmly on the bed plate of the machine to ensure shearing action and also to prevent drag.
3 All guards should be kept in place. They are required to be there by law in the interests of your safety.
4 When using a treadle guillotine make sure that any assistant or onlooker has not got his feet under the treadle.

Bending theory

Many metals and their alloys may be formed by bending. When considering metals with regard to their bending properties, the following facts are most important.

Behaviour of the metal in bending

In bending sheet metal to an angle, the inner fibres in the metal on the bend are compressed and given a compressive stress, while the outer fibres are stretched and given a tensile stress. The boundary line in the metal thickness between the two areas of stress is called the **neutral axis**. The position of the neutral axis may vary with the properties of the metal and its thickness.

Calculation of the bending length

The formula for calculating the bending allowance is shown in Figure 5.9. An approximate workshop method for calculating the bending allowance is $(r/2) + T$.

Neutral line or axis

Inner surface of the metal under compression stress

Metal stretched on the outside is subject to tensile stress

Figure 5.7   Bending action

1 Radius R on neutral line
2 Bending radius r
3 Thickness of the metal T

Bending allowance $= \frac{\pi}{2}\left(r + \frac{1}{3}T\right)$

Figure 5.8   Bending allowance

Bending machines

Angle bender (clamp folder)

The essential factors in producing a clean bend on sheet metal are that the edge or blade over which the metal is bent should be straight, smooth and fairly sharp and that the pressure applied to bend the metal over this edge should be equal throughout the length of the bend. The most generally used machine for bending sheet metal up to 1.6 mm is the angle bender or clamp folder. This consists of a clamp to which is attached the blade around which the metal is bent and is operated by a hand lever fastened to the end of the machine. The clamping action applies necessary pressure to hold the work to the bed or base of the machine in order to prevent movement while the bending is in progress. The actual bending of the metal is done by swinging up the front part of the bed, which carries the bottom blade on a hinged centre with the top blade.

The following procedure should be used:

1 Mark the bend lines on the metal.
2 Open the upper blade by pushing the clamping handles backwards.
3 Place the sheet of metal between the upper blade and the bed of the machine with the bend line directly under the edge of the top blade, then clamp the metal by pulling the clamping handle forwards again. Some adjustment to obtain the correct clamping pressure may be necessary.
4 Raise the lower bending blade to a position that will produce the desired angle of bend. Move the blade slightly beyond the required angle to allow for springback of the metal.
5 Drop the bending blade to its normal position and open the clamping handles, which will release the work.

Edge folder

As the name implies, this machine is used for turning the edges on sheets to make hooks for grooving seams or for wire insertion or for making small flanges, but generally only in mild steel of 1.00 mm or less.

Open-end folder or bending machine

In these machines the end frames in which the clamping beam is fixed have small gaps so that folds can be made in sheets of unlimited length, provided that the depth of the bend is below the width or height of the gap, which is usually about 50 mm. The maximum length of the blades in this type of machine is 2 m, and the maximum gauge which can be bent is about 1.6 mm. The lift of the clamping beam and top blade is about 38 to 50 mm in height. The clamping beam and top blade are operated by an eccentric shaft running across the front of the machine; this shaft has handles at each end which lift the blade up and down. The bending blade is lifted by a handle at the front of the machine. In larger machines there is provision for counter-balance weights to be fitted to the bending beam. The bending machine is only suitable for single bending operations.

Box and pan bending machine (universal bending machine)

It is possible to form boxes with more than one bend using the clamp folder if there is sufficient clearance between the clamping beam and the top blade to allow the workpiece to be removed. However, most box forming is done using the box bender, which differs from the clamp folder in the design of the upper blade (Figure 5.11); whereas the clamp folder has a single solid bending blade, the box bender has a series of individual blades known as fingers. The advantage of this type of machine is that any number of finger blades can be removed, permitting a great variety of bends to be made. The actual operation of the machine is the same as the clamp folder. Both machines are used to advantage in bodywork, as they are capable of bending to shape many articles suitable for both new and repair work.

Figure 5.9 Pinch rolls

> **Safety note**
> When working bending machines you must:
>
> - Before locking the clamping blade down, make sure that your fingers and those of your assistant are clear.
> - Check on the path of the swinging counter-balance weight, warning anyone working close by of your intentions to use the machine.

Rolling machines

Bending rolls

Bending rolls of 1 m capacity are well suited for panel work. They are used mainly to form curves in panels and to roll complete cylinders, but they can also be used for breaking the grain of coated sheet metal, such as tin plate or galvanized sheet, before it is worked. Machines are made in all sizes: bench rollers for light tin plate work, hand-powered rollers for general sheet metal, and motor-driven rollers for heavy-gauge sheet steel. All rolling machines comprise two front rollers, which lightly grip and draw the metal through, and a free roller at the rear to set or bend the metal to the desired radius. They can, however, differ in their mode of operation, and bend rolls fall into two main types: pinch rolls and pyramid rolls.

Pinch rolls

The rollers are of the same diameter and are usually geared together to give the same feeding speed and rotation. The two front rollers are adjustable for different thicknesses of metal; the third roller at the rear is also adjustable to provide for various degrees

of bend required. A swinging-type end support can be opened, permitting the top roller to be pivoted upwards so that completed cylinders can be removed. Nearly true cylinders can be formed using pinch rolls; the slight flat which occurs on the leading edge of the plate can be reduced by initially reversing the rollers and feeding the flat plate back into the rollers a little way, which will pre-form the edge ready for rolling. Pinch rolls grip or pinch the metal between the first two rollers and give good contact during bending, but if these rolls are too tightly adjusted the edges of the metal will be stretched, and the result will be an uneven edge. Pinch rolls are used for rolling all types of thin-gauge sheet metal up to 2.00 mm.

Pyramid rolls

This machine has two rollers of equal diameter mounted side by side which rotate in the same direction. The third roller, which is situated above the other two, is of larger diameter and is adjustable in the vertical plane in order to control the radius of bend. It pivots at one end to permit easy removal of the finished work. Pyramid rolls tend to leave more flat surfaces than pinch rolls. Again this may be reduced by rocking the plate between the

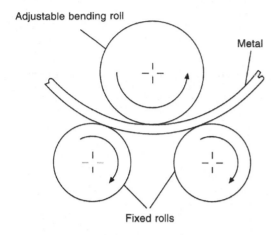

Adjustable bending roll

Metal

Fixed rolls

Figure 5.10 Pyramid rolls

rollers, but it may be necessary to pre-form the edges of the metal by hand before rolling. This type of machine is best suited for the heavier gauges of metal 3 mm and upwards.

As already stated, common metals used for panel work have a certain degree of elasticity. This means that before a metal sheet can be formed into a curve or cylinder, sufficient force must be applied to deform the structure of the material beyond its elastic limit so that the sheet assumes a permanent curved shape. Some heavier material may require pre-setting by hammering to provide a lead over the rear roller and also to prevent flats at the ends of the rolled surfaces. However, small flats may occasionally prove to be of assistance in the welding of thin sheet metal. The slight ridge formed by the flats tends to act in a similar manner to a swage and has the effect of stiffening the area to be welded. Excessive distortion is checked by the flats and, after welding, the smooth curve of the cylinder can be restored with a mallet. Bending rolls are sometimes used to repair flat aluminium panels by rolling them, then repairing them in this state, then rolling them flat again.

Cone rolling machine

Cone rolling machines are mostly used for rolling small conical-shaped fabricated articles in light-gauge metals. They consist of a pair of conical rolls, geared and mating together and acting as pinch rolls. There are no special sizes. They are usually made to the particular requirements of the company.

Safety note
Safety measures for the operation of rolling machines:

1 Loose clothing and long hair can easily be drawn between the rollers.
2 Care should be taken when putting the panel into the rollers that the operator's fingers are kept clear.

Wheeling machines – commonly called the English Wheel

These machines can be used to form flat sheets of metal into double-curvature shapes such as are found in automobile bodies. They are also used for removing dents, buckles and creases from sheets or panels previously shaped, for reducing the thickness of welds, and for planishing or smoothing panels which have been pre-shaped by hand. The main frame of the wheeling machine is of large C form. The size of the gap in this frame usually determines the machine's specifications, which are from 0.75 m to 1 m in width and about 0.6 m in depth; the wheel diameters are approximately 90 mm. The machine consists simply of an upper wheel which is nearly flat and a second lower wheel which is curved or convex in shape; the two wheels meet at a common centre. The lower wheel runs free on a spindle carried by a vertical arm, which may be raised or lowered by screw movement through a hand wheel, which regulates the pressure that can be applied to the work. Some machines have a quick pressure-release mechanism attached to the spindle. There are three standard shapes of faces for the bottom wheel – flat, small radius and full radius – and the choice of wheel depends on the required shape of the finished panel. The upper wheel is carried on a horizontal shaft which is allowed to rotate freely in bearings. The top wheel and housing can be swivelled round through 90° to make it more accessible when wheeling certain shaped panels. This also applies to the housing for the bottom wheel, enabling it to move in conjunction with the top wheel. Up to three times as much lift or stretch is obtainable with aluminium as with steel, and consequently aluminium can be shaped more by wheeling than steel. When wheeling aluminium, care must be taken not to apply too much pressure by the bottom wheel, as this could have the effect of overstretching the material.

This machine is one of the few machines that has been used in the panel beating trade since its infancy. The art of wheeling a panel to the correct shape and contour requires a highly skilled craftsman, who is known in the trade as a wheeler.

Swaging machines

These machines can be used on sheet metal blanks to carry out a large number of different operations such as swaging, wiring, joggling, flanging, beading and many other edge-type treatments. Hence the machine is also called a jennying, swaging, burring or beading machine, despite the fact that the same basic machine is used to perform the different operations. In bodywork the machine is used in the stiffening and strengthening of panels; for decoration in the form of beading or swaging; for the edge preparation of panels, such as wiring; and for making joints between panels as in lap and joggle joints. It can also be used to give rigidity to large flat panels to prevent 'drumming'.

The basic machine consists of a machine frame carrying two horizontal shafts which are geared together. The shafts are mounted one above the other, one of them in a fixed housing and the other in a housing which has some vertical adjustment with reference to

1 Wiring rolls for forming U-seating for wire
2 Closing rolls for closing metal around the wire
3 Jenny rolls (flanging)
4 Panning down rolls
5 Bottom closing rolls
6 Swage rolls
7 Collars or distance pieces
8, 9 Swage rolls
10 Necking rolls used for joggling

Figure 5.11   Swaging attachments

the fixed shaft. It also has some horizontal adjustment controlled by a gauging device known as the stop. In hand machines the shafts are turned by a handle which is fixed to the end of one shaft; in power-driven machines they are turned by an electric motor which is connected through gearing to the shafts. The shafts are usually arranged so that they have a 1:1 rotation. The actual swaging is done by attachments fitted onto the ends of the shafts, and the shape and form of these attachments can be varied as is shown in Figure 5.11. Attached to one end of the shafts are male and female rollers shaped to produce a swage of the desired form. The top section of the frame carrying the upper roller and shaft is hinged at the back and kept in upward tension by a flat spring. The wheels are brought into position vertically by the operation of a small hand-screw. A small lever situated near the gear wheel provides horizontal adjustment so that the wheels can be set to match exactly. An adjustable guide is provided to ensure that the swaged impression will be true and parallel with the edge of the panel.

To set up the swaging machine for operation, first select a pair of wheels which will give the required section. Fit these wheels to the shafts and line them up by slackening the locking screw and adjusting the small lever at the rear of the frame until the wheels engage centrally with each other. Next set the stop to the distance required between the centre of the moulding and the edge of the sheet. Adjust the top screw until the wheel forms a depression in the metal and then run the metal through the wheels. Give another half turn on the hand-screw and repeat the operation until the wheel 'bottoms', at which stage the swage will be fully formed. The object of forming the swage in gradual stages is to avoid strain on the metal, which may result in splitting or distorting the panel.

Brake presses

Brake presses are devices for bending sheet metal quickly and accurately, and their rapid development in the past 20 years has resulted in a very wide capacity range of up to 1500 tonnes. The bulk of the machines for bending thin-gauge metal are of 3 m or less, with capacities starting at 20 tonnes and going up to about 200 tonnes. Brake press capacities are usually given either in tonnage terms or in maximum bending of a certain metal thickness and are based on V-bending pressures. The tonnage specifications are usually determined by using a V-die opening of eight times the stock thickness, with a corner radius to the bend not less than the metal thickness. There are two important dimensions in brake press specifications: one is the maximum bending length over the bed, and the other is the distance between the housing. The width between the housing must permit the work to pass through the machine, and the brake press length is the overall bending length of the machine.

Brake presses comprise **a frame**, **a ram** or bending beam, and means for moving the beam. The frames are always of all-steel construction with parts welded or bolted together. Obviously these frames should be as rigid as possible to resist deflection – a fact of the highest importance. There are usually three main parts to the frame – a bed and two side frames. The top of the bed has a location slot into which the bending tools slide. The bending beam is usually a steel plate which works by simply sliding in the main frame. The load on the beam acts as near to the centre of the beam as possible to avoid side strain. The bottom of the beam face is made to receive the top bending tool or die, generally by means of a side plate to hold a tongue formed on the tool or die. The general methods of operating the beam are mechanical or hydraulic. The mechanical means comprise a crankshaft rotating in a phosphor bronze bush in the main frame. The working strokes are usually between 50 mm and 150 mm long, and this length can be adjusted to suit varying conditions. The motor drives a flywheel through a multi-plate friction clutch and single or double gearing. A brake is fitted to work in conjunction with the clutch and is of the drum-brake shoe-operated type. Its function is mainly to hold the weight of the moving parts when the clutch is disengaged.

The V-type blade and interchangeable die are used extensively for forming light mild steel sheet components, from simple bends to complex multi-bends in panels or sheet metals. Not only can it produce

Flywheel

Top die tool

Bottom die tool

Foot control to operate clutch

Control panel to start

Crankshaft housing

Acute angle dies   Flattening dies   Three-way and four-way dies

Bottoming dies   Air bend dies   Gooseneck dies

Figure 5.12 Brake press dies and applications

straight, sharp bends, but the tools can be interchanged to give curves or radius bends. Curved sections, ribbing or stiffening sections, notching plates, corrugating sheets, and punching holes in plates can also be formed. The machine is mostly used in sheet metal industries, but it does find a use in the manufacturing side of commercial bodywork where large and complicated panels are to be formed.

Forming and drawing

During forming, one area of a sheet metal blank is held stationary on a die while a punch forces the other area to assume a new contour or shape. The force applied is sufficient to stress the metal beyond its elastic limit so that the change in shape is permanent. Forming has the distinct characteristic of stressing the metal at localized areas only; in the case of bending, for instance, this localized stress occurs only at the bend radius, resulting in a reduction of the thickness of the metal at the bend. It is only in these localized areas that any structured change occurs within the metal itself. This type of change in shape of the metal, with little structural change, is known as **metal movement**. In drawing, however, total stretching of the metal occurs, with a correspondingly large amount of structural change within the metal itself. This structural change within the metal as a result of applied forces is known as *metal flow* (Figure 5.14). Many irregularly shaped panels are formed by drawing, but the simplest drawing operation, that of cupping, more easily illustrates the theory of drawing. During cupping, metal flows through an opening provided by a clearance between a punch and a die which is in a cup shape. The punch exerts a force on the bottom of the cup so that metal flows away from the bottom. Owing to the compressive forces on the outer edge of the blank, metal tends to flow into this region. These compressive stresses could cause wrinkles at the edge of the blank and cup, and to prevent this a blank holder is added around the punch. Pressure is applied to the blank holder by springs, air or an outer ram of a press. If the blank holder pressure is too high, metal flow will be restricted and excessive stretching will cause the cup side walls to break. If the pressure is too low at the start of the drawing, wrinkles will occur before the pressure can build up. Thus the blank holder pressure must be low enough to allow the metal to move or flow underneath it and high enough to prevent wrinkling from occurring. This pressure cannot be reduced or wrinkles will occur, and therefore a lubricant must be applied to reduce the friction. Nearly all drawing operations require some lubricant for this reason.

Sheet metal drawing operations

Drawing operations are classified according to the shape of the part drawn, as follows: cupping, box drawing, panel drawing shallow and panel drawing deep.

Cupping

Cupping is the drawing of parts having a circular or cylindrical shape. The cup may be perfectly cylindrical, be composed of several cylinders of different diameters or have tapered walls with spherical bottoms or flat bottoms. Changing the flat blank to the cup is called the **drawing** operation. Reducing the cup to a smaller diameter with greater height is called **redrawing**. If the cup is turned inside out during reduction, the operation is called **reverse redrawing**.

Box drawing

The drawing of square or rectangular shapes is called box drawing. Greater blank holder force is required in box drawing than in cupping to prevent wrinkling at the corners of the box. Box drawing consists basically of forming at the straight sections and drawing at the corners. Some flow of metal does, however, occur at the straight sections.

Figure 5.13 Box drawing

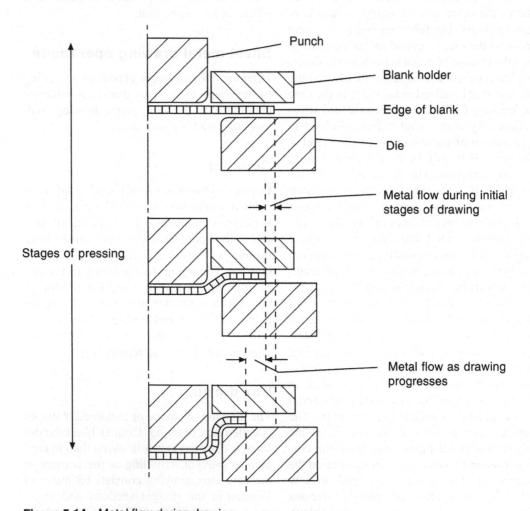

Figure 5.14 Metal flow during drawing

Draw   Redraw   Reverse Redraw

Figure 5.15 Metal punching

Panel drawing

Drawing of irregular-shaped parts is called panel drawing. The contour of the parts changes irregularly, causing a wide variation of stresses; for example, wrinkling or excess of metal may occur in one section while the metal tends to cause tears in other sections. It is very difficult to predict these stresses, as wrinkling is not limited to the flange area as in cupping, and tearing or failure is not limited to the bottom of the panel as in cupping. Hence panel drawing dies must go through a try-out period before production is possible. Adjustment in blank holding pressure, surface contour and pressure, draw radii and blank size are made during the try-out period to correct for variations in stresses from the predicted values.

If the panel height is great the operation is called **deep drawing**; low panel heights indicate **shallow drawing**. Shallow panels often incorporate a great deal of stretch forming of metal. Deep-drawn panels are used for automobile front wings, rear wings, rear quarter panels and instrument panels. Shallow-drawn panels are used for automobile roofs, doors, bonnets and boot lids.

Sheet metal cutting for press work

There are four types of sheet metal cutting for press work: **shearing, cut-off, parting** and **blanking**. In shearing the cutting action must be along a straight line. Shearing does not use a press or die and is therefore not a stamping operation. It is used for the following purposes:

1   To cut strip or coiled stock into blanks
2   To cut strip or coiled stock into smaller strips to feed into a blanking or cut-off die
3   To trim large sheets, thereby squaring the edges of the sheets.

A **cut-off** is made by one or more single-line cuts. The line of cutting may be straight, angular, curved or of a broken design. It is important to note that the blanks must interlock perfectly before a cut-off operation is carried out so that no scrap is produced during cutting. The operation is performed on a press by a die and therefore may be classified as a stamping operation.

Parting is used when the blanks will not interlock perfectly. Some scrap is inevitable since the blanks do not fit together, and so parting is not as efficient as shearing or cut-off.

In **blanking** the cutting action cuts out the whole shape of the part in one operation. Excessive scrap is formed during blanking, and therefore it is a very inefficient operation. The scrap is in the form of a ribbon commonly called the 'skeleton'. Blanking is performed in a press by a die and is a stamping operation.

Skills and questions

Metal forming and the use of workshop equipment are the backbone skills of the vehicle body builder and repairer. The following are suggested as exercises to help you gain professional-level skills. Please remember that these skills take a few years to completely master, but a few days getting familiar with the materials and equipment will stand you in good stead for future development. Get your tutor or a senior work associate to help you where appropriate.

1 Carry out tests on various metals to get the feel of them. The following tests are suggested to start your experiments:
   a Using the grinding wheel watch the sparks from steels. You'll see that the colour of the sparks changes with the carbon content.
   b Carry out tensile, bending and compression tests. You'll need to get your tutor to take you into one of the laboratories to do this.
2 Carry out hardening and tempering processes on scraps of material. Test your results to check the effects of your work.
3 Write out safety procedure cards for two of the processes which you regularly use.
4 Use as much of the workshop equipment as you are allowed. As often as you can, get a senior colleague to demonstrate them first, and practice using scrap material.
5 Inspect one of the vehicles in the workshop and determine the properties of the materials used for each body part and see if you can see how you would make one in the workshop. Make sure that you have permission to do this.
6 Describe the forming and fabricating properties of the steels used in vehicle body manufacture.
7 With the aid of sketches, describe how a low-carbon steel can be shaped into four different forms to provide strength in body construction.
8 Which properties of a low-carbon steel allow it to absorb energy resulting from impact through collision?

9 Explain the following treatments that are associated with aluminium alloys: (a) solution treatment, (b) age hardening, (c) precipitation.
10 Explain the following heat treatments to carbon steel: (a) normalizing, (b) hardening, (c) tempering, (d) case hardening.
11 What is meant by the term 'work-hardened material'?
12 Explain the process of annealing.
13 State the important conditions to be observed when using a wheeling machine.
14 List three safety measures to be observed when using a treadle-operated guillotine.
15 What is meant by the principle of shearing metals?
16 Describe the differences between a bending machine and a rolling machine.
17 Explain the operation of the swaging machine.
18 Explain the three methods by which heat can be transferred from one body to another.
19 Define the importance of the property 'ductility' in a material.
20 Explain the changes which affect metal when heated and then cooled.
21 Describe the condition of a metal panel which has been excessively wheeled.
22 After sheet metal has been work hardened, describe which process will allow further working of the sheet metal.
23 What is the minimum carbon content required for steel which will be hardened by heat treatment?
24 Describe a simple workshop method which can be used to estimate the temperature when tempering carbon steels.
25 When working with carbon steel, what is the difference between normalizing it and annealing it?
26 List the properties contained in commercially pure aluminium.
27 What practical factors limit the width of sheet metal which may be formed on a wheeling machine?
28 By means of a sketch, draw one type of swage which could be produced on a flat piece of sheet metal.
29 What is meant by the upper critical point of steel?
30 Explain the blanking operation.

6

Marking out and measuring

Marking out

A body repair worker must have a good general knowledge of **engineering drawing** and surface developments in order to mark out replacement parts on sheet metal, as this requires accurate calculation of sizes and angles. While the majority of bodyworkers who work in the repair side of the industry do not have to perform a lot of difficult marking out from elaborate drawings onto sheet metals for the making of the articles, those employed in the manufacturing side of the industry (e.g. coachworks) have to mark out all units and assemblies from drawings, whether for a one-off item or for mass production. Consequently a thorough knowledge is essential of measurement and marking tools and also of how to check these tools so that a standard of accuracy can be achieved and maintained throughout.

Blueprints, as working or production drawings are generally termed, are photographic reproductions of an original drawing. They take their name from the colour of the finished print when the drawing is reproduced on ferro-prussiate paper. This paper produces copies of the drawing in white lines on a blue background. Several other types of coloured prints are also produced by industrial contact photo equipment, but in spite of their colour these are still generally known as blueprints. The practical bodyworker is rarely called upon to produce a finished drawing, but he will find it necessary in many cases to understand and work from both general and detailed drawings as set out by a skilled draughtsman. The main object of reading a drawing is to obtain a clear mental picture of what another person has represented on paper by means of a conventional arrangement of lines and symbols. To this end the value of hand sketching cannot be too fully emphasized, as well as the making of simple models, for a much clearer impression of form can be obtained when movement and touch are combined with sight. The skilled craftsman, particularly in the small workshop or garage, is generally responsible for seeing his work through all stages of manufacture, commencing with careful study and accurate interpretation of the drawing. Before actual manufacture can begin, it is necessary to set out the desired shape by marking the outline on the surface of the sheet of material. Automobile work demands a thorough knowledge of the methods of sheet metal work, and although the machine has displaced the man for certain operations there is an increasing demand for craftsmen who are capable of developing the large variety of work met in the industry. Much of the work done deals with double-curvature shapes which are in some cases fabricated by hand in sections. This type of work needs templates made to the exact curvature required so that the work can be checked at each stage of its progression.

The Repair of Vehicle Bodies, Andrew Livesey, 9780815378709, 2019, Taylor & Francis

Also, very accurate alignment jigs have to be manufactured to hold these parts, after they are formed to the template size and shapes, so that they can be welded together into the finished component. Marked-out templates, which are usually in thin metal sheet, cut and then filed to a very accurate size, are used when one is working to rolling radii so that the curve can be checked as it is rolled. Angle templates are useful when bending in the bending machine to check the angle of bends so that they are constant.

Manufacture of templates

These can be made first of a paper or hard card known as template paper, then using materials such as thin-gauge aluminium, tin plate and in some cases thin-gauge mild steel. These metals are used for their ease of cutting. The next stage in the manufacture of the template, if it is to be marked straight onto metal, is to take the sizes from the appropriate drawings and, using such instruments as scribes, dividers, straight edges, trammels and chalk lines where long sizes are involved, accurately to mark them directly onto the metal. When the template is fully marked out the profile is centre punched very carefully on the scribed lines at close intervals, and then cut, either by machine or by hand using tin snips, to the centre-punched marks. The metal is next filed very carefully to obtain the final finish and checked from the drawing for its dimensional accuracy. The template is now ready to be used either for checking or for marking out predetermined shapes directly onto sheets. Metal templates have the advantage that they can be kept and used again if necessary.

Basic marking-out and measuring instruments

Rules

A length may be expressed as the distance between two lines (line measurement) or as the distance between two faces (end measurement). The most common example of line measurement is the rule. Checking by way of comparing with a rule is called measuring. Measuring rules can be adjustable or non-adjustable. The adjustable type is usually of the folding variety, but there are also thin spring-steel rules which are referred to as steel tapes. Rules are made of hardened and tempered high-grade steel and are usually graduated in the English and/or metric systems of measurement. Rules should not be misused as feeler gauges, scrapers, screwdrivers or levers; the end of the rule particularly should be preserved from wear because it usually forms the basis of one end of the dimension. In use, a rule should be held so that the graduation lines are as close as possible to, preferably touching, the faces being measured. The eye which is observing the reading should be vertically above the mark being read; this avoids what is known as 'parallax', which results in an inaccurate reading due to the graduation lines being viewed from a very oblique angle. Rusting of the rule should be avoided by oiling or an occasional rub with metal polish. Chalk is good for cleaning the rule.

> **Tech note**
> A **rule** is the steel item used by an engineer where measuring starts from the very end; a **ruler** is a plastic version on which the measuring scale starts a little distance from the end.

Scribers

A scriber is used when marking out work and will leave a fine line on the metal surface. This will more easily be seen if the surface of the metal is first prepared by lightly rubbing it with white chalk or a special colouring agent. Scribers are made of

Figure 6.1   Marking out with a scriber and rule
(*Neill Tools Ltd (Eclipse)*)

Figure 6.2   Engineer's scriber (double-ended)
(*Neill Tools Ltd (Eclipse)*)

hardened and tempered high-grade steel, with a knurled body which facilitates handling. They are ground at one or both ends to a fine point, which should be kept sharp and protected when not in use. Some of the double-ended scribers have points which can be replaced if damaged or worn. A handy pocket scriber is also available. When using the scriber, it is inadvisable to mark any lines other than cutting lines on coated metals, thus avoiding the destruction of the protective coating, as this may result in corroding of the parent metal. It is also inadvisable when working on soft metals, such as zinc, aluminium and copper, to scribe along the full length of fold or bend lines, because this could result in cracking during the bending of the metal. Short marks at each end are sufficient for bend lines.

Straight edges

These are used to test that surfaces are level or that parts are in line one with another and are ideal for lining up and jig work. There are two types of straight edge. One, made from good quality hardened and tempered steel about 3.175 mm thick with one bevelled face, is made in lengths up to 2 m. The other type, made from cast iron, is ribbed along its length

Figure 6.3 Straight edges (with square or bevelled edges) (*Neill Tools Ltd (Eclipse)*)

Figure 6.4 (a) Centre punches (b) nail punches (c) parallel pin punch (*Neill Tools Ltd (Eclipse)*)

and has a flat face. This type can be more accurate than the steel straight edge because its weight and robust construction reduce the possibility of flexing. It is usually made in lengths of straight edge up to 2.5 m. To test the surface contour of a metal panel or sheet. The straight edge is held against the panel and then moved over the surface, and any distortion or unevenness is shown as gaps between the edge of the straight edge and the edge of the article. Straight edges can also be used in sets of three for sighting parts into line. A straight edge should never be used for any purpose other than that for which it is intended.

Punches

Centre punch
This cast-steel tool is driven into the metal with a hammer blow. For marking out work, locating centres for curves and defining profiles, the centre punch is struck only lightly. Sometimes lighter pattern punches, called prick or dot punches, are used for marking out. A centre 'pop' should be made as deeply as possible where a hole is to be drilled. It is as well to mark hole positions lightly at first, check them for accuracy, and then increase their depth by heavier hammer blows, holding the work on a firm foundation such as an anvil. Punches should be ground so that the grinding scratches lead to the point. The tapering point of the marking punch is usually ground to an angle of 40°, whereas the centre or pop punch has a more obtuse angle of 60°, which facilitates the application of a drill.

Automatic centre punch
This is hand operated, and a hammer should not be used. It delivers a punch when hand pressure is applied to the knurled head. The depth of impression can be adjusted by turning the knurled head.

Nail punch
This is designed to drive nail heads below the surface of wood (Figure 6.4b). The square head prevents rolling and provides a larger face for striking with the hammer, while the cupped and bevelled point limits skidding off the nail head.

Parallel pin punch
This is used to punch out pins and dowels on mechanical assemblies (Figure 6.4c).

Try square (engineer's square)

This is the most common tool for testing squareness and is used for internal and external testing to check whether work is square (Figure 6.5). It can also be used for setting out lines at right angles to an edge or surface. A try square consists of a stock and a blade, and these may be made separately and joined together, or the whole square may be formed from a single piece of metal. Squares are of cast steel, hardened and tempered and ground to great accuracy. The size of the square is the inside length of the blade. Engineers' squares are made with a groove cut in the stock where the blade enters the stock; this allows for any burr on the edge of the metal being tested.

Before a square can be used, one surface of the workpiece must be made level and true. This surface is termed the 'master' or 'face' side. To use a square to scribe a line at right angles to the face side, the blade of the square is laid flat on the workpiece with the inside of the stock pressed firmly against this face side; the line can then be scribed along the outer edge of the blade at the position required. To check that an adjacent surface is at right angles to this face side, the square is held upright on the workpiece on the corner and the inside edge of the stock is pressed firmly against the face side; any discrepancy in the angle will be indicated by a gap or gaps between the inside edge of the blade and the surface being tested. Engineer's squares are available in three grades of accuracy: B (workshop), A (inspection) and AA (reference).

Bevel gauge

When two surfaces are at any angle other than 90°, the angle between them can be tested with an instrument called a bevel. Bevels consist of a stock with one or two blades which are adjustable and can be locked at any angle by means of a thumbscrew. There are three main types. First is the plain engineer's bevel, which has an offset slot for testing angles which are not possible with a straight slot. Second is the universal bevel, which has offset blades and can be set at any angle. The thumbscrew on the bevel is recessed into the stock, allowing it, if necessary, to lie flat on the work. Third is the combination bevel, which consists of stock, split blade and auxiliary blade. The split blade swings to any angle and, with the auxiliary blade attached, can be transferred from one surface to another, so work can be laid out or any desired angle checked and measured.

Bevel protractors

The alignment of planes or surfaces designed to form a definite angle may be checked by the universal bevel protractor. This is a graduated disc with a fixed adjustable blade and a stock with a straight-edge extension. The disc with the blade is graduated from 0° to 90° each way and rotates, together with its blade, through an entire circle in relation to the stock. For very accurate testing, a Vernier bevelled protractor must be used.

Adjustable protractors

When the edge of the workpiece forms an angle which deviates from the usual angles of 90° and 45°, this angle may be determined

Figure 6.5 Try square (*Neill Tools Ltd (Moore and Wright)*)

with the aid of an adjustable protractor. The protractor consists of a semi-circular segment covering 180°. The movable blade which indicates the angle reading is secured by a locknut which can be set and tightened at any angle, thus allowing angles to be read directly from the workpiece. Such a protractor may also be used for marking out when the marking lines do not form one of the usual angles.

Combination square

This is used to test that work is true, to **measure and check angles,** and to locate the centres of round bar or centre lines of tubes. It consists of a blade (Figure 6.6) which is usually marked in English and/or metric scales and which may be used in conjunction with any one of three heads. The various heads enable the tool to be used as a square, a protractor or a gauge. A central groove along the length of the blade enables each of the heads to slide along this blade and to be locked in any position on the blade. The three heads used on this square are:

1 The **square head**, which has two surfaces, one at 90° and the other at 45° to the base, so that it can be used as a normal or a metre square, and also as a depth or height gauge. This head is usually fitted with a spirit level and a scriber, which is contained in the head.
2 The **protractor head**, which is used to measure or check any angle up to 360° by simply reading off the scale.
3 The **centre head**, which is V-shaped with two internal surfaces. These form an angle of 90° and very quickly find the centres of round objects.

The complete set forms a very useful workshop accessory and fulfils many needs, but when a high degree of accuracy is required, a normal solid try square should be used instead of the square head of the combination set.

Figure 6.6 Combination square (*Neill Tools Ltd (Moore and Wright)*)

Centre square

This is a flat square made of hardened and tempered tool steel. It has two blades of different thicknesses which are designed for the accurate location of the centres of faces of round bar or discs. The centre square tool is used in conjunction with a scriber. It is held against the circumference of the round bar so that one blade touches the bar on its edge and the other blade lies across the surface, giving a diameter line which is then scribed. When two or more of these lines have been scribed the point of intersection is the centre of the bar.

Dividers

These are metalworkers' compasses, made of steel with hardened points. They are used for scribing circles and arcs and for marking off lengths. The most common type are spring dividers. These comprise a circular spring bow which joins the legs at the top and acts as a spring and pivot. The distance between the points is set by means of a fine adjustment screw mounted across the legs. In use, the centre point of the circle to be scribed is lightly marked and is used for the static position of the pivot leg, while the other leg is rotated to scribe the circle. The points of the divider must be kept finely ground and without making one

Figure 6.7 Spring dividers (*Neill Tools Ltd (Moore and Wright)*)

Figure 6.8 Trammel heads (*Neill Tools Ltd (Eclipse)*)

leg shorter than the other, and the points must be protected against damage when not in use.

Trammels (beam compasses)

Trammels are used to describe arcs and circles of large radii and mark off lengths which are beyond the range of dividers. They comprise two heads fitted with scribers which slide along a bar. The bar can be any length. To use a trammel, one leg is placed at the centre point of the required arc or circle. The other leg is then set and locked at the correct distance by the aid of a locking screw fixed in the head. Light pressure is kept on the pivot leg with one hand while the other hand describes the arc with the other leg. Additional legs can be supplied with curved tips and can replace the scriber legs for use as calipers.

Calipers

These are used to measure or compare distances or size. There are two designs: inside calipers, which have straight legs than turn outwards towards the points and are used for measuring bores and internal spaces, and outside calipers, which have curved legs that turn inwards towards the points and which are used for external diameters and other outside measurements. In their simplest form calipers are usually termed 'firm-joint' calipers and consist of two legs riveted together at the top. Also available are the 'spring-joint' type in which the legs are joined at the top by a circular spring band and a fine adjustment screw is fitted across the legs. In use, the legs of the calipers are first opened out to the approximate size and held square with the object being measured. The points are next screwed towards each other so that they just touch both sides of the part to be measured. The calipers are then

Figure 6.9 Calipers (*Neill Tools Ltd (Moore and Wright)*)

carefully withdrawn from the workpiece and the distance between the two points is measured.

Oddleg calipers

This measuring instrument is half caliper and half divider (Figure 6.10). It consists of two legs riveted together at the top, one of which is ground to a point while the other is turned inwards or outwards at its tip. In use, the curved tip is held against the edge of the workpiece and the line is scribed with the point of the other leg. These instruments are used for marking-out operations such as scribing parallel lines, following perimeters and drawing intersecting arcs on cylindrical objects.

Gauges

Whereas measuring is the establishing of the actual size of a component, gauging is the positive comparison of the size of a component with a standard. Gauges are normally made from a hard wear-resisting metal and find extensive use in the mass production of car bodies, where many components are made to standard sizes which require frequent checking to ascertain their accuracy. Handy 'limit gauges' are available for the routine testing of parts specified to be within certain limits, and come in various sizes and shapes such as plugs, rings, slotted plates and tapered plugs.

Figure 6.10 Oddleg calipers (*Neill Tools Ltd (Moore and Wright)*)

Depth gauges with reversible base or protractor

A depth gauge is designed to measure the depth of holes or recessed surfaces. It consists of a reversible head or a protractor base which sits on the top surface of the workpiece in use, and a graduated steel rule which can move vertically through the centre of the head. The rule can be locked in any desired position by a thumbscrew. When the head is placed on the workpiece the rule is lowered to the bottom of the recess, then locked in this position when the size or depth of the recess is indicated.

Feeler gauges

A feeler gauge is used to measure small clearances between two objects and in some cases to establish the amount of wear or distortion of a component part. Each tempered steel blade has a number indicating its thickness in millimetres. The thicknesses

(a)   (b)

Figure 6.11   (a) Reversible depth gauge (b) protractor depth gauge (*Neill Tools Ltd (Moore and Wright)*)

are carefully graded in order that various clearances ranging from 0.01 mm to 1.0 mm can be measured. The gauge illustrated in Figure 6.12 is imperial, but this tool is now mostly made in metric sizes.

Screw pitch gauge

A screw pitch gauge (Figure 6.13) consists of a metal case with two sets of flat pivoting blades, each having teeth cut on its edge to correspond to a screw thread form. In use,

Figure 6.12   Feeler gauges (*Neill Tools Ltd (Moore and Wright)*)

Figure 6.13   Screw pitch gauge (*Neill Tools Ltd (Moore and Wright)*)

the blades are placed in turn onto the profile of the thread to be checked until an exact match is found in the gauge. This matching thread form will give the correct name, size and pitch of the thread in question. The gauge illustrated in Figure 6.13 is imperial, but screw threads in common use are: Système International (SI) metric, International Organization for Standardization (ISO) metric, American National, Whitworth and Unified.

6.2.19 Radius gauges

These are used to measure small internal and external radii. They are supplied as a set having a number of pivoting flat blades each of hardened and tempered steel, cut to a specific radius size. They are used as a template to check radii sizes on the workpiece.

Drill gauges

The drill gauge is a flat piece of metal having a selection of holes of accurate diameters.

The size of a drill is found by inserting the drill in the holes until a hole is found which fits the drill exactly. These gauges are made in either fractional or decimal sizes, number sizes or lettered sizes.

Imperial standard wire gauges

A wire gauge can either be oblong or circular in shape. Slots cut into its outside edge give accurate wire gauge sizes, usually

Figure 6.14 Radius gauges (external and internal radii) (Neill Tools Ltd (Moore and Wright))

numbers 1 to 36 SWG or 0.2 to 10.0 mm. It is still used by the bodyworker and sheet metalworker, as metal sheet thickness may be given in SWG size. The gauge is used by slipping it over the edge of a metal sheet until the corresponding slot is found, and the gauge size is then read off.

> **Tech note**
> Although we use the word 'wire', the wire gauge is used for measuring the thickness of sheet material too. It originates from wire production, wire which was used for making nails in the Bristol area of the UK about 250 years ago.

Precision marking-out and measuring instruments

As vehicle body panel tolerances become tighter, a greater level of accuracy becomes necessary to carry out the various tasks. This section looks at some of the items available for accuracy down to 0.02 mm (0.001 inch).

Marking-off table

This is made of close-grained cast iron, rigidly constructed and supported on very substantial legs. The top surface is machined level and finally scraped to a degree of flatness which is determined by the use to which the table is to be put.

The poorest grade is used for marking out material appertaining to bodywork; it also supports the work during marking out and is often used as a datum from which all measurements can be taken. Marking-off tables must be used for no other purpose except marking out and measuring. When not in use they should be lightly oiled and covered against accidental damage and corrosion.

Figure 6.15   Drill gauge (*Neill Tools Ltd (Moore and Wright)*)

Figure 6.16   Wire gauges (metric) (*Neill Tools Ltd (Moore and Wright)*)

Surface plate

A flat surface is one of the fundamentals of engineering, and the flatness of a surface can be verified by testing it against the flatness of a standard surface, the surface plate. Surface plates are similar in construction and grading to the standard of finish of a marking table but are smaller in size and have two carrying handles. The back is strongly ribbed to give rigidity to the surface. In use, the surface plate is covered with a special marking compound and then lightly rubbed on the workpiece to be tested. The marking dye is transferred to the parts of the workpiece which actually come into contact with the surface of the plate, and if the surface of the workpiece is uneven, only the high spots will be marked. Any high spots must be scraped, and the workpiece must be retested until the whole surface comes in contact with the surface plate. Surface plates should be kept covered at all times when not in use.

Surface gauge or scribing block

There are various types of surface gauges, but in the main they consist of a very accurately machined base, an adjustable pillar fastened to the base and an adjustable clamp attached to the pillar for holding the scriber. The surface gauge is used in conjunction with a marking-off table or surface plate to mark out very accurate, parallel lines to a true surface. The machined base enables it to slide over the surface of the table while the scriber retains its accurately set position. The scriber can be set at various heights and angles in order to mark out different jobs or parts of jobs with horizontal lines at a fixed height above the surface of the marking-off table. Scribed lines show up more readily on a work surface that is first covered with a solution of copper sulphate or marking compound.

Figure 6.17   Universal surface gauges (*Neill Tools Ltd (Eclipse)*)

Figure 6.18   Vee blocks

Vee Blocks

These blocks are made in matching pairs and have to be used in conjunction with a marking-off table. They are made of cast iron with vees machined accurately to 90° into the top of the block to hold the work. Along each side of the block runs a groove into which a clamp can be fitted in order to secure round-shaped work in place during marking off with scriber, try square or scribing block. In order to make sure that you have a matching pair of vee blocks, before using them always check that each of the blocks is stamped with the same number.

Engineer's level

This is used for setting surfaces level and parallel to a marking-off table. The base is machined so that it can slide across the surface of the workpiece, and if the surface is level then a bubble in a tube across the top of the level lies at the centre of its scale. The base of the level has a concave groove along its length which allows it to lie on curved work.

> **Tech note**
> When assembling cars in the factory, **laser light** jigs are used to get the components such as doors and boot lids perfectly straight on their hinges.

Figure 6.20 Engineer's levels (*Neill Tools Ltd (Moore and Wright)*)

Vernier caliper gauge

These calipers are made in a variety of types, some being used for external measurements only and others for both external and internal measurements. Vernier calipers consist of a fixed jaw attached to a main beam which is graduated and another jaw which slides along the main beam. This sliding jaw is marked with the Vernier scale, which coincides with the main scale when the jaw is moved. The Vernier scale is based on the difference between measurement made on two scales which normally have one division difference. By careful use of this adjustment a reading can be taken to 0.02 mm (0.001 inch).

1 Spindle and anvil faces
2 Spindle
3 Locknut
4 Sleeve
5 Main nut
6 Screw-adjusting nut
7 Thimble-adjusting nut
8 Ratchet
9 Thimble
10 Steel frame
11 Anvil end

Figure 6.21 External micrometers (*Neill Tools Ltd (Moore and Wright)*)

Figure 6.19 Micrometer (*Neill Tools Ltd (Moore and Wright)*)

Figure 6.22 Vernier caliper gauge (*Neill Tools Ltd (Moore and Wright)*)

Micrometer

The micrometer is used to measure and record accurately to 0.02 mm. The size of an object is measured by the contact principle, which requires it to be positioned between the anvil and the spindle head of the micrometer. The micrometer has a C-shaped steel frame. An anvil with a flat face is attached to one end of the frame, and the other end accommodates the spindle and the barrel, which is graduated in linear measure. The bore of the barrel is screw threaded and engages the part of the spindle which is externally cut to the same thread. The outer end of the spindle has a ratchet stud and a thimble which is graduated around its bevelled circumference. When this thimble is turned it moves the spindle forward onto the object to be measured and in so doing moves along the linear scale on the barrel. A locking ring is usually provided in the frame at the point where the spindle emerges from the barrel, so that the spindle can be locked in position after a measurement has been taken. An object to be measured is held squarely in position between the anvil and the spindle, then the spindle is closed on the object by turning the ratchet stud clockwise until it starts to slip, indicating that there is a predetermined pressure on the object. If there is no ratchet stud the thimble is turned just until resistance is felt to the turning; great care must be taken to avoid over-tightening. Next the spindle is locked in position by turning the locking ring and the reading is taken. The working principle of the micrometer is based on the distance a screw moves forward for each turn made. The screw thread has a lead of 0.5 mm, while the thimble and barrel are graduated into divisions of 0.5 mm for the barrel and 0.01 mm for the thimble. Therefore, one revolution of the thimble moves it along the barrel for 0.5 mm. The thimble has 25 equal divisions, and as one revolution of the thimble is 0.5 mm, one of the thimble divisions will equal:

$$0.5 \text{ mm} / 25 = 0.02 \text{ mm}$$

There are a number of different types of micrometers made to serve a variety of uses in the field of precision engineering: internal micrometer, tube micrometer, screw-thread micrometer, adjustable micrometer, depth-gauge micrometer, tubular inside micrometer, bench micrometer and gear-teeth micrometer.

Skills and questions

To develop your measuring and marking-out skills, the following tasks are suggested. Carry them out using scrap material to save wastage. Get one of your senior colleagues to supervise you.

1 Mark out a circle of about 150 mm diameter and cut it out accurately Circles are one of the most difficult shapes to deal with.
2 Measure the thicknesses of a number of different panels and find out why different thicknesses are used for them.
3 Use a micrometer to measure the diameter of a range of screws and bolts. Note your results and the purpose of these items.
4 For something different, use your measuring skills to measure some engine components.
5 Using card, make a template in the shape of the outline of any vehicle you like at about 1/20 scale and use this template to make a decorative hanging to go over your bench, or put a prop on the back of it so that it sits like a greeting card.
6 When would a body repairer need to make templates?
7 Describe the making of a template and the types of materials which could be used for this purpose.
8 Make a list of measuring and marking-out instruments which would be used in vehicle bodywork.
9 Explain the type of steel that would be suitable to manufacture a scriber.
10 Suggest an appropriate repair situation which would require the use of a long straight edge.

11 Explain the difference between a try square and a centre square.

12 Explain the uses of a set of trammels.

13 Illustrate the two types of calipers which are used in the workshop.

14 State the reasons why gauges are a necessary piece of equipment in engineering.

15 Make a sketch, and state the operation of, a micrometer.

16 Explain the term 'datums'. Why are they used when marking out?

17 With the aid of a sketch, illustrate the principal parts of an engineer's combination square.

18 Explain the use of the vee block when marking out.

19 What is the purpose of a marking-off table and a surface gauge?

20 Explain the operation of a Vernier caliper gauge.

21 State the effects of parallax error when measuring with a steel rule.

22 Explain the purpose of a metre square when marking out.

23 Describe a method by which a surface can be checked to ensure that it is vertical.

24 With the aid of a sketch, show how to locate the centre between two points using dividers.

25 Explain the difference between a radius gauge and a drill gauge.

7

Methods of joining

Some methods of joining have been used for a long time and can be found on almost all makes of vehicle, while some are just used for a short time and found only on a small range of models. The methods of joining used by the vehicle manufactures are chosen on a number of criteria, typically:

- Cost – What is the cheapest method which will do the job commensurate with the quality of the vehicle
- Security and stiffness – How firmly will the panel be held
- Style – Frequently joints are used as design features. Is it suitable for the market that the car is designed for?
- Ease of manufacture and interchangeability with other models in the range
- Weight – Some super-sports cars use very light weight methods to bolster their image, and others look for lightness for environmental reasons.
- Serviceability – Some methods are essential to give accessibility to service items – like changing the engine oil.

Solid rivets

The strength of a riveted joint depends on several factors: the material of which the rivets and plates to be riveted are made; whether the holes for the rivets are punched or drilled; and the workmanship involved in the riveting of the joint. In light sheet, there is virtually no possibility of distortion. However, in heavier metal plate, riveted joints often have to be comparable with the surrounding metal, and so the rivets are inserted in the hot state, as this increases the strength of the joint.

Rivets used in hot riveting are made from steel and/or iron, while for cold working rivets made from copper, brass, aluminium and aluminium alloys are used. In bodywork, steel rivets are used on commercial chassis frames and also on fabricated sections such as sub-frames and structural load-bearing members. In addition aluminium rivets are used on alloy frame construction in the body-building industry.

Rivets are classified according to the **shape of the head** and the **diameter** and the **length** of the rivet. The shape of the rivet head is selected according to the intended use of the workpiece to be riveted. The diameter is selected according to the required strength and thickness of the component to be riveted. Most important, the length of the rivet must correspond to the total thickness of the components to be joined together with their British Standard proportions.

The snap or round head rivet is used where high strength is required. The pan head, similar to the snap head, is also used where strength is required. The mushroom is used in thin sheet metal which would be weakened by the use of countersunk rivets. Flat-head rivets are used for riveting flat bar

The Repair of Vehicle Bodies, Andrew Livesey, 9780815378709, 2019, Taylor & Francis

and angle sections to thin metal sheet, as the head is fairly flush and does not obstruct. The countersunk head rivet is used when the surface of the component must remain smooth.

Rivet holes

Rivet holes can be punched or drilled. Punched holes are generally slightly conical, and for an efficient joint the work has to be arranged so that the holes are brought together with their smaller ends adjacent. Punched holes also have ragged edges which must be smoothed; otherwise they will decrease the shearing resistance of the rivets. Drilling produces holes with smooth edges and has the further advantage that it can be carried out with the plates in position so that there is no damage due to badly aligned holes. A joint formed with drilled holes is about 8 per cent stronger than a joint made with punched holes.

Riveting dimensions

The **diameter of a rivet** is usually determined by $D = 1.2\sqrt{t}$, where D is the diameter of the rivet and t is the thickness of the plate.

The **allowance for riveting** is the allowance for the head of the rivet and is the amount of rivet showing above the plate before riveting.

The allowances are: snap head = $1.5D$; countersunk head = $0.15D$; flat head = $0.5D$.

The **pitch** of a rivet is the spacing between the rivet centres. It can be calculated from the principle that the part of the plate between each pair of holes should be the same strength as one rivet. Therefore if the pitch is less than $3D$ (Figure 7.3) the plate between the rivets will be too weak. The pitch can then be extended to $8D$ in special cases, but above that the plate will tend to buckle.

The **width of lap** is the distance from the centre of the rivet to the edge of the plate. It is generally taken as $1.5D$, so that a single-riveted lap joint would have an overlap of $3D$ and a double lap joint an overlap of $5D$.

Figure 7.1 Small rivet heads (BS 641)

Types of riveted joint

The simplest form of riveted joint is the single lap joint, in which the plates overlap a short distance with a single row of rivets along the centre of the lap. The double-riveted lap joint has two rows of rivets arranged either in line or staggered alternately. The single-strap butt joint consists of two plates butted edge to edge with a strap covering the centre of the butt and riveted down each side. The double-strap butt joint is the same as the single-strap butt joint with the addition of a

D = Diameter

Figure 7.2 Single-riveted lap joint

Figure 7.3 Butt joint

second strap on the opposite side. A single-riveted joint has an efficiency of only about 55 per cent. In a double-riveted joint this is raised to about 70 per cent, whilst in a treble-riveted joint 80 per cent or greater efficiency may be attained.

Riveting procedure

Plates to be riveted should be clamped together with their rivet holes in alignment. If hot rivets are being used, they should be at forging temperature and the operation should be completed before they become 'black' hot. For snap-head rivets, a punch or tool (snap) having a half-spherical cavity similar to the rivet head is used to support the rivet head, while the other end of the rivet is riveted over by holding a similar

punch to the rivet end and striking it with a hammer, thus forming a second cup head on the other end of the rivet. The aim in riveting is to swell the body of the rivet until it completely fills the holes and to complete the process as quickly as possible while the rivet is still very hot so that there is maximum contraction of the rivets after riveting to pull the plates together. Pneumatic or air-operated hammers are used extensively for closing over rivet heads. They are designed to deliver the right weight of blow at the correct speed to form the rivet head speedily and accurately. For cold riveting the closure of the rivet is similar to that of a hot rivet except that the metal is not as plastic, and greater difficulty will be encountered in swelling the rivet shank to fill the hole. Cold riveting plates cannot be as tight as with hot riveted because there is no contraction of the rivets to pull them together.

Care should be taken that the rivet head is spread evenly in all directions and not bent over in one direction only. This is often helped by giving the rivet a few preliminary blows with the ball end of the hammer, thus spreading the rivet a little before using the riveting tool. Countersunk rivets should be supported with a flat-headed punch. The initial spreading is done with the ball end of the hammer, and the head is finished with the flat end of the hammer. When riveted joints are large, it is advisable not to start at one end of the workpiece but to rivet the extremities first and next the centre, then the rest of the joint. This eliminates creeping of the plates and misalignment of the holes.

Bifurcated, tubular and semi-tubular rivets

These rivets have the outstanding advantage that there is no swelling of the solid portion of the shank during the closure operation. They are used extensively for the joining of soft materials such as plastic, rubber, leather and/or brake and clutch linings to metal. Only a small percentage is used in the body-building industry.

1 Rivet set used to bring plates together
2 Flat pein hammer used to spread rivet and fill up
3 Roughly shaped head with ball-pein hammer
4 Finish head to shape with rivet snap

Figure 7.4   Riveting procedure

Blind rivets

The blind rivet was originally developed in the 1930s for use in the aircraft industry, but since then it has been adopted in the light engineering industries. It is now employed as a standard sheet metal fastener in cars, buses and all forms of commercial vehicles. Blind rivets, as their name implies, are rivets which can be set when access is limited to only one side of a structure. They are invaluable both in car construction work, where many of the panel assemblies are of the double-skinned type so that accessibility would be impossible for conventional riveting methods, and also in body repair work, as they eliminate any unnecessary stripping of interiors. Blind rivets are manufactured by several companies under brand names, the most famous of which is POP. POP is the registered trademark of Tucker Fasteners Limited, who were the pioneers of blind riveting.

Blind rivet types

POP rivets

The POP rivet comprises a hollow rivet assembled to a headed steel mandrel or stem. It is inserted into a pre-drilled hole of the correct size in the workpiece, and a special tool containing a gripping device is applied to the mandrel or stem of the rivet. When the tool is operated, either manually or automatically, the mandrel head is drawn into the hollow rivet, expanding the end of the rivet which is on the blind side of the structure and at the same time pulling the material together. When and only when a tight joint has been formed, the mandrel breaks at a predetermined position, so that the mandrel head is left as a plug in the bore of the rivet (Figure 7.6). The spent portion of the mandrel is then ejected from the tool. POP rivets are manufactured from aluminium alloys, Monel, mild steel, copper and stainless steel. They are supplied with domed or countersunk (90°, 100° or 120°) heads in diameters of 2.4 mm, 3.2 mm, 4.0 mm, 4.8 mm and 6.4 mm, for riveting thicknesses up to 12.7 mm. Shear strengths range from 400 N for 2.4 mm diameter aluminium alloy rivets, to 5400 N for 6.4 mm diameter Monel rivets. Owing to the high clenching action of the rivet, tensile strengths are in excess of these figures.

Standard open rivet is subdivided into the break-head mandrel type, which leaves the mandrel head free to fall away from

Figure 7.5 POP blind rivet: standard open type (*Tucker Fasteners Ltd*)

Figure 7.6 POP blind rivet, standard open type: setting sequence (*Tucker Fasteners Ltd*)

the rivet and the break-stem mandrel type, which retains the broken portion of the mandrel within the set rivet, thus sealing the rivet to a certain extent. The latter type is intended for use in all normal blind riveting situations where the materials to be fastened do not present structural problems. This is a hollow rivet, preassembled onto a headed pin or mandrel. The mandrel is designed to fracture at a predetermined point during the setting operation, when the materials to be fastened have been drawn closely together and the joint is tight. *Sealed rivet* consists of a tubular rivet with a sealed end containing a steel or stainless steel mandrel. The riveting sequence is similar to that of a POP rivet, but this type has the advantage that in setting the rivet it is both compressed beyond its elastic limit and expands radially, thus ensuring a joint which is airtight and watertight up to 34 bar. The rivet is available with two alternative mandrels: one, called 'short break', fractures immediately under the mandrel head, which is retained in the blind end of the set rivet; and the second, known as the 'long break' type, fractures at a point outside the rivet and can be finished off to result in a flush finish. The sealed rivet is a fastener of high shear and tensile strength and vibration resistance. Owing to its high rate of expansion in setting, it cannot be recommended for use in very soft or brittle materials. It is designed for use where the fastening to be used has to be pressure- or watertight.

LSR rivet This range of aluminium rivets is designed to offer two particular advantages in joining soft friable or brittle materials. Its controlled setting gives outstanding reliability and it produces a neat uniform appearance. The LSR is suited to plastics, wood, GRP laminates and thin-gauge materials. It has an important role to play in such applications as the assembly of caravans and trailers. **MGR rivet** has been designed for use in situations where hole sizes are inconsistent. It also offers a multi-grip facility. The rivet is available in aluminium 2.5 per cent magnesium alloy with a carbon steel mandrel. MGR is valuable where components

Figure 7.7   Types of POP blind rivet (*Tucker Fasteners Ltd*)

Figure 7.9   POP blind rivet, sealed type: setting sequence (*Tucker Fasteners Ltd*)

Figure 7.8 POP blind rivet: sealed type (*Tucker Fasteners Ltd*)

are supplied to the user with holes already punched or drilled, or where the sheets to be joined are folded or curved, making it difficult to match up two holes of the correct size.

Grooved rivet has been developed for use in thick sections of soft or brittle materials such as hard board, plywood, glass fibre, asbestos board, concrete and brick. The POP grooved rivets gain their name from the series of grooves around the shank which engage into the workpiece on setting and set inside the material rather than against the rear face. Construction and setting action is similar to the standard open type. The body is of aluminium alloy with a carbon steel

Figure 7.10   POP blind rivet: grooved type (*Tucker Fasteners Ltd*)

Figure 7.11   POP blind rivet, grooved type: setting sequence (*Tucker Fasteners Ltd*)

carbon steel mandrel. On setting, the rivet body is split into four petals by the action of the mandrel head, producing a large blind-side bearing area capable of withstanding high a pull-out load.

ELF rivet is an aluminium 3.5 per cent magnesium alloy with an aluminium alloy mandrel. The body of this rivet splits and folds on setting to form three neat leaves which spread clamp-up load over a wide area. It is a totally sealed rivet designed for use on roofing but is suitable for materials like composite board, GRP, hard rubber and laminates, and it is used in commercial vehicle construction and caravans. Earth terminal rivets are designed to provide an effective earth connection to pre-painted sheets without damaging the finish. Tab rivets are

Figure 7.12   POP blind rivet, peel type: before setting (*Tucker Fasteners Ltd*)

mandrel. When set, this rivet is capable of withstanding very high pull-out loads.

Peel rivets are specifically developed for fastening soft or friable materials. They will secure blow-moulded or glass-reinforced plastic, rubber and plywood to metal panels or sections up to 13.5 mm thick. The rivet has an aluminium alloy body and a special

Figure 7.13 POP blind rivet, peel type: after setting (*Tucker Fasteners Ltd*)

Figure 7.14 Special POP rivets (*Tucker Fasteners Ltd*)

designed to perform two or more functions, such as fastening the tops of rechargeable batteries and at the same time providing a tab to serve as a connector. Tamper-proof rivets are set in the workpiece with a standard tool; a stainless steel pin is then tapped

into the bore of the rivet, and its domed head locates in a recess in the rivet head, making the fastening virtually impossible to remove without obvious traces. T-rivets are designed for exceptional strength and versatility; they can be reliable and vibration-proof in holes as much as 0.8 mm oversize.

Blind rivet nut is a threaded insert system and is a remarkably simple fastening system for use in all kinds of assemblies. It provides a straight, reliable threaded insert to which other components can be attached. It can be used as a blind rivet for permanently fastening one or more panels or sections together and provides an anchorage with the holding strength of at least six full thread turns. It can be installed at any stage without damaging the workpiece finish, even after painting. It can be used in materials with thicknesses varying between 0.25 and 7.5 mm. There are a wide range of sizes available in steel, aluminium or brass with thread sizes from M4 to M10. (M is the metric screw thread designation and is followed by a number, e.g. M4 is a metric thread of 4 mm diameter.) Sealed or open types are available with flat or countersunk heads. They can be used in drilled or punched holes with normal tolerances.

Well nut This demountable blind screw anchor is designed to provide a vibration-resistant fixing for engineering and automotive structures. It consists of a captive brass nut in a resilient neoprene bush which can be installed from one side of the workpiece using only a screw and a screwdriver. It is suitable for materials from light-gauge metal panel, too thin for self-tapping or set screws, to plastic sheet up to 16.5 mm thick. The flanged ends prevent electrolytic corrosion at metal-to-metal assembly points.

Repetition rivets

Repetition riveting systems are widely used on vehicle body construction, especially commercial vehicles. Some examples are as follows:

Chobert riveting system operates on an entirely different principle from other blind

Figure 7.15 Blind rivet nut, installation sequence

1 POP nut screwed onto tool
2 POP nut inserted into hole
3 Tool operated: mandrel retracts into tool, and unearthed part of nut expands on blind side of workpiece
4 Tool mandrel unscrewed
5 Assembly completed

riveting systems. Instead of a break-stem type of mandrel, there is a mandrel with a tapered hardened steel head larger in diameter than the tapered bore of the rivet. The mandrel is drawn through the rivet towards the head and is thus part of the placing tool rather than the component. It is possible to extend the mandrel length to accommodate a charge of up to 100 rivets and to achieve a rate of placing up to 2000 per hour. The Chobert rivet is made in a wide range of materials and sizes. Head forms currently available include snap, mushroom, flat and countersunk, and sizes range from 2.38 mm to 6.35 mm diameter. Materials include a wide range of aluminium alloys and zinc-plated steel. They are supplied in tube loaded form, each 317.5 mm long and containing up to 100 carefully aligned rivets. The tubes can

be used in special tools which can be either manually or pneumatically operated. If additional strength or watertightness is required the rivets can be pinned, giving in effect a solid rivet.

Chobert Grovit system has been developed so that the widest range of materials (such as wood, plastic, aluminium) may be joined. The Grovit has grooves on its shank which bite firmly into the material when the rivet is expanded. The rivets can also be used in the high-speed Chobert system. These types of riveting systems are ideal for mass production and are invaluable in the bodybuilding industry, where speed and economy are essential.

Briv system is a high-speed, high-clench riveting system for use with a wide range of assembly materials. Briv is installed from one side of the workpiece with the use of a mandrel loaded into the Avdel power tool. The action of the mandrel draws through the rivet, expanding the shank to fill the hole. The applications range from body component assembly to fixing electronic control panels.

Figure 7.16 The Chobert hollow rivet (*Avdel Ltd*)

Figure 7.17 The Chobert grooved rivet (*Avdel Ltd*)

Blind riveting tools

The majority of blind riveting tools have been designed on the assumption that the operator will be moving around a stationary structure, as in body building. However, some stand tools are available.

The range of tools has been designed to cater for all riveting conditions, whether they are used on high-speed assembly production lines or for occasional use. The basic types of tools consist of:

1 Manually operated plier types
2 Manually operated lazy tongs
3 Portable pneumatically operated guns
4 Portable pneumatic hydraulic guns
5 Pneumatically operated stand tools

In addition various heads are available for attaching to either manual or power-operated

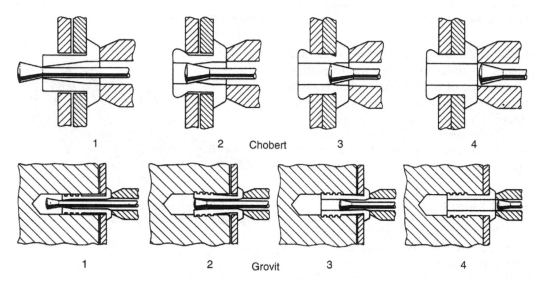

Figure 7.18   The Chobert and Grovit riveting systems (*Avdel Ltd*)

1 Place Chobert rivet or Grovit in prepared hole
2 Draw steel mandrel, which has opposite taper to rivet, through from tail end of rivet, expanding rivet tail around rear side of hole, to form shoulder. (For the Grovit, annual convolutions are forced into materials.)
3 Continue to pull the mandrel through rivet, which symmetrically expands shank to fill hole
4 The rivet has good bearing in hole and a parallel bore is left in the rivet

Figure 7.19   Briv placing sequence (*Avdel Ltd*)

1 Magazine-loaded rivet is placed into hole
2 Tool draws mandrel through rivet, compressing rivet
3 Mandrel expands shank of rivet to fill hole and clamp materials
4 Completed rivet. Tool reloads automatically.

Figure 7.20 Hand-operated riveting tool, plier type (*Tucker Fasteners Ltd*)

Figure 7.21 Hand-operated riveting tool, lever type (*Tucker Fasteners Ltd*)

tools which can gain access to restricted places. In order to change the rivet diameter or type of rivet being used, it is necessary merely to change the nosepiece or the gripping jaws which hold the mandrels. It must be remembered that certain special rivets and fasteners can only be used by their appropriate setting tool.

Rivet selection

The following factors must be considered when selecting the type of rivet to be used on a particular job.

Rivet diameter In some cases this may be decided by the size of an existing hole. When designing new work the main factor is usually the shear or tensile strength required from the riveted joint. In load-bearing joints the diameter of the rivet should be at least equal to the thickness of the thickest sheet but should not be greater than three times the thickness of the sheet.

Hole size – Drill sizes for each diameter of rivet are usually specified by the manufacturers. These are usually designed to give a clearance of 0.05 mm to 0.13 mm between the rivet and the hole, and in the case of rivets capable of only limited radial expansion, these recommendations must be followed if maximum efficiency is to be achieved.

Rivet length – Manufacturers also specify the correct rivet length to be used on a given thickness of material. It is often advantageous to use rivets which are longer than

Figure 7.22 Hand-operated riveting tool, lazy tongs type (*Tucker Fasteners Ltd*)

Figure 7.23 Automatic riveting system (*Tucker Fasteners Ltd*)

those recommended, as they will set satisfactorily on thinner materials, and the longer length, when expanded, increases the grip of the rivet, but the higher cost for longer rivets should be taken into account before doing so.

Pitch of rivets – In load-bearing joints the distance between rivets in the same row should not exceed six times the rivet diameter. Even when the joint is not load bearing, the rivet pitch should not exceed 24 times the thickness of the thinnest sheet in the joint.

Edge distance – In lap or butt joints which are likely to be subjected to shear or tensile loads rivet holes should not be drilled within a distance equal to 2 rivet diameters from the edge of the sheet but should not exceed 24 rivet diameters.

Rivet material – The choice of rivet material will normally be related to the strength required from the riveted joint. Usual rivet materials are steel, aluminium alloy, copper and Monel. Other factors affecting the choice are weight, high temperatures and corrosion resistance, especially electrolytic corrosion, which can occur when different metals are joined together.

Mandrel type – The break-stem mandrel is usually chosen where the rivet is to act as a waterproof plug and also where it would be inconvenient to eject the mandrel head on the blind side of the enclosed structure. The break-head mandrel is used where weight is a factor and where a comparatively clear hole is required through the rivet set.

Structural fasteners

Avdelok system

This is not a blind fastener, but has a number of advantages over the conventional rivet or nut and bolt which it replaces. The Avdelok is a two-piece high-strength fastener consisting of a bolt and collar made from carbon steel, stainless steel and aluminium alloy. It gives a high clench action and a positively locked nut or collar which is proof against vibrations. It can be placed very simply by manual or pneumatic tools. The Avdelok bolt is placed through the prepared hole, and the collar is slipped onto the other side of the joint. The nose of the tool is then pushed over the tail of the bolt and the trigger is pulled. This draws the bolt tight, clamping the sheets together. The pull continues until the nose of the tool is drawn over the collar, compressing the collar until it is swaged into the grooves of the pin. The tail of the bolt breaks off at the deep breaker groove, flush with the collar. No finishing is needed. This system is employed in such applications as chassis cross-bearer brackets on commercial vehicle bodies.

Figure 7.24   The Avdelok fastener (*Avdel Ltd*)

1   2   3   4

Figure 7.25   The Avdelok system (*Avdel Ltd*)

1 Insert Avdelok pin through pre-drilled hole and slip collar over pin tail
2 Place tool nose over pin tail and press trigger
3 Swage collar into locking grooves of pin
4 Tail of pin breaks flush with collar and is ejected

Hemlok system

The Hemlok is a blind fastener which comprises a steel stem and shell. Both components are zinc plated and gold passivated to ensure good corrosion resistance. The fastener is supplied in 6.4 mm diameter in various lengths. It is installed from one side of the workpiece, thus making it a blind fastener. It is designed with a chamfer at both ends of the stem; this facilitates easy insertion into the placing tip of the installation equipment and into the hole in the workpiece. The installed fastener breaks flush with or below the surface of the low-profile Hemlok head, leaving a clean finish and appearance. Hemlok provides strength and good clamp-up, and the well-formed fastener tail achieves this without deforming thin sheet material. This fastener is designed to be used on thin-gauge materials for high-strength joints suited to automotive work.

Threaded inserts

Steel threaded inserts (such as Nutsert) of high strength can be placed by any of three specially designed tools. The Nutsert is placed on the threaded drive screw of the placing tool and inserted into a prepared hole in the workpiece. The drive screw turns clockwise, drawing the tapered nose portion of the Nutsert back into its outer shell and thus causing the shell to expand into the preformed hole. The drive screw is then withdrawn, leaving the Nutsert permanently and tightly placed in the hole and ready to receive either a screw or a bolt. It can be

Installed fastener

Figure 7.26 Hemlok fastener (*Avdel Ltd*)

1   2   3

Figure 7.27 Hemlok placing sequence (*Avdel Ltd*)

1 Hemlok inserted into power tool and workpiece
2 Tool-operated: Hemlok tail deforms over material
3 Tail fully formed: fastener stem breaks off

used on all types of metals and also plastic and timber and is made in a wide range of thread sizes, both imperial and metric.

Monobolt system

Monobolt is a precision engineered two-part component. It is a high-performance flush-break fastener designed to build in strength, security and quality. It consists of a body and a stem and is supplied as a one-piece assembly. A sealing on the Monobolt stem

Figure 7.28   Standard Nutsert (*Avdel Ltd*)

provides exceptional resistance to moisture, an essential requirement for many applications in vehicle bodywork. The fasteners are available in aluminium alloy, carbon steel and stainless steel, in diameters of 4.8 mm and 6.35 mm. Fast and simple to install, Monobolt is a blind fastener, ideal for original structural assembly and to replace conventional fastening methods, particularly when access to the workpiece is difficult. A major feature of the Monobolt construction is the visible locking element, which allows easy, visible inspection after placement.

Avtainer system

This system has been designed for the joining of composite panels of plywood and glass-reinforced plastic to metal framing. This fastener has the ability to join firmly but without cracking or pulling the bolt right

(a)   (b)

Figure 7.29   (a) Standard Nutsert placement (b) Thin sheet Nutsert with bolt in place (*Avdel Ltd*)

Figure 7.30   Monobolt and placing sequence (*Avdel Ltd*)

through the composite. It is a two-piece fastener consisting of a zinc-plated carbon steel pin with a nylon seal and a collar which is pulled tight from the other side of the workpiece, thus causing no damage to the material during installation. One of its main applications is in the building of commercial vehicles.

Screws and bolts

Wood screws

Joining by wood screws is an essential part of the construction of vehicle bodies which incorporate timber. Screws are available in steel, brass and bronze, with a variety of

Figure 7.31 Avtainer and placing sequence (*Avdel Ltd*)

1 Drilled hole
2 Avtainer in position
3 Shell placed over pin tail

4 Tool applied: fastener completed. Seal protects against ingress of moisture

alternative finishes. Screw sizes are given as length and diameter in a number size

Metal screws (self-tapping screws)

Often there is a need for a fastening which will fix two or more parts together securely for indefinite periods yet can be dismantled if necessary.

Screws are available in stainless steel and in steel with various finishes. Sizes are given in length and diameter on a numerical scale. The types of heads available are countersunk, round head, raised countersunk, pan head, mushroom head and hexagon head. Most heads have a single slot for insertion of a screwdriver; some, known as Supadriv heads, have a star-type slot and must be placed with a special type of screwdriver.

Self-tapping screws are particularly useful to the body engineer and trimmer since no access is needed to the rear side of the work, it being necessary simply to drill the requisite size of hole and drive the screw home. The screw will stay secure until removal is necessary and can be removed and replaced many times without becoming slack if reasonable care is taken. Typical applications are the mounting of door-pillar trim pads, carpeting or rubber matting, parcel shelves, glove boxes, window cappings, ashtrays and kick plates.

A more recent development is the Taptite high-performance thread-forming screw which has a thread like that of a conventional machine screw. The thread on the Taptite screw has a tri-lobal cross-section instead of the customary circular form. As the screw is rotated the lobe profile rolls the material out of its path to form the thread. Not only does this eliminate the need for an additional tapping operation, but it also produces inherent resistance to vibrational loosening, thereby obviating the need for separate locking components. This is achieved by an increase in prevailing torque which comes from the radial pressure interference produced at and around the major diameter and not present on conventional fastenings because of the need for a clearance fit. These factors mean that in the assembly shop the operators (and drivers) can insert more screws in less time and with less fatigue. In maintenance

Figure 7.34 Supadriv Taptite thread forming screw (*European Industrial Services Ltd*)

Countersunk   Round

Raised countersunk

Figure 7.32 Unhardened wood screws (*European Industrial Services Ltd*)

Figure 7.33 Fixing self-tapping screws (*European Industrial Services Ltd*)

Figure 7.35 Plastite screw (*European Industrial Services Ltd*)

applications, a standard machine screw can be substituted if necessary.

The AB screw combines the benefits of Taptite in a heavy-duty thread-forming fastener and is available in sizes M6 and above. Another version of the Taptite screw is CA Taptite, designed in response to the increasing use of thinner-gauge materials. Fastening thin sheet poses problems with conventional self-tapping screws, and Taptite offers two solutions. Either a standard Taptite screw can be driven into holes which have been rim extruded to provide more metal, or, where this is not acceptable, the CA Taptite can be used; this has a gimlet point capable of self-extruding in thin sheet sections.

As plastic materials are increasingly used to replace aluminium, zinc die castings and other metal parts, so new fasteners have been designed capable of joining plastic components together. A screw can be driven directly into thermoplastics, making it a very cost-effective way of making positive joints. The screw can be removed and reinserted into the same hole many times and will always pick up the same thread, thus maintaining a strong joint. Polymate is a round-bodied, twin-threaded self-tapping screw with a closely controlled critical helix angle and thread pitch. The torque required to drive the screw is low, while the clamping load achieved is very high.

Steel hammer-drive screws

These sheet metal fasteners are used only where permanent fixing is required, as they are difficult to extract once placed. They are made in a very hard steel, and the head shapes available are round head and countersunk. The screws are inserted in predetermined holes of the correct size and hammered so that the spiral thread is forced to cut into the material. The cutting action of the thread reduces the size of the hole after the pilot point has passed through.

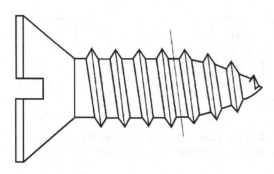

Figure 7.36 AB type self-tapping screw (*European Industrial Services Ltd*)

Figure 7.38 Polymate screw (*European Industrial Services Ltd*)

Figure 7.37 CA Taptite screw (*European Industrial Services Ltd*)

Figure 7.39 Hammer-driven screw (*European Industrial Services Ltd*)

Figure 7.40   Screw nail (*European Industrial Services Ltd*)

Screw nails

These are made of hardened steel and are used for fastening thin sheet metal onto wood structures. They are hammer driven through light-gauge sheet metal, taking care not to bend or break them, so that the hardened spiral thread cuts into the burr in the sheet metal and then worms its way into the wood. This makes a secure metal-to-wood joint which is ideal for panelling in the construction of vehicle bodies. Screw nails are available with countersunk, flat or round heads.

Bolting

Bolts are used extensively in the manufacture of motor vehicles, as components which are built in sections and then bolted together can easily be dismantled for repair or replacement. One example of this type of construction is the car body, which has certain panels which are made separately and later bolted in position so that they can be replaced as individual units if damaged. Another application of bolting is in cases where fabrications are too large for workshop assembly; then the article is made in sections which are assembled on site.

A bolt is a cylindrical rod having a head on one end and a thread cut along part of its length from the other end. Bolts are available in mild steel, alloy steel, stainless steel and brass and can be made of other metals and alloys for special applications. A nut and bolt is the most common means of joining. The unthreaded portion of the shank of the bolt is intended for fitting into clearance holes, and a washer can be used to minimize the effect of friction between the head of the bolt and the parent metal. A spring washer placed under an ordinary nut assists in preventing the slackening of the nut when the joint is subject to vibrations.

Tech note

There are frequently arguments in the workshop about the names of bolts and screws. Generally speaking (and there will always be variations), a bolt has a shank or unthreaded portion between the thread and the head, a set screw is threaded for its entire length, and a coach bolt has a square section below the head and uses a square nut.

An ISO metric bolt or screw made of steel and larger than 6 mm in diameter can be identified by either of the symbols ISO M or M embossed or indented on top of the head. In addition to marks to identify the manufacturer, the head is also marked with symbols to indicate the strength grade, e.g. 8.8; 10.9; 12.9; 14.9. As an alternative, some bolts and

Figure 7.41   (a) bolt, (b) Set-screw, (c) Coach bolt

Figure 7.42   Bolt identification

screws have the M and strength grade symbol on the flats of the hexagon.

The following types of head are available on bolts:

- *Hexagon head* –This is the most frequently used shape, and it is suitable for all spanners.
- *Square head* – The square-headed bolt is used mainly for heavy engineering and structural work.
- *Countersunk head* – A bolt of this type has a head shaped to fit into conical recesses in the work surface so that it lies level and flush. A slot is provided in the head for insertion of a screwdriver.

Screws (or machine screws)

A set screw is similar to a bolt but has the whole of its cylindrical shank threaded. Set screws are used either to fix two pieces of metal together or to adjust the distance of one piece relative to the other. They are placed by passing each through a clearance hole in the top piece of metal and screwing it into a hole tapped out with a thread in the lower piece of metal. Countersunk, round, cheese and pan heads are available, with either slotted or star recesses.

Nuts

A nut is a shaped block of metal internally cut to form a thread. It is usually made from mild steel, although other metals and alloys can be used for special purposes. Common nuts are:

- *Hexagonal plain nut* – Used in all classes of engineering.
- *Square nut* – Used in heavy engineering and structural work and also in coach building.
- *Castellated nut and slotted nut* – Used where the nut must always remain tight and slotted to take split pins which act as a locking device.

Figure 7.43   Machine screw head types: slotted and (a) countersunk (b) round (c) cheese (d) mushroom (e) pan; Pozidriv and (f) countersunk (g) pan; socket and (h) cap (i) countersunk (j) cup point (*European Industrial Services Ltd*)

Figure 7.44   Nut identification

- *Lock nuts* – These are thinner than the normal nut and are fitted above the main nut to act as a locking device.

Aero-tight nut

This nut is designed for use in conditions of severe vibration. On the top of the nut there are two threaded cantilever arms which are deflected downwards and inwards. When the bolt is passed through the nut these arms are forced out towards their original position, thus causing them to grip the thread very tightly.

A nut with an ISO metric thread is marked on one face or on one of the flats of the hexagon with the strength grade symbol 8, 12 or 14. Some nuts with a strength grade 4, 5 or 6 are also marked, and some have the metric symbol M on the flat opposite the strength grade marking. A clock face system is used as an alternative method of indicating the strength grade. The external chamfer of a face of the nut is marked in a position relative to the appropriate hour mark on a clock face to indicate the strength grade. A dot is used to locate the 12 o'clock position and a dash to indicate the strength grade. If the grade is above 12, two dots identify the 12 o'clock position.

Aero-tight nut

This nut is designed for use in conditions of severe vibration. On the top of the nut there are two threaded cantilever arms which are deflected downwards and inwards. When the bolt is passed through the nut these arms are forced out towards their original position, thus causing them to grip the thread very tightly.

Clinch nut

This is a captive nut used for blind hole fixing in thin sheet metal. It is so called because a threaded nut is held securely in position by an annular rivet, the head of which is hammered flat around the hole to form a secure anchorage.

Figure 7.45 Clinch nut (*European Industrial Services Ltd*)

Figure 7.46 Clinch nut fixing procedure (*European Industrial Services Ltd*)

1 Insert shank of nut into predrilled hole.
2 Rivet by closing spigot down with hammer. A convex punch may be used for large sizes to spread the spigot, but a flat tool must be used for the riveting operation.
3 An ordinary screw or bolt is then used for making an attachment to the clinch nut.

Nyloc self-locking nut

This is a self-locking nut with a built-in moulded nylon insert made of smaller diameter than the nut's normal thread. When the

(a)   (b)

Figure 7.47 (a) Nyloc (b) Cleveloc nuts (*Forest Fasteners*)

nut is screwed onto the bolt it runs freely until the end of the bolt meets the nylon insert. Further tightening of the bolt forces a thread onto the nylon insert, making the nut grip the bolt more tightly and thus creating a shock- and-proof nut.

Cleveloc self-locking nut and flange nut

The locking collar is an integral part of the Cleveloc nut, but in the flange nut the locking element is integral within the nut and has no collar. They are precision formed to a geometric ellipse and provide two locking elements of uniform shape and thread contour. These locking elements utilize most of the threads that come within the depth of the collar (on the Cleveloc nut) or the element (on the flange nut) and distribute locking pressure over wide areas. This ensures high fatigue life, with the flexibility necessary to give consistent performance and dependability in service.

There are no thread interruptions or pitch errors, and there is no deformation of the nut body or its hexagon form. The nut runs freely on the bolt threads until contact is made with the locking elements, each of which has a predetermined area of contact. These locking elements are designed to create gradually increasing areas of friction on the bolt threads.

Further tightening forces the locking elements to engage more fully with the bolt threads, and this increasing resistance to the entry of the bolt brings the full length of the nut threads into close contact with the working faces of the bolt threads. Further friction is created, and both forces combine to give a smooth, progressive and increasingly self-locking action.

Fastening devices

Spring steel fasteners directly contribute to an ever-increasing degree of automation and have probably saved the motor industry millions of pounds by enabling massive economies of labour and time in assembly. Industry data points to the fact that more than 50 per cent of the total cost of a motor body is in the area of assembly. A large portion of this arises from the time and labour expended in picking up and putting together a number of parts, some of them relatively small, and this is where spring steel fasteners achieve marked savings compared with earlier methods. Another ancillary industry affected by the advent of spring fasteners is the repair industry; here economies are less dramatic but spring fasteners still make valuable savings in time, labour and handling. All of these types of clips are made from a specially treated spring steel which retains its original spring locking power so that the clips are reusable and hence economical. Other advantages of these clips are that they are easy to apply, they eliminate the use of washers and nuts, and glass, plastic and other materials can be fastened together without fear of damage.

One of the firms that manufacture these clips is Forest Fasteners, who have developed the Spire speed nut (see below) which is used considerably in the car industry.

Spire speed nut

This nut has a double-locking action which operates by means of an arched base and arched prongs. Since the introduction of this nut, hundreds of different fastenings have been introduced over the years, and in some vehicles more than 200 spring steel fastenings are used.

The flat Spire speed nut has replaced the threaded nut plus lock or plain washer. It is available in a wide variety of sizes to suit machine screws and sheet metal self-tapping screws. First a hole is drilled in the appropriate panels, then the self-tapping screw is placed through the panels and pushed into the spire nut or clip, making sure that the prongs point outwards. When the nut is tightened it is locked both by the self-energizing spring lock of the base and by the compensating thread lock as the arched prongs engage the thread. These free-acting prongs compensate for tolerance variations, and the combined forces of the thread and spring locks are claimed to eliminate any risk of loosening by vibration.

Captive nut, U type

This nut is widely used for fastening blind assemblies in the motor industry. It can be assembled to the panels by hand, and no welding or riveting is required. The nut allows a certain degree of float, which facilitates speedy assembly. It remains captive to the panel, anchored by means of a sheared tongue on the lower leg, which drops into the mounting hole and holds the nut in the screw receiving position. The range available covers many panel thicknesses and screw sizes.

As the nut is pushed on the panel edge and over a predrilled hole, the locking tongue seats into the hole, thus holding the nut in position. The second panel is aligned, and the screw is driven through this panel into the captive nut, which holds it in place. This type of fastener has the advantage that it can be fitted before or after panels have been painted, because there is no danger of clogging during any spraying operation.

Captive nut, J type

This is similar in concept to the U-nut but with a shorter leg designed to snap into a clearance hole. The J-nut is easily started

Figure 7.48 Spire speed nut (*Forest Fasteners*)

1 Pre-locked position
The two arched prongs move inwards to engage and lock against the flanks of the screw thread. They compensate for tolerance variations.
2 Double locked position
A self-energizing spring lock is created by the compression of the arch in both the prongs and base as the screw is tightened, and vibration loosening is eliminated.

Figure 7.49 'U' type captive nut (*Forest Fasteners*)

1 Nut pressed into position: locating tongue snaps into predrilled hole and locks nut into position.
2 Assembly completed with screw.

over the edge of a panel and pressed into position with the thumb. A typical application for J-nuts in the motor industry is the replacement of reinforcing rings and blind, bushed-on headlamp assemblies. The short leg on the front side of the nut is clipped into screw receiving positions on the wing aperture, and ensures a good seal between gasket and wing, thus precluding mud leakage.

Nut grip

The nut grip is used in a square hole and replaces costly welded cage nuts and similar fastenings. It is installed into the panel by hand, where it remains captive. The fully threaded nut in the cage is permitted to float slightly to overcome the problem of

Figure 7.50 'J' type captive nut (*Forest Fasteners*)

1 Snap nut into position.
2 Drive screw into nut.

Figure 7.51   Nut grip (*Forest Fasteners*)

1 Insert one spring steel leg of nut into hole and snap in other leg using a suitable tool.
2 Complete assembly with bolt.

misaligned holes and is fitted after the finishing process at any convenient point on the production line. Three sizes of cage, covering threads from M3 to M10, are available.

Cable clip

This is a special clip used for carrying cables. It is fixed by clipping into predrilled holes, thus eliminating the need for screw fixings.

Push-on clips

In addition to spring steel fasteners that are associated with threaded members, there is a wide variety of push-on clips. However, these clips, unlike the nuts, are not pitched to follow the helix of a thread, and the two sheared arms are of equal height. As the push-on clip is forced over a plain stud, the fixing legs bite into the surface and, on finally depressing the arched base, which of course reacts as soon as the pressure is released, the fixing legs are given a strong upward and inward pressure which firmly holds the fastener in position. This has the effect of drawing the assembly together and removing any possibility of rattle. When using the push-on clip, it is important to ensure that the tolerance on the stud diameter is held to within reasonable limits; to get the best results from this type of fastener a tolerance of 0.05 to 0.08 mm is recommended.

Push-on clips take many forms, from the simple sheared type, which is still widely used, to the more modern blanked types, the multi-pronged clips for rectangular studs and now the plastic-capped type. To fix this

Figure 7.52   Cable clip used for latching: sectional view shows clip in position in a panel (*Forest Fasteners*)

(a)

(b)

Figure 7.53 (a) Simple push-on clip showing clip in position (b) tubular clip, showing fitting sequence (*Forest Fasteners*)

1 Clip enters hole: cut-outs register with panel, allowing clip to expand and hold.
2 Second part of assembly is lined up and rivet and integral stub inserted.
3 Rivet pressed home, engaging turned-in end of clip, which expands to bite on rivet.

type of fastener it is, of course, necessary to have access to the back of the panel, but if assembly is possible from one side only, a tubular type can be used which consists of a small spring steel split tube. This is pushed into a hole in the panel, where it remains captive and ready to receive the studs of the badge or nameplate. Such tubular clips permit the assembly of a badge to the grille as a final operation. A removable version of this clip is also made for applications which have to be taken apart from time to time.

Non-metal clips

The use of plastics has given rise to the development of new types of fastenings. The common fault with the strip or wire clip is that they become rusty and break, either on removal or on replacement, and plastics have

the advantage of being rustproof. Tough rubber has great possibilities for certain types of fastenings. Some time ago, a tough rubber strap was offered as an alternative to the traditional metal clip for retaining loose cables and wires. It is still widely used in this way, and rubber with a synthetic content is sometimes used for door check straps; this demonstrates the strength of the material, which clearly lends itself to many applications as yet unexploited.

Plastic captive nuts and trim panel fasteners

Plastic captive nuts are available in a large range of shapes and sizes for vehicle body applications. They can be used in high-corrosion areas and are designed to snap into a suitably punched hole in a metal panel, where they will remain captive until final

Screw

Panel

Hole

Fitted condition

Figure 7.54 Plastic capture nut (*TRW United-Carr Ltd*)

assembly. The shallow head gives highly desirable minimum clearance between panels. The combination of design and material permits the recommended screw to form its own thread by displacement, resulting in high torque load and good anti-vibration characteristics. Screws may be removed and the nuts reused, providing the screws are the same size and thread form. This makes them ideal for applications such as inspection covers and service access. They are manufactured from glass-filled nylon 66, which permits higher torque loadings. All other parts are manufactured from polypropylene.

Push-in panel fasteners (W-buttons) are designed to hold board, plastic, rubber or other soft materials to secondary panels. They are equally suitable for assembling components to panels. No special tools are required, as they assemble on the principle of a push-in rivet. Simply push the W-button through the aligned panel holes so that it contracts on entry and then relaxes as it snaps through the total panel thickness, securing the panels under tension. Many applications exist for W-buttons where a positive, attractive, light-weight non-corrosive fastening solution is required: to assemble painted or PVC-covered trim boards, to fasten instruction badges or to blank off holes. In addition, a special feature which makes the buttons particularly suitable for the automotive industry is the various styles of textured and colour-moulded heads to match adjacent panels (see Figure 7.61a).

Christmas-tree buttons are a variation of push-in fasteners, and when they are pushed into the hole they take up great variations in panel thickness. These parts have a good tolerance of poor-quality hole conditions. Although it is possible to remove the buttons, they are not normally reusable.

Trim panel fasteners are used principally in the automotive industry to attach door and tailgate panels to inner door panels. They are blind fixes, ensuring a neat, attractive assembly appearance.

They allow panel removal for repair, service or inspection purposes. Easily fitted without the use of special tools, they are positioned in a keyhole slot in the trim panel, then firmly pushed home into the inner door panel mounting hole. Various types and sizes are available to suit different conditions, and most have a flexible skirt which seals the mounting hole against entry of dirt or water.

Retainers – Where, for a technical or styling reason, a keyhole slot is not feasible, assembly can be simply achieved by using a separate retainer. With its own integral keyhole slot feature, it can be attached to the trim panel either with a suitable adhesive or by heat staking. This will provide a secure, attractive attachment with no evidence of fastener location.

Snapsacs provide increased retention and sealing for trim panel fasteners, and their smooth surface eases assembly and removal.

Quick-release fasteners (quarter turn) consist of a stud, a cam and an optional retaining washer. A selection of lengths and head forms is available to suit a variety of assembly conditions. The washer allows the

Figure 7.55 Plastic push-in panel fasteners: (a) W button (b) Christmas tree button (*TRW United-Carr Ltd*)

stud to be positively and securely located prior to and away from the final assembly point. Where the assembly is subject to routine servicing, the stud, washer and cam stay securely mounted, avoiding any risk of component loss.

Edge fasteners of the D type are lightweight fasteners which are ideal for assembling leather-cloth, soft plastics and fabrics to metal, rigid plastic or fibreboard panels. Fitted without the use of special tools, they are pushed or tapped onto a convenient

Figure 7.56   Trim panel fasteners (*TRW United-Carr Ltd*)

Figure 7.57   Quick-release fasteners, quarter turn (*TRW United-Carr Ltd*)

Figure 7.58   Edge fasteners, D type (*TRW United-Carr Ltd*)

panel edge using a light hammer. The barbs retain the assembly securely in position on one side, and the flat D side provides a neat, flush appearance on the other. The friction grip version offers good retention without paint damage where corrosion resistance is paramount. They are manufactured from austempered carbon steel and normally supplied in phosphate and black finish.

Tech note

When tightening or fixing any fastening you should think about the **torque** or **force** which you are applying. Always tighten to the manufacturer's torque setting and do not apply excessive force to any item.

Adhesives

Adhesives are to be found almost everywhere in the modern world. There are natural adhesives as well as man-made adhesives: for example, spiders use adhesive to spin and stick their webs and to catch their prey, and limpets and shell fish use adhesive to anchor themselves to rocks.

A bond between two surfaces may be regarded as a chain of three links, the strength of the bond being that of the weakest link. The central link is the adhesive film between the surfaces, and the outer links are formed where the adhesive film meets the bonded surface. To form a strong surface linkage, the adhesive must thoroughly wet the surface whilst liquid and when dry must adhere by penetration into the pores or fibres. In the case of smooth, non-porous surfaces such as glass or metal, a strong film of adhesive is best formed between the surfaces if these are joined at the appropriate stage in the drying of the applied adhesive. In this process the molecules of the adhesive and substrate are brought close together, and a variety of forces operate; the most important are atomic forces, which can be likened

to magnetic attraction. Many variations on this simple theme are possible and practicable, but the character of the bond in a particular case will always be governed by how closely the technique used approximates to this ideal. Almost everything that is manufactured or made will use adhesives. A good example is the motor car. Paint has to adhere to the metal on which it is placed. Inside the car, the upholstery will generally be stuck together with adhesives. All the brake shoes on the car are bonded together. Finally, some panel assemblies are bonded with adhesives.

Safety note

All adhesives require special precautions and procedures both when in use and when being stored. General precautions are:

1 Wear appropriate PPE.
2 Keep away from the skin and especially the eyes.
3 Do not breathe in the fumes.
4 Store in a locked cupboard with access to authorized personnel only.
5 Read the manufacturer's instructions before use.

Adhesive types

To facilitate the selection of adhesives, it is convenient to classify them into the following major groups:

*   *Anaerobics* – Often known as sealants or locking compounds. Acrylic based, they normally set in the presence of metal and absence of air (or to be exact, atmospheric oxygen). They are generally used to lock, seal and retain all manner of turned, threaded and fitted parts and are often used to seal flanges.
*   *Cyanoacrylates* – Also based on acrylic resins. Unlike anaerobics, they require

surface moisture as the vital catalyst in hardening. Generally, they harden in seconds. Often used in car trim applications to bond rubber trim to metal, rubber to rubber and also rubber to plastic.

- *Epoxies* – Based on an epoxy resin, which is mixed with a hardener. This allows great variety in formulation. Their strength is often employed in bonding larger components. Single-part epoxies (ESP) are a development in epoxide technology. Resin and catalyst are premixed, so they give high performance without mixing by the user. In the motor vehicle they see a great deal of use as a supplementary bond to welding: the process is known as weld-bonding.
- *Hot melts* – A refined form of the earliest adhesive, hot wax. They are convenient for assembling small, lightly loaded items for use in less severe environments. They are too viscous for use on the smallest parts but are often used for the assembly of motor vehicle trim.
- *Phenolics* – One of the earliest types of structural adhesive. Their use often involves specialized equipment and complex procedures. Nevertheless, they perform well in severe environments. The phenolics are hardly ever seen in motor vehicles, but they do have a significant residual use in the assembly of brake shoes.
- *Plastisols* – Based on PVC dispersions. They cure only at elevated temperatures and are generally used only on large-scale work or where there is access to a heat source intended for another purpose.
- *Polyurethanes* – Like epoxies, they offer variety in formulation. However, polyurethanes are difficult to handle and usually require specialized mixing equipment. Generally used for load-bearing applications in dry conditions, as they are prone to moisture attack. They can be successfully used for bonding painted metal surfaces and polyester components together.

Bulk of adherend 1
Surface layer of adherend 1
Adhesive
Surface layer of adherend 2
Bulk of adherend 2

Figure 7.59 Simplified cross-sectional representation of a typical joint (*Permabond Adhesives Ltd*)

- *Solvent-borne rubber adhesives* – Based on a rubber solution, where the solvent evaporates to effect bonding. Not suitable for loaded joints or harsh environments.
- *Tapes* – Adequate for bonding small components, but cannot support heavy loads. Some will withstand quite harsh environments.
- *Toughened adhesives* – Toughened variants are hailed as a breakthrough in adhesive technology. They incorporate low-molecular-weight rubbers that build in exceptional resistance to peel and impact forces. Acrylic-based, epoxy and single-part epoxy adhesives can be toughened in this way. Toughening reinforces the best features of these adhesives with the unique shock absorption and strength of the rubber matrix. Toughened adhesives are used in much the same way in motor vehicles, trucks and vans, but they may be used on unpainted metal. They are extremely durable in poor operating environments.

Selection of adhesives

Joint types involved

There are three basic joint types:

- *Co-axial joints*, where one part fits into another, usually require an anaerobic. Other adhesive types may be too viscous or lack the appropriate grades of strength.
- *Plain lap joints*, where the adhesive is primarily loaded in either shear or compression. As a rule, cyanoacrylates are better for unloaded and toughened variants for loaded lap joints.

- *Butt joints*, where the adhesive is in tension. These joints are very susceptible to peel and cleavage forces, which toughened adhesives withstand exceptionally well.

Joint function

Where a coaxial joint needs dismantling for maintenance, the weaker grades of anaerobic should be used.

Service conditions

It is possible to indicate conditions using a three-point scale:

- *Benign* – This is at normal room temperature, about 20°C where components are not expected to suffer high loads or impacts.
- *Normal* – This is at normal room temperature, with occasional excursions to about 80°C. Loads could be heavy, and components may experience occasional shocks.
- *Severe* – At high humidity and/or temperature with components heavily loaded, with heavy impacts during their use or assembly.

Curing times

This has important consequences for later production processes. Cyanoacrylates set in seconds and reach full strength in minutes. Anaerobics, epoxies and the toughened variants take longer to harden. However, all three can be made to cure rapidly using a variety of techniques.

Achieving good results with adhesives

Many applications require no preparation. For example, anaerobics need no pre-treatment unless contamination is excessive. This is because the locking or jamming mechanism of the coaxial joint plus adhesion is sufficient to hold components together. Similarly, cyanoacrylates are almost always used without surface preparation. On rubber and plastic they are capable of penetrating

surface debris. The toughened acrylic adhesives are even more hardy, and some will actually bond through a film of oil (so too will the heat-cured epoxies). Therefore, surface preparation can safely be omitted, or restricted to simple degreasing, on a wide range of jobs. However, where maximum performance is required or conditions are particularly severe, it is worth spending time on surface preparation.

Preparation involves degreasing (to remove oil, grease and contaminants) and abrading or etching (to increase surface grip).

- *Degreasing* – Best done with a chlorinated solvent, except for sensitive materials.
- *Abrasion* – Metals are best abraded by grit blasting. However, all the following give satisfactory results: abrasive discs, belts and cloths; medium-grit emery paper; and wire brushes. Plastics, when bonded with cyanoacrylates, generally need no abrading. When using epoxy-based systems, abrade lightly with abrasive discs, belts or cloths, or use medium-grit emery paper.
- *Chemical etching* – Some materials require chemical treatment to ensure optimum performance. After chemical etching, wear clean gloves to handle materials; even a fingerprint can contaminate the etched surface and so weaken the bond.

Tech note
The principles of good adhesive joint design are:

- Use an adequate overlap, as this gives a stronger joint.
- Choose a rigid adherence where loads are carried.
- Form joints from thick, rigid sections where possible.
- Avoid butt joints.
- Refer to the manufacturer's instructions for specific adhesives.

Adhesives and the automotive industry

Non-structural adhesives

The primary function of non-structural adhesives is to attach one material to another without carrying functional loads on the components. Loads are usually light, particularly for those used inside vehicles. Some interior applications include fabric on trim and door panels, headlinings, carpets and sound-deadening panels. Exterior applications include moulding, wood grain, striping and vinyl roof. The materials used for these bonds include pressure-sensitive tapes, hot melts, epoxy spray, vinyl plastisol, elastomer solvent, silicon, solvent-based rubber (used for large gap sealing)

and butyl rubber (formerly used to mount windscreens).

Hot melts

The advantages of hot-melt adhesives and sealants are their ability to bond to both pervious and impervious substrates and the speed with which the ultimate bond strength is attained, the latter being particularly valuable in high-rate production. There are, however, several other significant advantages in these solvent-free substances. With other types of adhesive, full bond strength is not achieved until all the solvent or liquid carrier has evaporated. A hot-melt adhesive, on the other hand, relies on mechanical keying to the surface roughness of the substrate surfaces and the adhesive's chemical affinity with the substrate.

Figure 7.60 Adhesive joint design (*Permabond Adhesives Ltd*)

Apart from the polyamide and EVA-based products, there are a number of other hot-melt products that are useful to the car industry. The thermoplastic rubber-based materials have many of the properties of vulcanized rubber systems yet can be applied as hot melts; they are especially useful in trim applications as derived pressure-sensitive adhesives. Where sealing is most important, such as to prevent moisture penetration into spot-welded seams, fairly soft butyl-based materials may be used; these do not liberate corrosive hydrochloric acid as a decomposition product when welded through. Another development is the hot-melt systems that cure after application, either by reaction with moisture or by the application of heat.

For assembling trim components at the end of an assembly line, hand or robot guns can be employed. For continuous rapid assembly, wheel or roller applicators dipping into reservoirs of the heated adhesive can be most effective. Machines with one or more applicator heads, and developed for high rates of production, are now available. Hot-melt adhesives may even be sprayed. There is also available a hot-melt adhesive in the form of a strip about 0.5 mm thick, impregnated in a synthetic fabric scrim made up into a composite tape backed with aluminium foil. This has been used to join pieces of carpet butted together. The heat is generated by passing an electric current through the foil. Light rolling ensures that wetting is uniform over the whole area.

Toughened structural adhesives

These materials span both the acrylic and epoxy technologies, where a specific technique is used to prevent catastrophic crack propagation when joints are overloaded. Improved performance of the adhesive film is brought about by the introduction of a rubbery distortable phase within the load-bearing matrix of the body of the adhesive. It is this physically separate but chemically linked zone which absorbs fracture energy and prevents crack propagation. In this manner the resistance of the adhesive to catastrophic failure is enhanced considerably, and such adhesive compositions show a marked resistance to peel, cleavage and impact forces.

Since the introduction of high-performance toughened structural adhesives, many designers and manufacturers are turning to bonded structures in order to reap the economic and technical benefits given by this technology. Materials used for vehicle manufacture are often chosen for strength, rigidity or light weight, and in some cases surfaces will be prefinished to enhance appearance. Bonding will not only overcome some of the assembly problems which such materials often present but will give a stiffer and stronger structure than could be obtained from either riveting or welding. In the correct circumstances, satisfactory bonds can be made on many metallic and non-metallic materials.

Adhesives may be used alone or in a secondary role to supplement welding, brazing

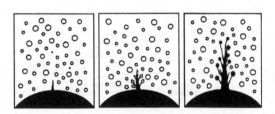

Figure 7.61 Toughened adhesive: when overloaded, crack propagation is stopped by the dispersed rubbery phase (*Permabond Adhesives Ltd*)

or riveting or in a primary role, complemented by welds or rivets. Weld bonding and rivet bonding are already standard procedures in many types of vehicle construction. Intermittent spot welding or riveting is a useful technique for pinning components during periods when the adhesive is uncured or temporarily softened from passage through a stoving oven.

A number of car manufacturers are already researching the possibility of a totally bonded car, especially those who want to use plastic or alloy bodies. If motor engineers are to build extremely lightweight vehicles, they will have to consider the application of adhesive engineering. Motor vehicle manufacturers are now extensively using high-performance epoxies (often toughened) to supplement welded joints in their vehicles. This has resulted in a major reduction in the number of welds required and the gauge of metal needed and at the same time has increased body stiffness and reduced corrosion.

Typical applications of structural adhesives include the following:

- *Toughened acrylic* – aluminium or steel fitments bonded to GRP roof sections; patch repairs to metal or plastic panels; internal steel fixtures bonded to steel
- *Toughened epoxy* – aluminium floor sections bonded to wood; inner and outer door skins bonded together
- *Cyanoacrylates* – rubber seals and weatherstrips

Adhesives used in vehicle body repair

- *Weatherstrip adhesive* – bonding rubber weatherstripping to door shuts, boot lid
- *Fast-tack adhesive* – repairs to trim fabrics, head-linings and carpets
- *Disc pad adhesive* – bonding paper discs to backup pads
- *Tape adhesive* – trim applications
- *Adhesive/sealant* (polyurethane-based compound) – structural bonding and sealing of replacement windscreen

Skills and questions

These tasks are suggested to help you gain knowledge and skills related to vehicle joining. Check them out with a senior colleague or tutor before commencing them.

1 Remove and refit a number of panels, noting the fastenings used. Pay particular attention to doing this without damaging or marking the vehicle in any way. Do this under the supervision of your tutor.
2 Use a rivet system, noting any particular safety points or procedures.
3 Write a paragraph explaining, as you would to a customer, how an adhesive is used. Pay particular attention to the various safety point and procedures and the need to keep the adhesive away from other parts of the vehicle.
4 Investigate how a premium-quality vehicle which uses aluminium alloy panels and bonding techniques for assembly is made, and state any special needs which must be observed if it were to be repaired. You may wish to use the internet as a starting point, along with vehicle repair manuals.
5 With supervised access to your parts department, investigate the different types of fasteners kept in stock and available to order.
6 Sketch four types of solid rivet.
7 What is meant by riveting allowance and rivet pitch?
8 Name three types of metals used in making solid rivets, and state a property which they all have in common.
9 Sketch and name the type of joint that would be used in solid riveting.
10 Name and sketch the type of rivet required when fastening a panel to a frame where the riveting is only possible from one side.
11 Describe four types of blind rivets and their placing procedure.
12 Explain the types of materials which are used in the manufacture of blind rivets.
13 Explain the use of a blind rivet nut.
14 Explain the advantages of the Hemlok system of structural fasteners.

15 Describe the advantages of the Monobolt system of structural fasteners.

16 What is meant by a permanent fastening method?

17 Describe the difference between a set screw and a bolt.

18 Explain what is meant by a self-tapping screw.

19 Explain the advantages of the Taptite screw system.

20 Describe the use of a screw nail in coachwork.

21 What purpose do nuts and bolts serve in vehicle bodywork?

22 List and describe four types of self-locking nut.

23 Explain the principle of the Spire speed nut and state where you would expect to find them on a vehicle body.

24 Explain the use of captive nuts in vehicle assembly.

25 What is the function of a spring washer when used as a securing device?

26 Name three types of plastic trim panel fasteners.

27 Sketch a pull-on panel fastener used in trim work.

28 Name the classified groups of adhesives.

29 Explain how an adhesive can be selected for use.

30 Explain what is meant by a non-structural adhesive.

31 Describe the importance of toughened structural adhesives.

32 List the applications where toughened adhesives would be used.

33 Explain how crack propagation is achieved in structural adhesives.

34 List the types of adhesive that could be used by the body repairer.

35 What type of adhesive will resist heat, water and acid and is used to join metal?

36 What purpose do nuts and bolts serve in vehicle body repair work?

37 List and describe four types of self-locking nuts.

38 What is the purpose of a spring washer when used with a nut and bolt?

39 With the aid of a sketch, illustrate and name a type of self-securing joint.

40 Give the advantages of blind riveting when compared with solid riveting.

41 With the aid of a sketch, name a type of joint that would be used when solid riveting.

42 Explain where quick-release fasteners would be used on a vehicle body.

43 Explain the principle of the Nyloc and Cleveloc nuts.

44 Name three types of blind riveting tools that could be used in assembly work.

45 Explain the purpose of using coach screws in vehicle body assembly.

46 Describe two ways of locating a bonnet stay.

47 How are two bars safely attached to a vehicle body?

48 How can you ensure that road wheels are fitted securely?

49 Why are cross-head screws popular?

50 State one advantage of self-tapping screws.

8

Soft and hard soldering methods

In this chapter we are going to look at soft and hard soldering, this is often referred to as **soldering and brazing** when discussed in a non–technically accurate way. At the end of this chapter you'll be able to state the differences in the processes and understand the different applications. **Soldering is a very useful skill** which has many applications, especially in building classic and custom vehicles using tubular frames. It is also essential for wiring and specialist sheet metal work.

Comparison of fusion and non-fusion jointing processes

The jointing of metals by processes employing fusion of some kind may be classified as follows:

- *Total fusion.* Temperature range: 1130 to 1550°C approximately. Processes: oxy-acetylene welding, manual metal arc welding, inert gas metal arc welding.
- *Skin fusion.* Temperature range: 620 to 950°C approximately. Processes: flame brazing, silver soldering, aluminium brazing, bronze welding.
- *Surface fusion.* Temperature range: 183 to 310°C approximately. Process: soft soldering.

In **total fusion** the parent metal and, if used, the filler metal are completely melted during the jointing. Thin sheet metal edges can be fused together without additional filler metal being added. Oxy-acetylene welding and manual metal arc welding were the first processes to employ total fusion. In recent years they have been supplemented by methods such as inert-gas arc welding, metal inert-gas (**MIG/MAG**) and tungsten inert-gas (**TIG**) welding, carbon dioxide welding and atomic hydrogen welding. Welding is normally carried out at high temperature ranges, the actual temperature being the melting point of the particular metal which is being joined. The parent metal is totally melted throughout its thickness, and in some cases molten filler metal of the correct composition is added by means of rods or consumable electrodes of convenient size. A neat reinforcement weld bead is usually left protruding above the surface of the parent metal, as this gives good mechanical properties in the completed weld. Most metals and alloys can be welded effectively, but there are certain exceptions which, because of their physical properties, are best joined by alternative methods.

In **skin fusion** the skin or surface grain structure only of the parent metal is fused to allow the molten filler metal to form an alloy with the parent metal. Hard solders are used in this process, and, as these have greater shear strength than tensile strength, the tensile strength of the joint must be increased by increasing the total surface area between

The Repair of Vehicle Bodies, Andrew Livesey, 9780815378709, 2019, Taylor & Francis

the metals. The simplest method of achieving this is by using a lapped joint in which the molten metal flows between the adjoining surfaces. The strength of the joint will be dependent upon the wetted area between the parts to be joined. Skin fusion has several advantages. First, since the filler metals used in these processes have melting points lower than the parent metal to which they are being applied, a lower level of heat is needed than in total fusion, and in consequence distortion is reduced. Second, dissimilar metals can be joined by applying the correct amount of heat to each parent metal, when the skins of both will form an alloy with the molten hard solder. Third, since only the skin of the parent metal is fused, a capillary gap is formed in the lap joint and the molten filler metal is drawn into the space between the surfaces of the metals.

In **surface fusion** the depth of penetration of the molten solder into the surfaces to be joined is so shallow that it forms an intermetallic layer which bonds the surfaces together. The process employs soft solders, which are composed mainly of lead and tin. As these also have a low resistance to a tensile pulling force, the joint design must be similar to that of the skin fusion process, a lapped joint.

Soft and hard solders

In spite of the growing use of welding, the techniques of soldering remains one of the most familiar in the fabrication of sheet metal articles, and the art of soldering still continues to occupy an important place in the workshop. While soldering is comparatively simple, it requires care and skill and can only be learnt by actual experience.

Soldering and brazing are methods of joining components by lapping them together and using a low-melting-point alloy so that the parent material is not melted. Soldering as a means of joining metal sheets has the advantage of simplicity in apparatus and manipulation, and with suitable modifications it can be applied to practically all commercial metals.

Soft soldering

The mechanical strength of soft-soldered sheet metal joints is usually in the order of 15 to 30 MN/m^2, and depends largely upon the nature of the solder used; the temperature at which the soldering is done; the depth of penetration of the solder, which in turn depends on capillary attraction, that is on the power of the heated surface to draw liquid metal through itself (Figure 8.1); the proper use of correctly designed soldering tools; the use of suitable fluxes; the speed of soldering; and, especially, workmanship.

Solders

Soft solder is an alloy of lead and tin and is used with the aid of a soldering flux. It is made from two base metals, tin and lead. Tin has a melting point of 232°C and lead 327°C, but the alloy has a lower melting point than either of the two base metals, and its lowest melting point is 183°C; this melting point may be raised by varying the percentage of lead or tin in the alloy. A small quantity of antimony is sometimes used in soft solder with a view to increasing its tenacity and improving its appearance by brightening the colour. The small percentage of antimony both improves the chemical properties of the solder and increases its tensile strength without appreciably affecting its melting point or working properties.

Solder

Capillary attraction pulling
the solder through the joint

Figure 8.1 Capillary attraction through a soldered lapped joint

The following solders are the most popular in use today:

- *95 to 100 per cent tin solder* is used for high-quality electrical work where maximum electrical conductivity is required, since the conductivity of pure tin is 20 to 40 per cent higher than that of the most commonly used solders.
- *60/39.5/0.5 (tin/lead/antimony) solder,* the eutectic composition, has the lowest melting point of all tin-lead solders and is quick setting. It also has the maximum bulk strength of all tin-lead solders and is used for fine electrical and tinsmith's work.
- *50/47.5/2.5 solder,* called **tinman's fine**, contains more lead and is therefore cheaper than the 60/40 grade. Its properties in terms of low melting range and quick setting are still adequate, and hence it is used for general applications.
- *45/52.5/2.5 solder,* known as **tinman's soft**, is cheaper because of the higher lead content but has poorer wetting and mechanical properties. This solder is widely used for can soldering, where maximum economy is required, and for any material which has already been tin plated so that the inferior wetting properties of the solder are not critical.
- *30/68.5/1.5 solder,* known as **plumber's solder,** is also extensively used by the car body repairer. Because the material has a wide liquidus–solidus range (about 80°C), it remains in a pasty form for an appreciable time during cooling, and while in this condition it can be shaped or '**wiped**' to form a lead pipe joint or to the shape required for filling dents in a car body. Because of its high lead content, its wetting properties are very inferior, and the surfaces usually have to be tinned with a solder of higher tin content first.

Fluxes

The function of a flux is to remove oxides and tarnish from the metal to be joined so that the solder will flow, penetrate and bond to the metal surface, forming a good, strong soldered joint. The hotter the metal, the more rapidly the oxide film forms. Without the chemical action of the flux on the metal, the solder would not tin the surface, and the joint would be weak and unreliable. As well as cleaning the metal, flux also ensures that no further oxidation from the atmosphere which could be harmful to the joint takes place during soldering, as this would restrict the flow of soldering.

Generally, soft soldering fluxes are divided into two main classes: corrosive fluxes and non-corrosive fluxes.

Corrosive fluxes

These are usually based on an acid preparation, which gives the fluxes their corrosive effect. They are very effective in joining most metals. If the flux is not completely removed after use, corrosion is set up in and around the joint, and the risk of this happening prevents the use of these fluxes in electrical trades and food industries.

The following substances are corrosive fluxes:

- *Zinc chloride (killed spirits)* – This is made by dissolving pure zinc in hydrochloric acid until no more zinc will dissolve in the acid. This changes the acid into zinc chloride – hence the name killed spirits. As an all-round flux for soft soldering, zinc chloride is without equal, but it has one disadvantage for some purposes: its corrosive action if the joint is not afterwards cleaned thoroughly with water.
- *Hydrochloric acid* – Although hydrochloric acid is not a good substitute for zinc

chloride, it is nevertheless used with excellent results on zinc and galvanized iron. It can be used neat, but it is better to dilute it with at least 50 per cent zinc chloride.

- *Ammonium chloride (salammoniac)* – This may be used as a solution in water in much the same way as zinc chloride, but it is not quite so effective for cleaning the metal.
- *Phosphoric acid* – This is effective as a flux for stainless steel, copper and brass and does not have the corrosive effect of other acid types of flux.

Tech note

Some fizzy drinks contain phosphoric acid – if you drop a dirty coin in the drink the acid will clean it.

Non-corrosive fluxes

These prevent oxidation on a clean or bright metallic surface during soldering. In general non-corrosive fluxes are not so active in cleaning the metal and serve chiefly as a measure of protection against further oxidation, when the material is hot. Therefore these fluxes should only be used when the metal has been cleaned prior to soldering.

The following substances are non-corrosive fluxes:

- *Resin and linseed oil* – Resin, finely powdered and dissolved in linseed oil, forms a good flux where non-corrosion is important. It is necessary that the parts to be soldered should be quite clean before applying oil and resin as a flux.
- *Tallow and palm oil* – Tallow and palm oil is often used as a flux for soldering

Table 8.1 Soft solders, fluxes and their method of application for different sheet metals

Sheet metal	Flux	Solder constituents (%)					
		Tin	Lead	Other	Used with		
Aluminium	Stearin	85	–	Zn8, Al7	Soldering bit	or	blowpipe
Brass	Zinc chloride or resin	65	34	Sbl	Soldering bit		
Copper	Zine chloride, ammonium chloride or resin	65	34	Sbl	Soldering bit		
Galvanized steel	Dilute hydrochloric acid	50	50	–	Soldering bit		
Lead	Tallow or resin	40	60	–	Soldering bit	or	blowpipe
Monel	Zinc chloride	66	34	–	Soldering bit	or	blowpipe
Nickel	Zinc chloride	67	33	–	Soldering bit		
Pewter	Olive oil, resin or tallow	25	25	Bi5o	Soldering bit	or	blowpipe
Silver	Zinc chloride	70	30	–	Soldering bit		
Stainless steel	Phosphoric acid + zinc chloride	66	34	–	Soldering bit		
Terne steel	Zinc chloride	50	50	–	Soldering bit		
Tin plate	Zinc chloride	60	40	–	Soldering bit		
Tinned steel	Zinc chloride	60	40	–	Soldering bit		
Zinc	Zinc chloride or hydro-chloric acid	50	50	–	Soldering bit		
Iron	Zinc chloride or ammonium chloride	50	50	–	Soldering bit		
Steel	Zinc chloride or ammonium chloride	60	40	–	Soldering bit		

lead. The surface of the lead must be first scraped clean before the flux is applied and the joint soldered. Tallow is a popular flux with plumbers for the purpose of wiping joints in lead pipes.

- *Olive oil* – A thin oil such as olive oil is generally used as a flux for soldering pewter. The soldering temperature for pewter is rather lower than for most soldering operations, hence the use of a thinner oil as a flux. As a flux, olive oil is sometimes called Gallipoli oil.

Soldering tools

Soldering irons are made in different sizes and shapes. The head is nearly always made of copper, although for soldering aluminium a nickel bit is necessary. Electrolytic copper is the best, as it gives longer life and holds the solder well on the working faces, and forged soldering bits are far better heat retainers than cast copper bits, as they are less liable to crack at the tips. A small soldering tool is not suitable for soldering any heavy or comparatively large sheet metal articles, because the heat loss by conduction is too fast to allow an even temperature to result, which will allow the solder to flow freely and sweat into the seams or joints. Soldering tools heated by gas, electricity and oxy-acetylene are now available to speed up the process of soldering in mass production work.

Soft soldering process

Soldering is a process of joining two lapped metal surfaces together by fusing another metal or alloy of a lower melting point in between them in such a way that the melted metal bonds firmly to the other two. A soldering iron or copper bit is usually used to apply the solder, although sometimes a blowpipe is used to sweat the solder into the joint. A liquid flux such as zinc chloride or resin or linseed oil is generally used to assist the solder to flow and run smoothly into the joint.

1. Chisel point bit
2,3,4. Hatchet soldering bits used for soldering the bottoms of deep containers

Figure 8.2 Soldering bits

1 Flux solution lying on the surface
2 Boiling flux removing oxides
3 Bare metal oxides removed
4 Liquid solder replacing flux
5 Tin reacting with metal to form alloy compound
6 Solder solidi

Figure 8.3   Soldering process

The soldering process comprises the following steps:

- *Choose the right materials* – The choice of the soldering iron and its shape is governed by the size and accessibility of the material to be soldered; in general always use as large a soldering iron as is practicable. Some means of heating the soldering iron is necessary, together with a quantity of selected solder, file, emery cloth and a tin of flux, either corrosive or non-corrosive, to suit the work in hand.
- *Clean the soldering iron* – Solder will not adhere to or bond to a dirty or greasy soldering iron, and whether the iron is new or old it must be clean and bright on its working surfaces, that is, approximately 20 mm up from the point on each face. An emery cloth could be used for this purpose, but generally a file is preferred. It is important to see that just sufficient copper is removed to get rid of the pitting or scale and leave a clean, bright surface.
- *Heat the soldering iron* – A clean flame such as gas is best for this purpose. Care must be taken not to allow the bit to become red hot, as overheating of the bit causes heavy scaling of the surface due to oxidation; this will mean re-filing and re-cleaning the bit and hence unnecessary wear and a shorter service life.

When the flame turns green around the soldering bit it is at the correct temperature; if the colour is allowed to change into a bright green the copper bit will begin to become red hot. With an electric soldering iron the heat is automatically controlled at the right temperature because the tool has a built-in thermostat for controlling the temperature.

- *Tin the soldering iron* – Dip the heated copper bit into flux to obtain a complete coating of flux on the surface faces of the bit, then rub the fluxed portion of the bit on a piece of solder to obtain a film of solder over the copper surface or working faces of the soldering iron. This operation is referred to as tinning the bit and makes it easier for the bit to pick up solder and then discharge it on the workpiece. Tinning also protects the bit against further oxidation, thus increasing its life.
- *Clean the surface of the workpiece* – All metals have a covering of oxide on their surface, although it may not be visible to the naked eye, and this oxide film will prevent the solder from bonding to the metal to be soldered and therefore create a weak joint. First clean the surface to a bright finish with coarse emery paper or steel wool, then immediately apply the flux. The flux helps to clean the surface chemically so that the molten solder can flow and penetrate into the metal, forming a strong joint. Also it prevents oxides reforming on the surface of the work as the soldering is carried out.
- *Reheat the bit* until a green flame forms around it, again taking care not to overheat it and destroy the tinning on its surface. Dip the bit into flux, then hold the tinned face of the bit against the solder stick until all the face that is tinned is covered with molten solder. The soldering iron is now ready for use.
- *Apply the soldering iron* loaded with solder to the face of the workpiece or joint which has previously been smeared with flux. The metal surrounding the iron is heated to the melting point of the solder by conduction of heat from the soldering

iron, and the solder will start to flow. Draw the iron slowly along the face of the joint, allowing solder from the bit to flow into the joint. A good joint has only a very thin film of solder, as too much solder weakens the joint. The length of joint that can be soldered before the bit needs recharging with solder depends upon the size of the bit, its temperature and the size of the job to be soldered.

The most important points in soft soldering are:

1 A perfectly cleaned joint.
2 A soldering iron that has been tinned and heated to the correct temperature.
3 The correct flux for the particular job in hand.
4 A good quality tin–lead solder.
5 Allow the heat from the soldering iron to penetrate into the metal before moving the iron along the joint; this will give the solder a chance to flow into the joint.
6 The correct type of joint must be used. Soft soldering can only be used on metals that are lapped one over the other to form the joint. This allows for capillary attraction of the solder

Lap seam

Corner seam

Butt

Folded seam

Grooved seam

Cable terminal soldered

Brass union joint sweated to tube

Figure 8.4   Types of soft soldered joints

The strength of a soft soldered joint is not governed by the amount of solder between the plates. The more solder, the weaker the joint; therefore a good joint has only a very thin film of solder between the metal plates, which forms a surface alloy by using the tin in the solder.

Hard soldering

Brazing

Brazing is used extensively throughout the panel beating trade as a quick means of joining sheet metal panels and other automobile parts. Although a brazed joint is not as strong as a fusion weld, it has many advantages which make it useful to the panel beater. Brazing is not classed as a fusion process and therefore cannot be called welding, because the parent metals are not melted to form the joint but rely on a filler material of a different metal of low melting point which is drawn through the joint. The parent metals can be similar or dissimilar as long as the alloy rod has a lower melting point than either of them. The most commonly used alloy is of copper and zinc, which is, of course, brass. Brazing is accomplished by heating the pieces to be joined to a temperature higher than the melting point of the brazing alloy (brass). With the aid of flux, the melted alloy flows between the parts to be joined due to capillary attraction and actually diffuses into the surface of the metal so that a strong joint is produced when the alloy cools. Brazing, or hard soldering to give it its proper name, is in fact part fusion and is classed as a skin fusion process.

Brazing is carried out at a much higher temperature than that required for the soft soldering process. A **borax** type of powder flux is used, which fuses to allow brazing to take place between 750 and 900°C. There is a wide variety of alloys in use as brazing rods; the most popular compositions contain copper in the ranges 46 to 50 and 58.5 to 61.5 per cent, the remaining percentage being zinc.

The brazing process comprises the following steps:

1 Thoroughly clean the metal to be joined.
2 Using a welding torch, heat the metals to a temperature below their own critical or melting temperature. In the case of steel the metal is heated to a dull cherry red.
3 Apply borax flux either to the rod or to the work as the brazing proceeds to reduce oxidation and to float the oxides to the surface.
4 Use the oxy-acetylene torch with a neutral flame, as this will give good results under normal conditions. An oxidizing flame used for materials having a high percentage of brass content will produce a rough-looking brazed joint, which nevertheless is slightly stronger than if brazed with a neutral flame.
5 Use only a small amount of brazing rod; if too much is used this weakens the joint.
6 The two pieces of material to be brazed must be either lapped or carefully butted after edge preparation and must fit tightly together during the brazing operation. Iron, steel, copper and brass are readily brazed, and metals of a dissimilar nature can also be joined. Typical examples are as follows:

   • Copper to brass
   • Copper to steel
   • Brass to steel
   • Cast iron to mild steel
   • Stainless steel to mild steel.

   Also, coated materials like zinc-plated mild steel can be better brazed than welded.
7 Carefully select the types of metal to be joined.

Although dissimilar metals can be joined by hard or soft soldering, corrosion may occur due to the electrolytic action between the two dissimilar metals in the presence of moisture. This action is an electrochemical action similar to that of an electric cell and results in one or the other of the two metals being corroded away.

> **Safety note**
> When heated, zinc-plated steel (galvanized) gives off very toxic fumes, so full preparatory equipment must be used. It is better to avoid this hazard if possible.

The main advantages of brazing are:

1 The relatively low temperature (750–900°C) necessary for a successful brazing job reduces the risk of distortion.
2 The joint can be made quickly and neatly, requiring very little cleaning up.
3 Brazing makes possible the joining of two dissimilar metals; for example, brass can be joined to steel.
4 It can be used to repair parts that have to be re-chromed. For instance, a chromed trim moulding which has been deeply scratched can be readily filled with brazing and then filed up ready for chroming.
5 Brazing is very useful for joining steels which have a high carbon content or broken cast iron castings where the correct filler rod is not available.

Silver soldering

Silver solder probably originated in the manufacture and repair of silverware and jewellery for the purpose of securing adequate strength and the desired colour of the joint, but the technique of joining sheet metal products and components with silver solder has come into wide usage in the automobile industry. The term 'soft soldering' has been widely adopted when referring to the older process to avoid confusion with the newer hard-soldering process, known generally as either silver soldering or silver brazing. The use of silver solder on metals and alloys other than silver has grown largely because of the perfection by manufacturers of these solders, which makes them easily applicable to many metals and alloys by means of the oxy-acetylene welding torch.

Table 8.2 Copper–phosphorus brazing alloys

BS 1845 ref.	Nominal composition (wt%)									Melting range (°C)	Tensile strength (N/mm²)	Elongation (%)	Hardness (HV)	Notes
	Cu	Zn	Mn	Ni/Ag	P	Bi	Si	Sn	Others					
CP1	Balance	–	–	14–15	4.3–5.0	–	–	–	For details of impurities see British Standards	645–800	670–700	10	187	Fluxless brazing; High strength, good ductility
CP2	Balance	–	–	1.8–2.2	6.1–6.9	–	–	–		645–825	490–560	5	195	Good strength
CP3	Balance	–	–	–	7.0–7.8	–	–	–		710–810	490–550	7	192	Good strength
CP4	Balance	–	–	4.5–5.5	5.7–6.3	–	–	–		645–815	490–550	7	192	Good strength
CP5	Balance	–	–	–	5.6–6.4	–	–	–		692–825				Good strength
CP6	Balance	–	–	–	5.9–6.5	–	–	–		710–890				Good strength

This process is employed for joining metal parts when greater strength is required than can be obtained by soft soldering, when the parts have to withstand a temperature that would cause soft solder to melt and in cases in which the high temperature developed by welding would seriously distort the metal parts. Vehicle parts which are manufactured from light-gauge sheet brass, stainless steel or very thin mild steel and sheet products or components fabricated from nickel, bronze or copper can be very effectively joined by silver soldering.

Solders and fluxes

Silver solders are more malleable and ductile than brazing rods, and hence joints made with silver solder have a greater resistance to bending stresses, shocks and vibration than those made with ordinary brazing alloys. Silver solders are made in strip, wire (rod) or granular form and in a number of different grades of fusibility, the melting points varying between 630 and 800°C according to the percentages of silver, copper, zinc and cadmium they contain.

As in all non-fusion processes the important factor is that the joint to be soldered must be perfectly clean. Hence special care must be taken in preparing the metal surfaces to be joined with silver solder. Although fluxes will dissolve films of oxide during the soldering operation, sheet metal that is clean is much more likely to make a stronger, sounder joint than when impurities are present. The joints should fit closely, and the parts must be held together firmly while being silver soldered, because silver solders in the molten state are remarkably fluid and can penetrate into minute spaces between the metals to be joined. In order to protect the metal surface against oxidation and to increase the flowing properties of the solder, a suitable flux such as borax or boric acid is used.

Silver soldering process

In silver soldering the size of the welding tip used and the adjustment of the flame are very important to avoid overheating, as prolonged heating promotes oxide films, which weaken both the base metal and the joint material. This should be guarded against by keeping the luminous cone of the flame well back from the point being heated. When the joint has been heated just above the temperature at which the silver solder flows, the flame should be moved away and the solder applied to the joint, usually in rod form. The flame should then be played over the joint so that the solder and flux flow freely through the joint by capillary attraction. The finished silver soldered joint should be smooth and

Figure 8.5   Joint design for brazing, showing the brazing equivalents to welding

regular in shape and require no dressing up apart from the removal of the flux by washing in water.

When making a silver solder joint between dissimilar metals, concentrate the application of heat on the metal which has the higher heat capacity. This depends on the thickness and the thermal conductivity of the metals. The aim is to heat both members of the joint evenly so that they reach the soldering temperature at the same time.

The most important points during silver soldering are:

1 Cleanness of the joint surfaces
2 Use of the correct flux
3 The avoidance of overheating.

Aluminium brazing

There is a distinction between the brazing of aluminium and the brazing of other metals. For aluminium, the brazing alloy is one of the aluminium alloys having a melting point below that of the parent metal. For other metals, the brazing alloys are often based on copper–zinc alloys (brasses – hence the term brazing) and are necessarily dissimilar in composition to the parent metal.

Wetting and fluxing

When a surface is wetted by a liquid, a continuous film of the liquid remains on the surface after draining. This condition, essential for brazing, arises when there is mutual attraction between the liquid flux and solid metal due to a form of chemical affinity. Having accomplished its primary duty of removing the oxide film, the cleansing action of the flux restores the free affinities at the surface of the joint faces, promoting wetting by reducing the contact angle developed between the molten brazing alloy and parent metal. This action assists spreading and the feeding of brazing alloy to the capillary spaces, leading to the production of well-filled joints. An important feature of the brazing process is that the brazing alloy is drawn into the joint area by capillary attraction: the smaller the gap is between the two metal faces to be joined, the deeper is the capillary penetration.

The various grades of pure aluminium and certain alloys are amenable to brazing. Aluminium-magnesium alloys containing more than 2 per cent magnesium are difficult to braze, as the oxide film is tenacious and hard to remove with ordinary brazing fluxes. Other alloys cannot be brazed because they start to melt at temperatures below that of any available brazing alloy. Aluminium-silicon alloys of nominal 5 per cent, 7.5 per cent or 10 per cent silicon content (Table 8.4) are used for brazing aluminium and the alloy of aluminium and 1.5 per cent manganese.

The properties required for an effective flux for brazing aluminium and its alloys are as follows:

1 The flux must remove the oxide coating present on the surfaces to be joined. It is always important that the flux be suitable for the parent metal, but especially so in the joining of aluminium-magnesium alloys.
2 It must thoroughly wet the surfaces to be joined so that the filler metal may spread evenly and continuously.
3 It must flow freely at a temperature just below the melting point of the filler metal.
4 Its density, when molten, must be lower than that of the brazing alloy.
5 It must not attack the parent surfaces dangerously in the time between its application and removal.
6 It must be easy to remove from the brazed assembly.

Many types of proprietary fluxes are available for brazing aluminium. These are generally of the alkali halide type, which are basically mixtures of the alkali metal chlorides and fluorides. Fluxes and their residues are highly corrosive and therefore must be completely removed after brazing by washing with hot water.

Table 8.3 Silver solders

BS 1845 ref.	Nominal composition (wt%)							Melting range (°C)	Tensile strength (N/mm²)	Notes
	Ag	Cu	Zn	Mn	Ni	Cd	Sn			
Cadmium-containing alloys										
AG1	49–51	14–16	14–18	–	–	18–20	–	620–640	470	Low melting point, very fluid, high strength, general purpose
AG2	41–43	16–18	14–18	–	–	24–26	–	610–620	470	Lowest melting point, very fluid. General purposes, especially small components
AG3	37–39	19–21	20–24	–	–	19–21	–	605–650	–	Very fluid and strong
AG9	49–50	14.5–16.5	13.5–17.5	–	2.5–3.5	15–17	–	635–655	480	Limited flow, useful for fillet joints
AG11	33–35	24–26	18–22	–	–	20–22	–	610–670	–	
AG12	29–31	27–29	19–23	–	–	20–22	–	600–690	485	
Cadmium-free alloys										
AG5	42–14	36–38	18–22	–	–	0.025	–	690–770	400	Cadmium-free for food equipment, etc.
AG7	71–73	27–29	–	–	–	0.025	–	780	–	Fluxless brazing of copper
AG7V	71–73	27–29	–	–	–	0.025	–	780	–	High-purity alloy for vacuum assemblies
AG8	99.99	–	–	–	–	0.25	–	960	–	Pure silver
AG13	59–61	25–27	12–16	–	–	0.025	–	730–695	–	
AG14	54–56	20–22	21–23	–	–	0.025	1.7–2.3	660–630	450	
AG18	48–50	15–17	21–25	6.5–8.5	4–5	0.025	–	705–680	360	For carbide brazing
AG19	84–86	–	–	14–16	–	0.025	–	960–970	–	
AG20	39–41	29–31	–	–	–	0.025	1.7–2.3	710–650	–	
AG21	29–31	35–37	–	–	–	0.025	1.7–2.3	755–665	–	

Brazing method

When the cleaned parts have been assembled, brazing flux is applied evenly over the joint surface of both parts to be brazed and the filler rod (brazing alloy). The flame is then played uniformly over the joint until the flux has dried and become first powdery, then molten and transparent. (At the powdery stage care is needed to avoid dislodging the flux, and it is often preferable to apply flux with the filler rod.) When the flux is molten the brazing alloy is applied, preferably from above, so that gravity assists in the flow of metal. In good practice the brazing alloy is melted by the heat of the assembly rather than directly by the torch flame. Periodically the filler rod is lifted and the flame is used to sweep the liquid metal along the joint; but if the metal is run too quickly in this way it may begin to solidify before it properly diffuses into the mating surfaces. Trial will show whether more than one feed point for the brazing alloy is necessary, but with proper fluxing, giving an unbroken path of flux over the whole joint width, a single feed is usually sufficient.

Bronze welding

Bronze welding is carried out much as in fusion welding except that the base metal is not melted. The base metal is simply brought up to tinning temperature (dull red colour), and a bead is deposited over the seam with a bronze filler rod. Although the base metal is never actually melted, the unique characteristics of the bond formed by the bronze rod are such that the results are often comparable with those secured through fusion welding. Bronze welding resembles brazing, but only up to a point. The application of brazing is generally limited to joints in which a close fit or mechanical fastening serves to consolidate the assembly and the joint is merely strengthened or protected by the brazing material. In bronze welding the filler metal alone provides the joint strength, and it is applied by the manipulation of a heating flame in the same manner as in gas fusion welding. The heating flame is made to serve the dual purpose of melting off the bronze rod and simultaneously heating the surface to be joined. The operator in this manner controls the work: hence the term 'bronze welding'.

Almost any copper-zinc alloy or copper-tin alloy or copper-phosphorus alloy can be used as a medium for such welding, but the consideration of costs, flowing qualities, strength and ductility of the deposit have led to the adoption of one general-purpose 60–40 copper-zinc alloy with minor constituents incorporated to prevent zinc oxide forming and to improve fluidity and strength. Silicon is the most important of these minor constituents, and its usefulness is apparent in three directions. First, in the manner with which it readily unites with oxygen to form silica, silicon provides a covering for the molten metal which prevents zinc

Table 8.4 Aluminium filler alloys for brazing

BS 1845 ref.	Nominal composition (wt%) (balance aluminium)						
	Si	Cu (max.)	Mn (max.)	Zn (max.)	Ti (max.)	Mg (max.)	Melting range (°C)
4004	9.0/10.5	0.25	0.10	0.20	–	1.0/2.0	555–590
4043A	4.5/6.0	0.30	0.15	0.10	0.15	0.20	575–630
4045	9.0/11.0	0.30	0.05	0.10	0.20	0.05	575–590
4047A	11.0/13.0	0.30	0.15	0.20	0.15	0.10	575–585
4104	9.0/10.5	0.25	0.10	0.20	–	1.0/2.0	555–590
4145A	9.0/11.0	3.0/5.0	0.15	0.20	0.15	0.10	520–585
4343	6.8/8.2	0.25	0.10	0.20	–	–	575–615

volatilization and serves to maintain the balance of the constituents of the alloy; this permits the original high strength of the alloy to be carried through to the deposit. Second, this coating of silica combines with the flux used in bronze welding to form a very fusible slag, and this materially assists the tinning operation, which is an essential feature of any bronze welding process. Third, by its capacity for retaining gases in solution during solidification of the alloy, silicon prevents the formation of gas holes and porosity in the deposited metal, which would naturally reflect unfavourably upon its strength as a weld.

It is essential to use an efficient and correct flux. The objects of a flux are: first, to remove oxide from the edges of the metal, giving a chemically clean surface onto which the bronze will flow, and to protect the heated edges from the oxygen in the atmosphere; second, to float oxide and impurities introduced into the molten pool to the surface, where they can do no harm. Although general-purpose fluxes are available, it is always desirable to use the fluxes recommended by the manufacturer of the particular rod being employed.

Bronze welding procedure

1 An essential factor for bronze welding is a clean metal surface. If the bronze is to provide a strong bond, it must flow smoothly and evenly over the entire weld area. Clean the surfaces thoroughly with a stiff wire brush. Remove all scale, dirt or grease; otherwise the bronze will not adhere. If a surface has oil or grease on it, remove these substances by heating the area to a bright red colour and thus burning them off.
2 On thick sections, especially in repairing castings, bevel the edges to form a 90° V-groove. This can be done by chipping, machining, filing or grinding.
3 Adjust the torch to obtain a slightly oxidizing flame. Then heat the surfaces of the weld area.

4 Heat the bronzing rod and dip it in the flux. (This step is not necessary if the rods have been pre-fluxed.) In heating the rod, do not apply the inner cone of the flame directly to the rod.
5 Concentrate the flame on the starting end until the metal begins to turn red. Melt a little bronze rod onto the surface and allow it to spread along the entire seam. The flow of this thin film of bronze is known as the tinning operation. Unless the surfaces are tinned properly the bronzing procedure to follow cannot be carried out successfully. If the base metal is too hot, the bronze will tend to bubble or run around like drops of water on a warm stove. If the bronze forms into balls which tend to roll off, just as water would if placed on a greasy surface, then the base metal is not hot enough. When the metal is at the proper temperature the bronze spreads out evenly over the metal.
6 Once the base metal is tinned sufficiently, start depositing the proper-size beads over the seam. Use a slightly circular torch motion and run the beads as in regular fusion welding with a filler rod. Keep dipping the rod in the flux as the weld progresses forward. Be sure that the base metal is never permitted to get too hot.
7 If the pieces to be welded are grooved, use several passes to fill the V. On the first pass make certain that the tinning action takes place along the entire bottom surface of the V and about halfway up on each side. The number of passes to be made will depend on the depth of the V. When depositing several layers of beads, be sure that each layer is fused into the previous one.

Body leading/body soldering

Sometimes called body leading, sometimes called body soldering. This used to be a popular way of repairing damaged bodywork; then it fell out of use. It has now become popular again, especially in the classic and custom car sectors of the industry. The advantage of using lead is that it gives a harder surface than paste filler and therefore cannot be noticed when the car is finished. There are two types of body solder available. One contains lead, while the other doesn't.

The main world-wide supplier of body leading materials and equipment is a firm called the Eastwood Company – eastwood. com. If you are a fan of classic films, you might know the name.

The one containing lead requires much tighter health and safety procedures – extraction equipment, breathing mask and skin protection – as lead is a poison. The content is 30 per cent tin and 70 per cent lead; the UTS is 6140 psi (42.3 NM/m²); the plastic-state operating range is 183°C to 254°C. The lead-free body solder is much stronger, with a UTS of 9000 psi (62 NM/m²) and an operating range of 280°C to 350°C. This also makes the lead-free solder the best choice if you are powder coating the finished component. Powder coatings tend to cure at about 200°C, which is above the start of the plastic state for lead-based solder.

Body soldering tools

Figure 8.6 Body soldering tools

Paddle – made from hard wood, the curved shape keeps your hands away from the hot working area.
Brush – to apply the paste flux.
Paste flux – to prevent the component being soldered from oxidizing
Body solder sticks – type to suit the application
Body file – to smooth the soldered surface

> **Safety note**
> When shaping body solder containing lead, only use a body file; do not use any sanding material. Using sanding – either mechanical or hand – produces lead particles which are extremely small and highly poisonous. It also gives a risk of metal particles entering the lead, which can lead to the finishing problems.

Body soldering procedure

- Clean the surface thoroughly and ensure that it is free of all contaminants.
- Warm the surface using a blowtorch – a soft propane flame is good, not too hot.
- Apply the paste flux with a suitable-size brush.
- Melt the solder onto the surface.
- With the paddle in one hand and the blowtorch in the other ,work the solder to the shape which you want.
- Apply additional solder if needed.
- When the shape is right for you, leave to cool slowly.
- Use the body file to finish the shape required.

> **Tech note**
> When body soldering, use only a non-acidic paste flux. Acidic, or killed spirit, fluxes can cause corrosion of the component.

Health and safety and the environment

The materials used in soldering can be very dangerous if not used in a safe way. As well as the normal workshop precautions, the following should be especially noted

- **All soldering and brazing must be carried out in an area with suitable fume extraction.**

- **Fluxes are mainly corrosive and should be handled accordingly.**
- **Always wash your hands after soldering or brazing.**
- **Use PPE as directed by your company safety guidelines.**
- **All materials must be stored securely and not be accessible to unauthorized personnel.**

Skills and questions

The following exercises are suggested to improve your soldering skills. You must ensure that until you are fully competent, you are fully supervised by your tutor or a senior colleague. Also ensure that all the health and safety and environmental conditions are fulfilled. Prevent waste by practising using scrap materials where possible.

1 Describe the differences between the following classifications:
   (a) total fusion
   (b) skin fusion
   (c) surface fusion.
2 Carry out soft and hard soldering procedures using scrap materials.
3 Make up a poster giving a guide to others on safe soldering.
4 Explain the reason why corrosive and non-corrosive fluxes are used in soft soldering as if to a customer whose vehicle you are repairing.
5 Investigate a vehicle and state where soldering or brazing may be used in its construction or repair.

1 Describe the differences between the following classifications: (a) total fusion, (b) skin fusion, (c) surface fusion.
2 Which three conditions are essential to achieve a good soldered joint?
3 What are the main functions of a flux in soft soldering?
4 Name two types of flux used in soft soldering, and state a typical application for each.
5 Explain the reason why corrosive and non-corrosive fluxes are used in soft soldering.
6 List the step-by-step procedure for the process of soft soldering.
7 Explain the sequence of operations required to tin a soldering iron.
8 What is the difference between bronze welding and brazing?
9 State the advantages of brazing.
10 Explain the reason why certain joints are brazed in preference to being soldered or welded.
11 Name the flux and detail the composition of the rods used in the brazing process.
12 Name one non-fusion welding process, and give a detailed description of the process.
13 Name the constituents of a brazing rod suitable for brazing low-carbon steel.
14 Define the term 'capillary attraction'.
15 Explain the type of application in which the process of silver soldering would be used.
16 Describe the differences between the processes of aluminium welding and aluminium brazing.

9

Gas welding, cutting and plasma arc

Overview of gas welding

Acetylene is made from a combination of calcium carbide and water. When mixed with the correct proportion of oxygen it burns at a temperature of 3200°C. This is hot enough to melt any metal; steel becomes fluid at about 1500°C. Welding is a process of fusing two pieces of metal together when they are molten – that is in a fluid state. When welding steel body panels it is normal to use welding rod, sometimes called filler rod – or sticks to provide extra metal to make up the joint. This is discussed further in the text.

Gas welding is used mainly in the repair of classic cars and the fabrication of components. It tends not to be used on most current vehicles because heating the steel to high temperature changes the steel structure and can cause defects.

Figure 9.2 A typical oxy-fuel gas welding and cutting system

Figure 9.1 High-pressure gas welding equipment in use (*Murex Welding Products Limited*)

Oxy-acetylene welding equipment

High-pressure welding equipment comprises:

1 Cylinder of dissolved acetylene
2 Cylinder of oxygen
3 High-pressure welding torch with separate oxygen control and acetylene

The Repair of Vehicle Bodies, Andrew Livesey, 9780815378709, 2019, Taylor & Francis

control and various-sized tips for different thicknesses

4 Acetylene regulator with gauges
5 Oxygen regulator with gauges
6 Two 5 m lengths of 10 mm high-pressure rubber canvas hose with clips and fitted with hose check valves (red for acetylene and blue for oxygen)
7 Acetylene flashback arrester and oxygen flashback arrester
8 Keys to fit cylinders and spanners for gland nuts and for fixing the regulators to the cylinders
9 Pair of welding goggles
10 Cylinder trolley.

Acetylene supply

The acetylene is supplied in steel cylinders painted maroon and containing between 0.57 m^3 and 8.73 m^3 of gas when fully charged. These cylinders contain a porous filling which acts as an absorbent for the acetone used to dissolve the acetylene. Acetone has the remarkable property of being able to dissolve large volumes of acetylene gas. At normal charging pressures (about 17.7 bar) the acetone will dissolve approximately 25 times its own volume. To prevent the interchange of fittings between cylinders, the acetylene cylinder valve into which the regulator is fitted has a left-hand thread, while the oxygen cylinder valve has a right-hand thread. The actual cylinder valves are all opened anti-clockwise and closed clockwise in the usual way.

Most acetylene cylinders supplied by the manufacturers are protected by at least one safety device. The majority of cylinders have a bursting disc on the back of the cylinder valve opposite the spindle. The disc operates at a pressure of about 90 bar. Others have either two fusible plugs fitted on the shoulder near the cylinder valve or a bursting disc. These plugs melt at about 100°C and relieve any excess pressure if the cylinder is subjected to undue heat or catches fire. Care is necessary in handling the cylinders, as any leak is liable to lead to fire, especially in confined spaces.

Figure 9.3 Cylinder valve (*BOC Ltd*)

Oxygen supply

The oxygen supply comes in seamless steel cylinders containing between 0.68 m^3 and 9.66 m^3 of oxygen and painted black. As this is a non-combustible gas, the cylinder has a right-hand threaded valve outlet. Oxygen is supplied at a maximum pressure of 200/230 bar at 15°C.

Cylinder labelling

To comply with the Classification, Packaging and Labelling of Dangerous Substances Regulations 1984, BOC cylinders now appear with a label designed to give more safety information. The format is shown in Figure 9.5 and the various sections are identified. Never use gas from a cylinder which has no label; return the cylinder to the supplier. If a cylinder is involved in an accident, withdraw it from service and set it aside clearly marked; contact the supplier.

Gas cylinders are designed and constructed in accordance with Home Office specifications and/or British Standards. These standards define the material of which the cylinder is made, the method of construction, its test pressure, the maximum permissible filled pressure and the method of regular testing.

All cylinders containing gas at high pressure are fitted with a cylinder valve which *must not* be removed or tampered with at any time except to tighten the gland nut when necessary. Some cylinders have a security cap over the cylinder valve, indicating that they have been filled and checked. This cap is removed by rotating the hexagon nut in either direction using a regulator spanner: this will cause the cap to split for easy removal. To open the cylinder valve, rotate the spindle anticlockwise using the special spindle key. Some cylinders are fitted with hand wheels, obviating the need to use a spindle key. These cylinders, together with others which are fitted with valves, normally have valve guards or valve protection caps. Valve guards should not be removed. Valve protection caps should always be replaced after use.

Only the owner of a cylinder can authorize its scrapping, and before this is done the cylinder must first be destroyed as a pressure vessel.

Figure 9.4   Typical cylinder label (*BOC Ltd*)

Figure 9.5   Identification markings on gas cylinder necks (*BOC Ltd*)

Welding torches

The welding torch comprises two needle valves (one for regulating the oxygen and the other the acetylene), a mixing chamber and an interchangeable welding nozzle, all incorporated in a shank. The torch, sometimes known as the blowpipe, mixes acetylene and oxygen in the correct proportion and permits the mixture to flow to the end of the welding tip or nozzle, where it is burned. Blowpipes vary to some extent in design, but basically they are all made to provide complete control of the flame during the welding operation. The equal-pressure torch is designed to operate in the high-pressure system and will operate when acetylene is supplied from cylinders. The acetylene and oxygen are fed independently to the mixing chamber and then flow through the tip. On the rear end of the torch there are two female unions for connecting the relevant hoses; to eliminate any danger of interchanging the hoses, the oxygen union has a right-hand thread and the acetylene union a left-hand thread.

Interchangeable welding nozzles

To make possible the welding of different thicknesses of metal, welding torches are equipped with an assortment of different-size heads or nozzles. The size of the nozzle is governed by the diameter of the opening at the welding end. The most common sizing system uses numbers which are marked on the nozzle, and the larger the number, the greater is the diameter of the opening in the nozzle; numbers 1 through 35 are for use with thicknesses up to 10 mm, and numbers 35 through 90 are for 10 to 25 mm.

Regulators

There are two types of regulator, the two-stage and the single-stage. The oxygen and acetylene pressure regulators perform two functions: they **reduce the cylinder pressure** to the required working pressure, and they **produce a steady flow of gas** under varying cylinder pressures. With the two-stage regulator the gas flows from the cylinder into a first chamber, where a predetermined high pressure is maintained by means of a spring and diaphragm. From the high-pressure chamber the gas passes into a second, reducing chamber where the pressure is governed by an adjusting screw. The principal advantage of the two-stage regulator is that the working pressure remains constant until the gas in the cylinder is exhausted, and little adjustment of gas flow is needed. In the single-stage regulator there is only one, low-pressure chamber; the gas from the cylinder flows into the regulator and is controlled entirely by the adjusting screw. The disadvantage of this regulator is that as the cylinder pressure drops, the actual working pressure also falls and thus needs continual readjustment.

Both types of regulator have two gauges, one indicating the pressure in the cylinder and the other showing the actual working pressure or line pressure at the torch. The content gauge on the oxygen regulator is calibrated up to 300 bar, whereas the working pressure gauge is calibrated from 0 to 6 bar. The content gauge on the acetylene regulator is calibrated up to 40 bar, whereas the working pressure gauge is calibrated from 0 to 2.5 bar (Figure 9.9). It is most important to remember to release (turn out) the adjusting screw on the regulator before opening the cylinder valve. Otherwise the tremendous pressure of the gas in the cylinder is forced into the working pressure gauge, and this could result in gauge damage. Before a

Figure 9.6   Welding and cutting blowpipe, high pressure (*Murex Welding Products Limited*)

regulator is attached to a cylinder, the valve of the latter should be 'cracked' by quickly opening and then closing it to blow out any particles of dirt that may be present; if the gas is acetylene, care must be taken to guard against any possibility of it becoming ignited. As previously mentioned, oxygen and acetylene regulators also have female unions, which have right and left threads respectively to prevent the risk of hose interchange. To distinguish the connections still further, the oxygen regulator gauges when new are painted black and those of acetylene regulators are red. An acetylene regulator may be used on a propane cylinder but not on a hydrogen or coal gas cylinder, as the initial high pressure would damage or burst the cylinder contents gauge.

Automatic flashback arrester

This is an automatic safety device for use in the high-pressure system with either oxygen or acetylene cylinders. An arrester is fitted to the pressure regulator outlet of a cylinder with its opposite end connected to the hose leading to the welding torch. If excess pressure builds up, the arrester acts by completely sealing off the gas supply. Only when the pressure has subsided to normal working level via the built-in escape valve will the arrester allow the main supply of gas to continue flowing. The increase in safety which this piece of equipment affords makes it an essential part of the high-pressure system of welding.

Figure 9.8 Acetylene regulator (*Murex Welding Products Ltd*)

Figure 9.7 Oxygen regulator (*Murex Welding Products Ltd*)

Figure 9.9 Flashback arresters for oxygen and acetylene regulators (*Murex Welding Products Limited*)

Gas hose

Gas hoses are of a special non-porous rubber, reinforced with canvas. They are designed to withstand the working pressures which transmit gas from the regulator to the welding torch. Hoses are available with internal diameters of 5 to 20 mm and are coloured red for acetylene and blue for oxygen to prevent the risk of hose interchange. The nut on the acetylene hose is distinguished from the nut on the oxygen hose by a groove that runs around its centre, indicating that it has a left-hand thread. A ferrule is used to squeeze the hose around the union shank to prevent it from working loose. The life of the hose may be prolonged by reasonable care in use, for example by keeping it away from heat, sparks, oil and grease and by preventing it from being crushed.

Hose check valve

This is a safety device in the form of a non-returnable valve which is fitted behind the welding torch and connected to the hose. There are check valves for both oxygen and acetylene connections. If the pressure or gas flow tries to reverse, the hose check valve immediately seals off the torch from the main supply, thus eliminating the possibility of flashback and backfiring of the welding torch through the hose to the regulator and cylinder.

Figure 9.11   Hose check valves (*Murex Welding Products Ltd*)

Cylinder key

This key is an essential part of the equipment and is used for turning the valves on the gas cylinders on and off. It should be always attached to the welding equipment or plant so that it is readily available in case of emergency.

Welding goggles

Goggles fitted with tinted glass of an approved type complying with the requirements set out in British Standard specification 679 should be worn to protect the eyes from the intense light of the flame as well as from sparks and small metal particles thrown up from the weld. As the goggles become pitted in time and obscure the work, it is usual to protect the tinted glass with a plain one which can be replaced at low cost.

Cylinder trolley

It is advisable to have all welding equipment portable and mobile when in use in a body repair shop, and for this purpose a cylinder trolley can be used. Both cylinders must be securely fastened but at the same time easy to replace. The trolley itself should be of sturdy construction and should be as narrow as possible so that it can pass through restricted spaces.

Figure 9.10   Hose fittings: nut, nipple, and O-clips (*Murex Welding Products Ltd*)

Figure 9.12 Types of welding goggles (*Murex Welding Products Limited*)

Assembly of high-pressure welding equipment

The cylinders must be kept upright, and all metal jointing surfaces should be free from oil or grease. Before fitting the regulators, blow out the cylinder valves to remove any dirt or obstruction. To ensure gas-tight connections between cylinder valves and regulators, first screw down the hexagon or wing nuts by hand, then give the regulators a twist to bed them down on their seats, and finally tighten the nuts properly, but without the use of excessive force. New hoses in use for the first time should be blown through to remove any grit that may be present. Attach the appropriate hoses to the regulators and the welding torch, and fit the latter with a suitable welding tip. Next slacken off the pressure, regulating screws, open the regulator outlet valves and turn on the gases with the cylinder key; this must be done slowly to avoid damage to the regulators. The cylinder valves must be opened at least two full turns to ensure that the flow of gases to the regulators is unrestricted. Next set the regulators to the required pressure, open the acetylene control valve on the welding torch, check working pressures when pure acetylene is coming out of the end of the nozzle, and then light the gas. Reduce or increase the acetylene supply by the welding torch valve until the flame just ceases to smoke, then turn on the oxygen by the welding torch control

valve until the white inner cone is sharply defined, and check working pressures again. The welding torch is now ready and is burning with a neutral flame, which is used on most welding operations.

On completion of the welding operation the following procedure must be carried out to render the equipment safe. Turn off the acetylene first by the welding torch control valve, then the oxygen. Close the cylinder valves. Open the welding torch valves one at a time to release the pressure in the hose; open the oxygen valve and shut it; and open the acetylene valve and shut it. Unscrew the pressure regulating screws on the oxygen and acetylene regulators.

Definitions of welding terms

The material of the parts to be welded is described as the **parent metal**. Any material that it may be necessary to add to complete the weld is known as **filler metal**, which, if in rod or wire form, is obtained from a **welding rod**. The surface to be welded is called the **fusion face**. That part of the weld where the parent metal has been melted, if filler is used and interdiffusion has taken place, is called the **fusion zone**, the depth of which is termed the **weld penetration**. Bordering on the fusion zone is the **zone of thermal disturbance**, consisting of that portion of the parent metal which, although not melted by the flame, has been heated sufficiently to disturb the grain structure; where the fusion zone and zone of thermal disturbance meet is known as the **junction**. A **bead** is a single longitudinal deposit of weld metal laid on a surface by fusion welding; a local deposit

Figure 9.13 Section through a welded joint

laid on a surface is termed a **pad**, and is usually formed by a series of overlapping beads. **Tack welds** are local welds used to hold parts in their correct relative positions ready for welding.

Most welded assemblies are made by **butt joints** in which the ends or edges directly face each other. In some cases the joining weld is made between these edges. Additionally or alternatively **fillet welds** may be used, which are of approximately triangular cross-section and lie externally to the parts joined. These fillet welds may consist of **T-joints or lap joints**; in the former the parts are usually set at right angles to one another, and in the latter the weld is made in the angle formed by the overlap. An alternative to the butt joint is the **edge joint**, in which the two metals are put together to form a corner.

Various terms are used to indicate the different parts of a weld. The **weld face** is the exposed surface of any weld, a **leg** is the fusion face of a fillet weld, and the **toe** is a border line where the weld face adjoins a welded part; along this line **undercut** or wastage of the parent metal in the form of a grooving may occur. The **root** is the zone at the bottom of a space provided for or occupied by a fusion weld, and the **throat** is the minimum depth of the weld measured along a line passing through the root. The condition which arises when the filler metal flows onto heated but unfused joint surfaces, and the interfusion of the filler and parent metals does not take place, is known as **adhesion**.

Welding rods and fluxes

Filler rods

Filler rods for use in oxy-acetylene welding are available in the following metals: mild steel, wrought irons, high-carbon steel, alloy steel, stainless steel, cast iron, copper, copper alloys, aluminium, aluminium alloys, hard facing alloys and zinc-based die cast alloys.

The metal from the rod has considerable influence on the quality of the finished weld. Good welding rods are designed to give deposited metal of the correct composition and have allowance in their chemistry for changes which take place in the welding process. Rods are obtainable in sizes ranging from 1.6 mm to 5.0 mm diameter. Some have a copper coating to keep their surfaces free from oxides or rust, but uncoated rods are equally efficient provided that they are clean. Sound welds, comparable in strength with the material welded, can be produced with satisfactory filler rods, but similar results cannot be expected with dirty or rusty filler rods or with any odd piece of wire that may come to hand.

Welding fluxes

It is impracticable to incorporate in welding rods all the elements necessary to overcome oxidation. Therefore it is necessary to use with them certain chemical compounds to act as deoxidizing agents or fluxes, which must be of the correct composition to ensure perfect welds. A flux *must* be used with the following metals: cast iron, high-carbon steel, stainless steel, copper, copper alloys, aluminium, aluminium alloys and magnesium alloys. In the majority of cases it is essential that the flux residues should be removed from the surface of the metal after the welding operation has been completed.

Flame control and types of flame

Temperature of flame

When acetylene is burned with an equal volume of oxygen, a maximum flame temperature of over 3200°C (about 2.5 times hotter than the melting point of cast iron and steel) is obtained just beyond the inner luminous cone (Figure 9.15). Variations of the proportions of oxygen and acetylene can be made at the welding torch, and three distinct types of flame are obtained as follows.

Figure 9.14   Flame temperature

Carbonizing (carburizing) flame (excess acetylene)

Neutral flame (equal quantities of oxygen and acetylene)

Oxidizing flame (excess oxygen)

Figure 9.15   Regulation of welding flame

Carbonizing, carburizing or reducing flame

For this flame the gases are adjusted so that all the acetylene gas is not completely burned and there is an excess of acetylene or an insufficiency of oxygen. When the welding torch is lit, the acetylene is turned on first and ignited, giving a very smoky yellow flame of abnormal size which shows two cones of flame in addition to an outer envelope; this is an exaggerated form of the carbonizing flame, but it gives out comparatively little heat and is useless for welding. The oxygen is turned on, and the supply is gradually increased until the flame, though still of abnormal size, contracts towards the welding torch tip, where an inner white cone of great luminosity commences to make its appearance. The increase in the supply of oxygen is stopped before the cone becomes clearly defined and while it is still an inch or so long; this results in a carburizing flame, which is characterized by a dullish white feather surrounding a brilliant, clearly defined white cone. The flame can be used to advantage in the welding of high-carbon steel, aluminium, Inconel and Monel metal in the technique of stelliting, for stainless steel, and wherever excess of oxygen on metals would cause detrimental oxidation. However, in some cases the flame is detrimental owing to the fact that it deposits carbon; for example, if it is used on mild steel the weld deposit will be higher in carbon and therefore becomes hardenable material, and cracks may result.

Neutral flame

As the supply of oxygen to the welding torch is increased beyond the point at which the carbonizing flame is formed, the flame contracts and the white cone assumes a definite rounded shape. At this stage approximately equal quantities of acetylene and oxygen are being used and combustion is complete, all the carbon supplied by the acetylene being consumed and the maximum heat given out. This flame is known as the neutral flame, and it is the one most extensively used by the welder. It does not oxidize or carburize the material, and it is used on mild steel, copper and magnesium.

Oxidizing flame

A further increase in the oxygen supply will produce an oxidizing flame in which there is more oxygen than is required for complete combustion. The inner cone will become shorter and sharper, and the flame will turn a deeper purple colour and emit a characteristic slight hiss. With this, the oxidizing flame, the molten metal will be less fluid and tranquil during welding and excessive sparking will occur. An oxidizing flame is only used for special applications and should always be avoided when welding steel, because it causes rapid oxidation of the metal. It is, however, used in the welding of brasses, bronzes and copper-silicon alloys, and in the process of bronze welding, the degree of oxidation required being determined experimentally.

Power of flame

The power or heat value of the flame is governed by the size of the orifice in the tip used. The power required depends on the thickness, mass, melting point and heat conductivity of the metal to be welded. Manufacturers of welding torches provide information regarding the size and rated gas consumption of tips for different thicknesses of metals and alloys. The power of a welding torch is measured by the number of litres or cubic feet of acetylene which are consumed in an hour with the flame perfectly regulated, and this figure is sometimes marked on the welding torch itself. The heat value of the flame must be adjusted by changing the tip and not by unduly increasing or decreasing the pressure and volume of gases used. The pressure of gases used for any size of tip determines the length of the inner cone, which should be 3 to 3.5 times as long as its breadth at the base. Too low a pressure gives too short a cone, which may cause lack of penetration and fusion; it also causes frequent backfiring and will be deflected by particles of metal and slag thrown up from the weld. On the other hand, too high a gas pressure giving too long a cone causes overheating and lack of control of the molten metal, resulting in adhesion and over-penetration.

Flame adjustment and gas pressure

The adjustment of a good welding flame is affected by the pressure of the gas entering the welding torch. If the pressure from either the acetylene or the oxygen regulator is excessive, the result will be a fierce or harsh flame which is very difficult to manipulate, especially when welding panel steel with the neutral flame. Excessive pressure also increases the tendency to blow the edges of the steel away as they are melted. On the other hand, if the pressure of either gas is too low a backfiring of the welding torch will result.

Methods of welding

Leftward or forward welding is the technique or method of welding in which the welding torch flame is directed towards the uncompleted joint. When the flame is directed towards the completed weld, this is termed **rightward**, **backward** or **backhand** welding. The positions in which welding is performed are termed respectively **flat**, **downhand** or **underhand** when the weld is made on the upper side of a horizontal or flat surface, **overhead** when on the underside, and **vertical** when on an upright or vertical surface.

Figure 9.16 Methods of welding (a) the leftward method and (b) the rightward method

Leftward welding

This is used on steel for flanged edge welds, for bevelled steel plates up to 3.2 mm, and for bevelled plates up to 4.8 mm. It is also the method usually adopted for cast iron and non-ferrous metals. The weld is started on the right-hand end of the joint, and welding proceeds towards the left. The welding torch is given a forward motion, with a slight sideways movement to maintain both edges melting at the desired rate, and the welding wire is moved progressively along the weld seam. The sideways motion of the welding torch should be restricted to a minimum. The advantages of the leftward welding technique are that it is faster because the flame, in preceding the weld, has a preheating effect; the molten metal is easily controlled; and penetration (complete fusion of the edges) is easily obtained over the range of application of the technique. One of the important features of a good weld is complete penetration. A disadvantage of the leftward method of welding is that if incorrect manipulation of the flame occurs, molten metal will flow ahead of the molten pool at the bottom of the weld and adhere to the comparatively cold plate, thus causing poor penetration.

The characteristics of the leftward welding technique are as follows:

- *Position* – Flat
- *Direction of welding* – Flame points away from the finished weld
- *Angle of welding torch* – 60°–70°
- *Movement of welding torch* – Straight along seam, with side to side movement reduced to a minimum
- *Gas consumption of the welding torch* – 86 l/h (litres per hour) and 3 ft³/h (feet cubed per hour) for 1.6 mm thickness
- *Type of flame* – Neutral
- *Position of filler rod* – Precedes the welding torch
- *Angle of filler rod* – 30°–40°
- *Movement of filler rod* – Slight movement along seam in and out of the molten pool

- *Size of filler rod* – Approximately equal to the plate thickness up to 2.4 mm, and then 3.2 mm for thickness above 2.4 mm
- *Plate preparation* – Up to 1.00 mm: edges flanged. From 1.00 mm to 3.2 mm: square edge preparation. From 3.2 mm to 4.8 mm: 80°V preparation

Rightward and all-position rightward welding

Rightward welding is recommended for steel plates more than 4.8 mm thick. Plate edges from 4.8 to 7.9 mm need not be bevelled. Plate more than 7.9 mm should be bevelled to 30° to give an included angle of 60° for the welding V. The weld is started at the left-hand end of the joint and the welding torch is moved towards the right, the welding torch preceding the filler rod in the direction of travel. The wire is given a circular forward action, and the welding torch is moved steadily along the weld seam. It is quicker than leftward welding and consumes less gas. The V-angle is smaller, less welding rod is required and there is less distortion. The advantages of the rightward technique are as follows: the direction of the flame holds back the molten pool, preventing any tendency to adhesion; the flame, which points directly towards the root of the weld, causes a hole to form as the edges melt, thus ensuring complete penetration; no preparation is necessary on thicknesses up to 7.9 mm.

The characteristics of the rightward technique are as follows:

- *Position* – Flat
- *Direction of welding* – Flame points towards finished weld
- *Angle of welding torch* – 40°–50°
- *Movement of welding torch* – Straight along seam, with sufficient side swing to ensure complete fusion of the edges
- *Gas consumption of the welding torch* – 140 l/h which is 5 ft³/h for 2.6 mm thickness
- *Type of flame* – Neutral

- *Position of filler rod* – Follows the welding torch
- *Angle of filler rod* – 30°–40°
- *Movement of filler rod* – Circular forward action
- *Size of filler rod* – Half the plate thickness up to a maximum of 6.4 mm
- *Plate preparation* – From 4.8 mm to 7.9 mm: square edge, with gap of half thickness. From 7.9 mm to 15.9 mm: 60° bevel, gap of 3.2 mm, and weld made in one pass. From 15.9 mm upwards: as above but step weld, using **multi-pass** technique

In the rightward and leftward techniques it is seen that edge preparation becomes necessary at 7.9 mm and 3.2 mm, respectively, in order to obtain complete fusion without fear of adhesion. This involves extra cost and additional filler material. If therefore the bevel edge preparation can be obviated, the cost of the welding is reduced.

Edge preparation and types of joint

In order to produce a satisfactory butt weld it is essential for the plate edges to be joined throughout their entire thickness; this necessitates complete melting of the edges. With thin materials this result can be achieved with square-edged plates. However, the same procedure applied to thicker metal is unlikely to result in complete fusion, and the fusion faces must be bevelled to enable the torch flame to be directed into the root of the weld in order to obtain complete fusion.

Square edge preparation is used for material up to 3.2 mm thickness. The edges are left square and separated by a distance equal to one-half the thickness of the sheet used. Such joins are termed **open butt joint**.

Single V preparation consists of bevelling the edges of each plate so that a V is formed when they are brought together. For thicknesses between 3.2 mm and 4.8 mm an 80° bevel is used, leaving a small gap at the bottom edges. For material more than 7.9 mm thickness the angle of V should be

60°. It is not necessary to bevel materials up to 7.9 mm thick if the rightward method of welding is used.

Double V preparation is used for material thicker than 15.9 mm, which must be welded from both sides of the plate; for this reason a V must be provided on each side. The top V should be 60° and the bottom V 80° and the edges separated by a gap of 3.2 to 4.0 mm.

Control of distortion

A proper understanding of the problems of distortion or buckling associated with the welding of sheet metal is of the utmost importance to the panel beater working on thin sheet metal. This distortion results from the shrinkage which occurs when molten iron passes from the liquid to the solid state. In consequence the welded seam tends to shorten in length, causing the parent metal to buckle. If the welded seam is so placed that it may be planished then the hammering along the weld will correct the distortion. As iron shrinks approximately 10 mm per metre when solidifying from the molten condition, the tendency to produce distortion is considerable. Where the work to be welded is of thick steel plate, the plate itself may be sufficiently strong to counteract or minimize some of the distortion, but where the work is of thin mild steel the distorting stresses take full effect in buckling the sheet. In the process of welding it is necessary to consider and make allowances for this distortion which, unless controlled, may buckle the work to such an extent that it is useless. When the joint edges are heated they expand, and as welding proceeds, contraction of the deposited weld metal takes place owing to the loss of heat by radiation and condition. The rate at which cooling takes place depends on various factors such as the size of the work, the amount of weld metal and the speed at which it is deposited, the thermal conductivity of the parent metal and the melting point and specific heat of the weld metal.

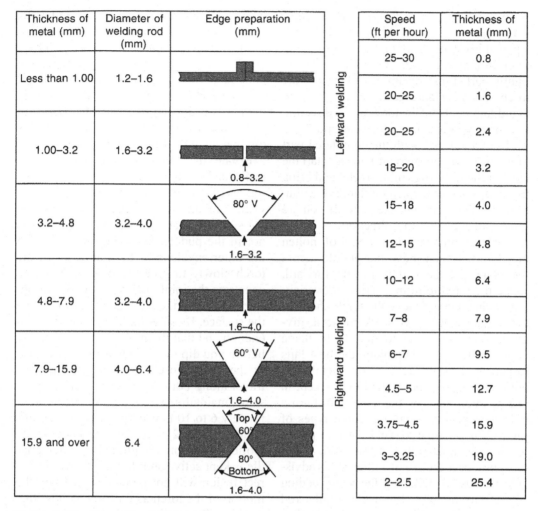

Thickness of metal (mm)	Diameter of welding rod (mm)	Edge preparation (mm)		Speed (ft per hour)	Thickness of metal (mm)
				25–30	0.8
Less than 1.00	1.2–1.6		Leftward welding	20–25	1.6
				20–25	2.4
1.00–3.2	1.6–3.2	0.8–3.2		18–20	3.2
		80° V		15–18	4.0
3.2–4.8	3.2–4.0	1.6–3.2		12–15	4.8
				10–12	6.4
4.8–7.9	3.2–4.0	1.6–4.0		7–8	7.9
		60° V	Rightward welding	6–7	9.5
7.9–15.9	4.0–6.4	1.6–4.0		4.5–5	12.7
		Top V 60°		3.75–4.5	15.9
15.9 and over	6.4	80° Bottom		3–3.25	19.0
		1.6–4.0		2–2.5	25.4

Figure 9.17   Edge preparation

Some methods of controlling distortion are as follows:

1 Efficient tacking or clamping, which maintains the edge positions of the plates
2 Tapered spacing of plates so that they pull together the process of welding
3 Use of intermittent weld technique and backstep weld technique
4 The offsetting and pre-setting of plates so that they are pulled correct by the contraction of the welds
5 The use of chilling bars and chemical foam barriers

6 The use of planishing when the weld is in the cold state
7 The use of jigs and fixtures.

Welding technique: butt joint in mild steel

To master the skill of welding with an oxy-acetylene torch, you will have to practise a series of operations in a definite order. The torch may be held in either one of two ways, depending on which is the more comfortable for you. When welding light-gauge metal, most operators prefer to grasp the handle of the welding torch with the hose over the outside of the wrist, which is the way in which a pencil is usually held. In the other grip, the torch is held like a hammer, with the fingers lightly curled underneath. In either case the torch should balance easily in the hand to avoid fatigue. Hold the torch in the direction in which you are going to weld and at an angle of about 65° with the completed part of the weld. If you are right-handed, start the weld at the right edge of the metal and bring the inner cone of the neutral flame to within 3 mm of the surface of the plate. If you are left-handed, reverse this direction.

Hold the torch still until a pool of molten metal forms, then move the puddle across the plate. As the puddle travels forward, rotate the torch to form a series of overlapping ovals. Do not move the torch ahead of the puddle, but slowly work forward, giving the metal a chance to melt. If the flame is moved forward too rapidly, the heat fails to penetrate far enough and the metal does not melt properly, but if the torch is kept in one position too long the flame will burn a hole through the metal. On some types of joints it is possible to weld the two pieces of metal without adding a filler rod, but in most instances the use of a filler rod is advisable because it builds up the weld, adding strength to the joint. The use of a filler rod requires coordination of the two hands. One hand must manipulate the torch to carry a puddle across the plate, while the other hand must add the correct amount of filler rod. Hold the rod at approximately half the angle of the torch, but slant it away from the torch.

It is advisable to bend the end of the rod at right angles, since this permits holding the rod so that it is not in a direct line with the heat of the flame.

Melt a small pool of the base metal and then insert the tip of the rod in this pool. Remember that the correct-diameter rod is an important factor in securing perfect fusion. If the rod is too large the heat of the pool will be insufficient to melt the rod, and if the rod is too small the heat cannot be absorbed by the rod, with the result that a

hole is burned in the plate. As the rod melts in the pool, advance the torch forward. Concentrate the flame on the base metal and not on the rod. Do not hold the rod above the pool, as if you do the molten metal will fall into the puddle, combining with oxygen in the air as it falls so that part of it burns up and will cause a weak, porous weld. Always dip the rod in the centre of the pool.

Rotate the torch to form overlapping ovals and keep raising and lowering the rod as the molten puddle is moved forward. An alternative torch movement is the semi-circular motion. When the rod is not in the puddle, keep the tip just inside the outer envelope of the flame. Work the torch slowly to give the heat a chance to penetrate the joint, and add sufficient filler rod to build up the weld about 2 mm above the surface. Be sure that the puddle is large enough and that the metal is flowing freely before you dip in the rod. Watch the course of the flame closely to make sure that its travel along both edges of the plate is the same, maintaining a molten puddle approximately 6 to 10 mm in width. Advance this puddle about 2 mm with each complete motion of the torch. Unless the molten puddle is kept active and flowing forward, correct fusion will not be achieved. Keep the motion of the torch as uniform as possible, as this will produce smooth, even ripples and so complete the weld.

Welding various metals

- *Mild steel* – Select the appropriate method and always use a neutral flame. Fluxes are not necessary.
- *Carbon steel* – A flux must be used and the flame kept in a neutral condition. After welding is completed the metal should be cooled slowly to avoid the weld metal becoming brittle.
- *Alloy steels* – It is very important that the correct welding rod is used for the appropriate alloy. The metal should be first preheated and then welded. It is essential to cool the metal slowly after welding.

- *Stainless steel* – A welding rod having the same composition as stainless steel should be used. A flux should be used and the welding flame kept in the neutral position. Do not interrupt the welding sequence, and carry out the weld as quickly as possible. On completion of the weld allow it to cool slowly. Remove all oxide and scale from the weld when it is finished.
- *Cast iron* – A silicon cast iron welding rod should be used, together with a cast iron welding flux. The metal must be preheated to dull red before the welding commences. It is very important that the cast iron cools very slowly, or the metal will crack.
- *Aluminium* – For pure aluminium sheet, either paint both sides of the metal with flux or dip the hot rod into the flux and allow the flux to coat the rod like varnish. Before welding, remove all traces of oil or grease and brush the edges to be welded with a wire brush. Tack the weld at frequent intervals using a neutral or slightly carburizing flame. The welding technique is leftward and should be carried out more quickly than when welding mild steel. On completion of the weld, wash it thoroughly to remove the remaining flux, as this could be harmful.
- *Aluminium castings* – Use cast aluminium alloy rods with silicon and flux. Preheat before welding. Melt the rod well into the weld, using the welding rod to puddle the molten metal. On completion cool slowly.

Gas cutting

The oxy-acetylene process is widely used to cut metal, especially in the body-building industry, where heavy sections have to be cut to special shapes for construction, and also in the repair side of the industry, where special nozzles have been developed for cutting away damaged parts of sheet metal body sections. Cutting may be done by means of a simple hand cutting torch or by a more complicated, automatically controlled cutting machine.

Flame cutting

Flame cutting, known also as oxygen cutting, is made possible by the fact that oxygen has a marked affinity for ferrous metals which have been previously heated to their ignition temperature.

Thus the cutting of iron and steel merely involves the direction of a closely regulated jet or stream of pure oxygen onto an area that has been previously heated to ignition temperature (1600°C, or when the metal has reached a bright cherry-red colour). As the iron is oxidized, the oxygen jet is moved at a uniform speed so that a narrow cut is formed. Since only the metal within the direct path of the oxygen jet is acted upon, very accurate results can be obtained if close control is exercised; when hand cutting 150-mm-thick steel, the working tolerance is about 1.5 mm.

Cutting torch

This differs from the regular welding torch in that it has an additional lever for the control of the oxygen used to burn the metal. It is possible to convert a welding torch into a cutting torch by replacing the mixing head with a cutting attachment. The torch has conventional oxygen and acetylene valves, and these are used to control the passage

Figure 9.18 Flame cutting process (*BOC Ltd*)

of oxygen and acetylene when heating the metal. The cutting tip has an orifice in the centre for the oxygen flow, surrounded by several smaller holes for a preheating flame which generally uses acetylene, propane or hydrogen.

The preheating flame has two purposes.

1 **To provide sufficient heat to raise a small area of the steel surface to the ignition temperature**
2 **To transmit sufficient heat to the top surface of the steel to offset the thermal conductivity of the metal.**

A wide variety of cutting torches is available, but in essentials they are all similar. Oxygen and a fuel gas for the preheat flame, for example acetylene, enter the torch separately and are mixed in the body of the torch or in the nozzle. Their respective flow rates are adjusted by two hand valves on the torch. The cutting oxygen stream is bled off inside the torch through a lever-operated valve

Figure 9.19 Cutting torches (*Murex Welding Products Limited*)

Preheat oxygen valve

Cutting oxygen lever

Fuel gas valve

Figure 9.20 Types of cutting torches (*BOC Ltd*)

(Figure 9.22). Two basic methods of gas mixing are employed.

In the first type, the two gases enter the torch at approximately the same pressure and are mixed in a separately designed chamber either in the body of the torch or in the nozzle. This is termed the equal-pressure or high-pressure cutting torch. Alternatively, oxygen enters the torch at a very much higher pressure than the fuel gas and sucks the fuel gas in through an injector which also mixes the two gases. These are termed injector cutting torches. Both designs are equally good for flame cutting.

Nozzles

A number of different designs of nozzles are available to suit the various combinations of fuel gases and torch design. In principle, however, nozzles are the same: they have a central orifice for the cutting oxygen stream, surrounded by a ring of orifices for the preheat flame.

Cutting torches in which oxygen and the fuel gas are mixed in the body of the torch (injector type) have two tubes leading to the nozzle, one for cutting oxygen and one for mixing oxygen and fuel gas. A so-called two-seat nozzle fits this design. Some torches are designed for mixing oxygen and fuel gas in the nozzle itself. These torches have three tubes leading to the nozzle and require special three-seat nozzles.

A typical oxy-acetylene nozzle has circular pre-heat holes surrounding a central circular cutting oxygen orifice and is frequently of a one-piece construction, made of copper and often chrome plated. On the other hand a liquid propane gas (LPG) or natural gas nozzle is mainly of two-piece construction: it has fluted ports for the preheat flame, and the central part of the nozzle is recessed. Nozzles should not be interchanged between different fuel gases.

The nozzle is one of the main keys to good-quality, efficient cutting. There are different sizes of nozzle for different metal thicknesses, and the nozzle manufacturers will state the correct size on their data sheets.

Figure 9.21   (a) Injector cutting torch and two-seat nozzle (b) three-seat nozzle (*BOC Ltd*)

One-piece acetylene
nozzle
(a)

Two-piece LPG
nozzle
(b)

Figure 9.22   (a) One-piece acetylene nozzle (b) two-piece LPG nozzle (*BOC Ltd*)

Hand cutting procedure

First remove any oxide or scale from the line of the cut. Set the preheating flame to neutral and hold the torch with two hands, one to act as a steady and the other to control the oxygen flow, and position the cutting torch so that the white cone is 6 mm from the work surface. When the metal reaches a bright red colour, switch on the oxygen cutting supply, and move the cutter steadily and at a speed which produces a smooth cut. Keep the white cone just clear of the work surface, and ensure that the cut is penetrating the surface. Whenever possible, the operator should draw the cutter towards him.

Machine cutting

Much oxygen cutting is done with machines, particularly if the cuts are long. Machine cutting has many advantages over hand cutting and results in greater accuracy and better edge finish, particularly when used for making single- and double-V-edge preparation. The principles of machine cutting are similar to those for hand work. A wide variety of types of machines are available, including stationary, general-purpose or universal models, multi-burner machines, straight-line and circle-cutting and joint-cutting machines. Some incorporate pantograph or electronic devices, enabling profiles to be accurately copied from templates or direct from drawings.

Gases: characteristics and colour coding

Safety note
You must remember that all gases are dangerous if not handled and stored correctly. Always follow the manufacturer's instructions.

The following is a summary of gas characteristics and cylinder colour codes.

Oxygen

Cylinder colour: black

Characteristics: no smell. Generally considered non-toxic at atmospheric pressure. Will
not burn but supports and accelerates combustion. Materials not normally considered
combustible may be ignited by sparks in
oxygen-rich atmospheres.

Nitrogen

Cylinder colour: grey with black shoulder.
Characteristics: no smell. Does not burn.
Inert, so will cause asphyxiation in high
concentrations.

Argon

Cylinder colour: blue

Characteristics: no smell. Heavier than air.
Does not burn. Inert. Will cause asphyxiation
in absence of sufficient oxygen to support life.
Will readily collect in the bottom of a confined
area. At high concentrations, almost instant
unconsciousness may occur followed by
death. The prime danger is that there will be no
warning signs before unconsciousness occurs.

Propane

Cylinder colour: bright red and bearing the
words 'Propane' and 'Highly flammable'

Maintain correct distance

Keep at right angles

90°

Remove rust and scale
before commencing

Adjust pressures correctly
at regulators

Make sure edge of plate is at
correct temperature

Figure 9.23   Cutting technique (*BOC Ltd*)

Characteristics: distinctive fish-like offen-
sive smell. Will ignite and burn instantly
from a spark or piece of hot metal. It is
heavier than air and will collect in ducts,
drains or confined areas. Fire and explosion
hazard.

Acetylene

Cylinder colour: maroon

Characteristics: distinctive garlic smell. Fire
and explosion hazard. Will ignite and burn
instantly from a spark or piece of hot metal.
It is lighter than air and less likely than propane to collect in confined areas. Requires
minimum energy to ignite in air or oxygen.
Never use copper or alloys containing more
than 70 per cent copper or 43 per cent silver
with acetylene.

Hydrogen

Cylinder colour: bright red

Characteristics: no smell. Non-toxic. Much
lighter than air. Will collect at the highest
point in any enclosed space unless ventilated
there. Fire and explosion hazard. Very low
ignition energy.

Carbon dioxide

Cylinder colour: black, or black with two
vertical white lines for liquid withdrawal

Characteristics: no smell but can cause the
nose to sting. Harmful. Will cause asphyxiation. Much heavier than air. Will collect in
confined areas.

Argoshield

Cylinder colour: blue with green central
band and green shoulder

Characteristics: no smell. Heavier than air.
Does not burn. Will cause asphyxiation in
absence of sufficient oxygen to support life.
Will readily collect at the bottom of confined
areas.

Safety measures

General gas storage procedures

1   Any person in charge of storage of compressed gas cylinders should know the regulations covering highly flammable liquids and compressed gas cylinders as well as the characteristics and hazards associated with individual gases.
2   It is best to store full or empty compressed gas cylinders in the open, in a securely fenced compound, but with some weather protection.
3   Within the storage area, oxygen should be stored at least 3 m from fuel gas supply.
4   Full cylinders should be stored separately from the empties, and cylinders of different gases, whether full or empty, should be segregated from each other.
5   Other products must not be stored in a gas store, particularly oils or corrosive liquids.
6   It is best to store all cylinders upright, taking steps, particularly with round-bottomed cylinders, to see that they are secured to prevent them from falling. Acetylene and propane must *never* be stacked horizontally in storage or in use.
7   Storage arrangements should ensure adequate rotation of stock.

Acetylene cylinders

1   The gas is stored together with a solvent (acetone) in maroon-painted cylinders, at a pressure of 17.7 bar maximum at 15°C. The cylinder valve outlet is screwed left-handed.
2   The hourly rate of withdrawal from the cylinder must not exceed 20 per cent of its content.
3   Pressure gauges should be calibrated up to 40.0 bar.
4   As the gas is highly flammable, all joints must be checked for leaks using soapy water.
5   Acetylene cylinders must be stored and used in an upright position and protected from excessive heat and coldness.

6   Acetylene can form explosive compounds in contact with certain metals and alloys, especially those of copper and silver. Joint fittings made of copper should not be used under any circumstances.
7   The colour of cylinders, valve threads or markings must not be altered or tampered with in any way.

Oxygen cylinders

1   This gas is stored in black-painted cylinders at a pressure of 200/230 bar maximum at 15°C.
2   Never under any circumstances allow oil or greases to come into contact with oxygen fittings, because spontaneous ignition may take place.
3   Oxygen must not be used in place of compressed air.
4   Oxygen escaping from a leaking hose will form an explosive mixture with oil or grease.
5   Do not allow cylinders to come into contact with electricity.
6   Do not use cylinders as rollers or supports.
7   Cylinders must not be handled roughly or knocked or allowed to fall to the ground.

Safe usage and handling of gas cylinders

Leaks

As a matter of routine, check for leaks. Test with a solution of 1 per cent Teepol in water; apply with a brush. Never use a naked flame to trace leaks. In hoses, leakages, cuts or local surface damage may be repaired by cutting out faulty sections and inserting an approved coupling. Worn ends should be cut back and refitted with the appropriate hose and connections and clips. Discard hoses that show signs of general deterioration.

If an acetylene valve shows a minor leak and it cannot be stopped by closing the valve or tightening the gland nut, eliminate all sources of ignition, move the cylinder outside to a safe area and contact your supplier.

If a cylinder is leaking and on fire, and you suspect the valve is damaged, do *not* attempt to extinguish the fire yourself. Call the fire brigade and evacuate the area. Use a fire extinguisher to put out any fire caused.

Fire: action to be taken

Gas cylinders involved in a fire may explode. Therefore the key actions to be taken are as follows:

1 Evacuate the surrounding area to a minimum distance of 100 metres.
2 Call the fire brigade immediately.
3 Advise persons between 100 and 300 metres from the cylinder to take cover.
4 Any attempt to fight the fire should be done from a protected position using copious quantities of water.
5 When the fire brigade arrives, inform them of the location, the number of gas cylinders directly involved in the fire and the names of the gases they contain.
6 Cylinders which are not directly involved in the fire and which have not become heated should be moved as quickly as possible to a safe place, provided this can be done without undue risk to personnel. Make sure the valves of these cylinders are fully closed.
7 As soon as possible, inform your gas supplier of the incident.
8 Be aware of the fact that, even after the fire has been extinguished, some cylinders which have been heated can **explode**, particularly acetylene cylinders. Therefore do not approach any cylinder until the key actions given in what follows have been taken.
9 When the cylinder content are unknown, treat as acetylene cylinders.

Acetylene cylinders in a fire: key actions

1 Never approach or attempt to move cylinders.
2 From a safe position, drench the entire surface of all cylinders with water for at least one hour after the fire has been extinguished. Do not use a jet of water of such strength that it would knock over a free-standing cylinder.
3 Check visually from a safe position. If steam is seen to be coming from the surface of the cylinder when water spray is interrupted, continue spraying with water. Then check at hourly intervals until it is seen that steaming has ceased.
4 Once steaming has stopped, observe from a safe distance whether the surface of the cylinder remains wetted. If patches dry quickly, continue to cool with water and observe again after half an hour. Repeat this operation until all surfaces remain wetted after water spray is stopped. Pay particular attention to the centre cylinders of manifold cylinder pallets (MCPs) and to any cylinder where some difficulty has been experienced in maintaining a good supply of cooling water.
5 Once all cylinder surfaces remain wetted after the water spray is discontinued, check using the bare hand that the cylinder remains cold for 30 minutes. Wait a further 30 minutes and check again: if any part of the cylinder feels warm to the touch, reapply the cooling water for 30 minutes and repeat the procedure until cylinders remain cold for one hour.
6 When you are satisfied that the entire surface has remained cold for one hour, submerge the cylinder in water carefully, avoiding shocks and bumps. Normally after 12 hours' immersion the cylinder will be safe to be disposed of by the gas supplier.

Ventilation and oxygen-enriched areas – Whenever cylinder gases are used, constant and thorough ventilation should be maintained. This is particularly important when they are used in confined spaces.

The normal content of air is 21 per cent. If this becomes enriched to 25 per cent there is an increase in the speed at which materials will burn. At 30 per cent the typical characteristics of an oxygen-fed fire are apparent. The fire is in two phases: an initial flash fire, followed by local burning at a

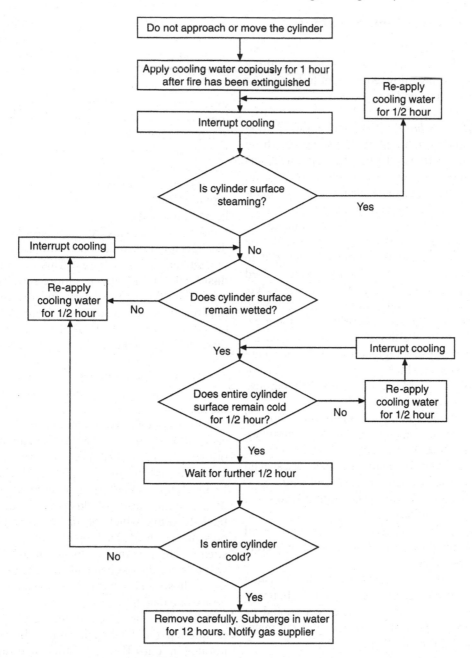

Figure 9.24   Acetylene cylinders in fires (*BOC Ltd*)

number of points. Fires in oxygen-enriched atmospheres are very difficult to extinguish and can spread rapidly across combustible materials from a single point source such as a spark from a cigarette.

Nitrogen, argon and carbon dioxide, if allowed to replace the oxygen in the atmosphere, can cause asphyxiation. The dangers occur typically when gas is released in a confined space. It is especially important to beware of argon and carbon dioxide; both are heavier than air and will displace air in confined spaces or spaces with no ventilation at floor level.

General equipment safety

All equipment should be subjected to regular periodic examination and overhaul. Failure to do so may allow equipment to be used in a faulty state and may be dangerous.

- *Rubber hose* – Use only hose in good condition, fitted with the special hose connections attached by permanent ferrules. Do not expose the hose to heat, traffic, slag, sparks from welding operations or oil or grease. Renew the hose as soon as it shows any sign of damage.
- *Pressure regulators* – Always treat a regulator carefully. Do not expose it to knocks, jars or sudden pressure caused by rapid opening of the cylinder valve. When shutting down, release the pressure on the control spring after the pressure in the hoses has been released. Never use a regulator on any gas except that for which it was designed, or for higher working pressures. Do not use regulators with broken gauges.
- *Welding torch* – When lighting up and extinguishing the welding torch, the manufacturer's instructions should always be followed. To clean the nozzle, use special nozzle cleaners, never a steel wire.
- *Fluxes* – Always use welding fluxes in a well-ventilated area.
- *Goggles* – These should be worn at all times during welding, cutting or merely observing.
- *Protection* – Leather or fire-resistant clothing should be worn for all heavy welding or cutting. The feet should be protected from sparks, slag or cut material falling on them.

Backfiring and flashbacks of welding torch

A welding torch is said to **backfire** when it goes out with a loud pop and then relights itself immediately, providing the heat of the job is sufficient to ignite the acetylene. Backfires result from defective equipment, incorrect pressures or incorrect lighting up or from careless use of the welding torch, permitting the nozzle to touch the work, overheating the nozzle tip, or working with a loose nozzle. Usually the backfire is arrested at the mixer or injector of the welding torch. If prompt action is taken, turning off first the oxygen then the acetylene valve, no damage occurs, and the welding torch may be relit provided that the cause of the trouble has been eliminated.

Sometimes a backfire may pass beyond the injector and travel back into either the oxygen or fuel gas hoses. This is termed **flashback**, and its effect is more serious, as immediate damage to hoses and regulators may result; there is also a risk of injury to the operator. A flashback should be suspected if there is a squealing or hissing noise, sparks coming from the nozzle, heavy black smoke or if the welding torch gets hot. If the flame burns back far enough, it may burst through the hose.

With oxy-fuel gas equipment, flashbacks can and do occur because the recommended pressures and procedures have not been observed and because of nozzle blockage, faulty equipment or leaking equipment. One of the main causes of flashback is due to **back-feeding**, which occurs when higher-pressure gas feeds back up a lower-pressure stream. The presence of hose check valves will prevent the oxygen and fuel gas mixing in the hose and subsequently causing fire, injury and damage.

A flashback arrester is a device designed to quench the flashback. When it incorporates a cut-off valve, this will automatically shut off the gas flow. These multifunction devices afford an additional safeguard, particularly in locations where a fire following flashback could not be tolerated, such as garages, bodyshops and other workplaces with flammable or hazardous materials.

Plasma arc cutting

Plasma cutting process and equipment

The plasma cutting process relies on the fact that if a gas or mixture of gases, such as air, is subjected to a very high temperature it becomes ionized: negative electrons are separated from the atom, which is then positively charged. This ionized state of the gas is called 'plasma', and in this state the gas is electrically conductive.

The high temperature necessary to create the plasma is achieved, in the case of plasma cutting, by a standing electric arc. This is constricted by forcing the plasma through a small nozzle which increases the temperature of the arc to more than 24,000°C and concentrates it on a very small area. When this plasma is directed at a conductive material, the arc is transferred through the plasma (transferred arc operation) to the material. The high energy of the arc melts the material, which, as long as it is within the cutting range of the equipment, will be displaced by the gas flow.

In order to initiate the standing arc it is necessary to produce an ionized path in the gases. This is achieved either by applying a very high voltage at a high frequency between the electrode and the tip and work, causing a high-frequency spark, or by momentarily touching the electrode and nozzle together and then quickly breaking the contact, causing an arc between the electrode and the nozzle. As soon as the gas between the tip and the nozzle is ionized, the main arc will ignite.

Plasma cutting and welding torches are designed with a small orifice which constricts the arc. Gas flows under pressure through the arc and is subsequently heated to an exceptionally high temperature. The heated gas cannot expand owing to the constriction of the nozzle, and it is forced out to form a supersonic jet hotter than any known flame or a conventional electric arc. The jet

Figure 9.25 Simplified diagram of plasma arc cutter (*Motor Insurance Repair Research Centre*)

melts down and partially vaporizes the base material, and the force of the jet blasts away the molten metal. The plasma jet is controlled by adjusting current, gas velocity and type of gas.

The distance between electrode and workpiece is made electrically conductive (ionization) by an auxiliary arc (pilot arc) between electrode and plasma nozzle. As the intensity of the high-voltage impulses of the pulse generator is not sufficient, part of the cutting current (limited by a resistor) is admitted for the sufficient ionization of the pilot arc distance. When the ionized gas jet contacts the workpiece, the main circuit is closed and the cutting process is introduced.

Clean, dry workshop air can be used when severing thin sheet steel, but it is recommended to use an argon/hydrogen mix or nitrogen when cutting gauges in excess of 5 mm, and for stainless steels, as these gases produce a higher thermal conductivity and will transfer more heat to the material to be cut. Reference to the manufacturers' recommendations is essential, as the correct cutting speed must be maintained to ensure the quality of the cut.

The equipment can be grouped as follows:

* *High-output units* (approximately 50 amperes) operating on input voltages of 380/415 V three phase, fitted with water-cooled torches and operating with workshop air or special cutting gases.

Figure 9.27 Plasma cutting unit: single phase, 30 amperes (*SIP (Industrial Products) Ltd*)

Figure 9.26 Plasma cutting unit: three phase, 40 amperes (*Olympus Welding Supplies Ltd*)

• *Medium-output units* (approximately 30 amperes) operating on input voltages of 220/240 V single phase. These cutters are normally available with air-cooled torches using clean workshop or bottled air (see Figures 9.29 and 9.30).

There are two types of electrode available for the plasma cutting torches. The first is a tungsten electrode intended for cutting with nitrogen or an argon/hydrogen mix. The second is an electrode intended for cutting with air.

Setting up the equipment for use

1 Select the electrode type according to the cutting gas used.
2 Check that the gas nozzle and electrode of the torch are not damaged, and set the electrode distance as required for the selected electrode type using the nozzle tool. For the best cutting results, the electrode distance should be regularly checked and adjusted.
3 Before performing any work in the torch, always disconnect the main voltage.

4 Check that the correct pressure (approximately 5 bar or 75 psi) is set on the regulator and that any valves ahead of the regulator are open.
5 Connect the earth lead, ensuring as good a contact as possible to the workpiece (if necessary, clean off any surface coating rust). The welding clamp must not be moved during cutting; the machine should be disconnected.
6 Set the main switch of the machine. The cooling unit and the fan will start, and the indicator 'on' lamp will light.
7 Check for correct gas flow by activating the switch on the torch handle. The gas valve will open, and the gas will start to flow from the nozzle (3.5–4.1 bar). After this gas test, reset the switch to its initial position.
8 Press the trigger button once only. Air should flow from the torch; if it does not, check the warning lamps on the machine. The red lamps should not be illuminated. The left-hand one indicates low air pressure and the right-hand one that thermal cut-outs have operated. If either is lit, then appropriate action should be taken. The green lamp should be illuminated, as it indicates that the machine is on. The green light is switched off when a red light goes on.

Figure 9.28 An air-cooled plasma cutting torch (*SIP (Industrial Products) Ltd*)
1 Handle
2 Electrode
3 Insulator bush
4 Spring
5 Plasma nozzle
6 Protection nozzle

Figure 9.29 Plasma arc cutting torch nozzle, showing position of electrode (*Motor Insurance Repair Research Centre*)

Figure 9.30 Plasma arc cutting machine in use (*SIP (Industrial Products) Ltd*)

9 Whilst the air is flowing, check that the pressure is 3.5 bar (50 psi). If not, adjust the regulator to give 3.5 bar.
10 Set the machine to either maximum or minimum according to the thickness to be cut. *Caution*: do not attempt to cut material beyond the range specified, as this will damage the torch.
11 Position the torch on the workpiece, ensuring that the tip has a good contact. Press the trigger button once and release it. Allow the air to flow for a few seconds. Press the button again and then, within one second, press and hold the button. The arc will strike and a hole will be cut in the metal.

Caution: If the arc does not penetrate the metal then stop, as the machine is either on the wrong setting or not sufficiently powerful for the thickness being cut.

12 Move the torch along so that the metal is cut right through. If the metal is not cut, reduce the speed of travel until it is. The torch head should be held at right angles to the work.
13 When the cut is complete, release the trigger button.

Caution: Do not switch off the machine until the air has stopped flowing; otherwise the torch will be damaged, as the air is required to cool the torch.
14 The operator must be constantly aware of the risk of fire and fumes, and therefore proper health and safety precautions must be adhered to at all times.

Cutting methods

The equipment can be used for contact or distance cutting. When cutting is started, the techniques differ between nitrogen and argon/hydrogen plasma cutting and air plasma cutting.

Contact cutting

Contact cutting is used for materials up to 5 mm in thickness. In this method the torch nozzle is in contact with the workpiece. The torch is tilted at an angle with the work surface to obtain plasma flow and gradually straightened up until it lies flush with the workpiece when cutting.

Distance cutting

This is used for materials more than 5 mm thick. With this method, the nozzle is not in contact with the workpiece, and an even distance must be maintained between them when cutting.

Process advantages

Plasma arc cutting is one of the most effective processes for high-speed cutting of many kinds of ferrous and non-ferrous metals. The quality of the cut is very high, with little or no dross and reduced distortion. Better economy is also a feature of plasma arc cutting as compared with oxy-fuel methods.

With plasma arc equipment, a saving in time and energy can be achieved in certain areas of repair, particularly in the removal of damaged panels and structural members. Plasma arc cutting will prove of benefit in a modern body repair shop as an alternative cutting technique to speed up damaged panel removal and to complement traditional mechanical methods.

Safety

General

1 Do not operate the machine with any of the panels removed.

2 Ensure that the machine is connected to the correct voltage supply, that the correct fuse is used and that the equipment is earthed.
3 Under no circumstances must the plasma nozzle be removed or any other work be carried out on the torch with the machine switched on. Ignoring this precaution could lead to contact with high DC voltage.
4 The torch should be kept free of slag at all times to ensure a free passage of air.
5 The tip and electrode will require changing when they have become pitted and the arc will not strike. The life of these components depends upon the current used and any misuse of the torch.

Fire

1 All inflammable materials must be removed from the area.
2 Have a suitable fire extinguisher available close by at all times.
3 Do not cut containers which have held inflammable materials or gases.

Glare and burns

1 The electric plasma arc should not be observed with the naked eye. Always wear goggles of the type used for oxy-acetylene work.
2 Gloves must be worn to protect the hands from burns.
3 Non-synthetic overalls with buttons at neck and wrist, or similar clothing, should be worn.
4 Greasy overalls should never be worn.
5 Do not wear inadequate footwear.

Compressed air

Compressed air is potentially dangerous. Always refer to the relevant safety standards for compressed gases.

Ventilation, fumes, vapours

Ventilation must be adequate to remove the smoke and fumes during cutting. Toxic gases may be given off when cutting, especially if zinc- or cadmium-coated steels are involved. Cutting should be carried out in a well-ventilated area, and the operator should always be alert to fume build-up. In small or confined areas, fume extractors must be used.

Vapours of chlorinated solvents can form the toxic gas phosgene when exposed to UV radiation from an electric arc. All solvents and degreasers are potential sources of these vapours and must be removed from the area being cut.

Skills and questions

Gas welding and cutting is a skill which takes time to learn, so practise as often as you can.

1 Describe the general precautions which should be taken in the storage and handling of oxygen and acetylene cylinders.
2 State the safety precautions to be taken when assembling oxy-acetylene equipment.
3 State how gas cylinders are visibly identifiable.
4 State the correct pressures for full cylinders of oxygen and acetylene.
5 Compile a list of safety measures which should be applied in the preparation and use of gas welding equipment.
6 What method should be used to find the location of a leak in an acetylene connection?
7 What safety precautions should be taken if an acetylene cylinder was to take fire internally?
8 State the reason why copper or high-copper-content alloy tubing should not be used for an acetylene connection.
9 Why is it dangerous to allow grease or oil to come into contact with the oxygen cylinder valve or fittings?
10 What is meant by the following terms: (a) high-pressure welding (b) low-pressure welding system?
11 What are the essential components of a high-pressure welding system?
12 Describe the reasons for both leftward and rightward welding methods, and state any benefits derived from these methods.

13 What is meant by an oxidizing flame?
14 What is meant by a carburizing flame?
15 Suggest two reasons for a welding torch backfiring.
16 Explain the meaning of the term 'penetration' in a welded joint.
17 Sketch and describe the best condition for a welding torch flame for welding low-carbon steel sheet.
18 When making gas-welded butt joints in sheet steel, which common faults should be avoided?
19 Sketch and name the type of flame suitable for welding each of the following: (a) aluminium (b) brass.
20 Describe the technique which would be most suitable for welding low-carbon steel up to 5 mm.
21 With the aid of a sketch, give a detailed explanation of the working and function of the hose-check valve on oxy-acetylene welding hoses.
22 List the procedure sequence necessary to shut down a gas welding plant high-pressure system.
23 Explain the difference between line pressure and contents pressure as shown on the regulator.
24 Describe the process of gas welding pure aluminium sheet.
25 Explain the precautions to be taken when welding aluminium.
26 State two disadvantages which limit the use of oxy-acetylene welding in vehicle body repair.
27 What principle makes possible the cutting of metal by means of oxy-acetylene?
28 Explain the basic principle of plasma arc cutting.
29 Explain the type of gases used in plasma arc cutting.
30 Describe how the plasma arc process can be used in vehicle body repair.
31 Name the oxy-acetylene flame that would burn equal quantities of the gases.
32 State the purpose of the two pressures that are shown on the gauge of an acetylene gas cylinder regulator.

33 Explain the safety device used on oxy-acetylene welding equipment to stop the risk of the flame travelling back down the torch supply hose.

34 With the aid of a sketch, outline two methods of limiting distortion when welding thin sheet metal.

35 Explain the difference in welding low-carbon steel and aluminium of equal thickness.

36 List the consumables used in plasma arc cutting.

37 State the welding process that is not recommended when welding HSLA steels.

38 What is meant by 'contact cutting' when using plasma arc?

39 What is the electrode made from in a plasma arc cutting torch?

40 What are the two gases that can be used in the flame cutting process?

10

Electric resistive welding

Resistance welding in car body manufacture

Resistance welding is a joining process belonging to the pressure welding sector. With its locally applied heat and pressure it has an obvious relationship with the forge welding technique practised by blacksmiths when joining metal. The resistance welding process was invented in 1877 by Professor E. Thomson of Philadelphia, USA, when an accidental short circuit gave him the idea for what was originally termed short circuit welding. From the beginning of the twentieth century it was used on a small scale in industry, but it was only after the Second World War that resistance spot welding had its real beginning in the automobile industry. It has since grown to be the most important method of welding used in the construction and mass production of vehicle bodies.

Resistance welding is extensively used for the mass production assembly of the all-steel body and its component sheet metal parts. Its wide adoption has been brought about by its technical advantages and the reductions in cost. Most mass-produced car bodies are assembled entirely by welding steel pressings together to produce an integral rigid chassis and body structure. Low-carbon steel thicknesses used in this unitary construction range from 0.8 to 1 mm for skin or floor panels to 3 mm for major structural pressings such as suspension brackets. Intermediate gauges such as 1.2 mm are used for hinge reinforcements, 1.6 mm for chassis structural members and 1.8 mm to 2.5 mm for suspension and steering members. With the introduction of high-strength steels (HSS steels), car manufacturers are producing body panels as thin as 0.55 mm and structural members with gauges of between 1.2 and 2 mm. This reduction in thickness can be made without loss of strength.

There are a number of resistance welding processes. Resistance **spot welding** is the most widely used welding process in car body construction; there are approximately 4500 to 6000 spot welds per body, which accounts for approximately 80 per cent of the welding used. A further 10 per cent is composed of other resistance welding processes: **seam, projection, flash** and **butt welding**. The remaining 10 per cent is divided between **MIG/MAG** welding and gas welding. Of the many welding techniques used in the mass production of car bodies, resistance welding dominates the field.

The fundamental principle upon which all resistance welding is based lies in the fact that the weld is produced by the heat obtained from the resistance to flow of electric current through two or more pieces of metal held together under pressure by electrodes which are made from copper or copper alloys. The engineering definition of **heat** (heat being the essence of all welding) is **energy × time**. This indicates a balance between **energy input** and **weld time**; therefore the faster the welding, the greater the clamping force. However, the engineering definition of resistance is such that the higher the clamping force, the greater the current needed to produce a constant heat. Heat is generated by the resistance of the parts to be joined to the passage of a heavy electrical current. This heat at the junction of the two parts changes the metal to a plastic state. When the correct amount of pressure is then applied fusion takes place.

There is a close similarity in the construction of all resistance welding machines, irrespective of design and cost. The main difference is in the type of jaws or electrodes which hold the object to be welded. A standard resistance welder has four principal elements:

- *Frame* – The main body of the machine, which differs in size and shape for robot, stationary and portable types
- *Electrical circuit* – Consists of a step-down transformer, which reduces the voltage and proportionally increases the amperage to provide the necessary heat at the point of weld
- *Electrodes* – Include the mechanism for making and holding contact at the weld area
- *Timing control* – Represents the switches which regulate the volume of current, length of current time and the contact period. Many now include adaptive in-process control units (pulsing weld timer).

The principal forms of resistance welding are classified as: resistance spot welding, resistance projection welding, resistance seam welding, resistance flash welding and resistance butt welding.

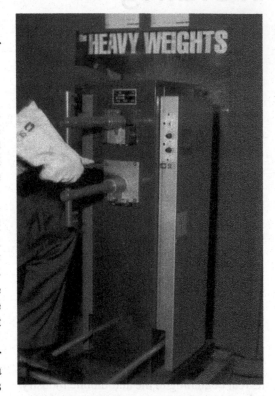

Figure 10.2   Stationary pedestal spot welding machine (*SIP (Industrial Products) Ltd*)

Figure 10.1   Robot spot welder on an assembly line (*Vauxhall Motors Ltd*)

Figure 10.3   Portable spot welding machine in use (*SIP (Industrial Products) Ltd*)

Resistance spot welding

Resistance spot welding is basically confined to making welds of approximately 6 mm diameter between two or more overlapping sheet metal panels. This type of welding is probably the most commonly used type of resistance welding. The art of production planning for spot welding is to simplify the presentation of panels in the region of mutual panel overlap. The limitation of spot welding is that the electrode assemblies have to withstand applied forces ranging from 2200 N to 4450 N for the range of sheet steel thicknesses used in vehicle construction and repair. Product design and planning must account, therefore, for the requirement that the electrodes, which are constructed from relatively weak copper alloys, need normal access to both sides of the overlapping panels to withstand such electrode forces.

The material to be joined is placed between two electrodes, pressure is applied, and the electric current passes from one electrode through the material to the other electrode. There are three stages in producing a spot weld. First, the electrodes are brought together against the metal and pressure is applied before the current is turned on. Next the current is turned on momentarily. This is followed by the third hold time, in which the current is turned off but the pressure continued. The hold time forges the metal while it is cooling.

Regular spot welding usually leaves slight depressions on the metal which are often undesirable on the show side of the finished product. These depressions are minimized by the use of larger-sized electrode tips on the show side. Resistance spot welding can weld dissimilar metal thickness combinations by using a larger electrode contact tip area against the thicker sheet. This can be done for mild steel having a dissimilar thickness ratio of 3:1.

There are three kinds of distortion caused by resistance spot welding which are relevant to car body manufacture and repair. The first is the local electrode indentation due to the electrode sinking into the steel surface. This is a mechanical distortion – a by-product of the spot welding process. Second, there is a small amount of thermal distortion which is troublesome when attempting to make spot welds on show surfaces such as skin panels without any discernible metal distortion. Last, there is the gross distortion caused when badly fitting panels are forced together at local spots. This is a mechanical distortion totally unconnected with the spot welding process; the same type of distortion would occur with rivets or with any similar localized joining method. A combination of all these distortions contributes to the general spot-weld appearance, which is virtually unacceptable on a consumer product.

(a)

(b)

Figure 10.4 (a) Resistance spot welding system (b) relationship between weld formation, current and pressure in welding

The technique on car bodies is to arrange for spot-welded flanges to be covered either with trims (door apertures) or with a sealing weatherstrip (window and screen surrounds). The coach joint is one of the features that distinguish a cheap, mass produced body from an expensive hand built one.

Spot welders are made for both DC and AC. The amount of current used is very important. Too little current produces only a light tack, which gives insufficient penetration. Too much current causes burned welds. Spot welds may be made one at a time, or several welds may be completed at one time, depending on the number of electrodes used. One of the most significant advantages of resistance spot welding is its high welding production rate with the minimum of operator participation. Typical resistance spot-welding rates are 100 spots per minute. To dissipate the heat at the weld as quickly as possible, the electrodes are sometimes water cooled. Although many spot welders are of the stationary design, there is an increased demand for the more manoeuvrable, portable type. The electrodes which conduct the current and apply the pressure are made of low-resistance copper alloy and are usually hollow to facilitate water cooling. These electrodes must be kept clean and shaped correctly to produce good results. Spot welders are used extensively for welding steel, and when equipped with an electronic timer, they can be used for metal such as aluminium, copper and stainless and galvanized steels.

Resistance projection welding

Projection welding involves the joining of parts by a resistance welding process which closely resembles spot welding. This type of welding is widely used in attaching fasteners to structural members. The point where the welding is to be done has one or more projections which have been formed by embossing, stamping or machining. The distortion on the embossed metal face is low and is negligible on heavy metal thicknesses, although in the case of dissimilar thicknesses it is preferable to emboss the projection on the thicker of the two sheets. The projections serve to concentrate the welding heat at these areas and permit fusion without the necessity of employing a large current. The welding process consists of placing the projections in contact with the mating fixtures and aligning them between the electrodes. The electrodes on the machine are not in this case bluntly pointed as in spot welding but are usually relatively large flat surfaces which are brought to bear on the joint, pressing the projections together. The machine can weld either a single projection or a multitude of projections simultaneously. The many variables involved in projection welding, such as thickness, kind of material and number of projections, make it impossible to predetermine the correct current. Not all metals can be projection welded. Brass and copper do not lend themselves to this method because the projections usually collapse under pressure.

Resistance seam welding

Seam welding is like spot welding except that the spots overlap each other, making a continuous weld seam. In this process the

Possible arrangement of projections

Figure 10.5 Projection welding system seam welding

Figure 10.6   Continuous resistance

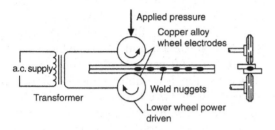

Figure 10.7   Seam welding (stitch welding)

Figure 10.8   Flash welding

metal pieces pass between roller-type electrodes. As the electrodes revolve, the current
supply to each one is automatically turned
on and off at intervals corresponding to the
speed at which the parts are set to move.
With proper control it is possible to obtain
airtight and watertight seams suitable for
such parts as fuel tanks and roof panels.
When spots are not overlapped long enough
to produce a continuous weld, the process is
sometimes referred to as roller spot or stitch
welding (Figure 10.8). In this way the work
travels the distance between the electrodes
required for each succeeding weld cycle. The
work stops during the time required to make
each individual weld and then automatically
moves the proper distance for the next weld
cycle. Intermittent current is usually necessary for most seam welding operations. Each
individual weld contains a number of overlapping spot welds, resulting in a joint which
has good mechanical strength but which is
not pressure tight.

Mild steel, brass, nickel, stainless steel
and many other alloys may be welded by this
method, although its use is largely limited

to operations on mild and stainless steel.
A maximum economy is obtained if the
combined thickness of sheets to be joined
does not exceed 3.2 mm. Before welding,
the surfaces must be clean and free from
scale, and this may be done by sand blasting,
grinding or pickling.

Resistance flash welding

In the flash welding process the two pieces
of metal to be joined are clamped by copper
alloy dies which are shaped to fit each piece
and which conduct the electric current to
the work. The ends of the two metal pieces
are moved together until an arc is established. The flashing action across the gap
melts the metal, and as the two molten ends
are forced together fusion takes place. The
current is cut off immediately this action is
completed.

Flash welding is used to butt or mitre
sheet, bar, rod, tube and extruded sections.
It has unlimited application for both ferrous
and non-ferrous metals. For some operations
the dies are water cooled to dissipate the heat
from the welded area. The most important
factor to be considered in flash welding is the
precision alignment of the parts. The only
problem encountered in flash welding is the
resultant bulge or increased size left at the
point of weld. If the finish area of the weld is
important, then it becomes necessary to grind
or machine the joint to the proper size.

Clamping electrodes

Controlled pressure

Forged pressure is applied
when welding point is reached

Figure 10.9   Resistance butt welding

Resistance butt welding

In butt welding the metals to be welded are
brought into contact under pressure, an electric current is passed through them, and the
edges are softened and fused together. This
process differs from flash welding in that
constant pressure is applied during the heating process, which eliminates flashing. The
heat generated at the point of contact results
entirely from resistance. Although the operation and control of the butt welding process
is almost identical to flash welding, the basic
difference is that it uses less current, has a
constant pressure and allows more time for
the weld to be completed.

Resistance welding
in body repair work

The object of car body repair is to put
damaged vehicles back into a pre-accident
condition. Today's chassisless bodies hold
engine, suspension and steering in the
right places and are designed to absorb
the impact of crashes by crumpling, thus
shielding the passenger compartment (and
its passengers) from shock and deformation. From the viewpoint of safety as well
as mechanical efficiency, proper welding is vital in this kind of repair. Car body
design demands careful choice of the sheet
metal, which was, until recent years, all
mild steel. Tensile strength and ductility,
which are good in mild steel, are vital to

'crumplability' (the ability to absorb the
impact energy), and this is why resistance
spot welds are used. The average body shell
is joined together by approximately 4500 to
6000 such spot welds. These remain ductile because the welding process does not
alter the original specifications of the steel.
Lighter body weight reduces the load on the
car engine and therefore has a direct influence on petrol consumption.

For weight- and fuel-saving reasons, high
strength steels have been introduced for
some panel assemblies on body shells. As
these steels have a different character from
low-carbon steel (mild steel), which still
accounts for 60 per cent of the body shell,
they cause repair welding problems. Higher-
strength steels have been made specifically
for motor car manufacturers to produce
body shells from thinner but stronger steel
sheet. These steels are less ductile and are
harder. Above all, they do not tolerate excess
heat from bad welding, which makes them
brittle or soft or can cause panel distortion.
Low-carbon steel tolerates excess heat well.
While older all-mild-steel bodies essentially
needed only to have the welding machine
set for the metal thickness, current new bodies can contain steel of up to four different
strengths, hardnesses and ductilities, some
coated on one side, some coated on both
sides and some uncoated. Zinc-coated sheet
materials require the use of heavier welding
equipment capable of producing a higher
current to penetrate the zinc coating, and the
electrodes must be constantly maintained
by cleaning to avoid zinc pick-up when
welding.

The repair of bodies incorporating low-
carbon steels and HSLA steels therefore
demands very different welding routines
from those for low-carbon steel alone. Low-
carbon steel bodies can be resistance spot
welded or gas (TIG) welded or arc (MIG)
welded; but higher-strength steels should not
undergo the last two processes, because they
involve nearly three times the heat of resistance spot welding. The temperatures generated are more than 3300°C for gas or arc

welding but only 1350°C for resistance spot welds for joints of similar strength. Higher-strength steels, however, with their higher tensile strength, limited ductility and greater hardness, are particularly vulnerable to heat and are apt to lose strength and change ductility when overheated.

Developments in welding equipment combined with the use of electronic controls have opened the way to new body-repair welding techniques that help to simplify the practical problems posed by bodies made from a mixture of low-carbon and high-strength low-alloy steels. The traditional resistance welding method has had to be improved to join higher-strength steels. Because the weld current flow is hindered by the steel's coating, these may require higher temperatures to break them down before a weld can be formed. For producing consistently good welds it is necessary to use two or three stages for welding, the duration of each stage being adapted to the nature of the steel and its coating.

Resistance spot welding of high-strength steels

The rigidity of the body and its ability to withstand high torsional and other stresses depend on the assembly method used to bring the various body panels together. Spot welding is used throughout the industry for two reasons: first, because it is the strongest and most reliable method of joining two pieces of metal and second, because of the total absence of panel distortion through the welding. In order to effect a satisfactory repair, vehicle welds having the same characteristics as the original are essential.

A comparison between an electrical spot weld and a forged weld shows that in both these processes a union is formed by an amalgamation of the metal molecules. These have been consolidated in each of the two pieces, despite the difference in the two processes employed. In the case of forge welding the pieces are heated in the forge furnace and then hammered until a homogeneous

material results. In the case of the spot-welding process the gun must have a pressure device which can be operated by the user and transmits the pressure to the electrodes and a transformer to enable current at high intensity to be fed to the electrodes.

Spot welding is the primary body-production welding method. Most production welding is carried out on metal gauges of less than 2.5 mm, although greater thicknesses can be spot welded. With the introduction of high-strength steels, car manufacturers are producing body panels as thin as 0.55 mm and using gauges of between 1.2 and 1.5 mm for structural members. By using the correct equipment, two-sided spot welding can be successfully accomplished on these steels.

Weld quality

Careful control of three factors is necessary to make a good-quality spot weld:

* Squeeze time
* Weld time (duration of weld flow)
* Hold time.

The main disadvantage associated with welding in general is that any visual inspection of the weld gives no indication of the weld quality. It is therefore essential that spot-welding equipment used in motor body repair takes outside circumstances automatically into account, such as rust, scale, voltage drop in the electric main supply and current fluctuations.

The quality of a spot weld depends on:

1 **The strength of the weld must be equal to the parent metal. This is a problem of tensile strength and homogeneity of the weld nugget itself.**

2 **The welding heat must not in any way alter the inherent qualities of the parent metal.**

3 **The welding pressure must prevent parent metal separation so as not to cause panel distortion and stresses in the assembly.**

Figure 10.10   Body panels made from high-strength/galvanized steels (*Motor Insurance Repair Research Centre*)

		Material			Material
1	Hood panel, outer	SGAC35R	20	Front pillar, outer, upper	SPRC35
2	Hood panel, inner	SGACC	21	Front Pillar, inner, upper	SPRC35
3	Upper frame, outer	SGACC	22	Centre pillar, inner	SPRC35
4	Upper frame, inner	SGACC	23	Centre pillar, outer	SPRC35
5	Front side member	SPRC35	24	Front floor, side sill, outer	SGACC
6	Front end gusset	SPRC35	25	Front floor, side sill, inner	SGACC
7	Front fender	SENCE	26	Side sill, outer rear extension	SGACC
8	Front skirt panel	SGACC	27	Quarter panel, inner	SENCE
9	Grille filler panel	SGACC	28	Rear pillar, inner	SPRC35
10	Front end cross member	SPRC35	29	Rear shelf panel	SPRC35
11	Side sill, inner front	SGACC	30	Trunk lid inner panel	SGACE
12	Front upper frame extension	SGACC	31	Trunk lid outer panel	SGAC35R
13	Cowl top outer panel	SGACC	32	Quarter panel, outer	SENCE
14	Front door outer panel	SGACC	33	Rear skirt panel	SGACC
15	Front door inner panel	SGACE	34	Rear floor cross member	SGACC
16	Rear door outer panel	SGACC	35	Rear floor side member	SGACC
17	Rear door inner panel	SGACE	36	Rear floor side sill	SGACC
18	Roof panel	SGACC	37	Roof reinforcement	SGAHC
19	Front pillar, outer, lower	SPRC35			

Legend:
SPRC is phosphorus added
SGACC, E and SGAHC are galvanized steel plate.
SGAC35R: phosphorus added (also galvanized)
SENCE: SPCE with electrogalvanized zinc-nickel coating
The numbers in the material codes indicate the tensile strength (kg/mm^2).

Heat application

With resistance welding, the welding time is controlled by either electronic or electro (electrical) devices. Hand-held pincer-type weld guns, as supplied to the car body repair trade, are usually controlled electronically with the minimum of pre-setting. Weld time can range from a fraction of a second for very thin gauges of sheet steel to one second for thicker sheet steel. Weld time is an important factor, because the strength of a weld nugget depends upon the correct depth of fusion. Sufficient weld time is required to produce this depth of fusion by allowing the current time to develop enough heat to bring a small volume of metal to the correct temperature to ensure proper fusion of the two metals. If the temperature reached is too high, metal will be forced from the weld zone and may induce cavities and weld cracks. In some grades of high-strength steels, cracking within the vicinity of the heat-affected zone has appeared after the welding operation.

Heat balance

In the resistance welding of panels of the same thickness and composition, the heat produced will be evenly balanced in both panels, and a characteristic oval cross-sectional weld nugget will result. However, in the welding of panels produced from steels of a dissimilar composition, that is conventional low-carbon steel to high-strength low-alloy steels, an unbalanced heat

rate will occur. This problem will also arise when spot welding two different thicknesses of steel; more heat will usually be lost by the thicker gauges, resulting in unsatisfactory welds. To compensate, the welder must either select an electrode made from materials that will alter the thermal resistance factor, vary the geometry of the electrode tip or use pulse equipment.

Electrodes

The functions of the electrode (chromium/copper) are to:

1 Conduct the current to the weld zone
2 Produce the necessary clamping force for the weldment
3 Aid heat dissipation from the weld zone.

The profile and diameter of the electrode face is dependent upon the material to be welded (Figure 10.14). The diameter of the face will directly influence the size of the nugget. The proper maintenance of the electrode tip face is therefore vital to ensure that effective current flow is produced. For example, if a tip diameter of 5 mm is allowed through wear to increase to 8 mm, the contact area is virtually doubled; this will result in low current density and weak welds. Misalignment of the electrodes and incorrect pressure will also produce defective welds. Electrode tip dressing is best carried out with a fine-grade emery cloth. Coated steels will often cause

Figure 10.11 Welding together panels of the same thickness (*Motor Insurance Repair Research Centre*)

Figure 10.12 Types of electrode profile (*Motor Insurance Repair Research Centre*)

Table 10.1 Recommended tip pressures according to the gauge of steel

Gauge of panel (mm)	Pressure (kg)	Length of arm (mm)
0.55 + 0.55	50	250
0.80 + 0.30	60	250
0.95 + 0.95	70	250
1.00 + 1.00	80	250
1.25 + 1.25	100	250
1.25 + 2.00	120	250

These pressures relate to HS steels: a reduction in pressure may be necessary for conventional steels.

particles of the coating to become embedded in the surface of the electrode tip face (pick-up). This necessitates frequent tip dressing; when the faces have become distorted, the profile or contour must be reshaped. Cutters are available to suit most profiles and tip diameters. Domed tips are recommended; they adapt themselves best to panel shapes and last longest. The adjustment of electrode clamping force is critical for any type of steel, particularly when long-arm arrangements are used and complex configurations of arms are used for welding around wheel arches. It should also be noted that the tip force is difficult to maintain with long arms fitted to hand-held pincer-type guns.

Factors affecting weld quality

The following factors can have a large influence on the final outcome of the spot weld:

1　The type of metal being welded (for example, coated metals)
2　Joint configuration
3　Current and timing settings
4　The type of welding equipment being used
5　Electrode maintenance.

Clearance between welding surfaces
Any clearance between the surfaces to be welded causes poor current flow. Even if welding can be made without removing

such gaps, the welded area would become smaller, resulting in insufficient strength. Flatten the two surfaces to remove the gaps, and clamp them tightly with a clamp before welding.

Metal surface to be welded
Paint film, rust, dust or any other contamination on the metal surface to be welded causes insufficient current flow and poor results. Remove such foreign matter from the surface before commencing welding.

Corrosion prevention on metal surfaces
Coat the surface to be welded with anti-corrosion agent that has a high conductivity (e.g. zinc primers). It is important to apply the zinc primer uniformly between the panel surfaces to be welded.

Minimum welding pitch
The strength of individual spot welds is determined by the spot weld pitch (the distance between spot welds). The bond between the panels becomes stronger as the pitch is shortened, but if the pitch becomes too small and the spot welds are too close together, this can allow current shunting, which in effect drains away the weld current towards the previous spot weld.

Positioning of welding spot from the edge and end of the panel
The edge distance is also determined by the position of the welding tip. Even if the spot welds are normal, the welds will not have sufficient strength if the edge distance is insufficient.

Inspection of spot welds
Spot welds are inspected either by outward appearance (visual inspection) or by destructive testing. A visual inspection is used to judge the quality of the outward appearance

Figure 10.13   Current shunting in resistance welds positioned in close proximity

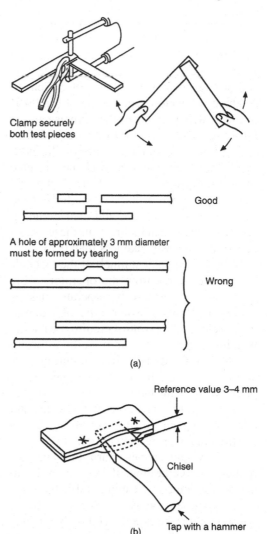

Clamp securely both test pieces

Good

A hole of approximately 3 mm diameter must be formed by tearing

Wrong

(a)

Reference value 3–4 mm

Chisel

(b)   Tap with a hammer

Figure 10.14   (a) Peel test (b) destructive test (*Motor Insurance Repair Research Centre*)

(tip bruises, pin holes, spatter, number of spots, spot positions), while destructive testing is used to measure the strength of a weld. Most destructive tests require the use of sophisticated equipment which most bodyshops do not possess. Consequently a simpler method called a peel test has been developed by general use in the workshop.

Destructive testing: peel test

After setting up the equipment a test weld should be made, using sheet metal of the same thickness and condition as the job to be done. When the test piece is completed, it should be torn apart. If the settings are correct, the joint will 'unbutton', that is the weld nuggets will be on one sheet and the holes on the other.

Weld tests for shear and peel for HSLA steels

When shear testing spot welds produced when welding together two HSLA steels, partial spot-weld failure may occur. If the partial failure does not exceed 20 per cent of the total nugget area, the strength of the weld should not be impaired.

ARO Spotrite Pulsa resistance welding system

For resistance spot welding in vehicle body repair there is a British invention called the Spotrite (made by ARO Machinery Co. Ltd). This is an adaptive (self-setting timing) in-process control unit which operates in several stages. Stage 1 is the monitoring stage for recognizing the material (low-carbon or HSLA steel) and for sensing through any protective coatings whilst judging how closely the clamped materials are fitting between the electrodes which will form the joint. Stage 2 is automatically linked and self-setting to provide a burst of current to produce the weld nugget. Stage 3 (and 4 if needed) will provide further precise timed pulses to bring the weld to the desired size, taking into account the number of sheets of metal and their thickness and the distance of the weld from the edge of the workpiece. The adaptive part of the control decides the number of stages needed and their duration. A weld cycle which includes all these stages added together will take less than 0.75 s.

With this equipment much of the technology is transferred from the operator to the controls, only because the operator has no means of knowing which type of material he is welding. The control monitors what the welding gun is doing and uses the information to take the decision by computing variables such as electrode tip wear, spot-weld pitch, voltage fluctuation, changes in metal thickness and the presence of scale, dirt and rust. Consequently quality control is no longer the operator's responsibility.

Audio-visual signals warn the operator if there is overheating, if the electrode tip needs reshaping or cleaning and if the weld arms are open too early. For difficult or unusual work, the experienced body repairer can switch to manual setting to control the unit's weld heat and time tolerance controls.

A pincer-type welding gun with a powerful weld transformer (to maintain the very short weld times) is fitted with a strong pressure system to deal with the majority of work, which can be approached from two sides in pincer-type fashion. A range of interchangeable weld arms and electrodes of different shapes and lengths can be used to convert the standard gun instantly into the special tool that a job may require. Special sets of arms with electrodes, specifically designed for use on individual makes of vehicles, have been designed in collaboration with most car manufacturers.

Figure 10.15 Modes of spot weld failure (*Motor Insurance Repair Research Centre*)

Figure 10.16 CEBORA spot welder (MIG/TIG/arc)

Figure 10.17 Electrode arm sets for different makes of vehicles (*SIP (Industrial Products) Ltd*)

Setting up equipment for use

The most important points to be considered in setting up the equipment are:

1 Thickness of the sheet metal to be welded
2 Accessibility of the parts to be welded
3 State of the surface of the sheet metal to be welded (this must be as clean as possible, and any rust, scale or paintwork should be removed).

The setting-up procedure is as follows:

1 With the equipment switched off, check that the electrode tips are aligned and correctly shaped. If domed tips are used (these generally give the best results) the tip radii should be 51 mm and 77 mm. When truncated cone tips are used, check that the diameter is correct for the gauge of material to be used: the diameter should be $5\sqrt{t}$, where t is the thickness of single sheet in millimetres. Check that the tip force is enough without bending the electrode arms or sliding the tip one upon another.

2 Switch on mains supply and check that the mains light (red) is on.
3 Switch function selection switch to resistance spot welding.
4 Switch on Spotrite control and check that the ready light is on.
5 Test earth leakage unit by pressing test button.

6 Set spot size to zero.

7 Make splashless welds by selecting the lowest heat and the longest time setting to make the desired size of weld nugget.

8 Make two welds approximately 51 mm apart. Peel test the second weld: tear the test piece apart and check whether the nugget is the size approximately to the weld strength needed. The nugget size should be $4.5\sqrt{t}$.

9 Increase the spot size setting from zero in single steps, and make five trial welds in each setting. If the reject alarm is heard (a high-pitched pulse note: red light illuminates) for more than three welds, reduce the spot size setting until the reject alarm is only activated two or three times.

10 Increase the heat step by step to reduce the actual weld time until splashing occurs. Then set back heat one step to stop splashing. The equipment is now fully set.

Welding procedure

1 Prepare the panel surface by removing any paint, primer and in general any insulating material covering the surface to be welded. This preparation is a key factor to good-quality spot welding. Paint must be removed by either paint remover or sanding. If rust is present on the panel surface, sanding is a better method of preparation, as it will leave the surface with a bright metallic finish which facilitates the flow of electric current.

2 Obtain the correct adjustment of welding gun electrodes and arms.

3 Determine a suitable weld pitch for the panel assembly to be welded to obtain maximum strength.

4 Make sure that the correct distance is set from the edge of the sheet metal panel to the nearest spot weld.

Single-sided spot welding

As a result of the operational limitations of conventional double-sided spot welding, single-sided spot welding equipment has become more widely used in the vehicle body repair industry. This system offers the benefits of conventional spot welding without the inherent problems of accessibility with double-skinned panel sections. Single-sided spot welding is therefore an alternative to conventional double-sided spot welding or can be used in conjunction with it. In the single-sided spot welding process the operator manually forces the single electrode against the panel, with the electrical circuit being completed by an earth clamp and the cable back to the transformer. This allows welds to be made in positions where access is possible only from one side.

The manufacturers are now using different types of steel to construct their vehicle bodies, the main three being low-carbon steel, galvanized steel and high-strength steel. This has led to confusion and difficulty in identification and welding. The main problem with high-strength steel is that it is heat sensitive and difficult to identify on vehicles without data (MIRRC Thatcham *Methods Manual*). In double-sided welding this problem is solved by using pulse welding, which keeps the heat in the weld as low as possible. Some single-sided equipment, especially that which operates at high current, does not require pulse-control welding. Instead these machines use a massive burst of power at 8000 A DC in a very short interval, which keeps the temperature of the spot weld below the recommended temperature. The other advantage of this system is that it does not need to identify whether the steel is high strength or low carbon. The problem with galvanized steel is that it has a very high contact resistance, making it difficult to weld; therefore the high-DC machines use a special preheat function for this type of steel. In the first stage of the weld the machine lowers the resistance of the steel by melting the coating, causing it to flow from the weld. Then, once the resistance is low enough, it will automatically carry out the second stage weld at the correct setting.

Single-sided spot welding equipment

This equipment ranges from 2500 to 9200 amperes as follows:

1 2500 A using two phases at 415 V
2 8000 A using three phases at 415 V
3 9200 A using three phases at 415 V.

The equipment can be used to carry out the following operations:

*   *Single-sided spot welding* – Ideal where access is difficult from both sides of the material. It is suitable for welding on wings, front panels and rear quarter panels.
*   *Two-electrode single-sided spot welding* – Suitable when unable to attach the earth clamp satisfactorily. It also allows two spot welds to be made simultaneously and is suitable for welding sill sections in place.
*   *Pulse-control roller spot welding* – This gives a continuous spot weld along an overlapped edge of metal. It is ideal for roof gutters and for welding patches to vehicle panels when dealing with corrosion repair without creating distortion.
*   *Welding copper ring washers for pulling* – This allows rows of washers to be welded to the panel surface. Ideal for pulling out large dints by using a slide hammer with a special hook attachment which fits through the rings, which are later broken off by twisting.
*   *Rapid puller* – This is designed for pulling out small dints quickly and effectively by welding the puller to the panel, pulling out the dint, then twisting the tool to release it from the panel.
*   *Copper shrinking of high spots* – This uses a copper tool for shrinking stretched panels which have been overworked by hammering.
*   *Carbon shrinking for over-stretched panels* – This uses a carbon pencil rod and is used for re-tensioning a panel surface that is only slightly stretched.

Figure 10.18 Single-sided and double-sided spot welding equipment (*Stanners Ltd*)

*   *Welding captive nuts to vehicle* – This piece of equipment can be used to weld captive nuts, trim studs and threaded bolts to panel surfaces when they need to be replaced.

Setting up equipment for single-sided spot welding

1 Set the weld timer dial.
2 Set the welding mode switch.
3 Set the welding power dial.
4 Connect the earth clamp plate to the negative side of the welding output cable holder and the single-sided electrode to the positive side of the cable holder.
5 Connect two earth clamp plates as close as possible to the area to be welded.

6   Ensure that both pieces of metal are clean and making good contact. Also make sure that the lower piece of metal is well supported to allow pressure to be applied to the electrode by the operator.

7   Press the electrode against the workpiece to be welded. Apply pressure to both pieces of metal and press the switch to carry out the weld.

8   It is important that the electrode is quenched in water after prolonged welding to prevent it becoming hot or overheating.

9   The electrode tip should maintain its original profile if good spot welds are to be achieved. Therefore, when necessary, the electrode should be redressed back to its correct shape.

Skills and questions

Different firms and training providers have different equipment; try to use as varied a range of equipment as you can. Visits to trade shows and equipment manufacturers' showrooms gives you the opportunity to expand the range of equipment which you know and understand and keeps you up to date with current developments.

1   What is meant by the term 'resistance welding'?

2   Name and describe three methods of resistance welding.

3   Give an example where resistance welding is used in vehicle manufacture.

4   Resistance welding uses various electrodes: select and sketch one type.

5   How is a workshop test carried out on a spot weld to test its strength?

6   Suggest one application of projection welding in the assembly of a vehicle body.

7   Illustrate types of joints that would be suitable for spot-welding applications.

8   Describe the three stages in the production of a spot weld.

9   Which important points should be considered when reshaping a resistance welding electrode?

10 Explain the importance of resistance welding in the repair of vehicle bodies.

11 Which metals are used for making electrodes in resistance welding?

12 What is the difference between butt welding and flash welding?

13 Name the principal parts of resistance welding equipment that would be used in the body repair shop.

14 State the qualities required of a good spot weld.

15 Explain the term 'robotic welding'.

16 What are the important factors to be considered when spot welding high-strength steel?

17 Explain the importance of the adaptive self-setting timing control unit when welding high-strength steels.

18 Explain the working procedure of a pincer-type welding gun.

19 Describe, with the aid of a sketch, the formation of a weld nugget during the spot welding cycle.

20 Describe a method of resistance welding that would be used on the assembly line to weld a roof panel into position.

21 What is the purpose of 'hold time' when carrying out the process of spot welding?

22 Describe typical repair situations that would use a pincer welding gun and a single-sided welding gun.

23 State where resistance seam welding could be found on a vehicle body.

24 Show, with the aid of a sketch, the principle of twin-spot welding.

25 Describe, with the aid of a sketch, the resistance projection welding process.

26 List the advantages of the resistance spot welding technique used in the repair of vehicle bodies.

27 Sketch one type of electrode arm sets used in vehicle repair.

28 With the aid of a sketch, explain how heat is generated to form the spot weld.

29 State the treatment that should be carried out before replacing a spot-welded panel.

30 State why water-cooled electrode tips are used in the manufacture of vehicle body shells.

11

Manual metal arc welding (MMA)

In the sphere of welding the electric arc has become an efficient and reliable means of welding sheet and metal plate. It is useful for welding the heavier-gauge plates used for commercial vehicle body building and also for the type of metal plate processes in which the metal ranges in thickness from 3 mm to 75 mm.

The use of arc welding depended naturally upon the development of electricity, and dynamos or generators were not developed until 1880. The first actual arc welding, meaning the melting of metal by means of electrodes and thus fusing them together, was developed by Bernardoz in 1885; he created a mechanism using a carbon electrode which produced an arc between the carbon and metal, melting the edges and thus performing a weld. The arc form of welding, using the metallic electrodes, was discovered by Slavinoff in 1892, but it had very little success because of the use of bare metal electrodes. However when Kjellberg, a Swedish inventor, developed the flux electrode in 1907, the success of the metallic electrode was assured. Progress accelerated as a result of the First World War, when productivity and speed of welding were of prime importance. However, it was not until the 1930s that good-quality joints could be reliably and consistently produced by the arc welding process. This was achieved by the development of coatings which gave adequate protection and improved arc stability whilst transferring metal between the electrode and the parent metal. From that time arc welding gradually displaced gas welding techniques, especially when joining heavy-gauge metal, although the major development in arc welding was due to the production of portable and automatic welding machines.

Although metal arc welding is only used a small amount in private car construction for heavy-gauge assemblies in cars which have chassis, it is still used in the commercial vehicle body-building industry for the assembly of the fully welded, trailer-type bodies in mild steel, stainless steel and aluminium. MMA is also used extensively in the construction industry for its suitability for welding thick metal plate and steel joists outside on site.

Principles of manual metal arc welding

Manual metal arc welding (MMAW or MMA) is used extensively in modern practice. It is based on the principle whereby intense heat is obtained from an electric current which creates an arc between a metal electrode and the plates which are to be welded. The heat produced is sufficient to fuse the edges of the plates at the joint, forming a small pool of molten metal. Additional molten metal from the tip of the electrode is deposited into the molten pool, and when solidified it results in a strong welded joint.

The Repair of Vehicle Bodies, Andrew Livesey, 9780815378709, 2019, Taylor & Francis

With this process the electrode from which the arc is struck is made of the same metal as the parent metal which is being welded; metal from the electrode is transferred to the weld, partly as drops under the influence of gravity and partly as high-velocity particles. The maintenance of a stable arc between a bare metal electrode and the workpiece is an unreliable procedure, and for this reason (among others) various coatings are applied to the electrode wire. Not only do these coatings help stabilize the arc, but they also perform the important functions of stopping oxidation of the heated electrode tip and molten workpiece. By their slagging properties they dissolve and segregate oxides and other impurities whose presence might otherwise adversely affect the quality of the welded joint. When the weld has cooled the slag forms a brittle coating which can be fairly easily removed by chipping and brushing. The electric supply for this type of welding may be either direct current (DC) or alternating current (AC); both systems possess certain advantages, depending on the purpose for which they are employed. Many different types and sizes of electric welding machines are now available, and two of the most generally used are the DC generator and the AC transformer.

Electrical terms used in arc welding

*   *Circuit* – A circuit is the path along which electricity flows. It starts from the negative (-) terminal of the generator where the current is produced, moves along the wire or cable to the load or working source and then returns to the positive terminal (+).
*   *Amperes* – Amperes or amperage refers to the amount or rate of current that flows through a circuit.
*   *Voltage* – The force (electromotive force or EMF) that causes electrons to flow in a circuit is known as voltage. This force is similar to the pressure used to make water flow in pipes. In the water system the pump provides the pressure, whereas in an electrical circuit the generator or transformer produces the force that pushes the current through the wires.
*   *Direct and alternating current* – There are two kinds of current used in arc welding: DC and AC. In DC the current flows constantly in one direction. In AC the current reverses its direction in the circuit a certain number of times per second. The rate of change is referred to as frequency and is indicated as 25, 40, 50, 60 cycles per second (hertz).
*   *Voltage drop* – Just as the pressure in a water system drops as the distance increases from the water pump, so does the voltage lessen as the distance increases from the generator. This fact is important to remember in using a welding machine, because if the cable is too long, there will be too great a voltage drop. When there is too much drop, the welding machine cannot supply enough current for welding.

Figure 11.1 (a) Principles of metallic arc welding (b) cross-section of a coated electrode in the process of welding

• *Open-circuit voltage and arc voltage –* Open-circuit voltage is the voltage produced by the welder when the machine is running and no welding is being done. After the arc is struck, the voltage drops to what is known as the arc or working voltage. An adjustment is provided to vary the open-circuit voltage so that welding can be done in different positions.

Metal arc welding equipment

DC generator

With a DC welding machine the electric current is produced by means of a generator, which is driven by a petrol or diesel engine (Figure 11.2) or alternatively by an AC or

1 Differential series field coils
2 Exciter armature commutator and brush gear
3 Exciter field coils
4 Main armature, commutator and brush gear
5 Spark-prevention interpole coils
6 Current regulator wound on laminated iron core
7 Main shunt-field coils
8 Prime mover: electric motor, petrol or diesel engine
9 Voltage control rheostat

Figure 11.2 Diesel-engine DC generator and circuit

DC electric motor. The motor-driven type of generator set is chiefly used for the type of welding work performed inside a workshop and is therefore often permanently mounted on the floor, but types are available for site work. The electric motor provides a good constant-speed drive for the generator and is not affected by the load imposed upon it. The generator is specially designed for welding purposes so that it generates about 60 volts on open circuit; this drops to about 20 volts immediately the arc is struck.

Generators are built in various current ratings ranging from about 100 to 600 amperes, and most modern types incorporate means for automatically adjusting the voltage to meet variations in the demand of the arc. Although the arrangement of the control unit may vary for machines made by different manufacturers, most consist of a wheel or lever control which selects the correct current for the right size of electrode and thickness of plate being welded. DC generators usually have a polarity switch which enables the welder to reverse the polarity, as is occasionally required when welding with special electrodes.

AC transformers

The AC welding machine employs a transformer instead of a generator to provide the required welding current. The AC transformer, as its name implies, is an instrument which transforms or steps down the voltage of the normal mains electrical supply to a voltage suitable for welding between 60 and 100 volts. In its simplest form an AC transformer consists of a primary and a secondary coil with an adjustment to regulate the current output. The primary coil receives the alternating current from the source of supply and creates a magnetic field which constantly changes in direction and strength. The secondary coil has no electrical connection to the power source but is affected by the changing lines of force in the magnetic field and, through induction, delivers a transformed current at a higher value to the welding arc. The output current is controlled either by an adjustable reactor in the output

Figure 11.3 AC transformers and circuit (*Murex Welding Processes Ltd*)

circuit of the transformer or by the adjustment of the internal reactance of the transformer. The current adjuster is operated by turning a hand wheel or crank. As the control handle is moved, a calibrated dial shows the current setting in amperes. Unlike the DC generator, the AC transformer has no moving parts and for this reason is usually referred to as a static plant. AC welding equipment has many advantages, namely:

1 Low initial cost
2 No moving parts and therefore negligible maintenance
3 Higher electrical efficiency
4 Ease of transport
5 AC can be converted to DC by means of a motor generator or rectifier, thus making both types available.

The disadvantages of the AC system are:

1 Coated electrodes must always be used.
2 Voltage is higher than in a DC system and therefore risk of shock is greater.

3 Welding of non-ferrous metals more difficult than in DC system.

Choice of system

As far as the quality of the weld is concerned there is little to choose between AC and DC systems. An AC system gives a smoother arc when using very high current, although a DC system is essential when welding certain non-ferrous metals. The availability of the mains supply is obviously a criterion.

Welding accessories

In addition to the actual welding generator or transformer, the following accessories should form a part of every welder's equipment:

• *Electrode holder* fitted with a length of flexible cable for connection to the plant. It should be of sufficient capacity to hold the largest electrode to be used, light in weight to reduce fatigue, well balanced and able to locate and eject the electrode easily. It should also be able to carry the maximum welding current without overheating. It can be a partially or fully insulated type, the latter being by far the safer.
• *Welding earth* – A length of cable which is flexible and connects the work to the plant. The diameters of these cables are governed by the voltage and distance to be carried from the machine. From the safety point of view it is essential that the welding circuit is efficiently earthed. The earth clamp which is fixed on the end of the earth cable should be as near to the work as possible. Its function is to keep the work at earth potential to safeguard personnel against any breakdown of the welding set. If the work is earthed and the return current lead is broken, no welding current can flow.
• *Head shield* or *face screen* fitted with special coloured lenses is an absolute necessity, because an electrode arc

Figure 11.4 Welding accessories (*Murex Welding Products Ltd*)

Figure 11.6 Welder wearing headshield and protective clothing (*Murex Welding Products Ltd*)

Figure 11.5 Welding accessories showing electrode holders and earthing clamps (*Murex Welding Products Ltd*)

produces a brilliant light and gives off invisible ultraviolet and infrared rays which are very dangerous to the eyes and skin. Never attempt to look at the arc with the naked eye. The helmet type of head shield fits over the head and leaves both hands free. The face screen provides adequate protection but needs holding by hand. The coloured lenses are classified according to the current used.

- *Gloves* are another important item for arc welding. These must be worn to protect the hands from the ultraviolet rays and from hot metal. Gloves, irrespective of the type used, should be flexible enough to permit proper hand movement yet not so thin as to allow any heat penetration.
- *Leather aprons* are often used by beginners in order to protect their clothes, although most experienced welders seldom wear an apron on the job except in situations in which there may be an excessive amount of metal spatter resulting from awkward welding positions.
- *Goggles* should be worn when chipping slag from a weld; this is a thin crust which forms on the deposited bead during the welding process. Whilst removing slag, tiny particles are often deflected

upwards, and without proper eye protection, these particles may cause a serious eye injury.

- *Cleaning tools* – To produce a strong welded joint, the surface of the metal must be free of all foreign matter such as rust, oil and paint. A wire brush is used for cleaning purposes. After a bead is deposited on the metal, the slag which covers the weld is removed with a chipping hammer. The chipping operation is followed by additional wire brushing. Complete removal of slag is especially important when several passes must be made over a joint. Otherwise, gas holes will form in the bead, resulting in porosity, which weakens the weld.
- *Welding booth* for use in production welding is designed to protect all other personnel from the arc glare. It comprises a steel bench, insulated from the booth, and a wooden duckboard to safeguard the welder from damp floors.

Electrodes used in welding: BS coding

Most modern electrodes are coated or covered and consist of a metal core wire surrounded by a thick coating applied by extrusion or other processes. The coating is usually a mixture of metallic oxide with silica which, under the heat of the electric arc, unites to form a slag which floats on the top of the molten weld metal. To a large degree the success of the welding operation depends on the composition of the coating, which is varied to suit different conditions and metals.

The principal functions of the coating are to:

1 Stabilize the arc and enable the use of arc welding
2 Flux away any impurities present on the surface being welded
3 Speed up the welding operation by increasing the rate of melting
4 Form a slag over the weld in order to protect the molten metal from oxidation

by the atmosphere, slow the rate of cooling of the weld and so reduce the chances of brittleness and provide a smoother surface.

In the past, each manufacturer employed different means of identifying the properties of his electrodes. To overcome the confusion arising from this, the British Standards Institution drew up a scheme for classifying all electrodes from a code number, which enables the user to know the main features of an electrode irrespective of the source of supply. The BS classification of an electrode is indicated by the following coding, in the order stated.

Strength, toughness and covering (STC) code

1 The letter E indicates a covered electrode for manual metal arc welding.
2 Two digits indicate the strength (tensile, yield and elongation properties) of the weld metal.
3 A digit indicates the temperature for a minimum average impact value of 28 J.
4 A digit indicates the temperature for a minimum average impact value of 47 J.
5 A letter or letters indicate the type of covering:
   E – – – – B basic
   E – – – – BB basic, high efficiency
   E – – – – C cellulosic
   E – – – – R rutile
   E – – – – RR rutile, heavy coated
   E – – – – S other types.

Tech note

The covering should be free from defects, such as cracks and abnormalities which would affect the operation of the electrode. The gripped end of each electrode should be free from covering for a minimum distance of 15 mm and a maximum of 40 mm.

Additional coding

The following additional coding must follow the STC coding:

1 When appropriate, three digits indicate the nominal electrode efficiency, which is the ratio of the mass of weld metal to the mass of nominal-diameter core wire consumed for a given electrode.
2 A digit indicates the recommended welding positions for electrodes:
   1 all positions
   2 all positions except vertical/down
   3 flat and, for fillet welds, horizontal/vertical
   4 flat
   5 flat, vertical/down and, for fillet welds, horizontal/vertical
   6 any position or combination of positions not classified above

3 A digit indicates the power supply requirement.
4 Where appropriate, a letter H indicates a hydrogen-controlled electrode.

Coding example

A typical classification for an electrode is therefore:

E51 54 BB [160 3 0 H]

The STC code is identified as follows: E51 strength (510–650 N/mm^2), 5 temperature for minimum average, 4 temperature for minimum average impact strength of 47 J (-30°C), BB covering (basic, high efficiency)

The additional code is as follows: 160 efficiency, 3 welding positions, 0 welding current and voltage conditions, H hydrogen-controlled impact strength of 28 J (-40°C).

Table 11.1 Designation for tensile properties (*British Standards Institution*)

			Minimum elongation (%)		
Electrode designation digit	Tensile strength (N/mm^2)	Minimum yield stress (N/mm^2)	When digit of Table 11.2 is 0 or 1	When digit of Table 11.2 is 2	When digit of Table 11.2 is 3, 4 or 5
E43— — —	430–550	330	20	22	24
E51— — —	510–650	360	18	18	20

Table 11.2 First digit for an impact value (*British Standards Institution*)

Digit	Temperature (°C) for minimum average impact value of 28 J using 4 mm diameter electrodes only
E— — 0 — —	Not specified
E— — 1 — —	+20
E— — 2 — —	0
E— — 3 — —	-20
E— — 4 — —	-30
E— — 5 — —	-40

Table 11.3 Second digit for an impact value (*British Standards Institution*)

Digit	Temperature (°C) for minimum average impact value of 47 J using 4 mm diameter and largest diameter electrodes submitted for classification
E— — — 0 —	Not specified
E— — — 1 —	+20
E— — — 2 —	0
E— — 3 —	-20
E— — — 4 —	-30
E— — — 5 —	-40
E— — — 6 —	-50
E— — — 7 —	-60
E— — — 8 —	-70

Table 11.4 Welding current and voltage conditions (*British Standards Institution*)

Digit	Direct current: recommended electrode polarity	Alternating current: minimum open-circuit voltage (V)
0	See manufacturer	Not suitable for use on AC
1	+ or -	50
2	-	50
3	+	50
4	+ or -	70
5	-	70
6	+	70
7	+ or -	80
8	-	80
9	+	80

Table 11.5 Nominal lengths of electrodes (*British Standards Institution*)

Diameter (mm)	Nominal length (mm)
1.6	200
250	
2	200
250	
300	
350	
2.5	250
300	
350	
3.2, 4, 5	350
450	
6 to 8	350
450	
500	
600	
700	
900	

Electrode and bundle identification

The STC code should be marked on the covering of each electrode as near as possible to the gripped end, except on electrodes of 1.6 mm and 2 mm diameter, where it is not practicable.

Each bundle or package of electrodes should be clearly marked with the following information:

1 The number and date of the British Standard Classification (STC code)
2 Name of manufacturer
3 Trade designation of electrodes
4 Size and quantity of electrodes
5 Batch number
6 Recommended current range, polarity and power supply
7 Recommendations for special storage conditions
8 Any other significant information on electrode characteristics or limitations on use
9 Health warnings.

Arc welding positions

Arc welding operations can be carried out with work in almost any position, but the degree of skill required of the welding operator will vary considerably depending on the position of the joint, which may be flat or downhand, horizontal, vertical or overhead.

Flat position (downhand)

Although welding can be done in any position the operation is simplified if the joint is flat. The speed of welding is then increased because the molten metal has less tendency to run, better penetration can be secured, and the work is less fatiguing for the welder in this position. Some jobs may at first glance appear to require horizontal, vertical or overhead welding, but upon examination it may be possible to change them to the easier and more efficient flat position, and for this reason positioning jigs are used in mass production. When welding in this position the electrode should be at an angle of 60 to 70° to the plate surface or horizontal.

Horizontal position

Occasionally the welding operation must be done while the work is in a horizontal position, which means that the welder must use

Figure 11.7 Arc welding positions

actually penetrating the metal. For horizontal welding, hold the electrode so that it points 5 to 10° and slants approximately 20° away from the depositing weld. When welding, use a narrow weaving motion, which will still further reduce the tendency of the molten weld pool to sag.

Vertical position

Vertical welding is done by depositing a weld in an upward or downward direction. Downward welding is very good for welding light-gauge metals because penetration is shallow, therefore forming an adequate weld without burning through the metal. Downward welding can also be performed more rapidly, which is important in production work. On heavier plates of 6 mm or more, vertical upward welding is more practical since deep penetration can be obtained. Welding upward also makes it possible to create a shelf upon which successive layers of weld can be placed. For vertical downward welding the electrode should be held at 10 to 15° tilted from the horizontal and starting at the top of the seam, moving downwards with little or no weaving motion. For vertical upward welding start with the electrode at right angles to the plate, then lower the rear of the electrode 10 to 15° with the horizontal.

Overhead position

Welding in an overhead position is probably the most difficult of operations. It is difficult because the welder must assume an awkward stance and at the same time work against gravity, which means that the molten pool has a tendency to drop, making it more difficult to secure a uniform weld and correct penetration. Heavily coated rods must not be used because of the continual dropping of the slag, and for this reason medium-coated rods are generally used. The most important points about overhead welding are to obtain the correct control of the current and keep a very short arc. Correct current control gives

a slightly shorter arc than for flat-position welding. The shorter arc will minimize the tendency of the molten pool to sag or run down and cause overlap. An overlap occurs when the pool runs down to the lower side of the weld and solidifies on the surface without a pool that is sufficiently molten to ensure good penetration. Hold the electrode at right angles to the seam, then tilt the rear of the electrode until it forms an angle of 10 to 15° to the horizontal.

Essential factors of arc welding

Correct choice of electrode

In selection, consideration must be given to position of weld, type of metal, size of electrode, type of joint, and current setting. The main group of electrodes include mild steel, high-carbon steel, special alloy steel, cast iron and non-ferrous aluminium, classified as bare and flux-covered. The bare electrodes have a very light coating which affords some protection against oxidation of the surface. The flux-covered or shielded electrode has a very heavy coating of several chemical substances which protect the molten metal from the oxidation and help to keep a steady arc. When this type of electrode melts, the coating produces a shield of gas around the molten metal, safeguarding it against the atmosphere. Some of the coating also forms a slag over the molten part, which also serves as a protection against oxidation.

Correct arc length

This will depend upon the type of electrode used and the nature of the welding operation. If the arc is too long, a wide, irregular bead is produced with insufficient fusion between the plate and weld metal. When the arc is too short, insufficient heat is generated to melt the plate properly, and the electrode will stick frequently, resulting in a very uneven bead.

1 This is how the beads of weld appear when the arc is too long
2 In this case the arc is too short
3 Here the arc is the correct length

Figure 11.8   Correct arc length

Correct speed of travel

This must be slow and even to ensure sufficient penetration without excessive build-up of bead. When the speed is too fast, the molten pool solidifies rapidly and impurities are retained in the weld. If the speed is too slow, excessive metal produces a high and wide bead.

Correct current

The correct current setting in accordance with the size of electrode used is most important. When the current is too great the electrode melts fast, causing a large pool of

1 Current, voltage and speed normal
2 Current too low
3 Current too high
4 Voltage too low
5 Voltage too high
6 Speed too slow
7 Speed too fast

Figure 11.9   Correct speed and current setting

Figure 11.10   Tapping method of striking the arc

Table 11.6 Current setting for electrode

Sheet thickness (mm)	Size of electrode (mm)	Current (A)
1.0	1.6	30
1.2	1.6	35
1.6	2.0	50
2.0	2.5	80
2.5	3.2	110
3.2	4.0	120–160
3.2–9.5	5.0–6.3	250–400
9.5 and over	8 mm	400–600

metal and excessive spatter. If the current is too low the heat generated will be insufficient to melt the plate and the molten pool will be small, resulting in lack of fusion.

Striking the arc

The successful step towards electric arc welding is learning to strike and maintain the arc and run a straight bead of weld metal. First set the control unit to the correct setting specified for the size of electrode being used. Then bring the electrode into contact with the plate by one of the following two methods.

Tapping method – The tapping motion method is as shown in Figure 11.10. The electrode is brought straight down on the plate and instantly withdrawn a distance of 3 to 5 mm, this distance being equal to the core diameter of the electrode.

Scratching method – The scratching method is where the electrode is tilted at an angle and is then given a slight circular movement similar to that of striking a match. As in the previous method, the electrode is promptly raised a distance equal to its diameter, otherwise it will stick to the plate.

Welding currents

Welding currents may vary from 20 to 600 A. For striking a DC arc an open circuit of 55 to 60 volts is required, whilst an AC set requires 80 to 100 volts. Once the arc is struck the arc voltage will drop to 20 to 25 volts. Before striking the arc the operator should have his head shield or face screen in position and observe the arc through the glass filter.

Weld defects

Undercutting
This is a condition that results when the welding current is too high. The excessive current leaves a groove in the base metal along both sides of the bead, which greatly reduces the strength of a weld.

Overlapping
This occurs when the current is set too low. In this instance the molten metal falls from

Figure 11.12 Undercutting

Figure 11.11 Scratching method of striking the arc

Figure 11.13 Types of arc welded joint

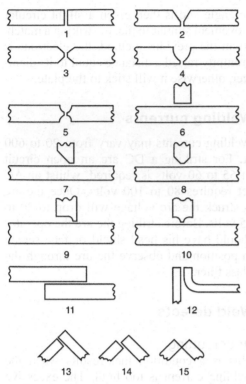

1 Square butt joint
2 Single-V butt joint
3 Double-V butt joint
4 Single-U butt joint
5 Double-U butt joint
6 Square-T joint
7 Single bevel-T joint
8 Double bevel-T joint

9 Single-U-T joint
10 Double-U-T joint
11 Lap joint
12 Edge joint
13 Flush corner joint
14 Half open corner joint
15 Full open corner joint

Figure 11.14   Overlapping

the electrode without actually fusing with the base metal, resulting in a defective weld.

Types of joint used in manual metal arc welding

The selection of the type of joint to be used for a particular application is governed by the following factors

1   The load and its characteristics
2   The manner in which the load is applied
3   The cost of joint preparation and welding.

Butt joints are of several types, each having a number of variations. However, they are generally classified as plain, single-V, double-V, single-U and double-U. T-joints also have several variations which can be applied in appropriate places. The other joints commonly used in manual metal arc welding are the lap joint and corner joints.

Welding joint design

Joint configuration
The three basic joint types are usually classified as fillet, butt or groove and lap These may be further subdivided depending on the joint detail and the joining process being used.

Strength and loading conditions
The achievement of the required static strength is a primary criterion in the selection of joint design. For fusion welding processes

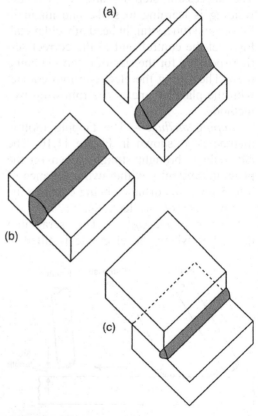

Figure 11.15   Basic joint types: (a) fillet (b) butt (c) Lap (*BOC Ltd*)

Figure 11.16 Static loading conditions: (a) butt weld (b) fillet weld, load carrying area shaded (*BOC Ltd*)

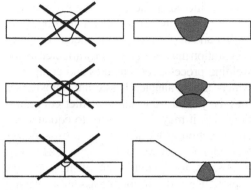

Figure 11.17 Acceptable and undesirable joint detail (*BOC Ltd*)

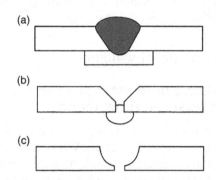

Figure 11.18 Plate preparation for limited access butt joints: (a) permanent backing strip (b) fusible insert (c) U preparation for root access (*BOC Ltd*)

which employ a filler material of composition matching that of the base material, it is normally only necessary to ensure that the cross-sectional area of the loaded joint is adequate. This is relatively easy to achieve where simple tensile or compressive forces are involved, but the effect of shear loading must also be considered in the case of lap joints. In dynamic loading situations the joint profile is more significant, and it is important to avoid discontinuities which may act as stress raisers. This is particularly important in joints subjected to fatigue.

Economic considerations

The cost of any weld connection will be influenced by the joint design. In general it is desirable to minimize the time taken to prepare and complete the joint to the required quality. This is usually associated with a reduction in the amount of material added to the joint. Narrow-gap butt welds, for example, give a reduction in weld time and, depending on the process, may reduce the complexity of plate preparation. Over-welding, that is the use of more weld metal than is needed to meet the strength requirements, may involve a significant higher cost.

Practical weld constraints

Some of the practical considerations which may affect joint design are:

1 Access
2 Welding position
3 Process/consumable/equipment availability
4 Dissimilar thicknesses.

If access is limited to one side of a butt joint it may be necessary to employ integral backing or complex plate preparation to ensure that adequate quality is achieved in the weld root (Figure 11.18).

In positional applications, non-symmetrical preparations may be necessary to improve joint profile and avoid excess weld metal. The joint preparation may be significantly affected by the welding process, consumables and equipment which are available. Dissimilar thicknesses may be joined using lap and fillet configurations, but it may be necessary to equalize the cross-section at the joint to perform satisfactory edge or butt welds. In some cases it may be possible for the joint to be self-jigging. This is particularly useful for site assembly, saving time and improving repeatability.

A range of approved joint designs is provided in international standards covering welding construction BS 5135.

Technique of welding

Welding mild steel plate in the flat position (downhand)

Prior to setting up the job it is necessary to clean the area to be welded. Where the material is heavier than 3 mm, V-ing of the welding edges is necessary to ensure penetration. Next make sure that the earth lead is securely fastened to the work. Painted or rusted surfaces provide a poor earth and should be scraped before connecting the earth clamp. Adjust the current setting of the transformer or generator in accordance with the electrode maker's specifications. Place the electrode in the holder, making sure that good contact is gained between the jaws of the holder and the bare end of the electrode. Care should be taken that the flux coating of the rod is clear of the electrode holder jaws. Never grip this holder in a tense manner; a firm but relaxed grip will produce a steady hand, which is essential while striking and maintaining the arc. The angle at which the electrode meets the plate being welded should be about 70°. The speed of travel should be such that a continuous penetrating bead of electrode is deposited. Welds on heavy plates are best made by making multiple runs rather than by excessive weaving of the electrode from side to side to build up the weld.

Welding light-gauge steel

The manual metal arc process offers a number of advantages for the welding of thin-gauge mild steels. The intensely localized heat input of the arc ensures that the minimum of disturbance occurs to the surrounding metal plate, therefore resulting in very little distortion. At present 1 mm is the thinnest gauge which can be readily welded by the MMAW process, and both butt and outside corner welds in this thickness can be produced with complete satisfaction. Inside corner fillet welds in this gauge are difficult to make because of the low heat input, which restricts the free flowing of the slag away from the depositing metal so that fusion is obstructed and the resulting weld is unsatisfactory.

Successful welding of light-gauge steel depends on three important conditions:

1   The close fitting of the edges to be welded
2   The right type and gauge of electrode
3   The accurate current adjustment.

Either DC or AC is suitable providing the supply is steady and capable of fine adjustment. The open-circuit voltage should not be less than 60 volts in the case of DC and 70 volts in the case of AC. Unless care is taken to see that edges or surfaces to be joined are in close contact along the seam, the heat conductivity of the weld area and the local heat will cause burning through wherever a gap occurs. Joint contact is maintained by tacking at 75 to 150 mm apart. The welding operation, which takes a little time to practise, consists of striking the arc and passing the electrode tip with the shortest possible arc rapidly along the seam to be welded at a uniform rate, without any sideways movement. If this is done correctly it will produce a satisfactory weld.

Welding stainless steels (austenitic group only)

The technique of welding stainless steel does not differ greatly from that of welding mild steel, but as the material being handled is very expensive, extra precautions and attention to detail at all stages of welding are desirable. In principle all stainless steel for high-class work should be welded with a short arc using a DC supply with the positive polarity applied to the electrode. For less important work an AC supply can be used with an open-circuit voltage of 100 volts. The current ranges recommended for different thicknesses will depend on the nature of the work and the size of the plate. Usually the lowest convenient current should be used; weaving should not be wider than twice the diameter of the electrode, and electrodes of like composition to the parent metal must be used. The edges of the welding should be free from scale. Clamps and jigs are advisable when welding thin sheet. Tack welds, particularly on thin sheets, should be placed much closer together than is the usual practice for welding mild steel.

Welding aluminium and its alloys

The equipment used is standard DC welding equipment. The operating voltage is about 25 volts, depending upon arc length and type of electrode, and over a current range of 40 to 450 amperes. The electrode coating on the rods usually consists of a mixture of chloride and fluorides, though other salts may be present. For manual metal arc welding, mixtures generally similar to those used for gas welding are supplied as the coating, but in this case the flux may include binders which prevent the coating from chipping and also help to stabilize the arc. The composition of the core wire is important and in general falls into three types: pure aluminium, aluminium silicon or aluminium manganese. If a wrought alloy is to be welded containing less than 2 per cent alloying element, then silicon-type electrodes are used. Pure aluminium should be welded with a pure aluminium electrode.

Vertical welding is possible with materials thicker than 5 mm, although downhand welding is preferable, with the electrode moving in a straight line along the seam without weaving. The speed of welding depends on the current

and the operator's skill, but is usually about three times that of mild steel. It increases as the electrode size is increased, and it should also be increased as the weld progresses in order to allow for the rise in temperature of the parent metal. Too-rapid welding on low currents does not give the required penetration, while welding that is slow or a current that is too high gives an excessive bead which may result in burning through the parent metal. When the arc is broken the coating tends to cover the tip of the electrode, thus obstructing the re-establishing of the arc. It is usually sufficient to tap the end of the electrode on the work to crack off this coating, although it may be necessary to cut the tip off. The slag on MMAW should flake off readily, especially if the weld is allowed to cool before the slag is removed. With all electrodes all traces of slag should finally be removed by scrubbing with a wire brush and washing.

Safety precautions for the welder

1 **Never look at a welding arc without a shield.**
2 **Always replace the clear cover glasses when they become pitted and encrusted due to metal spatter.**
3 **Examine the closed lenses in the helmet. If they are cracked, replace them immediately.**
4 **Wear goggles when chipping slag off a weld.**
5 **Always wear gloves and an apron when welding.**
6 **Use a holder that is completely insulated. Never lay it on the bench while the machine is running.**
7 **Welding processes should be carried out only in areas where there is adequate ventilation.**
8 **When welding outside a permanent welding booth, be sure to have screens around so the arc will not harm persons nearby.**
9 **Prevent welding cables from coming in contact with hot metal, water, oil or grease.**

10 Make sure that you have a good earth connection.
11 Keep the cables orderly to prevent them from becoming a stumbling hazard. Avoid dragging the cables over or around sharp corners. Whenever possible, fasten the cables overhead to permit free passage.
12 Do not weld near inflammable materials.
13 Be sure tanks, drums or pipelines are completely cleaned of inflammable liquids before welding.
14 Always turn off the welding machine when not in use.

Skills and questions

Welding takes time and practise. Spend as much time as you can practising your skills. Get your tutor or senior colleague to supervise you, and remember to take proper safety precautions.

1 What is the difference between alternating and direct currents?
2 Describe, with the aid of a line diagram, the principle of manual metal arc welding.
3 Name four factors which control the quality of manual metal arc welding.
4 Describe the process of manual metal arc welding and explain how the necessary heat is produced.
5 What is the function of the coating of a manual metal arc welding electrode?
6 What is meant by the term 'arc length'?
7 Give four essential factors in making a good arc weld.
8 In what way does a current affect a weld?
9 State the current in amperes and the size of electrode needed, for welding over 5-mm-thick steel plate.
10 Explain the function of an AC metal arc welding transformer.
11 What would be the resulting effect if the amperage is set too high in manual metal arc welding?
12 Why is it important not to look at an electric arc without proper eye protection?
13 Compare the use of a head shield to that of a hand shield for continuous welding operations.
14 What causes the defect known as undercutting, and how can this be avoided?
15 How is the weld affected when the arc is too short?
16 What happens when the welding arc is too long?
17 What determines the number of runs that should be made on a weld?
18 Why must a welder take into account the expansion and contraction of metal?
19 How are welding electrodes identified?
20 Why is it important to have adequate air extraction when welding in confined situations?

12

Gas shielded arc welding (MIG, MAG and TIG)

Development of gas shielded arc welding

Originally the process was evolved in America in 1940 for welding in the aircraft industry. It developed into the tungsten inert-gas shielded arc process which in turn led to shielded inert-gas metal arc welding. The latter became established in this country in 1952.

In the gas shielded arc process, heat is produced by the fusion of an electric arc maintained between the end of a metal electrode, either consumable or non-consumable, and the part to be welded, with a shield of protective gas surrounding the arc and the weld region. There are at present in use three different types of gas shielded arc welding:

- *Tungsten inert gas (TIG)* – The arc is struck by a non-consumable tungsten electrode and the metal to be welded, and filler metal is added by feeding a rod by hand into the molten pool.
- *Metal inert gas (MIG)* – This process employs a continuous-feed electrode which is melted in the intense arc heat and deposited as weld metal: hence the term consumable electrode. This process uses only inert gases, such as argon and helium, to create the shielding around the arc.
- *Metal active gas (MAG)* – This is the same as MIG except that the gases have an active effect upon the arc and are

not simply an inert envelope. The gases used are carbon dioxide or argon/carbon-dioxide mixtures.

Tech note
Gas tungsten arc welding (GTAW) – This is the terminology used in America and many parts of Europe for the TIG welding process, and it is becoming increasingly accepted as the standard terminology for this process.

Gases used for shielded arc processes

The shielding gases used in the MIG/MAG and TIG welding processes perform several important functions:

1 Protection from atmospheric contamination
2 Arc support and stabilization
3 Control of weld bead geometry
4 Control of weld metal properties.

It is necessary to prevent contamination of the weld pool by atmospheric gases which cause deterioration of the weld bead quality, by surface oxidation, porosity or embrittlement. In the consumable electrode process MIG/MAG, it is also necessary to consider the

The Repair of Vehicle Bodies, Andrew Livesey, 9780815378709, 2019, Taylor & Francis

Figure 12.1 TIG welding equipment AC/DC (*Murex Welding Products Ltd*)

Figure 12.2 MIG welding equipment (*Migatronic Welding Equipment Ltd*)

potential loss of alloying elements in the filler wire owing to preliminary oxidation in the arc atmosphere. In TIG welding, oxidation of the non-consumable tungsten electrode must be prevented. For these reasons most welding shielding gases are based on the inert gases argon (Ar) and helium (He). Active gases such as carbon dioxide (CO_2), oxygen (O_2) and hydrogen (H_2) may be added to the shielding gas to control one or another of the functions stated, but the gas chosen must be compatible with the material being welded.

The successful exclusion of atmospheric contamination depends on the ability to provide a physical barrier to prevent entrainment in the arc area, and in the case of some reactive metals such as titanium it may be necessary to extend this cover to protect the solidified weld metal whilst it is cooling. The gas cover depends on the shielding efficiency of the torch and the physical properties of the

Table 12.1 Chemical and physical compatibility of welding shielding gases and materials

Material	Gas	Compatibility
Plain carbon steel and low alloy steel	Argon, helium	No reaction
	CO_2, oxygen	Slight oxidation of alloying elements
	Hydrogen	Porosity and HICC risk
	Nitrogen	Porosity and loss of toughness
Austenitic stainless steel	Argon, helium	No reaction
	H_2	Reduces oxide
	O_2	Oxidizing
	CO_2	May cause carbon pick-up
Aluminium and alloys	Argon, helium	No reaction
	H_2	Gross porosity
	O_2	Oxidizing
Copper	Argon, helium, N_2	No reaction
	H_2	Porosity
Nickel	Argon, helium	No reaction
	N_2	Porosity
Titanium	Argon, helium	No reaction
	O_2, N_2, H_2	Embrittlement

Table 12.2 Density of common welding shielding gases

Gas	Density (kg/m³)
Argon	1.784
Helium	0.178
Hydrogen	0.083
Nitrogen	1.161
Oxygen	1.326
Carbon dioxide	1.977

gas. The higher the density of the gas the more resistant it will be to disturbance, and gases which are heavier than air may offer advantages in the downhand welding position.

The shielding gas used in MIG/MAG processes displaces the air in the arc area. The arc is struck within this blanket of shielding gas, producing a weld pool free from atmospheric contamination. The type of gas used determines the heat input, arc stability and mode of transfer, as well as providing protection for the solidifying weld metal. With any gas shielded arc process the type of shielding gas used greatly influences the quality of the weld deposit, weld penetration and weld appearance. The heat affected zone can also be influenced by the composition of the gas.

One of the important functions of the shielding gas is to protect the weld zone from the surrounding atmosphere and from the deleterious effects of oxygen, nitrogen and hydrogen upon the chemical composition and properties of the resulting weld. In this capacity the gas fulfils the major function of the fluxes used as electrode coverings or deposited as an enveloping layer during welding with other processes. The obvious advantages derived from the use of gas shielding are that the weld area is fully visible; that little, if any, slag is produced; and that the absence of abrasive flux increases the life of jigs and machine tools. In the MIG welding process, gas shielding enables a high degree of mechanism of welding to be achieved. Few gases possess the required shielding properties, however, and those that do with certainty – the inert gases, notably argon – are relatively expensive.

Argon

Argon is one of the rare gases occurring in the atmosphere and is obtained from liquefied air in the course of the manufacture of oxygen. Argon is an inert gas. It does not burn, support combustion or take part in any ordinary chemical reaction. On account of its strongly inactive properties it can prevent oxidation or any other chemical reaction from taking place in the molten metal during the welding operation.

The argon is supplied in steel cylinders coloured blue, a full cylinder having approximately 200 bar pressure, which is reduced by a regulator to approximately 15 litres per min when drawing for welding. Normally the cylinder should not be allowed to empty below 2 bar. This prevents air from entering the cylinder and helps to preserve the purity of the contents. Gas pressures are shown as bar: 1 bar = 14.505 lbf/in^2; 10 lbf/in^2 = 0.689 bar. Gas consumptions are in litres per hour (litre/h): 1 ft^3/h = 28.316 litre/h.

Argon has been more extensively used than helium because of a number of advantages:

1 Smoother, quieter arc action
2 Lower arc voltage at any given current value and arc length
3 Greater, cleaner action in the welding of such materials as aluminium with AC
4 Lower cost and greater availability
5 Lower flow rates for good shielding
6 Easier arc starting.

The lower arc voltage characteristics of argon are used to advantage in the welding of thin metals, because the tendency towards burn-through is lessened. Pure argon can be used for welding aluminium and its alloys, copper, nickel, stainless steel and also for MIG brazing.

Helium

Helium is a colourless, odourless inert gas almost as light as hydrogen. It is found in the United States in natural gas and is therefore more widely used in America than anywhere else. Helium has a high specific heat, so that a given quantity requires much more heat to raise its temperature by 1°C than does air; therefore the weld temperature is reduced and so distortion is minimized. A disadvantage of helium is that, because of its lightness, two and a half times the gas flow is required for a given performance as would be needed with argon. Helium is more favourable than argon where high arc voltage and power are desirable for welding thick material and metal with high heat conductivity such as copper. Helium is more often used in welding heavy than light materials.

Carbon dioxide

This gas is used as a shielding gas in the MAG welding process. It is not an inert gas: when it passes through the welding arc there is some breakdown into carbon monoxide and oxygen. To ensure that the oxygen is not added to the weld, deoxidants such as silicon, aluminium and titanium are included in the welding wire, which is specially made for carbon dioxide MAG welding. The deoxidant combines with the released oxygen to form a sparse slag on the surface of the completed weld. A gas heater with an electrically heated element is used to prevent freezing of the gas regulator after prolonged use with carbon dioxide gas. Since the development of the carbon dioxide process it has become widely used for the welding of plain carbon steels.

Argon mixtures

The gas mixtures that are suitable for vehicle body repair work consist of 95 per cent argon and 5 per cent carbon dioxide, or 80 per cent argon and 20 per cent carbon dioxide. These mixtures give smoother results with a better, cleaner and more attractive spatter-free weld. They also improve metal transfer and weld finish. These gases can be used to weld low-carbon mild steel, high-strength steel (HSS) and very low-carbon re-phosphorized steels.

Helium mixes

These are specially formulated for MIG welding of stainless steels. They are a mixture of high-purity helium and argon, with small controlled additions of carbon dioxide. They contain no hydrogen and are suitable for welding all grades of stainless steel including weldable martensitic grades, extra-low-carbon grades and duplex stainless steel.

Choice of shielding gases

Shielding gases must be carefully chosen to suit their application. The selection will depend on:

1 The compatibility of the gas with the material being welded
2 Physical properties of the material
3 The welding process and mode of operation
4 Joint type and thickness.

TIG welding

Principles of operation

The necessary heat for this process is produced by an electric arc maintained between a non-consumable electrode and the surface of the metal to be welded. The electrode used for carrying the current is usually a tungsten or tungsten alloy rod. The heated weld zone, the molten metal and the electrode are shielded from the atmosphere by a blanket of inert gas (argon or helium), fed through the electrode holder which is in the tip of the welding torch. A weld is made by applying the arc heat so that the edges of the metal are melted and joined together as the weld metal solidifies. In some cases a filler rod may be used to reinforce the weld metal.

Before commencing to weld it is essential to clean the surfaces that are to be welded. All oil, grease, paint, rust and dirt should be removed by either mechanical cleaning or chemical cleaners.

Striking the arc may be accomplished in one of the following ways:

1 Using an apparatus which will cause a spark to jump from the electrode to the work (arc stabilizer AC equipment); or
2 By means of an apparatus that starts and maintains a small pilot arc which provides a path for the main arc.

Once the arc is struck, the welding torch is held with the electrode positioned at an angle of about 75° to the surface of the metal to be welded. To start welding, the arc is usually moved in a small circle until a pool of molten metal of suitable size is obtained. Once adequate fusion is achieved, a weld is made by gradually moving the electrode along the parts to be welded so as to progressively melt the adjoining edges. To stop welding, the welding torch is quickly withdrawn from the work, and the current is automatically shut off.

The material thickness and joint design will determine whether filler metal in the form of welding rod needs to be added to the joint. When filler metal is added it is applied by feeding the filler rod from the side into the pool of molten metal in the arc region in the same manner as used in oxy-acetylene welding. The filler rod is usually held at an angle of about 15° to the surface of the work and slowly fed into the weld pool.

The joints which may be welded by this process include all the standard types such as square groove and V groove, as well as T and lap joints. It is not necessary to bevel the edges of material that is 3.2 mm or less in thickness.

Modes of operation

The TIG process may be operated in one of the following ways:

- *DC electrode negative* – In this mode the electrode remains relatively cool whilst the work piece is effectively heated. This is the most common mode of operation

for ferrous materials, copper, nickel and titanium alloys.
- *DC electrode positive* – With DC electrode positive there is a tendency for the electrode to overheat, and fusion of the workpiece is poor. The advantage of this mode of operation is the cathodic cleaning effect, which removes the tenacious oxide film from the surface of aluminium alloys.
- *AC alternating current* – This offers a good compromise between plate heating and the cathodic cleaning effect and is used with aluminium and with manganese alloys.

TIG spot welding

TIG spot welding is an adaptation of the main process. This method utilizes the heat from the tungsten arc to fuse the base material in much the same way as ordinary TIG welding. The tungsten electrode is set inside the argon shield to ensure that fusion takes

Figure 12.3 Basic principles of the TIG welding process (*BOC Ltd*)

Figure 12.4 Principles of the TIG welding process

place in a completely shrouded atmosphere. The sheets or parts to be joined are held in close contact by the manual pressure of the gun, and fusion is made from one side of the joint only. The equipment comprises a water-cooled torch with associated cables for argon, water and power, a standard AC/DC welding set and a timer and contact unit. The arc is struck by pressing a switch on the gun, activating the timer which carries the welding current to the tungsten electrode. The arc is struck automatically, and fusion takes place between the top and bottom components of the joint to be made. The depth of penetration through the component part is controlled by the current and time cycle.

The process is used for joining mild steel and stainless steel not exceeding 1.6 mm, but it is not suitable for aluminium and magnesium base alloys. The results of the spot welds differ according to the type and quality of the materials used, but providing the surfaces of the joint are clean and free from rust, scale, grease or dirt, satisfactory spot welds can generally be made.

1 Square edge close butt

2 Flanged edge close butt

3 Square edge open butt

4 Open butt inserted strip

70°–80°

5 Single-bevel close butt

70°–80°

70°–80°

6 Double-bevel close butt

Figure 12.5 Recommended edge preparation

Figure 12.6   Principles of the TIG spot-welding process

Equipment used in TIG welding

The equipment required for TIG welding consists of a welding torch equipped with a gas passage and nozzle for directing the shielding gas around the arc, a non-consumable electrode, a supply of shielding gas (argon), a pressure-reducing regulator and flow meter, a power unit and a supply of cooling water if the welding torch is water cooled. The individual components may differ considerably, depending upon power requirements and the type of work to be carried out.

Power unit

Standard AC and DC welding equipment may be used, but in most cases special welding units are used which are capable of producing AC or DC, have automatic control of argon and water flow and have fine current control switches for the stopping and starting of welding. The choice of welding current is determined by the material to be welded. Metals having a refractor surface oxide film, like magnesium, aluminium and its alloys, are normally welded with AC, while DC is used for carbon steels, stainless steels, copper and titanium. When direct current is used the electrode may be connected either to the positive or to the negative side of the power source, depending on the material to be welded. Usually the majority of general welding requires direct current with negative polarity, as the heat distribution and current loading are used to the best advantage, and the tungsten arc electrode can carry at least four times as much current, without overheating the electrode, as an equivalent positive arc. Practically all metals other than magnesium and aluminium are suitable for

Figure 12.7   Gas, water and power supply for TIG welding

this method, which gives deep penetration with a very narrow weld bead, whereas DC positive gives a shallow penetration with a wide bead.

Welding torch

Various types of torches are available to suit the different applications and current requirements. Torches may be water or air cooled: for currents below 150 A air-cooled torches are used, while from 150 to 400 A water-cooled torches are used. The air-cooled torches are used for welding light materials and the water-cooled torches for welding heavier materials, as the water cooling then prevents cracking of the ceramic shield at the tip of the torch. The torch is fully insulated electrically and has a quick release collet arrangement to facilitate convenient adjustment or changing of the tungsten electrodes. Tungsten having a melting point of 3400°C, is used as the electrode material owing to its refractory nature. It is almost non-consumable when used under ideal welding conditions. A ceramic shield, which is interchangeable, directs the gas so as to form a shroud around the arc and weld metal. The argon and the electric current are supplied to the torch through a combined cable and gas hose. In the water-cooled models a third cable is added to carry the water to and from the torch. A water flow switch can be provided to give complete protection to equipment and operator by shutting off the welding current if the water supply should fail.

To avoid contamination of either the electrode tip or the work, which would occur if the normal method of arc striking were employed, a high-frequency spark unit or an arc stabilizing device is used to stop the operator from having to touch the electrode on the surface of the work.

Gas supply

The inert gas, argon or helium, is supplied to the welding torch from the storage cylinder via a gas pressure regulator and a gas economizer valve, which may be a dual-purpose valve when cooling water is used, and a special flow

Figure 12.8 Alternative methods of DC connections (a) Theoretical distribution of heat in the argon shielded arc with the alternative methods (b) Average differences in arc voltages with equal arc lengths, using (left) negative polarity at the electrode and (right) positive (c) Relative depths of penetration obtainable with (left to right) DC positive, DC negative and AC

Figure 12.9 A water-cooled TIG welding torch

meter calibrated in cubic feet per hour or litres per minute of gas flow. The gas flow required varies with current setting, shroud size, material being welded and type of weld joint.

Filler metals

Filler materials for joining a wide variety of metals and alloys are available for use with the gas tungsten arc welding process. Among them are filler rods for welding various grades of carbon and alloy steels, stainless steels, nickel and nickel alloys, copper and copper alloys, aluminium and most of its alloys, magnesium, titanium and high-temperature alloys of various types. There are also filler materials for hard surfacing. Wherever a joint is to be reinforced, a filler rod is added to the molten puddle. In general, the diameter of the filler rod should be about the same as the thickness of the metal to be welded. For sound welds, it is important that the physical properties of the rod be similar to the base metal.

TIG electrodes

Normally pure tungsten electrodes are used, but to improve are striking and stability an addition of either thorium oxide or zirconium oxide is added to the tungsten. For alternating current welding zirconiated electrodes are used, while for direct current welding thoriated electrodes are used. The chosen electrode must be of a diameter which suits the current.

Improved performance can be obtained by alloying the electrodes lanthanum, yttrium and cerium, particularly in automatic TIG welding where consistency of operation is important. The electrode diameter is determined by the current and polarity. The angle to which the electrode is ground depends on the application. The included angle or vertex angle is usually smaller for low-current DC applications. In order to obtain consistent performance on a particular joint it is important that the same vertex angle is used.

Table 12.3 Recommended diameters and current ratings for TIG electrodes

Diameter (mm)	Maximum current (A) DC Electrode (-) thoriated	AC electrode (+) zirconiated
0.8	45	–
1.2	70	40
1.6	145	55
2.4	240	90
3.2	380	150
4.0	440	210
4.8	500	275
5.6	–	320
6.4	–	370

TIG welding techniques

The normal angles of the torch are 80 to 90° and of the filler rod 10 to 20° from the surface of the horizontal plate respectively. The arc length, defined as the distance between the tip of the electrode and the surface of the weld crater, varies between 3 mm and 6 mm, depending on the type of material and the current used. The filler rod is fed into the leading edge of the molten pool and not directly in the arc core and should be added with a slightly transverse scraping motion, with the tip of the filler rod actually making contact with the weld metal. This ensures that the rod is at the same electrical potential as the plate during transfer of metal to the weld and avoids any tendency of the rod to spatter. The heated end of the filler rod should always be kept within the influence of the shrouding argon gas in order to prevent its oxidation.

Appropriate vertex angles, degrees

Figure 12.10 Appropriate vertex angles of electrodes (*BOC Ltd*)

Figure 12.11 Recommended angles for torch and filler rod in TIG welding

Butt welds in thin-gauge materials are carried out with a progressive forward motion without weaving, but a slightly different technique is required when tackling medium- and heavy-section plate. As the filler rod diameter increases with increasing thickness of plate, there is a tendency for the end of the filler rod to foul the tungsten electrode. Contamination of the hot tungsten by particles of molten metal causes immediate spattering of the electrode and particles of tungsten may, according to the degree of contamination, become embedded in the weld pool. Loss of tungsten in this manner will cause the arc to become erratic and unstable, and the electrode will certainly require to be replaced before further welding is attempted. To avoid repetition of this occurrence, the arc length must be increased slightly to accommodate the insertion of the larger filler rod. This procedure cannot be taken too far, however, because there is a maximum arc length beyond which good welding becomes impossible. The upper limit is usually about 6 mm thick for aluminium, which does not allow for the free access of a 6 mm diameter filler rod. For heavier sections, therefore, a forward and backward swinging motion of the torch is employed. The weld area is melted under the arc; the torch is withdrawn backwards for a short distance from 6 mm to 3 mm along the line of the seam, and the filler rod is inserted in the molten pool (Figure 12.12).

Figure 12.12 Motion of torch and filler rod for TIG welding heavy sections

The torch is moved forward and the filler rod is withdrawn from the pool simultaneously. A rhythmical motion of both torch and filler rod backwards and forwards in a progressive forward motion melts down filler rod and plate without the filler rod entering the core of the arc and is recommended when welding plate in excess of 6 mm thick.

Application of TIG welding

This process has found an application in the body building side of the industry, where it is used to ideal advantage for fabricating components by welding in materials such as aluminium and stainless steel. The advantage of the method is that materials which would normally require flux can be welded without it; therefore cleaning of the weld is minimized and the effects of distortion are greatly reduced. This process is especially adapted for welding light-gauge work requiring the utmost in quality or finish because of the precise heat control possible and the ability to weld with or without filler metal. It is one of the few processes which permit the rapid, satisfactory welding of tiny or light-walled objects.

Among the materials which are readily weldable by this process are most grades of carbon, alloy or stainless steels, aluminium and most of its alloys, magnesium and its alloys, copper, copper-nickel, phosphor-bronze, tin bronzes of various types, brasses, nickel, nickel-copper (Monel alloy), nickel-chromium-iron (Inconel alloy), high-temperature alloys of various types, virtually all of the hard surfacing alloys, titanium, gold, silver and many others.

MIG/MAG welding

The development of the MIG/MAG processes is in some ways a logical progression from the manual metal arc and TIG processes. The effort throughout has been to obtain and maintain maximum versatility, weld quality, speed of deposition, simplification of the welding operation and lower operating costs. MIG/MAG welding has been adapted for many industrial applications, and over the past years it has become widely used for car body repairs. This method of welding is an electric arc process using DC current and a continuous consumable wire electrode without the addition of flux. On account of the absence of flux, gas is used to shield the arc and weld pool from atmospheric contamination. The process can utilize argon, argon/carbon dioxide mixture or carbon dioxide as the shielding gas, the choice being dependent upon the type of material being welded and the economics associated with the selected gas. If non-ferrous metals or stainless steels are to be welded, argon is the usual choice for shielding gas on grounds of compatibility. However when mild steel, low-alloy steels or high-strength steels are to be welded, argon/carbon dioxide mixture or carbon dioxide is generally used for reasons of overall efficiency, weld quality and economy.

Many types and grades of metal can be welded using this method: aluminium, aluminium alloys, carbon steel, low-carbon and alloy steels, micro-alloy steel, nickel, copper and magnesium. The success of this welding method is due to its capability of giving a consistently high-quality weld while also being very easy to learn. In addition it has the advantage of spreading very little heat beyond the actual welding point, and this helps to avoid distortion and shrinking stresses, which are a disadvantage in the oxy-acetylene process.

Principles of operation

Metal inert-gas or active-gas shielded arc welding (consumable) is accomplished by means of a gas shielded arc maintained between the workpiece and a consumable (bare wire) electrode from which metal is transferred to the workpiece. The transfer of metal through the protected arc column provides greater efficiency of heat input than that obtained in the TIG welding process. The resultant high-intensity heat source permits very rapid welding. In this process a continuously fed electrode passes through a gun, during which it passes through a contact area which impresses the preselected welding current upon the wire. The current causes the wire

to melt at the set rate at which it is fed. The shielding gas issuing from the nozzle protects the weld metal deposit and the electrode from contamination by atmospheric conditions which might affect the welding process. The arc may be started by depressing the trigger of the welding gun and scratching the electrode wire end on the work.

The equipment is designed so that the wire automatically feeds into the weld area as soon as the arc is established. Most MIG/MAG

Figure 12.13   Basic principles of operation of MIG/MAG welding (*BOC Ltd*)

Figure 12.14   Principles of the MIG/MAG welding process: argon, argon/CO_2 or CO_2

welding sets that are manufactured for the automobile repair trade are semi-automatic, the operator only being concerned with the torch-to-work distance, torch manipulation and welding speed. Wire feed speeds, power settings and gas flow are all pre-set prior to commencement of welding.

MIG/MAG spot/plug welding

An advantage of MIG/MAG welding is the ability of the process to be adopted for single-side spot welding applications, either semi or fully automatically. By extending the welding gun nozzle to contact the workpiece, one-sided spot welds may be performed using dip transfer conditions. Predetermined weld duration times may be employed, the gun trigger being coupled to a suitable timer and, if desired, fully mechanized. Unlike resistance spot welding, no pressure is required on the workpiece with MIG/MAG spot welding, and neither is a backing block. Mismatch of the sheets is permissible with a maximum gap equivalent to half the sheet thickness, the extra metal being provided by the electrode wire. Up to 30 spots per minute may be welded, which compares reasonably well with the 100 spots per minute from resistance welding techniques. The deep penetration characteristics of this welding process enable spot welding of widely differing metal thicknesses to be performed successfully, together with multi-sheet thicknesses.

MIG/MAG plug welding differs from MIG/MAG spot welding in that the outer metal panel has a predrilled or punched hole which is filled up with the weld metal to form the 'plug'. The hole sizes used are 5 to 6 mm with currents of 50–60 A for panel thicknesses of 0.75 to 1.2 mm. Care should be taken when MIG/MAG plug welding to avoid an excess build-up of weld metal and to reduce the necessity of dressing the finished weld. This method is used to advantage in the fitting of new panel sections, where the original was resistance spot welded and access to both sides is now difficult.

Higher-strength steels are used for selected panel sections, but as they are vulnerable to heat they are not as easily welded as mild steel. MIG seam and butt welding produce hard weld joints; they will tear from the sheet metal on impact. However, MIG spot welds and particularly puddle (or plug) welded MIG spots can be made ductile. The weld time has to be short to reduce the heat-affected zone around the weld.

Modern car body designs are constructed with deformation zones both front and rear, and the shear impact properties of the original welding have been carefully calculated to ensure that the energy of an impact is fully absorbed and contained within the zone. Changing the manufacturers' original welding specification could impair the safety of the vehicle. The crumplability (impact energy absorbing) design of car bodies makes new demands on welds. For its success the body design relies on the sheet metal to crumple or fold rather than tear in a collision. This protective design depends on the ductility of the metal, and the welds too have to be ductile. If they are soft they will break without forcing the assembly to crumple, but if they are hard the welds will unbutton and the assembly will fly apart instead of folding up slowly. Welds therefore must be ductile as well as large and strong.

Equipment used in MIG/MAG welding

The basic equipment required comprises a power source, a wire feed unit and a torch. The power sources commonly used have constant-voltage characteristics, and controls are provided for voltage and inductance adjustment. This type of power source is used in conjunction with a wire feed unit which takes the wire from a spool and feeds it through a torch to the arc. A control on the wire feeder enables the speed of the wire to be set to a constant level which will in turn determine the arc current.

The welding torch should be reasonably light and easy to handle, with provision for gas shield shrouding, control switch, easy wire changes and good insulation. The torch is connected to a wire feed and control unit by means of:

1   A flexible armoured tube carrying the welding wire
2   A plastic tube carrying the shielding gas
3   A pair of plastic tubes carrying cooling water, the return tube often carrying the welding supply to cool the welding cable (light-duty torches are air cooled)
4   The control wires for the switchgear.

The whole feed unit is bound together by a plastic sleeve and is 1.5 to 3 m long. The wire feed unit houses the drive unit for the wire feed, which can be varied in speed to suit current/voltage conditions. The wire reel is also carried in the unit and is loaded with some 15 kg of welding wire.

Power unit

DC power units are used as either rotary generators or rectifying units, which are specially designed to give full versatility of arc control. The equipment is either single phase (130–240 A) or three phase (250–500 A). The principal components of most machines are welding transformer, rectifier, choke coil, wire feed unit, gas solenoid valve and electronic control box.

The welding transformer is dimensioned so as to achieve optimal welding properties. The transformer is manufactured from materials able to withstand a working temperature of up to 180°C. By way of a further safeguard against overloading, there is a built-in thermal fuse which cuts out the machine at 120°C. The thermal fuse is automatically reconnected once the transformer has cooled down.

The rectifier is constructed from a fan, thyristors, diodes and a capacitor battery. During welding, or when the machine is hotter than 70°C, the fan cools the rectifier and the transformer. The rectifier is electronically protected against overloading in the event of any short-circuit between welding positive and negative, with the machine cutting out approximately 1 second after the onset of short-circuiting.

The transformer converts the high mains voltage into a low alternating voltage, which is rectified and equalized into DC voltage in the rectifier module. A choke coil reduces the peaks in the welding current and thus eliminates the cause of welding spatter. When the switch on the torch is depressed, the thyristors come on. The rectifier emits a DC voltage, which is determined by the remote control in the welding trigger and the gas/wire matching switch on the box. Simultaneously, shielding gas is turned on and the wire feed motor is started up at a speed also determined by remote regulation from the torch.

When the trigger is released the motor decelerates, and after a short time lag the gas flow and welding voltage are interrupted. This time delay is called burn-back and causes the welding wire to burn a little way back from the molten pool, thus preventing it from sticking to the workpiece.

Depending on the type of equipment selected, the following functions are available: seam, spot, stitch and latching (four-cycle).

* *Seam* – Welding starts when the switch on the welding trigger is actuated and stops when the switch is released. For use in short-duration welding and tacking.
* *Spot* – Welding starts when the switch is actuated and is subsequently controlled by the welding timer for a time between 0.2 and 2.5 s. It makes no difference when the switch is released. This function ensures uniform spot welds, providing the correct setting has been found.
* *Stitch* – The wire feed motor starts and stops at intervals which are set on the welding timer and pause timer. When welding is interspersed with pauses in this way, the average amount of added heat is reduced, which prevents any melting through on difficult welding jobs.
* *Latching* – Welding starts when the switch is actuated; the switch can then be released and welding continues. By re-actuating the switch, welding stops when the switch is released. For use on long seams.

A typical welding control panel has the following features:

1 *Selection switch* – This selects between the functions seam, spot, stitch and latch as described above.
2 *Power light* – This lights up when the machine has been turned on.
3 *Overheating warning light* – If this light comes on, the welding equipment is automatically switched off owing to overheating of the transformer. When the temperature is back to normal, the welding can be continued.
4 *Welding timer switch* – With this switch, the welding time is chosen when the selection switch is in the stitch or spot position.
5 *Adjustable pause time button* – With this button the pause time is chosen when the selection switch is in the stitch position.
6 *Burn-back button* – Pre-adjustment of the burn-back delay button indicates the time for stopping the wire feed until the arc is switched off. This varies between 0.05 and 0.5 s.
7 *Adjustment of wire speed switch* – This gives the adjustment of the wire feed from 0.5 to 14 m/min.
8 *Welding voltage switch* – This sets the welding voltage of the transformer. When set at gas test, the gas flows by pressing the switch on the torch handle.

The characteristics built into the welding power supply are such as to provide automatic self-adjustment of arc conditions as the weld proceeds. Depending on the relevant current and voltage used, metal transfer between electrode and work takes place in the following distinct forms, each of which has certain operational advantages.

Dip transfer (*short arc*)

This condition requires comparatively low current and voltage values. Metal is transferred by a rapidly repeated series of short circuits when the electrode is fed forward into the weld pool.

Metal dip transfer is the most suitable mode of metal transfer for welding on car repairs, as it offers good bead control and low heat input, thus cutting down distortion when welding in panel sections. This type of transfer will also permit the welding of thinner gauges of sheet steel, and is practical for welding in all positions.

The principle of the method (short-circuiting transfer) is briefly as follows. The molten wire is transferred to the weld in droplets, and as each drop touches the weld, the circuit is shorted and the arc is momentarily extinguished. The wire is fed at a rate which is just greater than the burn-off rate for the particular arc voltage being used; as a result the wire touches the weld pool and

Figure 12.15   MIG welding control panel (*Migatronic Welding Equipment Ltd*)

short-circuits the power supply. The filler wire then acts as a fuse, and when it ruptures a free burning arc is created. After the occurrence of this short-circuit, the arc re-ignites. This making and breaking, or arc interruption, takes place from 20 to 200 times per second according to the setting of the controls. The result is a relatively small and cool weld pool, limiting burn-through.

Free flight transfer

In this metal transfer a continuous arc is maintained between the electrode and the workpiece and the metal is transferred to the weld pool as droplets. There are three subdivisions of the system:

- *Spray transfer* – In this type the mode of transfer consists of a spray of very small molten metal droplets formed along the arc column, which are projected towards the workpiece by electrical forces within the arc and collected in the weld pool. No short-circuiting takes place; the welds are hotter, and the depositing of weld metal is faster. It is ideally suited for the rapid welding of thick sections in the downhand position and the positional welding of aluminium and its alloys.
- *Globular transfer* – This occurs at currents above those which produce dip transfer but below the current level required for spray transfer. The droplet size is large relative to the wire diameter, and transfer is irregular. This mode of transfer occurs with steel wires at high currents in carbon dioxide but is generally regarded as unusable unless high spatter levels can be tolerated. The use of corded wires gives a controlled form of globular transfer which is acceptable.
- *Pulse arc transfer* – In this mode the droplets are transferred by a high current which is periodically applied to the arc. Ideally, one drop is transferred with each pulse and is fired across the arc by the pulse. Typical operating frequencies are 50 to 100 droplets per second. A background current is maintained between pulses to sustain the arc but avoid metal transfer. Maximum control is obtained with this type of metal transfer, which utilizes a power supply to provide a pulsed welding condition in which the metal transfer takes place at pulse peaks. This leads to extreme control over the weld penetration and weld appearance.

Welding torch

Air cooled torches are available for the various welding applications ranging from 180 A to 400 A. The design of the torch is in the form of a pistol or is curved similar to the shape of an oxy-acetylene torch and has wire fed through the barrel or handle. In some versions where the most efficient cooling is needed, water is directed through passages in the torch to cool the wire contact means as well as the normally cooled metal shielding gas nozzle. The curved torch carries the current contact tip at the front end, through which the feed wire, shielding gas and cooling water are also brought. This type of torch is designed for small-diameter feed wires, is extremely flexible and manoeuvrable and is particularly suitable for welding in confined areas. The service lines consisting of the power cable, water hose and gas hose on most equipment enter at the handle or rear barrel section of the torch. The torch is also equipped with a switch for energizing the power supply and controls associated with the process. Some welding torches also have a current control knob located in the torch so that the welding current can be altered during welding. This ensures that the welding voltage is altered at the same time as any current alteration, allowing the welder to respond immediately to variations in weld gaps and misaligned joints. Welding characteristics are excellent. Fingertip control increases both productivity and weld quality.

Feed unit

The consumable electrode for welding ferrous and non-ferrous materials is supplied as a continuous length of wire on a spool or

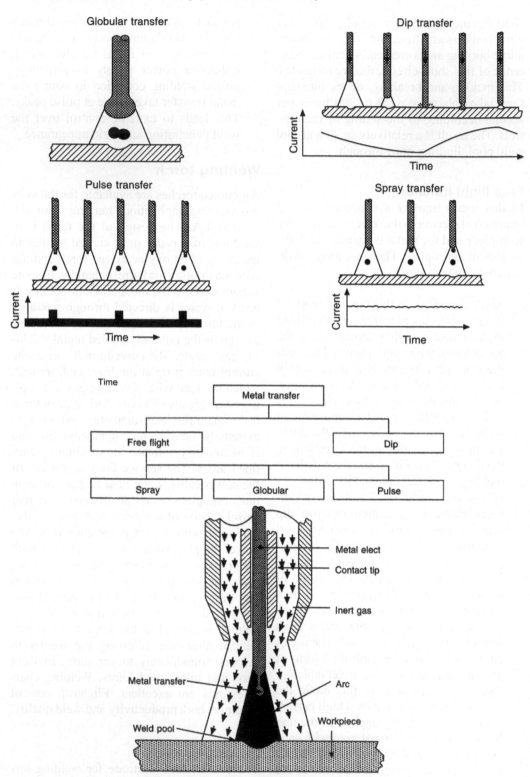

Figure 12.16   Metal transfer forms

reel, and the wire varies from 0.6 to 0.8 mm diameter. The feeding of the wire is achieved by the unit feed mechanism housing the necessary drive motor, gear box and feed rolls which draws wire from the adjacent reel, or by an integrally built motor drive connected to the torch which pulls the wire in the desired direction. The feed wire unit also houses the controls which govern the feeding of the wire at the required constant speed, plus the automatic control of current and gas flow. The wire from the feed rolls is fed along a carefully designed conduit system which also carries the welding current and shielding gas. The outer end of this conduit system is connected directly to the welding torch.

Figure 12.17 Air-cooled MIG welding torch (*Murex Welding Processes Ltd*)

Figure 12.18 Migatronic Dialog torch with current control in the torch handle (*Migatronic Welding Equipment Ltd*)

Figure 12.19 Wire feed unit (*Migatronic Welding Equipment Ltd*)

Gas supply

The primary purpose of the shielding gas used in MIG/MAG welding is the protection of the molten weld metal from contamination and damage by the oxygen, nitrogen, water vapour and other gases in the surrounding atmosphere. However, in use there are a number of other factors which influence the desirability of a gas for arc shielding. Some of these are the influence of the shielding gas on the arcing and metal transfer characteristics during welding, on the weld penetration, width of fusion and surface shape patterns and on the speed of welding. All of these influence the finished welds and overall results. Normal shielding gases in use are argon, carbon dioxide and mixtures of argon and carbon dioxide, although other mixtures such as argon and oxygen can be used.

Welding wire

The welding wires used for MIG/MAG welding are quite small in diameter compared with those used for other types of welding. Welding wire is 0.6 to 1.0 mm in diameter and is supplied in reels of 5 to 15 kg. Owing to the small size of the wire and comparatively high currents used for welding, the wire melts off very rapidly. Therefore the wires used must always be as long as possible, with suitable temper for being fed smoothly and continuously by motor-driven mechanical means through the welding apparatus. For this reason the wire is normally provided on convenient-size coils or spools.

In composition the wires are usually quite similar to those used by most other bare-wire processes. As a rule the composition as nearly matches that of the base material as is practicable, while also ensuring good weld properties and welding characteristics. Steel wires are often given a light coating of copper to prevent them from rusting and to improve the electrical contact and current pick-up when passing through the welding apparatus. Welding wires are available for joining most ferrous and non-ferrous metals:

- *Solid wire* – The composition of solid filler wire is usually chosen to match the parent metal being welded. Solid wire is the most common and widely used filler material for MIG/MAG welding car body replacement panels.
- *Cord wires* – Metal and flux corded wires are available. These materials are, however, more suitable for welding thicknesses in excess of 6 mm. Metal corded wires can be obtained with additions of alloying elements to meet special requirements (HS low-alloy steels).
- *MIG brazing wires* – There are a number of compositions of filler wire that can be used in MIG brazing, but usually they are of aluminium-bronze composition. When using these filler wires, pure argon must be used as the shielding gas.

MIG/MAG welding techniques

In the operation of a MIG/MAG welding machine there are two parameter settings which must match each other: the wire speed and the welding voltage. It is important therefore that the wire speed and welding voltage are adjusted to conform with each other. The wire speed gives the welding current, which

Figure 12.20 Incorrect settings for wire speed (*Stanners Ltd*)

must match the component panel section being melted. An increase in wire feed rate increases the welding current, at the same time reducing the length of arc, resulting in a shorter arc. When the wire speed is reduced, the current intensity is diminished, at the same time increasing the length of the arc. Raising the open-circuit voltage increases the arc voltage, though the current intensity remains almost constant. A reduced open-circuit voltage will produce a shorter arc length. A change in welding wire diameter will produce a change in current and voltage, since thinner

wires require higher voltage and higher wire speeds in order to produce the same current. If the limit values are exceeded a satisfactory weld cannot be obtained.

When the wire speed setting is too low, this will cause a long, drawn-out arc and spattering and a very unsatisfactory weld condition. When the wire speed becomes too great in relation to the voltage, knocking or stubbing will be felt in the welding torch, because the wire travels to the bottom of the molten pool. Welding under these conditions will normally produce adhesion defects owing to

FAULT		CAUSE
Porosity		Insufficient Si, Mn in wire Insufficient CO_2 shielding because of { flow rate frozen valve clogged nozzle draughts

Cracking		Dirty work – grease, paint, scale, rust
		  (i) Weld bead too small
		  (ii) Weld too deep, greater than 1.2:1
		  (iii) High sulphur, low manganese, slow cooling rate

Undercutting		Travel speed too high Backing bar groove too deep Current too low for speed Torch angle too low

Lack of penetration		Current too low – setting wrong
		  Wire feed fluctuating
		  Electrode extension too great
		Joint preparation too narrow
		  Angle too small
		  Gap too small

Lack of fusion		Uneven torch manipulation Insufficient inductance (short-circuiting arc) Voltage too low
Slag inclusions		Technique – Too wide a weave Current too low Irregular weld shape

Spatter – on work
       on nozzle
       in weld
		Voltage too high
		Insufficient inductance
		Insufficient nozzle cleaning

Irregular weld shape		Excessive electrode extension Wire temper excessive, no straightening rolls Current too high for voltage Travel speed too slow

Figure 12.21 Weld defects and their causes

the lack of fusion. If the voltage becomes too great in relation to the wire speed, large drops will form on the end of the wire. These will be reluctant to leave the end of the wire and will often settle as spatter beside the weld. When the ratio between the voltage and the wire speed is correct, a highly characteristic hiss or hum will be heard from the arc.

The mutual position of the welding torch and workpiece in relation to each other is of importance to the appearance and quality of welding. In practice many different combinations of inclination, torch handling and

welding position will be experienced by the body repairer.

MIG/MAG welding positions

Welding positions can be as follows: flat, horizontal, vertical, overhead and, owing to the complexities of panel shapes, a combination of these. In collision repair, the welding position is usually dictated by the location of the weld in the structure or panel assembly of the vehicle. Both heat and wire speed parameters can be affected by the welding positions.

	Thrusting weld	Drawing weld
Width of seam	wider	narrower
Upper bead	smaller	larger
Penetration	decrease	increase
Tendency to lack of fusion	greater	lesser

Pushing weld: torch sloped in direction of weld
Pulling weld: torch sloped away from direction of weld

Pushing weld   Pulling weld

Pushing weld sloping downwards   Pulling weld sloping downwards   Pulling weld sloping upwards

Figure 12.22 Welding position (*Migatronic Welding Equipment Ltd*)

Flat   Horizontal   Vertical   Overhead

Figure 12.23 Typical MIG/MAG welding positions

Flat welding is generally easier and faster and allows for better penetration. When welding a panel section that is off the vehicle, try to place it so that it can be welded in a flat position.

When welding a *horizontal joint*, angle the welding torch upwards to hold the weld puddle in place against the pull of gravity. When welding a vertical joint, the best procedure is usually to start the weld at the top of the joint and pull downwards with a steady drag.

Overhead welding is the most difficult. In this position the danger of having too large a weld puddle is that some of the molten weld metal can fall down, either into the nozzle, where it can create problems, or onto the operator. Therefore, always carry out overhead welding on a low voltage setting, which will cut down spatter while keeping the arc as short as possible and the weld puddle as small as possible. Press the nozzle of the torch against the work to ensure that the wire is now moved away from the puddle. It is better to pull the torch along the joint with a steady drag.

MIG/MAG basic welding processes

There are six basic welding processes used in collision repair work on vehicle bodies.

Tack weld

A tack is a relatively small temporary MIG/MAG weld that is used instead of a clamp or a self-tapping screw to tack and hold the panel in place while proceeding to make a permanent weld.

Like the clamp or self-tapping screw, the tack weld is always and only a temporary device. The length of the tack weld is determined by the thickness of the metal panel to be welded and is approximately a length of 15 to 30 times the thickness of the metal panel. Tack welds must be done accurately, as they are very important in maintaining proper panel alignment.

Continuous weld

A continuous weld is an uninterrupted seam or weld bead which is carried out as one steady progressive movement. While supporting the torch securely, use a forward movement to move the torch continuously at a constant speed, checking the weld bead frequently as you progress. The welding torch should be inclined between 10° and 15° to obtain the best weld bead shape, weld line and gas shielding effect. It is essential to maintain both the distance between the tip base and the metal surface and the correct torch angle; otherwise a problem can arise owing to the wire length being too long. When this occurs, penetration of the metal will not be adequate; therefore, to improve penetration and produce a better weld, bring the torch closer to the base metal. When the torch handling is smooth and even, the weld bead will be of consistent height and width with a uniform closely spaced ripple.

Figure 12.24   Tack weld

Figure 12.25   Continuous weld

Figure 12.26   Making a plug weld

Figure 12.27   MIG/MAG spot welding

Figure 12.28   Lap spot weld

Figure 12.29   Stitch weld

Plug weld

A plug weld is used to join two metal panels together. The outer metal panel is first pre-drilled or punched. Next the arc is directed through the hole to penetrate the inner metal panel. The hole is then filled with molten metal to form the plug.

Spot weld

In MIG/MAG spot welding the arc is directed to penetrate both metal panel pieces through the joint position while triggering a timed impulse of wire feed.

Lap spot weld

In the MIG/MAG lap spot weld technique, the arc is directed to penetrate equally between the edge of the upper metal panel and the top surface of the lower metal panel of the lap joint, and the puddle is allowed to flow onto the edges of the top piece. This is achieved by using the correct angle of the welding torch to the joint.

Stitch weld

A MIG/MAG stitch weld is a series of inter-mittent seam welds carried out at intervals along a lapped metal panel joint, where a continuous weld is not necessary or would cause distortion due to heat input.

Applications of MIG/MAG welding in vehicle body construction

The continuous search for increased effi-ciency in production techniques in the automotive industry has led to all major manufacturers adopting the use of MIG/MAG welding on a very wide scale. This welding process is ideally suited to flow line production processes owing to its inherent automatic characteristics. MIG/MAG weld-ing has numerous advantages over the major-ity of processes likely to be encountered in automotive production. The most obvious is its capability of being used either semi- or fully automatically, thus giving it virtually unrivalled versatility. In automobile con-struction it has almost totally replaced oxy-acetylene and manual metal arc welding by virtue of its high metal deposition rates, electrode efficiency and high duty cycles. Further advantages of the process are good penetration, even without edge preparation, the ability to handle all metal thicknesses and high weld strength (618 MN/m^2 being easily achieved).

MIG/MAG dip transfer welding is a low-heat-input process owing to the relatively low amperages required, and it is ideal for welding fine sheet thicknesses because of the absence of distortion with minimal jigging. With the absence of flux, further advantages accrue, the most obvious ones being the elimination of de-slagging and chipping back when second weld starts are required. The absence of flux also means that the welder has a clear view of arc opera-tion and weld bead appearance and is not subjected to excessive fumes. Spatter is min-imal and weld bead appearances enhanced owing to the improved wetting of the parent material by the molten metal on account of the various additives present in the wire composition.

The MIG/MAG welding process can in many cases replace existing welding techniques, usually manual metal arc and oxy-acetylene, to give superior welding conditions and economies. Semi-automatic and more particularly fully automatic MIG/MAG welding processes with a wide range of technical advantages have opened up new areas of production technique. Numer-ous companies in the United Kingdom have already adapted MIG/MAG welding to their own particular requirements. On a mod-ern mass-produced passenger car the body shell is a relatively low-stressed unit, high stresses imposed by road conditions being carried by sub-frame assemblies. However, where continuous welds are required or spe-cific components carrying high stresses are to be fitted MIG/MAG welding is employed using semi-automatic equipment. Thus com-ponents such as safety harness attachments, door hinges, jacking points, spring shackles, steering column brackets and seat frame guides are attached using the dip trans-fer technique, which enables welds of this type to be made on thin shell panels with no distortion of the panel and deep penetra-tion into the thicker workpiece. Where MIG/MAG welding has been introduced into automotive production lines, its value as a production process is well established.

Applications of MIG/MAG welding in body repair work

The semi-automatic MIG/MAG welding process which is used in the construction of vehicle bodies is also used for the repair of these bodies. Equipment is now specially produced for thin-gauge welding in the repair of vehicle bodywork.

Welding in the repair of car bodies was mainly carried out using oxy-acetylene and resistance spot-welding techniques. Both of these techniques have certain disadvantages. In the oxy-acetylene method the weld metal and surrounding panels are liable to distort owing to heat input. In spot welding if the repair is visible, the joints, which in most cases are lapped, must be solder filled to achieve an acceptable finish.

The advantages of MIG/MAG welding equipment in repair are first that it is a dual machine capable of welding in any position, and it can weld material from 0.5 mm to 4.4 mm in thickness. Second, it can be used for spot welding from one side of the panel only, which is an advantage when welding inaccessible panel assemblies. The equipment itself is very portable, consisting of a transformer, rectifier, built-in feed wire system, lightweight welding torch

Figure 12.30 CEBORA synergic pulse (MIG/TIG/Arc)

with interchangeable nozzle heads for conventional welding and spot welding, earthing cable and clamp and gas supply in steel cylinders, all of which are incorporated on a manoeuvrable trolley which lends itself to bodywork application in the workshop.

Most equipment used in motor body repair usually has the following controls and functions: a weld timer which can be set from 0.2 to (usually) 2.5 seconds; some form of wire speed control which ranges from 2 to 12 m/min; and a voltage control usually ranging from 0 to 10, the 0 setting giving the lowest welding current. On some machines the settings for wire speed and welding voltage are coupled together into one control, which can either be on the machine or positioned on the welding torch for instant adjustment. The following function controls are the most widely used: seam for continuous welding; spot, which makes use of a time setting to ensure uniform spot welds; and stitch, which is used through the weld timer and pause timer and provides intermittent welding so that the welding is interspersed with pauses, therefore reducing heat input and preventing melt-through on thin materials and gapped joints.

Preparing the equipment for welding

1 Select the correct dimensions of wire: 0.6 mm is best for welding thin materials and panels; 0.8 mm is used for general purposes; and 1.0 mm is used for aluminium only (using aluminium wire).
2 Select the tip for the type and dimension of wire being used and fitted to the equipment. Also select the torch nozzle for seam or spot welding application.
3 Mount the welding wire reel on the reel holder, roll off the tip of the wire and insert it into the inlet between the feed rollers of the wire feed unit.
4 Connect the machine to the main supply, set the wire speed selector, pull the welding torch trigger and feed the wire through the hose to the welding torch till

it appears at the end of the contact tip. (Some machines have a wire inching switch which is used when a new reel of wire is put into the machine and fed to the hose, and the wire is fed at a gentle speed to the torch.)

5 Connect the hose for the shielding gas to be used and set the gas flow regulator between 5 and 15 litres/min.

6 Set the wire speed and voltage control and select the mode of welding operation.

7 Connect the earth return clamp as close as possible to the point of welding, making sure of a secure connection to facilitate a good welding circuit.

8 Before commencing to weld, a visor-type or hand-held welding screen must be used for the protection of the welder. Then hold the welding torch down to the component to be welded, and an arc will be struck between the wire electrode and the workpiece, thus commencing the welding.

9 Before welding on a car it is essential that the electrical connections to the battery, starter motor and generator/alternator are disconnected in order to avoid damage to the electrical equipment of the car.

For the best results for seam, spot or stitch welding, the surface of the panel should be cleaned to bare metal condition, as paint or grease causes bad welds. In bodywork applications the heat input is limited on each side of the weld owing to the shielding gas; this has the effect of reducing possible distortion to a minimum so that adjacent parts do not get so hot as to damage trim, screen rubbers and windscreen glass. It can also be used to great advantage in repairing by sectioning or patching, as two panels can be butt welded and then the weld completely dressed off on one side with a sanding machine and finished off, using normal panel-beating hand tools, to a perfect finish. As this system of welding gives a perfectly controlled penetrating weld of great strength, it will be found ideal for replacing new panel sections

such as sub-frame and underframe members, where the welding is highly critical because of the stress factor present in load-bearing assemblies. The speed at which this welding can be carried out in the repair of the modern mono-constructed body shell is proving itself to be an economical asset.

MIG/MAG welding of low-carbon steel and HSLA steels

Until the mid 1970s most if not all vehicle bodies were constructed from plain low-carbon steel, but since then there has been an increase in the worldwide demand for corrosion resistance in body structures. This has led to the introduction of single-sided and double-sided coated galvanized steel, high-strength steel and aluminium. The Japanese, in particular, and the European industries have moved to the use of HSS steels in further attempts to reduce body weight. Body skin panels have been reduced from around 0.9 mm to 0.7 mm, with structural components reduced from 3 mm to between 1.2 mm and 2.0 mm.

The use of these different materials can lead to problems for the welder. Welding galvanized steels (zinc coated) can reduce rust-resistant properties around the weld areas if they are not correctly treated.

Cosmetic body panels are the thinnest panels on the vehicle and are generally repaired by the use of butt, fillet, lapped-edge or corner joints. It is therefore necessary to use as low a current as possible when using the MIG/MAG process on these thin panels.

The choice of shielding gas for the process is important, as it affects not only the quality of the weld but also the depth of penetration of the weld into the material and the amount of weld spatter that is formed around the finished weld. It also controls the temperature, which affects the suitability of the process for the thickness of materials to be welded. The choice of shielding gases commonly used can be seen in Table 12.4.

Table 12.4   MIG/MAG recommended operating conditions

Application	Metal thickness (mm)	Shielding gas	Wire dia. (mm)	Current (A)	Voltage (V)	Wire feed speed (m/min)	Welding speed (mm/min)	Gas flow (l/min)
Horizontal/ vertical fillet	1.0	Argoshield 5	0.8	66	14	3.4	900	12
	1.6	Argoshield 5	0.8	130	18	9.7	600	14
		Argoshield 5	1.0	160	17	5.9	563	14
	3.0	Argoshield TC	1.0	180	20	6.4	391	15
		Argoshield TC	1.2	180	17	3.8	430	15
	6.0	Argoshield TC	1.2	260	27	7.0	480	16
	10.0	Argoshield TC	1.2	290	27	7.4	430	16
Butt weld, flat position	1.0	Argoshield 5	0.8	66	14	3.0	750	12
	1.6	Argoshield 5	0.8	84	18	4.0	800	14
		Argoshield 5	1.0	125	16	4.0	640	14
	3.0	Argoshield TC	1.0	140	19	4.0	350	15
	6.0	Argoshield TC	1.2	300	28	7.6	650	16
	10.0	Argoshield TC	1.2	300	28	7.6	680	16
				320	28	7.6	580	16
Butt weld, overhead	1.0	Argoshield 5	0.8	50	15	4.0	360	12
	1.6	Argoshield 5	1.0	150	17	5.9	600	14
	3.0	Argoshield TC	1.2	180	19	4.3	520	15
Butt weld, vertical up	1.0	Argoshield 5	0.8	45	15	3.5	350	12
	1.6	Argoshield 5	1.0	130	17	4.7	410	14
	3.0	Argoshield TC	1.2	135	17	3.1	275	14

Under the right conditions, any of the gases in the table can be used for the range of thicknesses encountered in the bodyshop. However, the ideal for welding thin body panels is the colder mixture of argon, with 2 per cent oxygen and 5 per cent carbon dioxide. When welding structural components and chassis members the material tends to be thicker, up to 3 mm or even more, and so it is necessary to use a hot arc in order to penetrate the material. For this reason a mixture of argon with 2 per cent oxygen and 12 per cent carbon dioxide is used to give better heat flow, greater tolerance to variation in setting and low spatter generation. It is also a common practice to increase the wire diameter from 0.6 mm to 0.8 mm when welding these thicker panel assemblies.

MIG brazing

As well as being used in vehicle repair, this process is also used in vehicle production on some assembly lines, as it offers increased speed and a smooth flat bead requiring little dressing after brazing. The MIG brazing process can be used with the traditional brazing filler materials of aluminium bronze, phosphor bronze and silicon bronze in the form of continuous wire on a reel. With this process the choice of filler material and shielding gas is important if the necessary flat bead is to be achieved; otherwise excessive amounts of dressing would be required. The shielding gas used is normally pure argon. However, gases that are a mixture of helium and argon produce a more fluid weld pool, which again leads to a flatter bead than with argon. A 70 per cent argon and 30 per cent helium mix is generally preferred for thinner cosmetic panels, since the richer 80 per cent argon and 20 per cent helium mix can produce excessive penetration with an increased risk of distortion, although this can be used to advantage on thicker panel structures.

Traditional MIG welding wire melts at about 1450°C. The zinc or aluminium galvanizing coating on car body panels melts at 420°C and vaporizes at 906°C. This means that as soon as a conventional MIG arc is struck you are burning the coating and releasing toxic gases. The current recommendation is to use pure argon; but a 2 per cent carbon dioxide – CO_2 – is very popular. Bronze, that is copper and tin – CuSn – melts at 950°C, and the popular silicone bronze melts at 980°C. To prevent the parent metal coating getting to the point of vaporization a programmable synergic MIG is used set fairly low at 45 to 65 Amps. The synergic MIG works in pulses, almost laying down the bronze in droplets. To ensure that it works smoothly a Teflon gun liner should be used.

As this is not welding, that is not fusing, the two parts together one technique is to leave a gap of about 0.5 mm between the two work pieces so that the bronze filler can flow into the gap with capillary action.

MIG welding of aluminium

While most welding equipment is supplied primarily for the welding of ferrous metals, some can also be used for the welding of aluminium. Equipment with low amperages is really not suitable, although it can be used for short periods of welding. The larger-amperage machines (180 A and over) are better equipped to handle aluminium. The wire sizes used are 1.0 mm and 1.2 mm for larger machines for thicker aluminium. The torch contact tip must be the correct size for the wire to be used. When welding with aluminium wire a Teflon liner must be used in order to prevent the aluminium from sticking

Figure 12.31   MIG Brazing

and damage occurring to the wire itself. Also, pure argon must be used as the shielding gas owing to its total inert characteristics, and not argon mixes or carbon dioxide.

Welding stress and distortion in the MIG/MAG process

The MIG/MAG welding process and its effectiveness may be measured by its ability to produce joints of acceptable quality. Unacceptable joint quality may arise from a failure in the specific design or from the incident of weld defect. These problems, however, are preventable if their causes are understood and the correct operating procedures are followed.

The most common welding problems are:

- Residual stress and distortion
- Incorrect bead geometry and appearance
- Defects
- Loss of properties

Residual stress and distortion

The solidification and contraction of the weld bead will induce strain and consequently stresses in the joint. In the case of the contraction of the solid metal the stress will be equal to the change in temperature as the metal cools from its melting point to its ambient temperature. Under normal circumstances movement of the weld bead is restricted by the adjacent body panels in the vehicle structure, and the stress which could be generated is given by the Young's modulus of the material. This stress level often exceeds the elastic limit of the material, and plastic deformation takes place. The stresses locked in the material, which may reach levels up to the elastic limit of the material, are called residual stresses, and the deformation of the material is known as distortion.

Residual stresses

In a butt weld the weld bead will tend to contract longitudinally and transversely, and this will induce tensile stresses in the weld and also balancing compressive stresses in the sheet material. In the joining processes which rely on heating or fusion, it is difficult to prevent the formation of residual stresses. If the heat-affected zone is ductile and defect free (as in thin-sheet panel steel) the presence of some residual stresses may be acceptable. The most common technique used to relieve residual stress in thicker materials is post-weld heat treatment. This consists of uniform heating of the joint to a temperature at which the yield stress of the material is lowered and the residual stresses are relieved by plastic deformation.

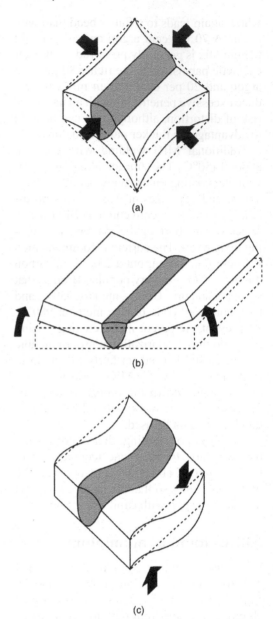

(a)

(b)

(c)

Figure 12.33 Types of distortion: (a) longitudinal and transverse shrinkage (b) angular distortion (c) out-of-plane buckling distortion (*BOC Ltd*)

Figure 12.32 Residual stress in a simple butt weld (*BOC Ltd*)

Distortion may result in unacceptable appearance (buckled body panels), prevention of subassembly fabrication or the inability of the structure to perform its intended function (alignment of body panels after welding).

Distortion

Distortion may take the form of a change in dimensions of the joint, transverse or longitudinal shrinkage, angular movement or out-of-plane buckling.

The following steps may be taken to minimize distortion:

1 Use the minimum amount of weld metal. Over-welding and excessive reinforcement should be avoided in fillet welds and flat butt welds. Intermittent or stitch welding may be used.
2 Use square edge on narrow gap procedures to reduce angular distortion when welding.
3 Use high travel speed and low heat input to limit heat build-up in the panels to be welded.
4 Use the back-step weld sequence or preset the joint.

Incorrect weld geometry and appearance, defects, loss of properties

Root concavity or suck-back

This may occur if the rear of the weld pool is too hot or large, and the combined effect of contraction and surface tension results in a root surface which is concave. The surface profile will often be smooth and less likely to produce stress than over-penetration or lack of penetration; however, it may result in unacceptable loss of cross-section. The cause of this defect is incorrect parameter selection, in particular the use of high current or slow travel speed.

Overfill

Overfill or excessive reinforcement may be described as the presence of weld metal which exceeds that required for the joint. This

Figure 12.34 Root concavity: suck-back (*BOC Ltd*)

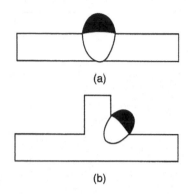

Figure 12.35 Excessive reinforcement in (a) butt weld and (b) fillet weld (*BOC Ltd*)

Figure 12.36 Root convexity: excessive penetration (*BOC Ltd*)

creates an unacceptable appearance or surface finish and would require weld dressing.

Root convexity

This results from excessive penetration on the underside of a full penetration weld. The cause of this defect is incorrect welding parameter selection (shielding gas or operating technique).

Underfill and undercut

If insufficient weld metal is deposited in the joint, the parent material may remain unfused and the joint may be underfilled. Alternatively, the weld area may be melted, but the combined effect of the arc force and the flow of the molten metal may prevent complete joint filling or produce a depression in the surface at the weld boundaries: this is known as undercut. Underfill is usually avoided by careful placement and operating technique. Undercut normally occurs at high speeds and high current and may be avoided by reducing both current and speed.

Figure 12.37   (a) Underfill and (b), (c) undercut (*BOC Ltd*)

Figure 12.38   Lack of penetration (*BOC Ltd*)

Figure 12.39   Lack of fusion (a) at sidewall and (b) inter-run (*BOC Ltd*)

Lack of penetration

This is classed as partial penetration and is not acceptable, particularly with respect to fatigue, loading and corrosion resistance on the joint, where full penetration is required. The defect results from poor joint preparation and unsuitable welding operating parameters, especially current settings.

Lack of fusion

Lack of fusion between successive runs or between the weld bead and the parent metal can produce serious crack-like defects. These defects reduce the load-carrying area of the joint and produce stresses and are therefore unacceptable in welds which are subjected to fatigue loading or static loading. The cause of the defect is usually incorrect operating parameters, high deposition rate or low heat input.

Spatter

Spatter takes the form of small particles of metal which are rejected from the arc or the weld pool and adhere to the surface of the metal being joined. Spatter may be unacceptable for purely cosmetic reasons: it may also inhibit the application of surface coatings (as in vehicle painting) and may initiate corrosion. Spatter usually results from either excessive voltage or insufficient induction in dip transfer or argon-enriched gas mixtures.

Porosity

Porosity results when bubbles of gas are nucleated in the weld pool and trapped during solidification.

The source of the gas may be a chemical reaction such as the formation of carbon monoxide from carbon and oxygen or the expulsion of gases which have been dissolved by the liquid metal but are relatively insoluble in the solid phase. The problem is best controlled by eliminating the source of the pore-generating gas. This may be achieved by effective shielding and the prevention of leaks, the ingress of air or moisture to gas lines and joint surface contamination from oil, grease, moisture or paint.

Inclusions

Non-metallic inclusions may be trapped in the joint owing to inadequate cleaning. Inclusions tend to be more angular than gas pores and can result in higher stress levels as well as forming a plane of weakness in the weld. The cause of this defect is usually insufficient inter-run cleaning, incorrect welding parameters or **incorrect gas shielding**.

Figure 12.40   Solidification cracks (*BOC Ltd*)

Solidification cracks (hot cracking)

Solidification cracking occurs when the solidifying weld pool is exposed to transverse shrinkage. The last part of the weld pool to solidify is usually rich in low-melting-point impurity elements, and liquid films are often present at the centre line of the weld; as a result the ductility of the weld metal is lowered. The transverse contraction strain can be significant and, if the low-ductility area along the centre line of the weld is unable to accommodate this strain, cracking takes place. The problem is usually avoided by careful selection of the weld metal composition.

Weld testing and inspection

Weld testing and inspection can be divided into two types. In non-destructive testing, the test samples are not destroyed in the process. In destructive testing, the test samples are destroyed in the process.

Tech note
There is a qualification known as **coded welder** that is gained by completing a series of tests as to weld quality. This qualification is needed for those working on pressure vessels and pipe work.

Visible examination

This is normally carried out during and after welding and prior to any other non-destructive or destructive test being used. This visual check will usually determine the following:

- Weld size
- Profile or weld face shape
- Surface defects in weld face
- Undercut and overlap
- Root defects
- Weld penetration
- Surface slag inclusions.

Non-destructive testing

Non-destructive testing of weld samples is normally carried out using the following methods.

Macro examination

This method uses a low-power magnification to examine weld specimens which have been levelled, polished and etched to detect the following: lack of fusion, lack of penetration, porosity, oxide inclusions, internal cracks.

Crack detection

By using dye penetrant, surface defects in both ferrous and non-ferrous metals may be detected. A solution of coloured dye is sprayed onto the weld and parent metal and allowed to soak. The dye is then washed off and the surface dried. A liquid developer is then sprayed onto the weld to give a uniform dry powder coating which is white in colour. The coloured dye oozes out of any crack in the weld into the white coating and can be seen in normal lighting conditions.

Magnetic particle method

Surface defects in mild steel and low-alloy steels may be revealed by this method. The test specimen is connected to a special power source. The magnetic test liquid is sprayed along the weld. The magnetic particles then collect along a line of crack when the current flows through the weld. Better test results are obtained if the surface of the weld has first been ground smooth. As these penetrants and magnetic test liquids give off harmful vapours, the work must be carried out under well-ventilated conditions.

Ultrasonic inspection
This method uses sound waves which are passed through the weld. They are transmitted as pulses by a probe connected to an ultrasonic test set. Defects reflect these pulses back to the probe through a flexible cable which is attached to the ultrasonic test set, where it is displayed on an oscilloscope screen as deflections of a trace. These deflections measure the location and size of the defect. Grease is used between the probe and the work to improve sound transmission. Also, light grinding of the weld or parent metal surface to remove spatter may be required before testing.

Radiography
This method uses penetrating X-rays or gamma-rays which are passed through the weld and recorded on photographic film held in a light-proof container. When the film has been processed, any defect in the weld will show up as shadows in the photographic image. Only trained and radiation-classified personnel may operate radiography equipment, as the equipment is extremely dangerous.

Destructive testing

Mechanical testing is a destructive procedure and therefore cannot be carried out on any component required for use. Representative test samples produced under similar conditions to the in-service components, for example welding procedure tests, are normally used and accurate comparisons made. The tests most frequently used to assess the properties of welded joints are as follows.

Tensile test
This test can be used to assess the yield point, ultimate tensile strength and elongation percentage of the weld specimen. Failure usually occurs in the parent material; therefore exact measurements are not usually obtained for the weld itself, although in this case the ultimate tensile strength (UTS) of the weld is higher than that of the parent metal. Test specimens are cut from the designated area of the weld assembly, the edges are smoothed and the corners are radiused. If an elongation value is required, then two centre punch marks often 50 mm apart are applied, one on either side of the weld; this is called a gauge length. The test equipment can vary, but the basic principles are that two sets of vice jaws are used to clamp the specimen, hydraulic power is applied to force the jaws apart, and a dial calibrated in tonnes or newtons records the load. As the load increases, the dial registers the amount of applied load until fracture occurs.

Bend test
Bend tests are carried out on butt welds and are used to determine the soundness of the weld zone. The test piece is the full thickness of the material, with the weld bead reinforcement dressed flush on both sides. It is common practice to take two specimens from a test piece: one can be bent against the face and the other bent against the root. A bend test machine consists of a roller or former connected to a ram and operated hydraulically. Unless otherwise stated, the diameter of the former should be four times the weld specimen thickness. The specimen is placed on support rollers, and the centre point of the former is brought into contact with the weld face or root. Pressure is then applied to bend the weld specimen through either 90 or 180°.

Nick-break test
In this test procedure the butt-joint weld must first be ground flush. Then two nicks (saw cuts) are made, one on either end of the joint in the weld specimen. The weld

specimen may be broken during bending, or alternatively the specimen may be placed in the jaws of the tensile test machine and struck with a hammer when tensioned.

Impact test

There are two main types of impact test, the Charpy V notch and the Izod test. Both tests employ a swinging pendulum which breaks accurately prepared specimens to which a notch has been applied. Both the tests have the same principle, which is to determine the energy measured in joules absorbed by the notched test piece at a specified temperature as it is broken from a single swinging pendulum. During fracture, energy from the pendulum will be absorbed by the specimen, more energy being absorbed by strong materials than by brittle materials. The distance the pendulum swings after fracture is measured by a pointer on a dial calibrated in joules. The lower the value indicated, the more brittle the specimen; conversely, the higher the reading, the greater the toughness.

Vehicle repair weld testing quality control

Those vehicle body repair companies wishing to obtain British Standards will have to conduct weld test procedures. These procedures will have to become an integral part of the company's policy on quality in the repair of vehicles while maintaining safe working practices. In addition regular testing will demonstrate the skill and ability of body technicians to maintain a high quality of weld standard in all types of crash repair work carried out by their company.

Equipment maintenance and safety

For the best results the gas and contact nozzle of the welding torch must be cleaned of welding spatter at regular intervals. Unless this is done, there is a risk of the welding current arcing between the contact nozzle and the gas nozzle to the workpiece, which could cause severe damage to the welding torch. When cleaning the gun of spatter and deposits the condition of the gas distributer should also be checked, and if it appears to be burnt and worn by spatter then it should be replaced. This is because in addition to distributing the gas it also acts as an insulator and is of vital importance to the service life of the swan neck of the torch. It is recommended that the machine should be cleaned with compressed air once yearly, or more frequently if it is used in dusty conditions. The feed mechanism and rollers require more frequent cleaning. Regular maintenance to the following parts is important: gas nozzle, gas distributor, contact nozzle and wire guide rollers. It is never advisable to weld with a worn-out contact nozzle and deformed gas nozzle.

Any exposure to a naked welding arc, even for a fraction of a second, is sufficient to damage the cornea of the eye (welding flash). Consequently a welding screen should always be used, preferably of the helmet and visor type. By using this type both hands are free for control over the welding equipment. The hand-held type of welding screen is of course fully adequate from a safety point of view but only leaves one hand free. This could result in a shakier movement of the gun and a loss of precision in following the joint.

Wear welding gloves and suitably fitting protective clothing, especially when welding overhead so that the welding spatter does not run down the neck or sleeves of the overall worn.

Another hazard is the build-up of fumes given off by the welding process, which could be dangerous to health when welding in a confined space. When welding in a normal workshop, there is no problem, but if welding in a tightly enclosed area, such as inside the boot of a car, make sure that the area is well ventilated.

Precaution must be taken when welding vehicle panels in case there are flammable materials attached to the other side of the panel which could ignite and burst into flames. Therefore always have adequate fire-fighting appliances available.

Skills and questions

Gas shielded arc welding is the most common type of welding in the vehicle repair industry, and as with the other skills, it needs practice. There is no alternative to practise.

1 Explain the term 'MIG/MAG welding'.
2 Explain the term 'TIG welding'.
3 What is an inert gas? Explain its function in welding.
4 What are the advantages of gas shielded arc welding?
5 Name three types of gases that could be used in MIG/MAG welding.
6 What is the difference between water-cooled and air-cooled torches in TIG welding?
7 What determines the size of the ceramic nozzle to be used for TIG welding?
8 Which material are the electrodes made from in the TIG welding process?
9 State the advantages of MIG/MAG welding when compared with oxy-acetylene welding, for the repair of vehicle panels.
10 When MIG welding aluminium and its alloys, which type of gas would be most suitable for use?
11 Describe the difference between plug welding and spot welding.
12 Explain the mode of metal transfer used for welding thin-gauge materials.
13 Describe the three functions that most MIG/MAG welding equipment is capable of performing.
14 Explain the term 'spray transfer'.
15 Describe the purpose of a contact tip in a MIG/MAG welding torch.
16 Show, with the aid of a sketch, a repair situation in which resistance spot welding can be substituted by a MIG/MAG spot weld.
17 With the aid of diagrams, describe the following defects made by MIG/MAG welding: (a) porosity, (b) incomplete fusion, (c) spatter on workpiece, (d) poor penetration.
18 What colours are used to identify argon and argon-mix cylinders?
19 Why is it important to disconnect a battery on a vehicle before commencing any MIG/MAG welding operation?
20 List the safety precautions to be observed before and during welding.
21 Explain the form in which the electrode is manufactured for MIG/MAG welding.
22 With the aid of a sketch, show a repair application which involves MIG/MAG spot welding.
23 Describe the major features of a TIG welding torch.
24 Which mode of metal transfer uses the short-arc process in MIG/MAG welding?
25 With the aid of a sketch, illustrate one method of metal transfer used in MIG/MAG welding.
26 With the aid of a sketch, explain how plug welding is carried out in repair work.
27 Describe the differences between the processes of TIG welding and MIG/MAG welding.
28 By using a sketch, illustrate a typical weld fault in MIG/MAG welding.
29 State the function of the controls of standard MIG/MAG welding equipment.
30 Explain the setting to be used when welding a large gap with MIG/MAG welding.

13

Craft techniques and minor accident damage

Panel beating: forming panels by hand

Essentially, panel beating is a hand method of producing hollow or double-curvature shapes by means of hammering. Nevertheless, the panel beater's craft still retains its place in bodywork and as yet is irreplaceable by more modern methods; in spite of the tremendous developments in recent years of mechanical methods of forming, panel beating remains an essential means of fabrication of special parts. Some metal shapes cannot be produced at all by mechanical methods and others only with great difficulty, and in such cases panel beating is used to finish the shape that has been roughed out by power processes. Often, too, the prototype of a component ultimately to be made in quantity by stamping or pressing is hand made to allow minor modifications to be studied before mechanical production begins; the part produced by panel beating is used as the pattern for press or stamp tools. Panel beating may also be used where a small number of components only are required and where the cost of press or stamp tools would be uneconomical. In body repair work, panel beating is used to advantage where sections which are either unobtainable or uneconomical to replace completely can be fabricated by hand either in part or as a whole. In many cases corroded areas can be repaired by fabricating new sections for replacement

purposes. In the body-building trade, panel beating is still used to a large extent where new vehicles are being built either in aluminium or mild steel. Many of the components for these vehicles are still made using the traditional hand shaping methods. Also a lot of the aluminium moulds used in fibreglass construction, where highly developed double-curvature shapes are needed, are made by hand, welded and dressed and planished or wheeled to a final finish.

Panel beating is essentially a hammering process, involving different kinds of blows that can be struck on sheet metal. It should be borne in mind that most metals used in bodywork possess high malleability and may be overstretched even with a wood tool. The types of blow that can be struck on sheet metal are three:

- *Solid blow*, where the work is struck solidly over a steel stake.
- *Elastic blow*, where either the head or the tool or both is made of a resilient material such as wood.
- *Floating blow*, where the stake is not directly under the hammer.

Each type of blow has its uses for particular purposes. A solid blow will stretch the sheet and may be necessary when forming a panel, bending a curved strip or angle, removing a loose or tight place in a sheet or throwing

The Repair of Vehicle Bodies, Andrew Livesey, 9780815378709, 2019, Taylor & Francis

an edge over when thickness is not a consideration. An elastic blow will form metal without undue stretching; indeed, metal can be thickened if desired, as in working out a tuck or pucker. The floating blow is given to the metal when it is held over a suitable head and hit 'off the solid', so forming 'dents' at the points of impact.

Shaping metal by hand

Thin-gauge mild steel and light aluminium sheet material up to 1.2 mm, and in some cases even 1.6 mm, can be satisfactorily hand beaten by wood mallets into double curvature forms. The usual practice is to beat the metal in a suitable recess in a wood block or upon a sandbag in what is known as the **hollowing** or blocking process. Alternatively, the metal can be hammered into a wood block hollowed out to the shape of the job.

Another method is that of *raising* the metal by means of floating blows over steel stakes or wood formers. This raising process can be speeded up considerably by 'taking in' or making tucks at the edges of the metal. This method is generally called **puckering**, and after being made each pucker must be eliminated by careful blows to drive the metal into itself, so thickening the work at the edges. Much greater advantage may be taken of puckering in aluminium alloy sheet than in most metals because of its malleability and ductility. The exact degree of shaping of the metal which is permissible depends upon the particular alloy being worked, its malleability and the ability of the craftsman.

Another method of shaping metal which can be used in conjunction with hollowing and raising is that of *wheeling*. A panel can either be wheeled from a flat sheet to the desired double-curvature shape, or it can be pre-shaped using either hollowing or raising techniques and its final shaping and smoothing completed by wheeling. The technique of wheeling is not entirely done by hand, as a wheeling machine is used. The finished result depends on the skill of the craftsman as he manipulates the sheet by hand in the wheeling machine.

An experienced panel beater uses the beating method suitable to the job in hand, complex shapes often being beaten up by using both hollowing and raising or wheeling methods. For light-gauge material too much hollowing or blocking is not to be recommended, as it tends unduly to thin the metal. The skilled panel beater is a craftsman who, to a great extent, relies on a good eye for line and form; this can only be cultivated by years of experience which, combined with dexterity in the use of hand tools, is the secret of his craftsmanship. Wood formers or jigs, upon which the beaten shape can be tried in order to obtain uniformity of shape for each workpiece, are used for many jobs. The shape of the job is retained in the vision of the panel beater, who can, by beating on the sandbag and raising over a suitable stake, obtain a very good approximation to the desired form. This only needs hammering lightly on the wood former to obtain the finished shape. Until the metal fits the former it is often necessary to check repeatedly to find any high or low areas which may prevent the workpiece from fitting the former correctly. Instead of using formers or jigs many panel beaters use templates only, which are cut to the shape of the various cross-sections of the workpiece. These templates are used towards the end of the beating process to check the finished shape of the job. After shaping, the surface of the finished panel has to be smoothed by the technique of *planishing*, using a steel bench stake, or by wheeling.

Figure 13.1 Shaping metal by hand

Hollowing

One of the methods of shaping metal by hand is that of hollowing. In this method the metals used for the purpose of shaping panels for bodywork applications are usually aluminium and its alloys and mild steel; aluminium is by far the easier of the two to shape owing to its higher ductility and malleability properties. This is a process of shaping a sheet metal blank into a double-curvature shape by beating the blank into a sandbag or hard wood block with the aid of a pear-shaped boxwood mallet, or for thin metal a rubber mallet, or for steel a hollowing hammer with a steel head.

First the blank which is going to be shaped is cut to size and the edges trimmed for any rag by filing. If this rag is not removed as the blank is shaped, it may tend to split from the edge of the rag. The next step is to place the metal blank, which should be held tilted, onto a sandbag, and to give it a series of blows around its outside edge, working towards the centre by means of a pear-shaped mallet. This hammering has the effect of sinking the metal into the sandbag, which is of course resilient to the blows. After each blow the disc is turned and the next blow struck near the first one, and so on until a series of overlapping blows is made round the circumference of the blank. The tendency will be for creases or wrinkles to appear on the edges of the metal, and these must be gently malleted flat to avoid overlapping of the metal and eventual cracking. The hammering of each course has to be done with steady and even blows to bring up a regular curved shape. On completion of the first course of blows, a second course is begun further in from the edge of the metal than the first set. This process of hammering in courses and rotating the blank is continued until eventually the centre of the blank is reached; by this time it will have taken on a double-curvature shape, but should greater depth be required the whole operation must be started again, working gradually from the edge towards the centre of the

Figure 13.2 The technique of hollowing

work until it is completely to the desired shape and size.

At this stage templates cut to the correct size and shape can be used to check that the panel being shaped is the correct size and curvature. Alternatively it is possible to use a jig constructed to the correct size and shape, usually in wood. It may be necessary during the beating-up process to anneal the workpiece to restore its malleability, because the hammering tends to harden the metal by work hardening. In some cases, instead of using a sandbag for the shaping to be carried out on, a hollowed-out recess in a wood block can be used. When the panel has reached the desired shape by hollowing it can then be smoothed to a final finish by hand planishing using a hand dolly or over a stake, or wheeling to obtain the final smoothness.

Raising

Raising is another method of shaping metal by hand into a double-curvature shape. The method of raising is carried out by drawing the metal in courses over a suitably shaped steel stake or wood block, using floating blows which are struck slightly off the stake

Figure 13.3 A pucker or tuck made at the edge of the panel to quicken the raising process

Figure 13.4 Raising

with a boxwood pear-shaped mallet. A series of blows is made in the metal starting at the centre, the blows being struck slightly off the stake. This has the effect of shrinking or reducing the circumference of the blank, thus forcing it down and around the stake. The disc or blank is rotated after each blow as in hollowing, but working from the centre in courses outwards towards the edges of the blank. The same process is repeated with frequent annealing of the metal until the final degree of raising is reached and the desired shape obtained. In the course of the raising, the edges of the metal around the circumference will be continually subjected to creasing, and care and skill are needed to avoid allowing these creases to become too sharp. If they are not worked out the edge will crack or fracture as the shape proceeds.

When the article is partly beaten up over the stake, a series of tucks can be made in the outer edges of the work in order to quicken up the beating or raising process by taking in or shrinking surplus metal. This method is to make a pucker or tuck at any point on the edge of the blank by bending the metal on a stake into an ear-shaped form. After the tuck is formed, the disc is placed on a steel stake and each side of the tuck is lightly malleted to stiffen the metal and hold the tuck in position. This is done in three or four places on the circumference of the blank, thus decreasing its diameter. Then working from the base or point of the tuck, this surplus metal is

malleted out over a stake towards the edge of the disc. This has to be done very carefully with the minimum of blows, and overlapping must not take place. As each tuck is formed and worked out, the blank deepens towards its final shape.

The processes of hollowing and raising in sheet metal are often applied together in the making of articles in the form of double-curvature shapes, bowl-like shapes and so forth. When the work is only slightly domed, the process of hollowing alone may be sufficient to complete the work. There are limitations to the depth which can be obtained by hollowing. This is governed by the diameter of the finished article and access for hand tools. Where the diameter is going to be small and the article deep, the raising method will have to be used to shape the work. Again as in the hollowing process, the final finish can be obtained by planishing the pre-shaped article to a smooth surface finish.

Wheeling

The craft of wheeling has been used for many years in the production of curved panel assemblies that are used to make up the modern vehicle body. Wheeling was a very skilled art when vehicles were coach built and hand methods were employed to make the component panels. With the advent

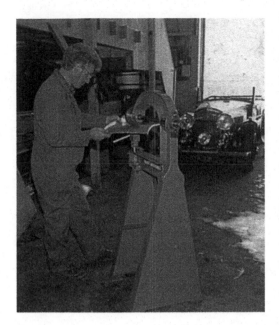

Figure 13.5 Wheeling a panel to restore a vintage vehicle (*Frost Auto Restoration Techniques Ltd*)

of mass production and the development of the motor car, speed of production became an essential factor. Consequently hand-made panels, which were usually made by wheeling, were replaced by pressed panels made on power presses with mating dies, which very speedily produced accurate panels having a good surface finish. Today wheeling is still used to produce panels for prototype vehicles; one-off assemblies which are specially built to order and small-volume production, in which the panels are prepressed to somewhere near the finished shape and then finished off by wheeling. The vehicle-building industry still uses wheeling in producing panels for commercial and private coaches, road transport vehicles and any assembly requiring a double-curvature shape in its production.

The art of wheeling lies in the operator's ability to handle the panel successfully in the wheeling machine. Wheeling is simply the stretching of metal between two steel rolls known as wheels. The upper wheel has a flat face and revolves freely on its own bearings.

The lower wheel also revolves but has a convex curved face and is made interchangeable so that different-shaped lower wheels, from nearly flat to full curved, can be used. The two wheels meet at a common centre at which the stretching takes place when the wheels are tensioned by applying pressure through the bottom wheel against the panel which is pressed onto the top wheel. As the metal panel is pushed through the wheels a stretched area, the length of the panel, is produced, and is known as the wheel track. By carefully allowing these wheel tracks to overlap on the entire surface of the flat panel being wheeled, a curve in one direction only develops. To create double curvature on the same panel, stretching must take place in two opposing directions; therefore a second set of wheel tracks must cross the first set transversely, and this is achieved by turning the panel through approximately 90° before making the second tracks. As the panel is pushed through the wheels the pressure can be increased gradually until the desired curvature is obtained. Therefore the skill of wheeling lies in the use of correct wheel pressures and careful manipulation of the panel through the wheels; this can only be learnt by experience. An alternative method of achieving double curvature on a flat panel is to form the first curve by using a rolling machine, and then to wheel it in one direction only, allowing the tracks to cover the panel in the opposite direction to the first curve, thus stretching the metal in such a way as to form the double curvature shape. Wheeling can therefore be used, first, for shaping a flat metal blank to a finished double-curved panel; second, to finish a pre-shaped panel which has been hollowed or raised to its final shape; and third, to smooth or planish a pre-shaped panel to its final finish.

The materials best suited for wheeling are aluminium, some of the aluminium alloys and mild steel, all of which possess the properties of malleability and ductility to a certain degree. When using the wheeling machine for wheeling aluminium and its alloys, care should be taken not to put too

Figure 13.6   Sports car panel being shaped by wheeling (*Autokraft Ltd*)

Figure 13.8   The technique of wheeling (*Autokraft Ltd*)

Figure 13.7   Sports car panel being shaped by wheeling (*Autokraft Ltd*)

much pressure on the work of raising the bottom wheel. Up to three times as much lift or stretch is obtainable with aluminium as with steel, and much more shaping by wheeling is possible in the case of aluminium than with harder metals like mild steel. Thus too much pressure could have the effect of over-stretching the particular panel or workpiece. The wheeling machine may be used simply to planish, producing a smooth surface by

the friction and rolling action derived from passing the sheet backwards and forwards between the wheels when these have just the right amount of pressure.

Where components of moderate curvature are to be produced by wheeling alone, the sheet or blank is placed between the two wheels at one edge or, in the case of a round blank, in the centre; pressure of the exact amount suitable for both the thickness and the type of material is applied and the sheet is wheeled. In the case of aluminium the wheeling lines are difficult to see, but by smearing the surface of the panel or sheet with mineral turpentine or very light oil these tracks become more easily seen and are therefore more easily lapped. Care should be taken at this stage when passing the metal through the wheels not to twist or jerk the panel, as this could result in ridged sections and an uneven surface. Movement of the sheet is varied until the desired shape is obtained, those parts of the panel which require to be only slightly curved receiving less wheeling than other parts which must be more curved.

To wheel a panel that has been pre-shaped by blocking in a sandbag, it is necessary to smooth its surface without altering its shape, and for this reason the panel should be pulled right through the wheels at every stroke. This edge-to-edge wheeling will result in an evenly stretched panel surface. When carrying out this smoothing work, or planishing as it could be termed, the pressure exerted on the work by the wheels should be very slight. Another common task that can be successfully carried out with the wheeling machine is to tighten up the loose wavy edges which sometimes occur when shaping a panel by hand. The method of overcoming this trouble is to wheel directly adjacent to the stretched edge. This stretches the area being wheeled and so tightens up the loose edge. The reverse sort of problem, that of a panel with a fullness or stretched area just in from the edge of the sheet, can also be overcome by wheeling. The method is to start wheeling from the centre of the fullness on a track parallel to the edge and to work right out to the edge of the panel. With most machines three standard wheels are supplied which are generally referred to as flat, medium and full curved. The widths of the wheel tracks are flat 25 mm (1 in), medium 10 mm (3/8 in) and full curve 5 mm (3/16 in). The best rule to follow when selecting a wheel is to use the flattest wheel possible for the job you are doing. This not only speeds up the wheeling but prevents to a large extent marking the surface of the panel with wheel tracks, which are really the evidence of an overstretched panel.

The most important points when using the machine are:

1 The wheels must be kept clean, free from dirt and in perfect condition.
2 The pressure exerted by the bottom wheel must be correct for both the thickness and type of metal being wheeled.
3 The right-shaped wheel must be used to suit the required shape of panel.
4 When wheeling, the panel should be held without tension, allowing it to move freely without twisting or jerking it.
5 The wheel tracks must be carefully overlapped to achieve a smooth curved surface.

Split-and-weld system of shaping metal

The introduction of welding into the panel beater's craft has led to the development of split-and-weld panel beating, which is at once less laborious and much quicker than the older methods of hollowing and raising. The system consists of making a pattern on a panel jig with pattern paper. The paper is held off the jig by tension at its edge. To allow the paper to drop onto the jig, the paper is slit at suitable points, the edges then opening out to let the pattern fall into position. It is obvious, then, that additional material is required at the slits. This may be obtained in the panel either by stretching the metal at these points

Figure 13.9 Standard wheel sets (*Frost Auto Restoration Techniques Ltd*)

Figure 13.10 Split and weld pattern

until enough is obtained to meet the requirements, or by welding in V-shaped pieces of metal. The final shape is then achieved by wheeling or planishing.

Planishing

The technique of planishing is a very old and established craft in the history of hand-fabricated metal articles. Basically planishing takes over from hollowing and raising, which shape the article, to smooth its entire surface and finalize its shape.

Planishing can be performed in three different ways. First, there is the technique which is used mostly by the panel beater in planishing new work. In this case a planishing hammer is used in conjunction with a steel stake, both having highly polished faces. The steel stake is mounted on a bench and is of suitable curved shape for the article being planished. The work is taken to the stake and planished over it to achieve the final finish. Second, there is the technique used by the body repair worker, where the planishing hammer is used in conjunction with a dolly block which is in fact a miniature stake or anvil, again with polished faces. The dolly block is held in place under the panel by hand, while the blows are directed onto the panel surface and transmitted through to the dolly block by the force of the blow being in direct contact with hammer face, work surface and surfaces of dolly

block. In this method the tools may be taken to the job and the work carried out on the spot. This fact makes planishing ideal for the repair of vehicle body panels. Third, there is the technique of planishing using the wheeling machine as a means of smoothing the work surface. This is accomplished by the friction and roll action of the workpiece as it passes between the two steel rollers. This method is normally used by panel beaters in smoothing and finalizing new work; it can also be used by body repair workers, but the difficulty arises that the panel has to be dismantled and removed from the body shell, and it is therefore an uneconomical proposition.

Consequently planishing using hand dolly and hammer is accepted universally as the best technique in the repair of panel surfaces by planishing.

In some cases the three techniques can be used together; for instance a panel can be planished using a stake and then finished off by wheeling, or a panel can be wheeled, fitted to the job, and then minor rectification carried out using hand dolly and block. All three techniques have one common feature: when planishing, the metal surface is slightly stretched because of the metal-to-metal contact between the working faces of the tools and the work face of the panel. The skill in this process lies in the fact that the craftsman has to merge, by careful hammer blows or wheel action, all these stretched blows into one to create a continuous smooth surface.

Where planishing hammers are employed, the process is carried out over a metal stake

Figure 13.11 Planishing using a hammer and dolly block

Point of impact

Point of support

Solid even blows smooth the surface without altering the shape of the work

Figure 13.12 Planishing using a steel stake

Figure 13.13 Planishing using the wheeling machine

or hand dolly. The planishing is carried out over the whole surface of the workpiece; the blows are light and given squarely; otherwise they will produce crescent marks difficult to eliminate. Each hammer blow produces a flat spot, and the blows are so directed that the spots merge imperceptibly into one another over the whole surface. Any low places or 'valleys' on the surface of the workpiece can be eliminated by careful hammering on the head, which slightly stretches the metal, causing it to rise to the correct contours. Both the hammer face and the steel stake must be kept scrupulously clean and perfectly smooth; otherwise it will be impossible to avoid marking the sheet. Planishing should leave the metal with a dead smooth surface. If this is not attained, small hammer marks can be removed by smoothing off with emery cloth glued to a piece of wood and used like a file.

General guide to the fabrication of hand-made panels

When a panel is required to be made by hand, the first essential is that a jig or former should be made to resemble the exact line and contour. This jig or former can be made in either wood or metal, but preferably in wood, as this is easily shaped to double-curvature shapes. Moreover, if the panel has to be fastened to the jig this can be done by putting small panel pins through the metal

into the wood jig. The holes left by these tacks can be later filled in by welding.

Once the former has been built, the next stage is the making of the blank template, which is the developed flat form for the total shape of the panel being made. Some shapes can be developed using geometric drafting methods, but where there is no recognized method only the trial-and-error basis can be used. This pattern can be made in strong brown paper or special template paper. At this point it is necessary to decide whether the panel is to be made in a one-piece construction or made in several pieces which can then be fabricated by welding. On very complicated shapes it is sometimes necessary to use joints; therefore the exact location on the panel should be considered very carefully, taking into consideration facts such as the length of the joint and its position in relation to accessibility for planishing with hand tools when assembled. Once the pattern draft has been developed it can then be marked from the pattern paper onto the surface of the metal from which the panel is to be made, the two most popular metals being aluminium and mild steel. The golden rule of hand-made panels is that an allowance should be made all the way round the developed size; in other words, the developed blank should be bigger than the pattern, as it is easier to cut off surplus metal on the finished component after shaping than to have to weld pieces to it.

The metal blank can now be shaped by any hand methods. Care must be taken to check continually the shaping against the jig or former to see that no part is over-shaped. In some cases where opposing curves meet or combine, small metal templates cut to the correct curvature are useful to check the relative position of individual curves as the panel is being shaped. As each piece is completed it is tried on the jig and made to fit it exactly by planishing. When all the shaping has been carried out, the next stage is to join these pieces by welding. The only accurate method of doing this is to fasten the appropriate pieces to the jig, making sure that the joints are butted and not overlapped together, leaving no gaps which may need extra welding filler rod which leads to later difficulties in planishing. Once the panels are secured using either clamps or nails, the appropriate sections should be tack welded together at intervals no larger than 19 mm. After each tack the assembly should be cooled for two reasons; first, because the former or jig is usually made from wood, and second, the smaller the heat input the greater will be the accuracy of the alignment of the job.

The work is then carefully removed from the jig, when welding of the joints can commence. The utmost care must be taken when welding to ensure that there is adequate penetration, but not so much as to leave unwanted surplus weld metal on the underside, while reinforcement on the face side should be slightly above the surface of the panel to allow the weld to be completely filed off without losing its strength. It is best to weld small sections at a time, planishing with a hammer and dolly whilst the weld is still hot; this allows the weld to be flattened easily, and it gives the weld inherent

Figure 13.14 Planishing a hand-made panel (*Autokraft Ltd*)

strength. Once the weld is completely finished and flattened in this manner the final finish can be obtained by further planishing and filing, which, if done correctly, should make the weld indistinguishable from the parent metal. The whole assembly is now tried back on the jig, and any small rectification required is carried out by further planishing methods. If this is then satisfactory it is trimmed to the correct size. Any flanges, wired edges or safe edges can then be formed by using either hand tools or flanging jigs or a combination of hand tools and swaging machine. The final finish can be achieved by using different grades of emery paper or a sanding machine with a very fine-grit sanding disc.

Edge stiffening of sheet metal

Edge treatment is a general term used to cover the many methods of forming the edges of sheet metal, panels and components. The bodyworker frequently has to increase the strength and rigidity of the edges of large unsupported metal panels to stop movement and vibration when the vehicle is in motion and to create resistance against buckling and twisting. This can be provided on the sheet itself or by adding stiffening agents. The various types of edge treatment are normally classified as follows:

- *Formed* – In this case the edge stiffening is formed from the metal panel itself.
- *Applied* – Here the edge stiffening is made up as a separate piece and then fixed to the panel edge.

Safe edges, flanges, wiring and swaging can all be classed as formed-edge treatments, while the attachment of strips, half-round beads, mouldings angle sections and false wire edges are classified as stiffening agents. Panel edges are treated in these ways for the following reasons:

1 To stiffen and strengthen the panel or component at its extreme edge

2 To act as a safe edge, as it is important that panels which are to be handled frequently should be effectively treated to avoid the risk of injury due to exposed raw edges

3 To ornament and decorate the panel.

Often edge treatment is used for more than one reason; an example of this may be found on the modern motor vehicle wing, where the edge treatment provides a stiffening effect, gives a safe edge and also has the effect of being pleasing to the eye.

Folding

Folding is the simplest form of edge treatment and is satisfactory when neatness and speed are the main factors. Folds may be creased to give a flushed effect on one side of the panel, or doubled over twice to increase the strength. These edges can be formed either with hand tools or with the aid of a folding machine.

Flanging

In flanging, the edge is formed at right angles inside the panel. In addition to imparting rigidity it can be readily cleared of road dirt, thus reducing the possibility of corrosion from moisture-retaining matter. One disadvantage of this type of edge treatment is that when the edge suffers a severe blow, the metal tends to crease badly, and the edge may crack. When repairing such a fracture, care should be taken to avoid rigidity at one point, which often results in further cracks appearing some distance from the repaired

Figure 13.15 Folded edges (a) single fold, (b) folded and creased edge, (c) double folded edge

Figure 13.16 Flanged edges

portion. In the case where the edge is formed outside, a plastic moulding is clipped over the protruding edge. Stiffening is provided by the formed edge, the plastic moulding being used for decoration and also to render the edge safe.

Flanging is the process of hammering the edge of a piece of sheet metal in such a manner that the required width of metal is worked into a position usually at right angles to its original form. This has the effect of strengthening the metal in the area of the bend. It is used to advantage on panel edges to stiffen, to make the edge safer and to improve its appearance and in many cases to perform the function of a flanged joint where one metal panel is joined to another. If the flange is on a straight piece of metal the flanging technique can be carried out either by using a stake and mallet and hammering the edge over the stake to form a right angle or by using a bending machine, which gives a more consistent edge.

Difficulty may be experienced when forming flanges on curved sections of panels depending on whether the flange is on an external or an internal curve, as these require different techniques of working. This type of flanging can be carried out by using a swaging machine fitted with flanging rolls, which turns the metal edge at right angles as it is fed through the rolls. Owing to the awkward shape of some panels, they cannot be flanged in the swaging machine and so must be processed by hand, using either a stake and mallet or a hand dolly and mallet, and finishing off with a planishing hammer. If the panel is of a complicated curved nature and requires flanges, it is sometimes necessary to make a flanging jig of two identically shaped pieces of wood. The panel is inserted between the two pieces and clamped, then the metal edge protruding is carefully hammered over. In some cases annealing has to be carried out where the corners are very sharp so as not to split the metal. When taken from the jig the panel retains the curved shape with the flange following its contour. To form a flange around the edge of a cylinder, it is placed against the edge of a stake so that the width of the metal to be flanged lies on the stake. A stretching hammer is then used to stretch the flange metal, working the cylinder steadily round and keeping the width of the flange constantly on the face of the stake. The maximum amount of stretching must take place on the outside of the flange, gradually diminishing to nothing at the inside. The metal must be kept flat on the top of the steel stake, so that the hammer strikes the metal hard on the stake top. In the case of a flange raised around the edge of a flat disc, a mallet is used instead of a hammer, and the metal is gradually drawn inwards or upwards by careful working round the edge over a curved end of a steel bench stake. By allowing the outer edge of the flange to be drawn in slowly a good deal of creasing is avoided; moreover, the creases which do occur around the edge are carefully worked out as they appear. In the working of a deep flange on a disc the metal may be annealed at intervals, except, of course, in the case of coated metals such as tinned or galvanized steel.

Swaging

A swage is a moulding or indentation raised upon the surface of sheet metal by means of male and female rollers, the rollers being made in a variety of contours. The machine to which the rollers are attached may be driven either by hand or by electric motor, the choice of machine often being influenced by the type or amount of work to be undertaken. Although swaging has many similar functions to that of wired edges, it is not

Figure 13.17 Common swaged edges: (a) ball swage – may be produced in various sizes according to the size of the wheels used; is used as a stiffener on sheet metal; (b) return curve swage – may be produced in various sizes; is used as a stiffener in bodywork; (c) joggle swage – used to produce a creased lap joint in sheet metal; (d) radius swage – used to produce radius corner joints; the small flange may be trimmed off thus forming a butt joint

confined to edge treatment but may be used some distance from the edge within the limits of the throat of the machine. In addition, as no extra allowance of metal is required, a saving in material with consequent reduction in weight is achieved with the use of swaging. Special composite swages are also available for use on panels; the rollers are made in sections so that their width may be adjusted. If it is found that the throat of the normal swaging machine will not accept curved panels or wings, the rollers may be fitted onto a wheeling machine which is virtually throatless.

The projecting shape of the swage above the surface imparts considerable strength to sheet metal articles. Panels which would otherwise be slack and lacking in rigidity can be stiffened by the use of swaging. Motor vehicle body panels are subject to vibrations and fluctuating stresses, and for this reason swages are used to obtain a suitably rigid body shell. The return curve swage is frequently used to strengthen the centre portions of cylindrical containers because of its high resistance to externally or internally applied forces. In addition to strengthening purposes, swaging is often used to relieve plain surfaces; this kind of decorative effect is a common feature on motor vehicle bodies.

Wiring

Wiring is the process of forming a sheet metal fold round thin wire to give extra strength to the edges of a panel and also to improve its appearance. The wired edge, although not so popular as in the past, is still used where impact strength is the most important requirement of the edge. Mudguards on earth-moving equipment meet conditions in service to which the wired edge lends itself most admirably, and repairs can be readily made, if necessary, with the aid of welding equipment. The wired edge is still found to be a useful form of edge treatment on the wings of public service vehicles,

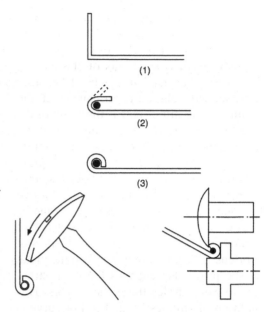

Figure 13.19 Wiring process:
1 Allowance marked off and sheet folded
2 Metal beaten over wire with mallet
3 Edge closed using wiring hammer or wiring rolls

particularly where the wings are not integral with the body.

To carry out the technique of wiring, the edge of the metal should be bent up in the folding machine or by hand, using a mallet and stake, at the corresponding bending line. The bend should not be too sharp, as the metal has to be worked round the wire. Next a length of wire is cut to size and placed in position, and the metal is beaten over by a mallet to hold the wire in position. The edges can be finally closed by using a wiring hammer or by passing through a swaging machine fitted with wiring rolls.

Techniques of damage rectification

Before a systematic approach to body repairs is possible, it is necessary to understand the characteristics of sheet metal as used for body panels. When a flat sheet of metal is bent to a wide arc or radius it will

Figure 13.18 Wired edges: (a) plain wired edge and (b), (c) creased wired edges

regain its former shape when released; that is, it is elastic or possesses elasticity. However, if this sheet is bent to a short arc or radius it exceeds the limits of elasticity or flexibility; the metal in the bend becomes stiff and will take on a permanent set and retain the curvature. This is the result of the stresses which have been set up at the bend, making the material work hardened. Before the sheet is formed in the press the grain structure is constant and the thickness uniform throughout. When the metal is formed to make the body panel it is bent beyond its elastic limit. The outer surface stretches or lengthens while the inner surface shrinks or shortens. The pressure exerted on the metal by the press also changes the grain structure to work harden the surface layers. This build-up of stresses in the bend or curve is an essential factor in the design of vehicle body panels which together form the body shell. A common feature in the design and manufacture of a motor vehicle is the many curved surfaces which are normally referred to as *crowns*. Vehicle body panels consist of flat or slightly curved areas, sometimes quite large and elastic in nature (low crowns), such as door panels; these are held in position by stiffened, rigid sharp bends and swages which are non-elastic in nature (high crowns), such as the cant of roof panels.

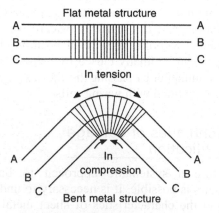

Figure 13.20 Change in grain structure during pressing (*Sykes-Pickavant Ltd*)

If a panel is damaged in an accident the buckled area, being sharply bent, will create additional stiffness in the panel, whether in an elastic or non-elastic area. The slope of the buckles surrounding the sharp creases will be fairly elastic, but a greater amount of effort will be needed to reshape the sections of the panel which are made rigid either in manufacture or through accidental damage. When a panel becomes damaged due to impact, the resulting force on the metal causes buckling in the form of creases or ridges which are created because the panel has gone beyond its elastic limits to become non-elastic, therefore establishing unwanted rigid sections within the damaged area on the panel. The characteristic stiffness of the ridges prevents the panel returning to its original shape unless additional force is applied to release the stress in the ridges in the damaged area. When these stresses in the unwanted rigid areas are released, the elastic areas will also be allowed to return to their original shape. It is important that these corrections be made in the right sequence on the individual panel; otherwise additional damage will be caused to the panels. Repairs must be performed using the reverse order and an opposing force to that of the original force which caused the damage. Consequently the correct sequence should be first to remove the last ridge which was formed and then to work towards the first point of impact of the damaged area.

Rectification of vehicle bodies, following a true assessment of the damage, can be divided into two stages: roughing out or straightening of the reinforced sections and panels to approximately their original shape and the finishing or preparing of the surface to a smooth appearance for repainting. Both stages are of prime importance, and many man hours can be saved if the job is processed correctly. In cases of damage where the body is distorted, the temptation is to use rough-and-ready methods depending on brute force to restore some resemblance of shape. Whilst this may speed up the first stage of a repair, it will be found that such

1 Low-crown, elastic areas
2 High-crown, rigid areas

Figure 13.21   Prepressed wing panel

methods result in additional marking of panels; considerably more time will be spent on the final stage of finishing than would be required if more thought had been given to the job in the first instance and better methods had been used to rectify distortion. Damaged panels should be restored by relieving the stresses which have been set up by the force of impact. The skill of all body repair techniques lies in the correct handling of the basic hand tools, in a variety of combinations best suited for the job in hand.

Hammering techniques

Unlike most other trades, where the hammer is used with a follow-through action from a combination of wrist, elbow and shoulder, in the skilled hands of a body repair worker the planishing hammer swing is a rhythmic action involving finger and wrist movement, producing a ringing blow. The hammer should not be held tensely, but during the complete cycle of movement it should be held loosely in the hand. This will achieve a higher degree of accuracy and at the same time help to reduce fatigue. This loose holding of the hammer applies equally to dolly

Figure 13.22   Swing of the planishing hammer

blocks, as it permits them to bounce back naturally and to assume the correct position for striking the next blow. With practice the wrist becomes strengthened, and consequently working in restricted places becomes easier where an even wrist action is impossible. The dolly should be allowed to lie naturally in the hand with the face to be used uppermost and, as with the hammer, should be held firmly but not tightly. Tap lightly at the dolly to obtain the feel of metal to metal, and check for control of force of blow; each blow will give a metallic ring which should be the same for each stroke of the hammer. When no metallic ring is heard the hammer is not hitting the metal in alignment with the dolly.

Roughing out damage

In minor repair work which can be carried out using hand tools, the first major operation is to reshape the damaged area back to its original contour. This is done by a technique known as roughing out, which must be carried out prior to any finishing process such as direct hammering or planishing. Roughing out is the reshaping of the area by hand with the aid of a heavy dolly, which forces back the damaged section to its original shape.

When repairing collision work, the normal method of correction is to reverse the process which caused the original damage. In a case of minor repair the point of impact is now the lowest part of the damage. To reverse the process this point on the underside of the panel should be struck using the same force as was originally directed against it.

If this spot is hit accurately with a roughing-out dolly using the same force, the panel will spring back almost to the contour it had prior to the damage. In some cases you will be able to correct the panel damage with a single blow which will spring the panel back to its original shape. In other cases, where the repair is larger, it will be found that several blows are necessary. Hold the roughing-out dolly lightly in the hand and strike the hardest blow at the centre of what appears to be the lowest point of the damaged area, then direct the blows around the first one and gradually work outwards, decreasing the force of the blows until all the damaged area has been roughed out. However, in most cases the damage will not be completely restored to its original contour, although it will be roughed out and can be straightened to its correct contour by direct hammering or by combination of direct hammering and indirect hammering.

The use of a heavy hammer for roughing out is not advisable, for this permits heavy blows which are concentrated in small areas and invariably results in stretching or otherwise distorting the metal, whereas a well-directed blow with a dolly that matches the original contour of the repair spreads the blow over a larger area, resulting in very little distortion of the metal. In some cases body repair workers use a boxwood mallet for roughing out, because there is less chance of stretching the metal. The technique is similar to that of using a dolly, as the mallet is used on the inside of the panel to hammer the damaged section back to its original shape; then the work is finished off by direct hammering using a panel hammer and dolly. A disadvantage in using a mallet is that on modern panel assemblies there is insufficient space to use a mallet for roughing out; therefore most bodyworkers find a dolly more useful.

Direct hammering

Direct hammering is in fact the process of planishing, and the body repair worker uses it as a finishing process after the work has been pre-shaped and roughed out. It is the essential practice to master and develops as a result of continuous experience.

Figure 13.23 The technique of roughing out damage

Before using the hammer and dolly together, it will be necessary to clean the underside of the portion of the wing or panel on which you will be working. Body panels and wings are covered with a sound-deadening material which must be removed before starting the work. If you fail to clean the surface of this material it will not only stick to your dolly but will to a large degree destroy its effectiveness.

It is most important to choose the correct dolly block for the job, because they differ in shape, curvature and weight. In repairing the high-crowned radius of a wing you will have to use a dolly block with a high-crowned radius. In repairing large body panels and door panels which are fairly flat it is necessary to use a dolly block with a low-crowned radius. In direct hammering, by having a dolly which matches the original contour under the damaged area and striking it with a hammer, you are pushing the

Figure 13.26 Using a high-crowned dolly

uneven displaced metal surface back to its original contour to give a smooth and level finish. The dolly provides support and prevents the undamaged areas that have been previously roughed out from being pushed out of place. If you do not strike squarely over the dolly, you will be hitting an unsupported area of the repair and will displace the metal, creating further damage that must be rectified later.

Direct hammering requires skill in directing the hammer blows and close observation of what you are doing so as not to hit the metal too hard, thereby displacing it. Perfect coordination between your two hands is necessary to enable you to move the dolly around under the damaged area and still continue to hit squarely over it with the hammer. Start hammering by using light blows; these will not do the job but will show you whether or not you are hitting squarely over the dolly. Do not forget to let the dolly just lie in your hand and to grip the hammer loosely. A true ring will be heard if you are directly over the dolly; otherwise the sound will be dull. Increase the force of the blow gradually until you have found just the right force to push the raised points of the roughed-out section back without flattening the surrounding metal. The hammer should bounce back of its own accord so that it is ready for the next stroke. Likewise the dolly will spring away from the surface, and the normal resilience of your arms will bring it back, striking a

Figure 13.24 Positioning of blows in the roughing-out technique

Figure 13.25 The technique of direct hammering

Figure 13.27   Using a low-crowned dolly

blow on the metal from underneath. These things will occur normally only if you hold both hammer and dolly loosely.

Figure 13.28 The technique of indirect hammering

Indirect hammering (off the dolly block)

Indirect hammering is another technique which uses hammer and dolly to level a panel surface. A low area can be raised by hammering round the outer edges in such a manner that the rebound action of the block tends to push the low area upwards to its original contour. This in fact is achieved by the sequence of hammering just off or at the side of the dolly block; hence the name 'indirect hammering'. This technique is used in conjunction with direct hammering or planishing to achieve a final finish on the panel surface. Metal that has not been excessively hammered, displaced or stretched will have a tendency to return to its original contour of its own accord. This is due to the internal strain imparted to the metal by the forming dies in manufacture. If the metal is prevented from springing back by other strains imparted to it by additional bends or creases that have been formed by accident, the metal can then be restored to its normal contour by relieving whatever new strain is holding it out of position.

In direct hammering a dolly block having the correct contour to match the original shape of the panel is held under the low spot, and a series of light blows are aimed around the outer edge of this low spot and slightly off the dolly block. The light blows will not displace the surrounding undamaged area, but the force of the downward blow will be transferred to the dolly block. As a result of receiving the hammer blows indirectly, the dolly will rebound, and the hand holding the block will automatically bring it back in place so that it imparts a light push upwards on the area. The centre of the damaged area will slowly rise until the original contour is restored.

Spring hammering

This is another technique of using hand tools to smooth and level a panel surface. In this case only a hammer is used, and it is not supported with a dolly block. The technique is used to reduce high spots which sometimes form as a panel is planished. In some cases these high spots can be reduced by careful, controlled hammering which spreads the force of the blow over the area of the metal, thus reducing the high spot. When a crown or curved surface is formed in a metal panel, it becomes strong in that it resists any change to its shape. The strength of this crowned surface can be used to support the surface being hammered without the use of a dolly. This type of hammering is called spring hammering, and it can be used to correct high spots on metal panel surfaces. To take advantage of a great amount of the natural support provided by the crown of the metal, the force of the hammer blow is spread over a larger area. Once the metal is

Figure 13.29 The technique of spring hammering (*Sykes-Pickavant Ltd*)

can strike the underneath of the low spot on the panel with one of its rounded corners. It must be noted that if the operator does not hit exactly in the centre of the low spot, he will raise metal in some unwanted place. Accuracy is therefore essential, and it can be achieved by holding a finger in the low spot and lightly tapping the underside of the panel with the rounded corner of the dolly until you feel that it is exactly beneath your finger, then strike a sharp blow and raise the metal at this point. After each low spot has been raised in this manner, these points can be filed to check that they are level with the surrounding panel surface.

The second common method of raising low spots is by pick hammering. Bringing up low spots with a pick hammer is more difficult than by the use of a rounded corner of a dolly block. With a pick hammer more accurate placing of the blow is required. Likewise greater control over the force of the blow is necessary. Start using the pick hammer in a manner similar to the dolly block. Hold the end of your finger in the low spot. Tap the under surface of the panel until the pick is directly below your finger. Then strike a light blow from beneath the panel of sufficient strength to form a pimple in

back to its original contour, additional hammering will cause the surface to sink below its original contour line, and it may not be possible to raise it readily. Always start with light blows, and as the repair nears completion, inspect the work after each blow. This will reduce the possibility of sinking the surface too low. Keep the surface of the hammer face clean and highly polished. Any marks on the surface of the hammer will be transferred to the surface of the metal and create additional work.

Removing low spots

Low spots can be removed in several ways, the two most common being the use of a pick hammer or a dolly block. When using the dolly block, start by holding it so that it

Figure 13.30 Pick hammering used to remove low spots

the low spot. Care must be taken to avoid overstretching the metal by using too hard a blow. These pimples represent stretched metal but in being formed also raise the surrounding metal. When all the low spots have been raised with a pick hammer in this manner, the pimples can then be lightly hammered level by direct hammering and finished by filing.

Figure 13.31   The technique of filing (*Sykes-Pickavant Ltd*)

Filing

Filing is one of the most important aspects of finishing a body panel. It is carried out using an adjustable file holder, fitted with flexible blades which can be adjusted concave or convex to suit most contours on the average vehicle body. Initially the file was used for smoothing off panels prior to sanding and locating high and low spots. With the introduction of body solder and later metal and plastic fillers, filing took on an even greater importance in the finishing of repairs on body panels. Filing indicates any irregularities in the repaired surface of a panel and is carried out as the panel is planished. First of all fasten the correct file blade to the file holder with the cutting edges of the teeth facing away from the handle or operator. Adjust the contour of the file holder so that it is almost but not quite matching the contour of the surface on which you intend to work.

One hand is used to hold the file handle, while the other grasps the knob at the opposite end. The file should be applied with long, straight strokes, pushing it away from you along the length of the panel. Short, jabbing strokes should never be used, as these will only scratch the panel and will not indicate low spots. If the file digs in, too much pressure is being applied, and hence a need for reduction is essential. At the end of the first stroke, raise the file and, without dragging it over the metal, bring it back to the starting position and make a second stroke. Repeat this procedure until the area has been covered, making the file marks parallel to one another. This is termed **line filing**, and

it indicates the levelness of the panel in the direction in which it has been filed. At this point both the high and low areas will show up. The high spots can be corrected by spring hammering and the low spots by direct hammering, pick hammering, or in some cases by using the corner of a dolly block. Line filing indicates curvature in one direction only, and as most panels are double curved the panel surface must be cross filed to give an accurate contour check. **Cross filing** means a change in the direction of the file strokes so that the file is moved at an angle between 45° and 90° over the previous file strokes, thus checking the accuracy of the curvature in that direction.

After filing, and prior to refinishing the panel, the damaged area is sanded using a fine-grit sanding disc which leaves a smooth, even surface ideally suited for painting.

Sanding or disc grinding

Use of grinder

Several general rules govern the use of the disc grinder. If these are observed they will enable the operator to become proficient very quickly in the use of the grinder. The rules are considered good shop practice and are directed towards the safety of the operator. In the first instance, if the device is electrically operated see that it is properly connected and earthed. Shop floors are usually of cement; they are generally moist and, therefore, relatively good conductors of electricity. If the grinder is not properly earthed it is possible to receive a fatal electric shock when the machine is in use. Always wear goggles to protect the eyes from flying particles of metal and from small abrasive particles that come loose from the grinding disc. Always replace worn discs as soon as a tear is noticed; torn discs may catch in the work and twist the grinder out of your hands.

Always maintain a balanced position when using the grinder. This position not only permits perfect control over the machine at all times, but it will also produce less fatigue over longer periods. When operating the grinder, hold it as flat as possible without permitting the centre connecting bolt to come in contact with the surface being ground. Hold the grinder so that only 40 to 65 mm of the outer edge of the disc is in contact with the surface being ground. The grinder must never be tilted so that only the edge of the disc contacts the surface. Failure to observe this will cause gouges or deep scratches in the metal which will be hard to remove. Move the grinder from left to right, overlapping the previous stroke with each new stroke. Make the cutting lines as clean and straight as possible. Move the grinder in the same manner whether using it for the removal of paint or for rough or finish grinding. For most grinding operations, finish grind in the longest direction possible on the repaired surface.

Sanding discs

The coated abrasive disc is the part of the sander that does the actual sanding, and selection of the right grit and coating for each job is important. There are five different minerals which are commonly used for manufacturing abrasives. These are garnet, flint, emery, aluminium oxide and silicon carbide. Aluminium oxide is the most important of these. Because of its toughness and durability it is used in the motor body repair trade, where it is chiefly used on metal. The abrasive is fixed on a backing which is either of paper, cloth or a combination of the two.

For dry grinding or sanding, high-quality hide glues are used for anchoring the abrasive grains to the backing. For wet sanding, resins are used as the bonding agent.

Coated abrasives fall into two additional classifications based on how widely the minerals are spaced. If the minerals are close together the abrasive is close coated. When they are widely spaced it is open coated. In close-coated abrasive discs, the abrasive is applied in such quantity as to entirely cover the backing. In open-coated abrasive discs, the backing is from 50 to 70 per cent covered. This leaves wider spaces between the abrasive grains. The open coating provides increased pliability and good cutting speeds under light pressures. Open-coated discs are used where the surface being ground is of such a nature that closely spaced abrasive materials would rapidly fill up and become useless. When grinding a body panel, always use an open-coated disc up to the time that the area being sanded is completely free of paint. Then use a close-coated abrasive to grind the metal to the point where the surface needs no further correction. The final finish is accomplished with a fine grit to get the surface smooth enough for re-finishing by the painting department.

Hot shrinking

One of the most important skills in the repair of damaged panels is that of hot shrinking. It is important because in most cases of collision causing the damage of body panels, stretching of the metal takes place. The actual process is carried out by gathering the stretched metal into a common centre or area and then by heating this section. The panel steel is then at its best condition to be hammered down, thus reducing the surface area and so making shrinking possible.

The oxy-acetylene flame is used as a means of heating the panel, and care must be taken to reduce the spread of heat to the surrounding area of the panel. This can be done by cooling the panel with water after every shrink. The advisable welding nozzle size when using for a 1.00 mm panel is a number 2 nozzle.

Tools required for hot shrinking

Before commencing to heat up the stretched section, it is essential that all tools and material needed to carry out the shrink should be conveniently placed so that they can be brought into use quickly. The tools required are as follows:

- *Wire brush* – The wire brush and scraper are used in preparing the panel prior to shrinking. With the scraper any anti-drum or underseal compound must be removed before applying the heat.
- *Mallet or shrinking hammer* – The mallet face has a soft surface much larger than that of a planishing hammer. If used together with a suitable dolly block it will bring the metal down to a level surface whilst hot and also avoid stretching the panel, as the mallet is made of wood and not metal like the panel hammer. Another tool that can be used is the shrinking hammer; this is similar to a planishing hammer, but the faces have cross-milled serrations, which reduce the tendency for the hammer to stretch the metal because of the very small contact area between the points on the serrations and the panel.
- *Planishing hammer* – This should be used to complete the shrink after malleting, as the hammer is better for levelling out the surface of the panel than the mallet.
- *Dolly block* – This should not be fuller in shape than the actual panel being repaired. Also it should be of a rather light weight so that it forms a relatively weak backing for the malleting, thus reducing the stretching during the levelling operation. A grid dolly has a serrated face to reduce the possibility of stretching in the panel.
- *Damp cloth* – This is helpful in checking the spread of heat, thus reducing the risk of panel distortion.

Hot shrinking process

Arrange your tools so that they are within easy reach, as it is necessary to change quickly from one tool to another when performing a shrinking operation. Locate the highest point in the stretched section of the panel with which you are working. Light the torch and heat the spot to approximately 10 mm diameter in the centre of the high spot or stretched area to a cherry red, using a circular motion when heating the spot.

Figure 13.32 The technique of hot shrinking

Be careful not to burn through the metal by over-heating. As soon as the spot is cherry red place the shrinking dolly (grid dolly), which is usually fairly flat, under the spot and strike several sharp blows with a mallet around the heat spot; this will force the surplus stretched metal into the heat spot. Then the blows are directed on the centre of the heat spot, which pushes the metal down while it is still red hot. At all times the dolly block must be held loosely against the underside of the panel, exactly centred under the heat spot. After four or five mallet blows, the last being on the centre of the heat spot, the heat spot will turn black and should be quenched immediately with a water-filled sponge. Repeat this operation, taking the next-highest spot in the stretched section of the panel until the bulge or fullness is finally shrunk down below the level of the surrounding surface. It can then be brought up to its correct level by planishing with a hammer and dolly and then finishing with a panel file.

The following points should be noted:

1   Never quench a red-hot heat spot; wait until the metal has turned black.
2   Never heat an area greater than that which can be hammered with mallet and dolly.
3   Never use anything but an oxy-acetylene welding torch for heating a stretched section.
4   Never attempt to shrink a panel until it has been roughed out.
5   Always hammer the stretched section outwards before applying heat.
6   If the stretched part of the surface is small, make a smaller heat spot.
7   It is possible to shrink metal without quenching each spot. However, the shrinking operation is much faster when each spot is quenched with water, as fewer heat spots are required if the heat expansion is drawn out by quenching than by additional spots.
8   In some panels it is possible to use a spoon for the backing tool for the hammering operation, especially on door panels or over inner constructions.
9   When performing a shrinking operation, take care to avoid over-shrinking the panel. This will cause the metal to warp and buckle both in and out of the stretched area, owing to overheating of the stretched section. If this does arise, heat a small spot in the area where the panel is buckling, apply a dolly block or spoon with enough pressure to hold the buckling section up, then allow the metal to cool. Do not use the mallet or water in this case. In extreme cases like this it may be necessary to repeat this operation in several different places in the buckled section.

Shrinking aluminium

Sheet aluminium can, with the application of heat, be subjected to shrinking. The process is similar to that used for sheet steel except that the work should be carried out faster than when shrinking steel. This speed is essential because aluminium is a good heat conductor, and the spread of heat must be prevented from distorting adjoining panels. Unlike steel, aluminium does not change colour when heated and, because of the melting temperature differences, great care must be taken not to melt holes in the panel while it is being heated.

Shrinking equipment

Shrinking can also be carried out using shrinking equipment, which can be either a specialized piece of equipment or an attachment to a MIG welder. Basically the equipment consists of a power source to which two cables are attached: one is to the shrinking torch fitted with a tungsten or carbon electrode (which is interchangeable), and the other is to the earth return clamp for completing the circuit.

To use the equipment, first position the earth clamp, making sure of a good connection. Switch on the power source, which in some cases has a built-in timer, and then apply the tip of the electrode to the highest point on the surface of the stretched panel to be shrunk. The resulting arc will produce an extreme concentration of heat to the small area on the panel surface. Heat until bright red or to the pre-set time, then immediately cool with water. This high concentration of heat in one area together with rapid cooling is extremely effective in reducing high spots by heat shrinking in vehicle repair.

Cold shrinking

Cold shrinking is another method of repairing stretched sections on vehicle body panels. In this case a hammer and special dolly block are used in conjunction with one another. This dolly block is a shrinking dolly and is shaped like the toe dolly but has a groove running along the full length of its top face. The dolly block is placed under the high spot formed by the stretched metal of the damaged panel, and the panel is hammered down into the groove with hammer or mallet to form a valley, care being taken not to make the valley any longer or deeper than is necessary to draw the stretched metal back to its original contour. The valley formed is then filled using body solder, which is filed to give a good finish. From the reverse side of the panel it will be seen that a rib has been formed, which will in effect give strength to the damaged area should this be required. This method is also most useful in reducing welds: the weld on the face of the panel is depressed into the channel of the dolly to form a valley below the line of the face of the panel, and the valley is then filled by the body soldering method.

Body soldering

Body soldering (or loading, filling or wiping, as it is sometimes termed) has become a general practice where owing to the structural design of the all-steel body, the use of normal panel beating methods of repair using hand tools is not possible. It is ideally used to hide a lapped joint in the construction of a body shell and is very useful to the panel beater repairing a windscreen or door pillars or when a dent or crease is backed by a bracket or is double skinned, which prevents the placing of a dolly behind the panel and beating the section out. Body solder was first used in the motor trade in the mid 1930s and was introduced into the repair industry by the car manufacturers, who were using it on the mass production of car bodies. It was during this period that the all-steel body was developed, along with rapid advances in streamlining. To finish a panel in the area of the production welds, which in most cases were lapped, spot-welded joints, body solder was used and found most useful. This method has been continued, and today solder filling plays a major part in the finish and repair of the modern car body.

Composition and requirements of body solder

Body solder consists of 68.5 per cent lead, 1.5 per cent antimony and 30 per cent tin. The solder has to possess the following characteristics:

1 It must remain plastic over a large temperature range so that it stays workable.
2 It must wipe with the solder stick and not crumble, thus providing a clean surface finish.
3 The lead and tin must not separate, as it is worked in vertical or overhead positions.
4 It must be capable of being reheated and reworked without forming hard spots.

Process of body soldering

During the repairs of body damage, some means of reproducing the normal contour of the damaged area is necessary when it cannot be restored by normal panel beating methods. This is a case in which body solder can be used to advantage so the area can be filled and then dressed down to a smooth and perfect finish.

Body soldering requirements are soldering blocks, which are usually hardwood blocks shaped to suit the panel being soldered; a suitable grease or light oil which is rubbed onto the surface of the blocks to stop them sticking to the solder, resulting in a smoother surface finish (tallow is the optimum substance for this purpose); a sanding machine, files and emery cloth for cleaning purposes prior to tinning the metal for soldering; a welding torch or similar heating appliance which produces a low-temperature flame (if a welding torch is used the flame should be feathered, which means slightly carbonizing, resulting in a soft flame); a quantity of body solder and a suitable tinning paste complete with a tinning brush for ease of application; and a clean rag for rubbing off the tinning paste. The operator must also use a self-contained air-fed mask together with a fume extractor for his own protection and to conform with health and safety regulations. Before commencing work make sure that the appropriate tools and other materials are close to the job and within easy reach, thus avoiding a delay during working operations.

The first step in body soldering is to clean the surface to be soldered to a bright metallic finish. This can be done by using a sanding machine with the right type of sanding disc until the area to be soldered is cleaned to the bare metal. Any small particles of paint which the sander will not remove can be cleaned off by using emery paper and a file; this pre-cleaning is very important, as the tinning paste will not tin the surface unless it is perfectly clean and free from paint. Apply the tinning compound by brushing it over the cleaned section so that an area slightly larger than that to be soldered is covered with tinning paste. Using the welding torch with the flame set in the carbonizing condition, which is slightly feathered and gives a very soft flame, heat the tinning compound until it becomes fluid. Then with a clean rag wipe the tinned area to spread the tinning over the cleaned surface, making sure that every part to be soldered is perfectly clean and completely tinned. An important point at

this stage is not to overheat the tinning paste when tinning or the surface will turn discoloured, usually blue, and the tinning will be burnt; solder will not then adhere to this surface unless it is re-cleaned and tinned again.

Hold the welding torch in one hand and a stick of body solder in the other and play the flame over the tinned section, heating it just sufficiently to cause the tinning to begin to flow. Whilst doing this, hold the stick of body solder near the work and apply the flame over both the tinned area and the solder stick so that the melting stage of each coincides. When the end of the solder stick begins to melt, press it against the tinned section, thus causing a quantity of body solder to adhere to the tinned surface of the work. After sufficient solder has been deposited on the surface of the work, select a suitably shaped solder block, which should have previously been greased or dipped in oil or tallow, and commence to push the body solder over the damaged section of the panel. From time to time the flame should be played over the solder to keep it in a plastic or movable state; then, using the solder block, the solder should be moulded to the general contour of the panel. The solder blocks must be kept continually coated with oil or grease to allow them to slide over the surface of the solder without sticking and picking up pieces of solder whilst it is in the plastic state. This coating also produces a very smooth surface on the face of the area being soldered. It is very important to make sure that the tinned surface is heated to the melting point as the solder is smeared across the area being built up; unless this is done a poor bonding between the solder and the panel will be the result and the solder may fall out. After the required shape has been formed and the solder built up to a level slightly above the existing panel, the final finish is gained by filing the body solder with a flexible panel file, being careful to ensure that the level does not fall below that of the surrounding area. After filing, the solder can then be rubbed down with emery paper to give a finer surface finish for painting. The sander should never be used for dressing down the solder except as a last resort, because it is too severe and tends to cut deeply and unevenly into the solder. The dust given off from the sander when using body solder is also injurious. When the final shaping and smoothing of the loaded area is complete it is essential to remove all traces of soldering flux, oils and grease which may have been used during the loading operation. If these were not removed they would have a harmful effect when the section is finally spray painted.

Safety points

1 **Avoid skin contact with fluxes.**
2 **Use applicators of some form when necessary.**
3 **Do not inhale fumes from heat/flux application.**
4 **The use of sanding machine in finishing produces injurious lead dust.**
5 **Health and safety regulations demand protection for the operator in the form of a self-contained air-fed mask.**
6 **A fume extractor reduces the risk of adjacent working areas being contaminated and protects the operator.**

Chemically hardening fillers (plastic fillers)

Although body soldering still provides the best quality of filling for a repair, there are cases in which an alternative method is required. Much research on this matter has led to the development of plastic fillers which would come up to the standard of body solder. These fillers are used in the body repair trade as an alternative method to body soldering. They are based on the polyester group of thermo-setting resins, and require a catalyst or activator to cure them. Therefore the fillers obtainable are of a duo-pack type containing paste and hardener. When they are mixed together a catalytic action takes place, resulting in the filler hardening very quickly. As this plastic filler does not require heat during its application it has certain clear advantages. Its use eliminates fire risk, especially when filling next to petrol tanks or any inflammable material. It also eliminates the problem of heat distortion which can occur when solder filling flat body panels. It is cheaper than body solder and much easier to apply. The first of these fillers developed were released some years ago, and many disappointments were associated with their use; problems encountered were poor bonding to panels, too much delay in hardening and too hard to file when dry. Several fillers now available have been vastly improved; they will harden in as short a period as 12 minutes provided the quantities are correct and the two chemicals thoroughly mixed. When hard these fillers have excellent bonding qualities and will feather out to a fine smooth edge. Filler materials can be dispensed either from wall-mounted, air-operated filler dispensers or from portable dispensers. These dispensers take a 10 kg tin of polyester filler together with a matching cartridge of benzoyl peroxide hardener.

Portable dispensers offer the user the advantage of being able to take the filler from repair to repair or from workshop to paint shop. In some bodyshops it is the normal practice to use a general-purpose filler for the main filling operation; then once shaped and flatted, it is surfaced with a fine filler (30 per cent unsaturated polyester resin plus 70 per cent inert filling material) to bring the repair up to pre-paint condition. Alternatively a bodyshop may require different types of filler such as a general-purpose filler, a glass fibre–reinforced filler (with added fibreglass strands, which provide great strength for bridging holes and strengthening weakened areas in metal and GRP), a fine surfacing filler (a filler with very fine surface finish to fill fine pinholes, shallow scratches, file marks or sanding marks) and, where repairs are being made to galvanized panel surfaces, a filler suitable for zinc (a filler possessing high adhesion to coated steel surfaces). Aluminium metallic fillers are body fillers containing aluminium. The aluminium content makes them particularly easy to mix and apply. Unless the bodyshop is very large it would not be cost effective to have many wall-mounted dispensers, and the portable dispenser offers a viable alternative.

The portable dispenser is operated by pushing the handle down; the filler is then extruded together with the correct amount of hardener for that quantity of filler. With most fillers this is normally 2 per cent by weight. Using a dispenser ensures that the correct proportion is added, making it impossible to over-catalyse the filler and thus eliminating or very much reducing the possibility of spoilt paintwork. With over-catalysed filler there is a risk, particularly with metallic paints, of the repaired area showing through owing to the action of the peroxide in the hardener bleaching the paint. Another major advantage of using filler dispensers is that the filler is kept clean and uncontaminated from the beginning to the end of the tin because there is no need to replace the lid every time the filler is used. This is a feat which is commonly accepted as impossible in most bodyshops owing to the rim of the tin becoming encrusted with filler as the applicator is cleaned off or simply owing to forgetfulness. In any event, even with the lid being replaced every time the filler is used, there is exposure to air and therefore a tendency for the filler to lose its styrene content and thus to become stiffer.

The technique of repair using this plastic filler is carried out by first roughing out the damaged area to as near the original shape as possible, or, if the area has been patched, by hammering the patch down below the original level of the panel. The area is then ground using a sander to remove all rust and paint present. A coarse-grained sanding disc should be used for this operation, as this will provide better adhesion for the plastic filler. Next mix the paste with its hardener on a flat surface using a stopping knife or flexible spatula, making sure that the materials are mixed in accordance with the maker's instructions. The filler can then be applied to the damaged area after making sure that the surface is absolutely clean and free from any trace of oil. If the area to be filled up is deep, make several applications, allowing each layer to dry before adding more filler. After the filler is applied and allowed to

set, it can be shaped to the contour of the panel using a plastic filler file with an abrasive paper backing or sanded using a sander.

Plastic filler should never be used over rusted areas without patching the area with new metal so that the filler is applied to a solid base. It should not be used in areas on a panel or body surface which are continually in direct contact with water. Never drill holes in a panel to give the filler a better grip, as the holes will allow water or dampness to work in between the filler and the metal surface, leading eventually to the separation of filler from the metal surface. The filler should not be applied in any great thickness, especially where excessive vibration occurs, or the surface will crack and fall out. A good standard of finish can be achieved using these plastic fillers if the correct working procedure is adopted.

Body jack (hydraulic)

Development and principles

The evolution in motor car bodywork design and construction has called for many changed methods of handling repairs. This in turn has demanded an increasing amount of repair equipment to augment the traditional hand tools and equipment of the body builder, panel beater and sheet metalworker. Modern equipment has been made necessary by the chassisless construction of modern mass-produced cars and vans, which require careful alignment of the complete structure following any serious impact in addition to panel damage rectification. An essential piece of equipment for repair work is the hydraulic body jack, which is used to push and pull body shells and component body parts into alignment following an accident. Blackhawk Automotive Ltd markets a hydraulic body jack under the name of Porto Power. A range of kits of various sizes is available to suit the needs and capital expenditure of various repair shops. Kits are made up in carrying cases, on wall-board storage, and on trolleys incorporating a small press or in the form of a bench rack. Attachments and fittings have been developed to use with the body jack equipment.

The use of hydraulic body jacking equipment is not new, nor was it invented suddenly. It has developed to its present sophisticated state over many years. At one time the only type of jack available was the ratchet or screw-type body jack. The use of hydraulic body jacking equipment developed from the use of an ordinary hydraulic jack for this purpose. The hydraulic hand jack had all the advantages of providing tons of closely controlled torque-free power for the minimum of effort by the operator. It soon became apparent that the hydraulic jack was ideally suited to repair work because it could be operated in any plane and controlled from outside the car. In the essential repair of collision work, a large percentage of the work will require the use of the body jack to push or pull large areas or sections back to, approximately, their original positions. Hydraulic body jacks can be extended to any desired length by incorporating a number of attachments which are available for pushing or pulling.

The outer skin, or panels, of a body is made from light-gauge metal, placed over a framework of heavier, stiffened metal which is reinforced with various types of supports and braces. In addition to damage as a result of a collision to the outer panels of the body, the inner construction which is attached to the outer panels also becomes damaged, which means that the surface of the outer panel is prevented from being restored to its original contour; hence the inner construction must be restored to its original shape and position either before or at the same time as the outer panel metal is corrected. In some instances it will be found that once the inner construction has been restored to its normal position, the outer one will have been corrected at the same time. Before any correction can be made of such damage it is necessary to restore this inner construction. This is generally done by applying pressure to the damaged member or members. Where sharp kinks or creases

Figure 13.33 Snap 4 tonne standard set supplied for carrying case (*Blackhawk Automobile Ltd*)

1	Manual pump with hose	7	Tube connector
2	4-T Bantam hydraulic hose, 18 m	8	Serrated saddle
3	4-T ram toe	9	Wedge head
4	Extension tubes	10	Plunger toe
5	Snap flex head	11	Spread ram
6	90° V base	12	Flat base

have been formed at any point in the inner construction, it may be advisable to use heat while the pressure is being applied, but only on low-carbon steel and not on high-strength steel. This permits the metal to return to its original shape with little danger of cracking.

In using the body jack, it is important to understand that pressure is being applied at both ends of the jack simultaneously; therefore there could be a danger of distorting adjacent undamaged panel assemblies during the jacking process if the pushing points are not carefully selected. Such pressure can if necessary be applied either locally or spread over larger areas by the introduction and use of pressure pads, which are usually hard wood blocks. The body jack is also useful for providing support or pressure at otherwise inaccessible portions of the outer panels, as well as applying controlled pressure in a higher degree than is possible with hand tools on the various panels.

Basic equipment

A body jack consists of three basic units: a pump, a flexible hose connecting pump to ram and the ram unit. The pump comprises

Figure 13.34 Body jack kit on a mobile trolley (*Blackhawk Automobile Ltd*)

1 Slip lock extension	13 Lock-on female and male connectors
2 Pump	14 Pull clamp
3 10 tonne ram	15 Wedgie ram
4 10 tonne short ram	16 Flat base
5 Wide-angle wedge head	17 Serrated saddle
6 Pull plates	18 Spreader plunger toe
7 Pull ram	19 Clamp toe lock pins
8 Midget ram	20 Pull chain
9 Pull rings	21 90° 'V' base
10 Lock-on extension tubes	22 Wedge head
11 Male connector	23 125 mm flex head
12 Lock-on tube connector	24 Trolley and press stand

reservoir, pump handle and hose and is controlled by a simple open and close release valve. The handle can be screwed into the pump in two different positions for ease of operation. The hose is connected to the ram by a simple quick-release coupler which needs to be only finger tight. Pressure is applied by closing the pump release valve and operating the pump handle. The pump will build up sufficient pressure only to overcome the external resistance against the ram. The need to apply excessive pressure to the pump handle indicates that the ram has reached the limit of its movement. The ram is designed with a snap-on system to enable extension tubes and attachments to be positioned to harness the hydraulic power for any desired type of application.

All collision damage repair work which makes use of the body jack equipment is carried out by using one or another of a number of simple set-ups or, in the case of more complicated repair, a combination of set-ups. The first important step therefore is to understand the set-ups that can be built with this equipment, the attachments required and their application; then it is a question of breaking down a job into its basic set-ups and applying the corrective force in the correct sequence. The corrective force should be applied as near as possible in the direction opposite to the force which caused the damage. The body jack set-up should be applied so as not to push at the deepest point on the damaged section; instead work round the outer edges in ever-decreasing circles,

1 Hydraulic pump	2 Release valve	3 Pump handle
4 High-pressure hose	5 Speed-coupler	6 Spread ram
7 4-ton midget ram	8 20-ton ram	9 10-ton ram
10 10-ton hollow ram	11 7-ton shorty ram	12 Wedgie

1 Thread cap protector
2 Protector ring
3 Spring protector

Fabric
Rubber
Interwoven wire
Rubber

A section of the
high-pressure hose

Figure 13.35 Hydraulic body jack (*Blackhawk Automotive Ltd*)

which will tend to spring the remaining damage into the final position. If the body jack set-up is applied directly against the lowest part of a damaged section without relieving the strain, then as the pressure is applied the metal surface may become kinked and stretched and require further attention to return it to its correct level.

Application of the body jack

Pushing

Pushing is the simplest operation of all and is achieved by inserting the ram between two points and operating the pump. The plunger extends until it touches the point at which the load is to be applied, and as pumping is continued pressure is built up to overcome the resistance of the metal at the point of application. Movement of the damaged area will take place as long as pumping is continued.

Care must be taken when selecting a ram anchor; for example, if a ram was placed between two chassis members and it was intended to push from the undamaged member to straighten the damaged member, the force applied would not rectify the damage but would distort the undamaged member! The first essential is to ensure that the pushing anchor point is stronger than the point which is receiving the corrective force. This can be done by attaching a base plate to the bottom end of the ram to spread or distribute the load, which can be spread over an even larger area by putting a piece of solid hardwood timber between the base plate and the pushing point. It is seldom that a pushing application can be achieved using the ram only, because of its limited travel; therefore there are available various combinations of extension tubes, couplings and pusher heads to cater for any repair requiring straightforward push.

1 Bantam pulling	10 10-ton clamping
2 Bantam pushing	11 10-ton pushing
3 Dozer pulling	12 10-ton pushing
4 Bantam spreading	13 10-ton pulling
5 10-ton pushing	14 Wedge spreading
6 Bantam clamping	15 Spread ram
7 Bantam pushing	16 Bantam pushing
8 10-ton pushing	17 Bantam pulling
9 10-ton spreading	18 10-ton pulling

Figure 13.36 Body jack combinations (*Blackhawk Automotive Ltd*)

Figure 13.37 Pushing with body jack (*Blackhawk Automotive Ltd*)

Pulling

Pulling is also a simple operation but uses a slightly different type of ram. With the standard equipment a pull converter set is available to enable the one push ram to fulfil both functions, but it gives an off-centre pull and is less convenient to use than the separate pull ram. With the direct set-ups using the pull rams, the tubes are under tension and there is no risk of bending. The thread sections of tubes and couplings are now under tension, and it is the threads themselves that carry the load, so it is essential to ensure that they are kept clean, free from dirt and damage and securely mated. The pulling combination obtainable makes use of the pull rams and direct pulling attachments, though it is also possible to use chain plates and chains for obtaining a pull with a push ram. The latter method is quite convenient in such cases as pulling across the width of a body from an undamaged door pillar which has been bowed outwards. It would be necessary to reinforce the undamaged pillar with timber to prevent distortion and to protect both pillars from marking by the chains with suitable packing material.

Spreading

Spreading is a similar application to pushing except that in the latter case there is sufficient room for access for the ram and extension between the two members to be moved apart to permit a direct push. When this is

not possible a means must be found of inserting jaws or attachments which are capable of applying an indirect thrust. The most obvious means of spreading is provided by the wedgies and spread rams. This is an off-centre load once more, and even under the most favourable conditions it is not possible to apply a force of more than about 7 tonnes with the 10 tonne ram, or 2.5 tonnes with the 4 tonne ram.

Stretching or tensioning

The technique of stretching or tensioning is another means of obtaining a pull using a push ram. It is different from the type of pulling previously described, which was a method of applying force to pull towards one another sections which have been forced apart. Here the reverse takes place, and an external pull is applied to pull apart or draw outwards areas that have been pushed or drawn in towards each other. The combination used to obtain this external pull employs toes, links and clamps on the end of the push ram and extension tubes. Two types of clamps are used, the pull clamps for attachment to flat edges and the wing clamps which have deep throats or lipped edges. Both clamps have an alligator type of action and are first tightened down on the centre bolt until the jaws are parallel and in contact with the surface of the panel to be gripped. Pressure is then applied by tightening the rear bolt, which cants the jaws forward and causes them to bite into the surface of the

Figure 13.38   Pulling with body jack (*Blackhawk Automotive Ltd*)

Figure 13.39   Spreading with body jack (*Blackhawk Automotive Ltd*)

Figure 13.40   Stretching with body jack (*Blackhawk Automotive Ltd*)

Figure 13.41   Clamping with body jack (*Blackhawk Automotive Ltd*)

metal. It may well be that the time taken to set up this combination is greater than the time to pull out the damage once it is in position. On flat panels where it is not possible to get at the edges or where there is not an edge to locate a clamp, the same result can be achieved by locating the toes in the bosses of a pair of solder plates which have been sweated onto the panel. A roof panel or car door is a typical example of this. To get at both edges of a car door it is usually necessary to remove the door, but this can be eliminated by using a clamp at the free edge of the door and a solder plate at the hinged edge. If properly used the solder plates will withstand a pull of 13,800 kN/m2. The plates should be sweated on with a layer of body solder about 3 mm thick, using the minimum of heat and quenching with a wet rag to prevent distortion. The same technique may be applied to boot lids, front wings, bonnets and rear quarter panels and can be the means of repairing a panel in a position where direct pressure from inside would not have achieved a satisfactory result.

Repair of component motor body panels

To repair damaged motor body panels requires great skill in the use of hand tools and repair techniques and an ability to assess the cause, extent and sequence of the damage sustained. Damage by accident and its subsequent repair covers a wide range of incidents from minor scratches and cosmetic damage to the write-off. The repair will only be approved by an insurance company after considering the car's age, condition and relevant market value.

Vehicle body repair work can be divided basically into two groups: minor accident repair work and major accident repair work. Body repair workshops vary in their opinions as to the dividing line between the two assessments. Generally they can be defined as follows.

Minor accident repair work

To be able to carry out minor repairs, a good level of ability with hand tools and all types of power tools, both electrical and air operated, is essential to the body repair worker. He or she needs a knowledge of body construction, an appreciation of materials and their properties, and a practical expertise in appropriate repair techniques, including the use of all types of welding equipment.

Minor repair work can be classified as the simple cosmetic repair requiring the use of hand tools only; the cutting out of damaged panel sections and their effective replacement; the replacement of complete panels such as wings, bonnets, boot lids, doors and sill panels; and the finishing of these panels to a point where paint can be applied. A minor repair can also involve adjustment to door locks and window mechanisms, the restoration of body apertures, and the use of hydraulic equipment to realign body panels and apertures to their original shape.

Major accident repair work

Major repair work can be classified as a repair which involves the use of manufacturer's replacement parts and the reconstruction of the body or parts of the body on an approved body jig. A major repair will normally involve all the features listed for minor repairs. In addition it will include the rectification of any misalignment of the underbody and body shell using conventional hydraulic pulling equipment, together with either a fixed bracket jig or a universal jig with measuring system; the realignment of the body shell, correcting damage with a combination of pulling and pushing equipment; and the cutting out of panels and reinforcing members which are damaged beyond repair; and the welding in of replacements.

The conclusion of all repairs must restore the vehicle to its original safe, roadworthy condition to the satisfaction of the owner and the insurance company. In order to achieve this the repairer needs to use his acquired skill and knowledge and, whilst each accident repair has its own individual features, he or she must be able to assess the damage and make the choice of appropriate methods of repair and the best equipment to effect a speedy first-class result.

Repairing rust damage

In this work the body repairer relies on his own judgement, skill and experience. Rust is a corrosion, known chemically as iron oxide, which occurs on the surface of iron and most of its alloys when they are exposed to air and moisture. The designs of all-steel body shells over the past years have provided many pockets which could be termed water or moisture traps, thus creating suitable conditions for the rusting process. In particular, rust attacks wings, sill panels, wheel arches, floors of luggage compartments and bottoms of doors, the worst being the sills because of their close proximity to the road. During the past few years some manufacturers have made improvements which have reduced the rusting of bodies. In some cases this has been achieved by good design and in other cases by treating the metal prior to painting.

When the body repair worker is faced with rust problems, he has a choice of two methods of repair, depending upon the extent of the damage. The best possible results are obtained by cutting away the corroded panel or section and replacing it with a hand-formed section or a factory-pressed panel. The replacement panels can then be fitted by welding them into position and finished by planishing, filing and sanding. In the case of the older vehicle where new panels are not available or where the vehicle's age or condition make it uneconomical to fit new panels, the method of fabricating or patching these areas, by cutting out and replacing the sections from flat sheets fabricated by hand, is adopted. In many cases if the original sections are carefully cut out and not too distorted, they can be used as an excellent guide to the manufacture of the new sections. Templates can also be made up before the old sections are cut out to check alignment and curvature when fitting the new section. Often double-curvature panels such as wheel arch fabrications must be made up in three or four parts for ease of patching onto the original body. The main difficulty experienced in this type of repair is the welding of the new metal onto the old; this is one of the reasons why it is important to cut away all the rusted section so that the new metal can be welded to a rust-free section. In cases in which the sections are very

badly corroded and welding is difficult, the section can be cleaned of surface rust and sometimes brazed to form the joint.

The second method of rust repair, which is only a temporary measure and not recommended for longevity, is to fill the corroded section with either body solder or a chemically hardening filler. If the corrosion is only very slight, giving a pinhole effect, the area can be sanded down to bare metal, hammered down carefully to below its original level and then filled up with body solder and filed to a finish. Plastic filler should not be used in this case because the moisture would seep through, parting the filler from the panel. Where the corroded area has turned to holes it can only be repaired by placing a patch over the corrosion and welding or brazing it in place. This patch must be tapped down below the panel level, then filled with body solder or plastic filler to obtain the final finish. In some cases where it is difficult to weld a patch, the perforated section can be reinforced by using glass-fibre matting impregnated with resin and bonded to the underside of the repair. The surface can then be filled with plastic filler and finished by filing.

Repair of minor accident damage

Repairs arising from minor accidents are usually of a relatively straightforward nature, as the damage is either a dent, a scratch or a scruff of the outer panel surface of the body and does not always involve structural distortion.

When repairing it is essential to know the nature or properties of the sheet metal used in body panels. When a flat sheet of metal is bent to a wide radius it will regain its former shape when released; in other words, the metal is flexible. If this metal is bent to a small radius it exceeds the limits of its flexibility, and the metal in the bend becomes stiff and retains its curved shape. These are known as rigid sections. Therefore a body shell and panels are a combination of rigid and flexible areas. The rigid areas are such assemblies as door posts, sill members, roof cantrails and flanges around wing edges and bonnets. The flexible areas are large areas of mainly flat or slightly curved sections such as bonnet tops, outer door panels, sections of wings and centres of roof panels. In the main after a collision if the stresses are removed from the rigid areas first, then the flexible areas (except where badly creased in the impact) will regain their former shape. If a panel is damaged in an accident then the buckled area, being sharply bent, will create additional stiffness in the panel, whether in an elastic or a non-elastic area. The slopes of the buckles surrounding the sharp creases will be fairly elastic, but a greater amount of effort will be needed to reshape the sections of the panel which are made rigid either in manufacture or through accidental damage. The theory of repair is that areas which are elastic in manufacture and remain so after damage can generally be reshaped by their characteristics once the rigid areas in the surrounding buckles are forced back into position.

In general the method of repair is to analyse the crash, establish the order in which damage occurred, and reverse the order when correcting the damage. The majority of repairs to minor accident damage require the skill of the body repair worker in the use of hand tools and general repair techniques. Although every panel assembly has its own individual repair procedure, the basic approach is the same for all minor repairs and is as follows:

1   Carefully inspect the damaged area and analyse the severity of the damage and its ease of accessibility for repair using hand tools.
2   Decide if the repair can be carried out using hand tools only or whether the hydraulic body jack is required.
3   Select the necessary hand tools and, if using the hydraulic body jack, decide which basic set-ups and attachments are best suited to the job.

4   If the repair is carried out using hand tools only, then the back of the damaged section under repair, if accessible, must be cleaned of anti-drum compound.
5   If the accessibility for the use of hand tools is difficult owing to the presence of brackets behind the panel or double-skinned sections, rough out the damaged area either by cutting out the back and using spoons to lever the damage out or by using a panel puller on the front of the damaged section to pull out the damage. The area should then be cleaned, filled with the appropriate material and finished.
6   When using hand tools only, the damaged area should be roughed out to its original shape, although this operation can also be carried out using hydraulic equipment to push the damaged section back to its original shape.
7   Once the damaged section has been reshaped, either by hand or hydraulic means, planish it using a combination of direct and indirect hammering and filing.
8   Then check the low spots, if any, by cross filing and raise them by further planishing or pick hammering. At this stage high spots may develop due to stretching of the metal, and hot shrinking of these points will be necessary.
9   Sand the panel surface over the damaged area using a very fine sanding disc.

Minor repairs to wings

Removal of a small dent from a wing is one of the simplest of repair jobs. The first step is to check the underside of the wing to ensure that it is free from road dirt and anti-drum compounds; then the damaged area is roughed out and the wing reshaped to its original contour. The reshaping is done with a rather heavy hand dolly of a similar shape to that of the contour of the damaged wing section. The blows should be heavy with a follow-through action which will speed up the roughing-out operation. It is very important, while roughing out damage, that full-shaped or very curved dollies and heavy ball pein hammers are not used, as they cause excessive stretching of the already damaged metal. Metal that is returned to its original pressed shape during roughing out will smooth out more quickly and evenly.

When the roughing out is complete, the next step is to planish the roughed area. In the case where the wing has been heavily coated with cellulose, it is advisable to remove this coating using the sander fitted with an open-coated disc suitable for paint, or with paint remover; the latter must be neutralized to stop the burning action on the paint. The first step in smoothing is to try to level out the uneven surface by working the high areas into the low areas. This can be done by holding the dolly block under the low area and tapping down the adjacent high spot with a planishing hammer. In some cases there may be no high spots and no adjacent low areas, in which case the high spots, if any, can be tapped down often with just the planishing hammer and without the dolly block backing. These high and low areas are found by passing the hand over the section under repair. As planishing continues, filing will show up any low and high areas, which will need correcting by tapping down high areas and lifting the low areas by further planishing with hammer and dolly or using a pick hammer. The final finish is achieved by cross filing the original file strokes; this is a double-check on curvature and smoothness of the panel surface.

Minor repairs to door panels

When repairing door panels the main difficulties arise through the fact that most door panels are nearly flat sheets of metal, which means that even the slightest stretching due to over-hammering will produce a fullness which will stand out after the repair has been painted. The greatest care must be taken.

The first step is to strip the door of interior trim and window. The amount of stripping necessary depends on the extent and position of the damage. With most door repairs it is quicker to remove the door from the vehicle by releasing the check strap and then taking out the set screws in the door side of the hinges, or, if the hinge is welded, by knocking the hinge pin out. Having prepared the door and the work bench, an inspection can then be made of the door frame to determine the accessibility for the use of hand tools and hydraulic equipment. When carrying out the repairs it is often necessary to cut the back of the door frame and bend it, thus permitting greater freedom of movement for the use of hand tools. The thorough removal of all anti-drum material is essential before any actual planishing is commenced.

The shape is gained by roughing out with the mallet and dolly block, after which the finish is attained with the planishing or pick hammer in conjunction with the file. In selecting a dolly it is advisable to choose one that is rather light and slightly flatter than the panel being planished. This will reduce the risk of stretching the panel. In some cases where the door panel is badly creased and is inaccessible for the use of hand tools, it would be better to rough out the damage with the aid of the body jack or spoon acting as a lever, smooth it over with hammer and dolly, then fill it with body solder and file the area to obtain the final finish.

Re-skinning door panels

This process affords considerable saving in time and cost compared with the practice of replacing complete door assemblies. The technique of this process is the same regardless of the shape or type of door to be re-skinned:

- *Removal of a damaged skin panel* – This is achieved by drilling out the spot welds around the flanged areas of the door using either a Zipcut or cobalt drill or Spotle tool and also, on some types of door, using a power saw to make joints across the pillars. The flange can be then eased up using a flange-removing tool, or in some cases the outer edges of the flange are ground through using a sander. If this method is used, care must be taken not to damage the door frame. The damaged panel skin can then be removed using a hammer and chisel and easing up the flange strips that are left.

- *Preparing door flanges* – The door flanges should now be straightened up using a hammer and dolly and the face of the flanges sanded to make a smooth surface for the attachment of the new skin panel.
- *Preparing the new surfaces* – The door flanges should now be painted with a zinc-rich interweld primer to stop any chance of corrosion between the door flange and the new skin.
- *Fitting the new skin panel* – The pre-painted skin panel is now placed over the frame and aligned using the door lock holes as a guide and MIG tacked onto the joint pillars if necessary. The edges of the skin are now hammered over the door frame using a hammer and dolly until the flange is folded over completely. During this operation great care is essential if the panel face is not to be marked.
- *Welding in the new skin panel* – The spot welds should now be replaced to fix the new skin panel to the frame, taking care to use the correct electrodes, especially on the face side of the panel. Adhesive is used on certain doors to fix the skin panel to the door frame. Where joints have been made in the pillars they must be MIG welded, ground flush and soldered.
- *Final cleaning and inspection* – The soldered joints must be filed and sanded with a DA sander. Finally, inspect the skin panel for any defects, which will need correcting by filing and sanding.

Pulling out double-skin panel damage

When inaccessible body panels are damaged, a method of repair for the removal of the damage, without leaving holes or having to remove interior trim, is to use a special pin welding gun. Also pin welding attachments can be fastened to spot welding equipment. The welding gun attaches pins, screws, washers or studs to the surface of any damage so it may be pulled out. There are two sizes of pins: small (2 mm), which are normally used for skin panel damage, and larger (2.5 mm), which are used for chassis legs or thicker-gauge metals.

First the damaged area must have its paint removed and be sanded down to a bare metal finish in order to give a good electrical contact between the pin and the welding gun. Insert the pin into the spring-loaded nozzle of the gun. The two points of contact are the tip of the pin and the outer copper ring on the welding gun, which should, when pushed down, make as even a contact as possible with the panel surface under repair. Press the gun to the surface of the metal, press the trigger for 1 to 2 seconds only and remove the gun from the pin, which now should be welded to the panel. The number of pins which need to be welded onto the panel will vary in accordance with the depth and extent of the damage, starting at the deepest section of the damage and working outward.

The dint is then gradually pulled out using a slide hammer, which tightens onto both sides of the pin by the rotating action of the chuck. The dinted area is then pulled out by using the slide hammer action, on one pin at a time, beginning at the deepest part of the damage and working outwards. The pins are removed by breaking them from the panel surface and sanding down their residue.

When the damage is a crease and very deep, ring washers may be used instead of pins to facilitate a stronger pull. A clamp may be fastened to these washers and attached by chain to pulling equipment which would then pull out the damage. Alternatively a hook is inserted into the slide hammer to pull out each washer individually.

When the damage has been completely pulled out the area may be filled using body solder or plastic filler according to the position and extent of the damage. The filled area is then sanded and filed down to an acceptable finish for painting.

Minor repairs to bonnets

The bonnet of a car is firmly fastened when closed. Many models have spring-loaded hinges so that the tension of the springs holds the bonnet firmly in place at the hinge side. At the point of locking a sturdy catch holds the bonnet down under spring tension. When open the bonnet is virtually unsupported, as it is only held in place by the hinges. This means that only minor repairs can be carried out with the bonnet in place, and if hydraulic power tools are necessary it is advisable to remove the bonnet from the car. Large areas of almost flat sections running into gradual return sweeps are incorporated in the designs of most bonnets. The flat panel and the return sweep are both shapes that need careful repairing to avoid stretching of the metal.

A rather common trouble that arises when repairing bonnets is the development of a loose edge. This looseness usually occurs along the back edge of the bonnet, which in most cases is a raw edge, and even the slightest stretch will cause a buckle. It is essential that this buckle be removed, and the best way of carrying this out is by hot shrinking the edge. Another characteristic of bonnets is that the edge flanges crack, and these should be welded as early as possible to prevent the extension of the crack and to keep the bonnet in its correct shape.

The repair procedure for minor, accessible dents is first to rough out the damaged area using a mallet to minimize stretching and then to planish the area using a beating file in conjunction with a dolly block, which will also reduce the possibility of stretching. If the repair is in such a position as to make the use of hand tools difficult because of the strengthening struts, the area should be roughed out using very thin body spoons and then filled with plastic filler, as the heat produced when soldering might distort the panel. The prominent position of the bonnet requires its finish to be perfect, because ripples and low spots can be easily seen when painted.

Minor repairs to sill panels

The sill panel on a car is attached to the door pillar and floor and is of double-panel box-like construction having both inner and outer panels. Because of its position it is one of the most likely places to receive damage.

Two methods of repair can be used for double-panel assemblies. First, cut away part of the inner panel, or in the case of the sill, along the top. Then bend the panel or top section upwards to make the outer panel accessible for the use of hand tools. Push out the damage using the spread ram which acts as a wedge, or force it out using body spoons as levers, then planish and file. Bend the top section or back panel back into position and weld. Care should be taken here to avoid overheating the repaired area.

The second method, which is less time consuming, can be used if the dent is small and not very deep. The area can be sanded clean to bare metal and filled using body solder. If the damage is very deep it will need pulling out, and this can best be done using a panel puller on the outer panel surface. The holes needed for using this tool should then be welded up and the damaged area should be sanded, filled with body solder and filed to achieve the final finish.

Minor repair of body panels using adhesives

With the advancement of adhesive technology, Permabond have developed an adhesive specially formulated for car repair applications and called Autobond. This is a two-part product which has been developed to meet the specific requirements of the car repairer. It is designed to replace panel welding with a minimum of surface preparation; it will also protect against corrosion. It provides substantial labour saving in avoiding the necessity to remove fuel tanks and soft trim as required before any welding repair. The adhesive not only bonds metal to metal but is also suitable for bonding GRP components.

For ease of use the adhesive is supplied as a layer of resin on top of a layer of hardener inside a plastic tub; the two layers will not react until mixed with the stick provided. The resin and hardener are coloured light grey and black respectively to provide a visual check that thorough mixing is achieved. After mixing is complete, setting begins at a rate depending on the ambient temperature. The best performance is achieved by heating the bond area to 40°C for 1 hour; however, at 60°C full hardening is achieved in 30 minutes. Curing at normal ambient temperature will give joints of reduced strength and is not recommended.

Removal and replacement of exterior and interior soft trim and hard trim

Body exterior mouldings are attached by weld stud–retaining plastic clips, weld stud–retaining plastic clips with attaching screws, an adhesive bond using either tape or sealant, a self-retaining spring clip or clinch-type clip or special attaching screws.

During repairs, observe the following:

1 Adjacent paint surfaces need protecting with masking tape to help prevent any possible damage.
2 Always use the correct tools for the job, and take great care with mouldings.
3 Water can leak into the interior of the body through any of the body panel holes made by screws, bolts and clips unless they are securely sealed.

Door trim

The door trim pieces are normally assembled in one-piece or two-piece trim pads. Removal of door trims is carried out in the following order: door lock handle and door locking knob, mouldings, window regulator handle and door latch handle, armrest assembly, mirror remote control bezel nut and door trim retaining screws (if any). Then lift the trim panel retaining clips from the door inner panel and disconnect all wiring so that the trim panel may be removed. Prior to reinstallation of the trim pad, always make certain that the watershield is in position and correctly sealed.

Quarter trim panel

This is the panel which fits over the interior of the quarter panel. Removal of the quarter trim panel is carried out in the following order: rear seat cushion, rear seat back, rear window handle (if any) and any screws or mouldings. Then unfasten the panel retaining clips and pull the trim panel away from the quarter panel.

Floor carpet

Carpet for car floors is made either in one piece which is moulded to fit the shape of the floor or in two pieces to fit the front and rear sections of the floor respectively. In order to remove the one-piece carpet it is necessary to remove everything in the way, including seatbelts, seats, centre console and door sill scuff plates. However, the two-piece carpet can be removed with the seats still in place because they are made with cut-outs to allow for easier removal. In some cases the floor of a car is covered with heavy rubber flooring instead of carpet, and this can be removed and replaced in a similar manner.

Head-linings

Head-linings are of two types. The first is soft and made from cloth or vinyl coated. The second is harder, being formed from moulded hardboard coated with foam or cloth which has a vinyl facing. The replacement of the moulded headlining is a simple, straightforward operation because of its one-piece rigid structure. The soft type is more complicated in its removal owing to the necessity of removing a large amount of trim in addition to the windscreen and rear window in some cases.

Aluminium panel repair

Differences

The repair of aluminium panels and other components is slightly different to repairing steel ones. Aluminium panel repair is not difficult; it is just different to repairing steel ones. Not greatly different, just that you should follow a set of procedures and remember the characteristics of aluminium. More and more vehicles are using aluminium every year. More correctly, we should say aluminium alloy. Let us have a look at the characteristics of the material, particularly its advantage over steel.

Aluminium is light in weight; strength for strength it is about a third of the weight. In an accident, it will absorb double the energy of a similar steel component. Given the correct environment it does not corrode. It is a very abundant material and easy to recycle. Because aluminium is easier to work than steel, and it can be worked in different ways, aluminium bodies can be made for a lesser number of parts than steel ones. The big environmental advantage is that by reducing the weight of the vehicle the fuel efficiency is increased; a 10 per cent weight reduction typically gives an 8 per cent reduction in fuel usage.

Galvanic corrosion

Aluminium is the fifth least noble metal, it is almost at the end of the anode–cathode scale, and it has an anodic index of 0.95. This means that whenever steel and aluminium come into contact there is likely hood that galvanic corrosion is going to take place. If there is dampness of any kind, such as when rubbing down or cleaning, then some form of corrosion is bound to take place. Although this might appear insignificant at first, the smallest pinprick of corrosion can

lead to a large rust patch. The smallest corrosion bubble will look unsightly if it is in the centre of the bonnet or other highly visible panel.

Steel particles from tools and abrasives can cause corrosion to aluminium, so careful consideration must be given to every possibility of cross-contamination between tools and panels to prevent corrosion taking place.

Contamination

Repairers specializing in working with aluminium bodied vehicles, such as Audi, Jaguar and Aston Martin, use specialist enclosed work booths for those vehicles – that is, a booth which is fully enclosed and heated and ventilated separately from the rest of the workshop. Also the tools and equipment are dedicated for use only on aluminium-bodied cars. The tools used, where possible, have non-ferrous working faces. That is, tools made from plastics materials or plastics-coated tools are used where possible.

However, this is not usually practical for the smaller repairer, or the general accident repairer. In which case the following is advised:

- **Position the aluminium-bodied vehicle in the workshop so that it is quite separate from steel-bodied ones, especially when welding, sanding, grinding or other intensive repair procedures are being carried out.**
- **To avoid cross-contamination by minute steel particles from steel bodies to aluminium ones use a set of tools for aluminium use only. That is, you should dedicate a new set of tools for aluminium repair use only. Colour code the tools by painting or plastic dipping their handles.**
- **Keep the tools for use with aluminium bodies in an enclosed tool box separate from those used on other vehicles.**
- **If possible, dedicate sanding and grinding tools for use only on aluminium bodied cars. Use the same**

colour coding system and store them separately under covers. If this is not possible, then clean the equipment with solvent before use on aluminium bodies and use new discs or abrasives to ensure that there is no cross-contamination.

- Always wear a clean set of overalls and gloves when working on aluminium-bodied cars.
- Always use new wipers and cloths when carrying out aluminium repairs.

Riveting

When drilling out steel rivets from aluminium panels be sure to collect any steel chipping to prevent contamination and possible corrosion. Punches are available to remove old rivets.

When replacing rivets be sure to use the correct type. Often coated rivets are used to reduce the risk of galvanic corrosion. The coating is usually anodizing, giving an identifiable colour.

Rivets are best applied with a rivet gun, however copper rivets can sometimes be used and closed using snaps which are hit with a hammer.

Oxidation

Oxidation on aluminium panels is invisible to the naked eye except as there may be a slight dulling of colour. As the oxidation layer will be over the whole panel, the dulling will not be distinguishable. A coating of oxidation is formed within a period of 2 hours even in a well-heated dry workshop. Therefore a primer coat should be applied within 2 hours of sanding or other preparation.

If the aluminium panel is left for a long period in damp conditions, then the oxidation will become visible as a coating of white powdery aluminium oxide. Aluminium oxide ($Al2O3$) is both infusible, difficult to melt without very high temperatures and amphoteric – can form a base or acid – which implies it can be dissolved in caustic soda. Be careful when using caustic soda as

Figure 13.42 Enclosed work booth for aluminium repair (*Courtesy of POWER-TEC ®*)

a cleaner; always wear rubber gloves, goggles and apron as the minimum PPE.

Immediately before welding aluminium panels the areas to be welded must be deoxidized clean – the best technique is to use a stainless steel wire brush which has been dedicated for use only on aluminium panel work.

Heating aluminium

Most aluminium body panels are hardened and tempered for strength and dent resistance. The aluminium will work harden on impact. Therefore if the panel is dented the damage will be firmly set in place. It will be necessary to heat the panel to soften it to remove the dent and bring the panel back into shape and contour.

Aluminium is a very good conductor of heat, so spot heating is difficult. That is, the whole panel becomes hot very quickly, as does the surrounding area.

Aluminium fuses (melts) at about 650°C, but the aluminium oxide will not burn off at less than 1250°C. Therefore it is important to clean off the oxide prior to welding to be able to keep the aluminium fluid during the welding process at 650°C.

When heating the panel to carry out a repair the temperature should not exceed 200°C. This is the threshold temperature for the aluminium. A usual temperature range for the repair of panels is between 110°C and 160°C. It is advisable to keep the temperature below 160°C.

Special attention must be given to adhesively fixed or bonded areas, whether it is the panel which is being repaired which is bonded or the bonding on an adjacent panel or nearby supporting member. Heat maybe applied to help disassemble an adhesively secured or bonded joint; but be careful not to apply any heat to a joint which should be kept intact. Always pay special attention to these areas both in terms of accidentally applying direct heat and allowing heat to be conducted to reach them. The use of heat soaks in the form of damp material is one way of doing this; but make sure that this is clean and thoroughly dried off afterwards to ensure that no corrosion can occur later.

Heat application

Aluminium panels can be heated in the same way as any other panel; but the working temperatures are very low compared to steel. Heat sources are:

- *Oxy/acetylene torch* – Be careful with the flame to ensure even temperature distribution.
- *Gas blowtorch* – Use a low to moderate setting.
- *Hot air gun* – Probably the best method, especially if it has a temperature control.

Temperature measurement

Aluminium does not change colour or otherwise give an indication of its temperature as it gets hot – that is, until it suddenly melts.

There are several ways of measuring the temperature. The choice depends on personal preference and required accuracy. The more accurate the better, especially when dealing with vehicle with large numbers of bonded joints.

- Infrared or laser 'point-and-shoot' digital thermometer – These are very accurate, being able to measure to a fraction of a degree. They allow quick measurement not only of the area which is being worked on but of any other areas, for example adjacent bonded joints which must be kept cool. The readings are instantaneous.
- Digital thermometer strips – These are placed on the panel, their plastic backing strip holding them to the panel and allowing reading to be observed whilst work is taking place. The readings take several seconds to change, as the

(a)

(b)

(c)

(d)

(e)

Figure 13.43 (a) Heating the panel with propane gas (b) Conventional bumping file being used to raise dent (c) Using a panel file (d) Threaded studs and loops for pulling repair from outside (e) High density nylon compound panel hammer (*Courtesy of POWER-TEC ®*)

Figure 13.44 Laser point-and-shoot digital thermometer (*Courtesy of POWER-TEC ®*)

material changes colour with the change in temperature to indicate the temperature. Several of these may be needed at any one time, and it may not be possible to secure them to bonded frame areas if there is not a suitable smooth surface. Accuracy is limited to the pre-set limits of the strip.

• Colour crayon – The area to be heated is given an outline mark with a crayon which changes colour at a given temperature. With crayons the correct one must be chosen for the job, as the accuracy is limited to pre-determined constituents of the crayon, and the speed is that of the chemical colour change. These are limited too in that it is not possible to track heat travel to adjacent panels.

Dent repair

Clean the whole area with soap and water to remove any road dirt then dry off the panel. Remove the paint coating using 80-grit or finer abrasive. Use only a light pressure when sanding, especially on the bare metal.

Wipe the panel with wax and grease remover and dry off with lint-free cloth or a wiper. Allow any vapours to evaporate.

Warm the area to be repaired, checking that the temperature does not exceed 160°C. Also check that adjacent areas, especially where there are bonded joints, are not getting too warm. Use an accurate thermometer for this; don't guess.

Work the damaged metal using conventional straightening techniques, keeping in mind how soft the aluminium panel is compared to steel.

Use the tools which have been dedicated for use only with aluminium bodies. Make sure that any metal tools such as hammers, dollies and spoons have smooth edges to avoid gouging the aluminium panel. Pick hammers are likely to be unsuitable because they are too sharp. The use of wooden bossing hammers and purpose-made high-density nylon hammers is recommended.

Remember to re-heat the panel continually throughout the repair process.

Miracle repairs

Lifting and repairing damaged panels from the outside without removing interior trim or on otherwise blind sections can be done using Miracle tools. The system uses treaded studs and a range of rings, pulling loops or collars and pulling tools. The stud is welded to the damaged area of the panel after appropriate removal of the paint coating. A suitable pulling loop or collar is screwed onto the stud. The panel is heated up if needed and the pulling tool applied to pull the dent out.

When the dent is out the stud can be snipped off and its remains filed or ground down level with the surrounding area. Check for low and high spots using a panel file, taking care not to remove too much metal. Aluminium clogs files, so keep the file clean using a stainless steel or brass wire brush.

Figure 13.45 'MIRACLE' pulling tool (*Courtesy of POWER-TEC®*)

Figure 13.46 Snip-off the studs (*Courtesy of POWER-TEC®*)

Applying a guide coat then using a long sanding block, as is used for filler, high spots can be identified without removing metal.

Heat shrinking

Applying heat to any dent, without any kind of manipulation, may reduce the size of the dent due to the thermal expansion of the aluminium.

Heat shrinking may also be used with hammer and dolly to manipulate the panel and reduce the dent.

Cutting aluminium

Aluminium, though it is soft, requires special care when cutting. The following points should be observed:

- Always ensure that the panel is fully supported on both sides. If it is not, it may sag and bend, causing damage.
- Use a reciprocating saw, such as an electric jig saw.
- Use a fairly coarse blade. Forrest Blades aluminium saws have alternate six- and four-sided teeth. Cutting fluid must be used with this tooth shape.
- Use the correct cutting speed, up to 5.5 m/s (1000 feet per minute) – much faster than for steel.

Aluminium welding

MIG and TIG are the normal welding processes for use with aluminium. Follow normal procedures using aluminium wire when MIG welding and adjusting the amperage accordingly. Clean the panel immediately before welding with a stainless steel wire brush. If possible pre-heat the panel before welding to give better weld quality and reduce the risk of distortion. Control boxes are available to convert conventional MIG welders to weld aluminium.

Figure 13.47 Aluminum saw (Courtesy Forrest Blades)

Figure 13.48 Control box to convert conventional MIG welder to weld aluminium (*Courtesy of POWER-TEC ®*)

Body electrical and electronic systems

There are very few body panels on current vehicles which do not have some form of electrical or electronic device attached or nearby. This means that the repair of any of the panels will probably require contact with an electrical component or the power supply to a nearby component. This section aims to look at the basics of body electrics as may be tackled by the personnel in a body repair shop. For more complex systems and system fault diagnosis the services of a vehicle electrical/electronics specialist should be sought.

Electrical terms

Amp
The unit of current is the amp and is the quantity of electricity which flows in a circuit. If it were water you would refer to it in gallons or litres. To measure current flow the ammeter must be connected in series with the circuit. The ammeter must be able to handle the expected current or it will be burnt out. To measure large current flow, such as in a starter circuit where the motor may be drawing 200 A, you will need an induction ammeter which clamps over the starter cable

without making electrical contact. The most common use of an ammeter is to check the charging rate of a vehicle's alternator; you'll need a meter which goes up to 60 A for this job. A centre zero scale is useful for measuring the current flow when working on vehicles, as it can show the direction of the flow, and you do not need to change the connection of the leads. Some ammeters which do not have a centre zero have a button to change the polarity to show charge or discharge.

Volt
This is the pressure of the electricity. The water pressure from a fire hose can knock you over; the electrical pressure from a mains socket can do the same. Any voltage over 50 V is potentially lethal. Vehicle electrical circuits operate at a safe 12 V or 24 V on HGV/PSV applications. The reason for the higher voltage on the large diesel vehicles is to reduce the amperage needed by the starter motor. It also gives better interior lighting in buses. The voltage in a circuit is measured by connecting the voltmeter red to positive and black to negative each side of a component, that is, in parallel with the component.

Watt
This is the unit of electrical power.

$$\text{Watts (W)} = \text{Volts (V)} \times \text{Amps (A)}$$

746 W equals 1 HP.

Tech note
Bulbs are usually identified by their wattage. Headlamps are typically 60 W.

Resistance
This is measured in ohms. If you have a narrow hosepipe it will restrict the flow of water, in the same way the diameter of the electrical wire will affect the flow of the electric current. As a long hosepipe will

reduce the flow of water, the length of the wire will increase the electrical resistance.

$$\text{Resistance (ohms)} = \text{Volts/Amps}$$

Resistance is measured with an ohmmeter. If there is no resistance the meter needle will move across the scale and read zero. Electrical cables and earth connections should show zero or very small amounts of resistance. An exception is plug leads, which have an inbuilt resistance for radio suppression purposes, typically 30,000 ohms per meter of length.

Analogue and digital meters
Analogue meters are ones which have a needle which moves across a scale; digital ones show numbers on an LCD (liquid crystal display) or illuminated display.

Resistor
A resistor is used to control the flow of electricity. It allows electricity to flow, but the flow is at a set rate.

Capacitor
A capacitor (also called a **condenser**) stores electricity, or absorbs current surges, for a limited period of time. The capacitor in a distributor absorbs the flow of electricity in the LT circuit when the CB points open.

Diode
This is a one-way valve for the flow of electrical current; zener diodes only allow electricity to flow when the voltage exceeds a set figure.

Transistor
This term covers a number of different solid-state devices, including diodes. The most common transistor has three connections, base, collector and emitter. By applying a small current to one terminal a larger current can be switched between the other two; remove the small current and the larger current is stopped. A **thyristor** is such that a small current starts the flow which remains after the small current is stopped. A thermistor is temperature operated. Its resistance may increase with temperature or it may be of the negative **temperature coefficient** (**NTC**) type where its resistance decreases as it gets warmer.

Series
This refers to a circuit where components are connected in line with each other, that is, positive to negative. As the number of items in the circuit is increased the resistance will increase, and the voltage across each component will decrease.

Parallel
A parallel circuit is one in which items are connected electrically alongside each other. The resistance decreases, and the voltage remains constant. Vehicle lighting is connected in parallel.

Earth return
This is the name used to describe the use of the vehicle body/chassis as a carrier of electricity. This system is used on most cars and light vans, all the circuits returning to the battery negative via the body/chassis. The battery must have a good earth connection. Tankers and buses use a two-wire system in which each component has a feed wire and a return wire to the battery negative.

Short circuit
When a live bare wire or terminal touches the chassis earth or another earth point it produces a short circuit. This may lead to the wire becoming hot and burning, even causing damage to the car.

Open circuit
When a wire breaks or becomes disconnected, an open circuit is produced. In this case the component involved will not operate.

Fuse

A fuse is a breakable link in a circuit. The fuse will blow instead of starting a fire in the case of a short circuit.

SAE J1930

The Society of Automotive Engineers (SAE International) have a task force to produce a list of standardized terms and acronyms for automobile electrical and electronic systems. The following table lists some of the common terms.

Standardized term	Standardized acronym
Accelerator pedal	AP
Air cleaner	ACL
Air conditioning	A/C
Charge air cleaner	CAC (intercooler)
Data link connector	DLC
Diagnostic test module	DTM
Diagnostic trouble code	DTC
Distributor ignition system	DI
Electronic ignition	EI
Engine control module	ECM (Engine ECU)
Engine coolant temperature	ECT sensor
Engine speed sensor	RPM sensor
Exhaust gas recirculation	EGR
Fan control	FC
Generator (alternator)	GEN
Ignition control module	ICM (Ignition ECU)
Intake air temperature sensor	IAT sensor
Knock sensor	KS
Malfunction indicator lamp	MIL
Manifold absolute pressure	MAP sensor
Mass air flow sensor	MAF sensor
On-board diagnostic system	OBD system
Open loop	OL
Park/neutral position	PNP switch/sensor switch/sensor
Scan tool	ST
Service reminder indicator	SRI
Throttle body	TB
Transmission control module	TCM (transmission ECU)
Turbocharger	TC
Vehicle speed sensor	VSS

Battery

The battery is the source of power for starting the vehicle. Most vehicle batteries are 12 volts. Before carrying out any dismantling which entails disconnecting any electrical or electronic components or connections it is good practice to disconnect the battery. Always disconnect the chassis earth lead first and re-connect the same lead last. It is good practice to remove the battery completely from the vehicle and put it on charge to ensure that it will re-start the vehicle when the repair is completed. The battery must always be disconnected before any arc/MIG/TIG welding is carried out to prevent damage to the electrical system by the high voltage of the welding apparatus.

Before disconnecting the battery you should ensure that the security codes are available for the radio/CD player. Also, if appropriate, any other security item codes or personal settings are stored.

Some special vehicles have two batteries, one battery for starting the engine and one battery for maintaining the supply to the electronic components that operates when the vehicle is not running. The second battery is usually a gel battery. This is used to maintain a power supply to the clock and the security system when the vehicle is parked. In this case, both batteries will need to be disconnected.

> **Tech note**
> The earth battery terminal is usually very short and can be seen bolted to the bulkhead or inner wing.

Alternator

The alternator generates the electrical power to charge the battery when the engine is running. The rectifier inside or, in some cases, near to the alternator contains a number of diodes. These diodes will be damaged if the alternator is not disconnected before carrying out any form of electrical resistance welding. Disconnecting the battery will usually also disconnect the alternator.

Electronic Control Unit (ECU)

Vehicles are fitted with at least one ECU. The ECU may also be called by other names including ECM or 'brain'. Typically ECUs are used for the engine management system, the ABS, the security system and electronically controlled transmission systems. ECUs are usually attached to the panel with two screws. They are connected to the wiring loom with a 32-pin plug. Pull back the spring clip holding the connector plug into the ECU then pull the 32-pin plug out gently without pulling on any of the wires, as this may pull the wires out of the connector.

Before replacing the plug ensure that the plug and ECU are both clean and that each of the 30 pins are aligned before pushing them into the ECU.

Central locking

Central locking may be operated with or without a key. The keyless type uses either an infrared or a wireless signal to a sensor on the vehicle from the key-fob control, or 'zapper'. The small mercury battery in the key-fob control should be replaced every two years. The central locking mechanism may be operated by either a centrally pumped pneumatic system or individual solenoids. Look out for the locking mechanism when working on vehicle doors. Always ensure that the connections are in place before re-fitting the door trim. It is good practice to ensure that the windows are wound down before testing the central locking mechanism.

Electric windows

Most electric windows are operated by a small electric motor and a continuous cable system. The cable is only about 3 mm in diameter and will break if subjected to normal rusting. The cable goes stiff and brittle when rusty. The windows may be operated by switches on either the centre console or the inside door trim. Some have two switches for each window arranged to work in parallel. You will need to disconnect the electric window power supply leads and the central locking wires from inside the door, after removing the door trim, before attempting to remove the door or repair the door skin panel.

Electric sunroof

Electric sunroofs operate in a similar way to electric windows. The control for these is usually a switch mounted just in front of the interior rear-view mirror. Access to the sunroof usually entails removal of the interior head lining (also called roof lining).

Electric seat

Electric seat controls are usually available for height, backwards and forwards position, rake and tilt. Some seats have memory settings available for different drivers, referred to as setting 1, 2 and 3 and so on. The connectors for these are usually underneath the seat cushion. Be careful when disconnecting and re-connecting the seat wiring, because seats also may incorporate an electrical earth cable to reduce the risk of travel sickness and connections for the pyrotechnic anti-submarine seatbelt pre-tensioner.

Lights

Most light fittings are screwed to the front panel or wing. Removal is usually simply unscrewing and unplugging the cable

Figure 13.49 Mercedes McLaren SLR tyre pressure sensor

Figure 13.50 Mercedes McLaren SLR door

socket. Re-fitting is the reverse procedure. However, the head lamp beams will need adjusting using a beam setter to ensure that they do not dazzle on-coming traffic and comply with the VOSA regulations.

Supplementary Restraint System (SRS)

Restraint systems to protect the driver and passengers in the event of an accident are designed to work in conjunction with the vehicle body crumple zones, the idea being to slow down the vehicle occupants in a controlled manner. That is, if the vehicle hits a wall, then the occupants will decelerate at as gentle a rate as possible in the time available. Typically, the amount of time is 120 ms (0.120 sec) – the time it takes to blink. One of the major causes of death in car accidents is damage to the brain caused by deceleration rates which are too high.

The standard three-point seatbelt is defined as the primary restraint system. Other features, namely seatbelt pre-tensioners and airbags, come within the category of SRS.

Airbags

The air bag is designed to protect the occupant's head and upper torso – not to take all the weight of the body. It is inflated by gas generated by the pyrotechnic device. The gas is usually nitrogen, which is inert and poses no after-effect problems. The air bag module, comprising the airbag, airbag cover, the pyrotechnic device and its fitting, must be treated with extreme care when working on the SRS or any other related system. For instance, if you are replacing the indicator or ignition switch, you will probably need to disturb the airbag module. The pyrotechnic device is ignited by a low-voltage electrical current, and small tablets fill the airbag with nitrogen in about 30 ms (0.030 sec).

Always carry or handle an airbag module so that it is kept in the upright position. If the module were to accidentally go off, the bag would inflate upwards, and the heavy part of the module would go down to the floor. Airbags cannot be repaired, and any tampering is probably on a par with taking the pin out of a hand grenade and hoping that the handle will stay in place.

Control unit

The control unit, sometimes called the diagnostic and control unit, is central to the operation of the SRS and is a form of ECU. The unit has three jobs: it monitors the system for faults, controls the operation of the airbag and seatbelt pre-tensioners and stores electrical charge for emergency use. The electrical charge is stored in capacitors within the control unit, so if the main vehicle battery is damaged or disconnected before or during impact, the capacitors can supply an electrical charge to ignite the pyrotechnic materials in the airbag and seatbelt pre-tensioner. As the electrical current from the control unit can operate the SRS independently of the battery, obviously precautions must be taken to avoid accidental operation before working on the SRS. Typically the procedure is to switch off the ignition and completely remove the key, then disconnect the battery – remember always to remove the earth lead first – and then wait about 10 minutes for the capacitors in the control unit to discharge themselves. The capacitors will re-charge when the battery is reconnected, before the ignition is switched on, so it is prudent to ensure that nobody is in the vehicle in case there is a fault which might accidentally operate the SRS when you are first re-fitting the battery terminals.

Release sensors

Most vehicles are fitted with two negative acceleration (deceleration) sensors wired in series. This is a form of logic gate so that the airbag and seatbelt pre-tensioners can be activated only if both sensors agree. The sensors may be fitted in a number of different places. Generally the main crash sensor is located close to or in front of the vehicle centre of gravity and is usually bolted to a longitudinal structural member of the body/chassis.

The crash sensor operates the SRS at a deceleration of about 15 g, which is like hitting a wall at 20 mph. The second sensor, called the safing sensor, will trigger at about a tenth of the crash sensor's deceleration, that is at about 1.5 g. If a sensor is faulty, the control unit should find it and illuminate the warning light.

Rotary coupling

The SRS rotary coupling, between the steering wheel and the outside of the steering column, is unlike any other electrical coupling. It is not a friction terminal. The SRS rotary coupling consists of a length of wound wire ribbon which is permanently connected at both ends so that there is no risk of a bad connection. The ribbon coils and uncoils inside the coupling housing as the steering wheel is turned. Before removing the coupling, the steering wheel must be centralized and the road wheels set to the straight-ahead position.

Wiring harness

The SRS wiring harness is a special construction; do not attempt to replace any damaged wires or change the plug terminals. The wiring harness can be damaged by heat, so if an airbag or seatbelt pre-tensioner is fired

the SRS wiring harness must be replaced too. Similarly, if the vehicle suffers from any fire damage, the SRS wiring harness must be replaced. The damage caused by heat cannot always be seen from a visual inspection.

Warning light

The warning light is usually located on the dashboard but sometimes can be found on the airbag casing in the steering wheel. When the ignition is first switched on, the warning light will illuminate for between 3 and 5 seconds to indicate that the system is functional and that the control unit is carrying out a self-test of the system.

The amount of time that the lamp is lit indicates the length of the system check. If the system is correct the light will go out. If the light stays on then the system has a fault. On some vehicles the warning lamp will flash to indicate a fault code – this may need to be triggered by a diagnostic tool. Some vehicles have an SRS OFF switch. This is useful when carrying babies or small children in seats which are protected by airbags, or the seatbelts are used as part of a child restraint system. Usually, a small white lamp adjacent to the switch is illuminated when the SRS is in the switched off mode.

Safety check

Do

- **Check all procedures in the workshop manual before working on the SRS.**
- **Ensure that all replacement parts are of the correct type and in good condition.**
- **Carry out any diagnostic procedures using the correct equipment.**
- **Carry and store the airbag with the cover facing upwards.**
- **Keep all SRS components dry.**
- **Store all SRS components in a locked safety-approved area.**
- **Ensure that nobody is in the vehicle when re-connecting the battery.**

Do not

- **Drop any SRS components.**
- **Ever rest anything on an SRS component.**
- **Place an SRS component near electrical equipment, any source of heat or flames or anything which gives off electromagnetic radiation.**
- **Carry SRS components inside the passenger compartment of a vehicle.**
- **Attach anything to or otherwise mark components.**
- **Attempt to tamper with, dismantle or repair any components.**
- **Attach any other wires to the SRS circuit.**

Crash sensors

As well as the release sensors for the SRS, many vehicles are fitted with a crash sensor which cuts off the fuel pump in the event of a serious impact. This can usually be reset by pressing a button on the actual sensor. The sensor is usually positioned behind the glove box.

Security and alarm systems

On most vehicles there are three different systems in operation:

- *Immobilization* – an electronic system, usually in conjunction with a key or a key pad that immobilizes the engine unless the correct key is inserted or the correct numbers are entered in the key pad
- *Perimetric* – a series of micro-switches on the doors, bonnet and boot which set off the alarm if they are opened when the system is alarmed
- *Volumetric* – a sensor inside the car which senses movement or changes in pressure if something moves inside the car or a door is opened.

Air conditioning

Air conditioning heat exchangers (or radiators) are often incorporated in or mounted adjacent to the cooling system radiator. Be very careful when dealing with air conditioning radiators and other parts. The fluid in the system is under pressure and should be safely removed before disconnecting any parts. After changing parts the system will need re-charging and re-setting – this requires special equipment. On older cars the air conditioning fluid was R12. This contains CFCs and must not be allowed to escape, as it causes damage to the ozone layer and can poison the operator. An expensive evacuator and re-filler is needed to re-charge the air conditioning system.

Fuses

You will find up to 20 or more individual fuses on vehicles. Do not replace a blown fuse unless you are sure that the circuit fault which caused the fuse to blow has been repaired.

Sensors

When carrying out body repair look out for sensors. Sensors for a wide range of uses are fitted, including outside temperature sensors, rain sensors and on the Mercedes McLaren SLR tyre pressure sensors on the wheel inner arches.

Checklist

After completing any repair it is a good idea to check all the electrical/electronic components. A PDI checklist for the particular vehicle is useful to do this job.

Skills and questions

This work is the core of your profession. You will develop your skills over the period of your apprenticeship or training course. You should try to develop all the areas of work discussed, but with the tendency not to carry out repairs to panels, you may have a limited chance to do everything. Whatever you do, do your best.

1 Describe, with the aid of sketches, the processes of hollowing and raising. List the necessary tools for these processes.

2 Describe the process of planishing.

3 Compare the differences between planishing and wheeling a panel to produce a smooth surface.

4 Explain the method and name the tools used in fabricating by hand an aluminium component with a high-crowned surface.

5 State two important conditions to be observed when using a wheeling machine.

6 After extensive wheeling on an aluminium body panel, the metal becomes work hardened. Describe a method used to anneal the panel and a workshop test which will indicate when the correct temperature is reached.

7 Describe, with the aid of sketches, the technique of split and weld for producing a hand-made panel.

8 What are the purposes of forming angles and flanges along the edges of sheet metal panels?

9 With the aid of sketches, describe three methods of edge stiffening a sheet metal body panel.

10 Explain the terms 'direct hammering', 'indirect hammering' and 'pick hammering' as applied to body repair techniques.

11 Describe the technique of cross filing.

12 Describe, with the aid of sketches, three different methods of using a portable hydraulic body jack in the realigning and straightening of vehicle body damage.

13 Explain each of the following stages of repairing a front wing which has received minor damage: (a) the analysis and assessment of the extent of the damage, (b) selection of the appropriate tools needed to carry out the repair, (c) the repair techniques required, (d) the alignment and final check.

14 Explain the process of body soldering when used in the repair of a damaged body panel, and list the necessary equipment.

15 Describe, with the aid of sketches, the technique of hot shrinking to restore a panel to its original contour.

16 Explain the difference between the hot shrinking and cold shrinking techniques.

17 When is it considered necessary to fill with body solder in preference to any other method of repair?

18 Describe a method of repairing a sill panel which is showing signs of corrosion.

19 What precautions must be taken, after body soldering a repair, to prevent subsequent paint defects?

20 Explain the importance of tinning a panel before commencing body soldering.

21 State the disadvantages associated with the use of hydraulic push rams when used on the inside of a vehicle.

22 Give one practical application of the body jack when being used for each of the following operations: pushing, pulling and spreading.

23 Describe a suitable method of removing the damaged skin during the repair process of reskinning a door panel.

24 Explain how a new door skin is secured to the door frame, and list the tools required.

25 Indicate the type of welding necessary when securing a new door skin to the door frame.

26 Why is it important to clean the underside of a damaged panel before commencing repairs?

27 Illustrate a tool which could be used in the removal of damage from a double-skinned section.

28 When and how would a pin welding gun be used in repair work?

29 What type of attachment would be used on the end of a body jack to spread the force of the load during a repair?

30 Explain the differences between a hydraulic push ram and a hydraulic pull ram.

31 The lower section of an A-post has suffered corrosion damage. Explain a suitable method of repair.

32 Describe a suitable process for the removal of spot welds from a body panel without distorting the flanges.

33 Describe the sequence of repair (not replacement) to a rear quarter panel which has received minor accident damage.

34 With the aid of a sketch, show how a body jack can put a door panel in tension.

35 A small corroded area of a wing needs a patch repair. Explain how this process will be carried out.

36 Describe, with the aid of sketches, how the hydraulic system of a body jack works.

37 With the aid of a sketch, show the techniques involved in the manufacture of a small double-curvature panel by hand.

38 Describe, with the aid of a sketch, a vector pull for correcting vehicle body alignment.

39 Explain why careful study of the accident damage is important before any stripping or repair work is carried out.

40 Explain the importance of correct alignment when fitting new body panels and state how this is achieved.

14

Major accident damage repair

Damage classification and assessment

Damaged bodywork is corrected by first observing the extent of the damage, then deciding how it was caused and the sequence in which it occurred. The reason for doing this is to help plan the repair procedure and correctly estimate the final cost. Your experience will teach you what to expect, especially in terms of hidden damage, that is the damage of inner panels and components in particular types of accidents. It is usually the concealed damage that is the most expensive to repair, as it usually involves considerable stripping to reach the damaged part. So let's have a look at this in more detail. Damage can be classified into two groups, these are:

- *Direct or primary damage* – This results from the impact on the area in actual contact with the object causing the damage. This will result in the largest area of visible damage and is the cause of all other consequent damage. Primary damage is identified by first determining the direction of the primary impact. This knowledge will help in the search for concealed damage.
- *Indirect or secondary damage* – This is usually found in the area surrounding the direct damage which causes it, although in certain cases it may be some distance from the actual point of impact. After the impact, internal damage is caused by the forced movement of objects and passengers towards the point of impact and can be seen in the form of damaged dash panels, broken seat frames and twisted steering wheels.

Direct and indirect damage can each be subdivided in two further ways:

- *Visible damage* – This is damage that can be readily seen in the area of actual contact, such as a vehicle having suffered frontal impact causing damage to the bumper, grille, bonnet and front wings. A detailed examination may discover distortion of the inner wing valances, which would indicate visible indirect damage.
- *Concealed damage* – This is indirect damage, but it is not easily detected by visual examination unless the vehicle has been partially dismantled to allow a detailed inspection. In most situations measuring equipment in the form of body jigs must be used to detect concealed damage, because complicated mono-constructed vehicles may hide further damage such as misalignment, which could therefore affect the steering and roadworthiness of the vehicle.

Direction of damage

Direction of damage, or line of impact, is particularly important to the body repairer. It is used to identify the sequence and direction in which the damage occurred and consequently the reverse sequence to be followed for the repair. Direct damage marks are usually scratch marks where the paintwork has been damaged; they are an excellent guide to what happened and indicate the possible location of any hidden damage.

By careful study of the damage sustained it should be possible to ascertain the direction and strength of the impact force, and this is always the preliminary stage of a detailed assessment.

Methods of describing major damage

As part of your professional vocabulary, there are terms which are used to describe damage in a generic way. These are as follows:

- *Parallel side damage* is caused by the impact object moving parallel to the vehicle and causing substantial damage along the full length of the side of the vehicle, for example wing, doors, rear quarter panel.
- *Direct side damage* is caused when the vehicle is struck at an angle to its side, causing substantial damage at the point of contact.
- *Front-end damage* is the result of head-on collision, collapsing panels from the bumper to the front bulkhead.
- *Three-quarter frontal damage* is the result of an angled front end collision, sustaining damage to one front wing, the grille and bonnet.
- *Rear-end damage* is the result of an impact, direct or slightly angled, to the rear end of a vehicle and causing substantial damage to bumper, rear panel, boot lid, boot floor panel and quarter panels.

Figure 14.1 Proportions of damage sustained in different directions in UK accidents (*Motor Insurance Repair Research Centre*)

- *Roll-over damage* is caused by the vehicle rolling completely over and returning to its wheels. In some cases this type of movement could cause damage to almost every panel.
- *Total write-off* is damage so extensive that the vehicle is either unrepairable or the total cost of repair would be greater than the value of the vehicle.

Tech note
You will find this vocabulary of descriptive terms useful when talking to colleagues and insurance company staff.

Pulling, alignment and repair systems used on major accident damage

Since the monocoque construction of car bodywork based on engineering principles took the place of coach-built bodies mounted on a separate chassis, the whole concept of repair work has undergone radical changes in both methods and the equipment employed. Repairs to damaged body shells mean a great deal more than the beating out of a few dents or even the cutting away of a damaged portion and welding in a new pressing.

The majority of motor cars on the road today are of unitary or mono-construction. The whole unit is constructed round a framework of light sheet metal box sections, which provide an overall rigidity of exceptional strength and twist resistance. Owing to this construction, collision damage usually is confined to the area of impact; the box sections tend to crumple or concertina, thus absorbing much of the shock and rarely permitting end damage to travel beyond the centre of the car.

In order to deal with the major damage which occurs to the mono-constructed vehicle, specialist equipment has been developed. The equipment consists of frame and body straighteners to pull the damaged body shell back into alignment. These are used in conjunction with body repair, alignment and measuring jigs and thus provide accurate location points where exact positioning becomes critical, especially when the panel assemblies carry major mechanical components and suspension units.

The new generation of pulling and pushing equipment began in the 1950s with the introduction of Dozer equipment, which enabled the repairer to use external hydraulic pulling to aid the repair. Then came the idea of repairing a vehicle on an alignment jig using attachable/detachable brackets, so that it was unnecessary to check the repair after every pull. Next came the universal measuring system or bracket-less jig, which involved a measuring bridge or in some cases laser measurement. Multi-pulling repair systems were introduced, which were either static or mobile and which used the vector pulling principle. Later came mobile jig benches and lift operated systems with pulling attachments. Finally, dual repair systems give the repairer the advantage of precision measuring allied to the use of brackets.

The types of pulling, alignment and repair systems include the following:

- Portable pulling equipment
- Stationary pull-post systems
- Stationary floor-mounted rack
- Stationary floor-mounted rails
- Floor-anchored or anchor-pot pulling systems
- Stationary bench-type jig-mounted pulling systems (universal or brackets)
- Mobile bench-type jig mounted pulling systems (brackets, mechanical measuring, electronic measuring, laser measuring)
- Mobile bench-type jig-mounted pulling system using universal brackets
- Centre-post hoist jig system with pulling equipment (brackets/measuring)
- Four-post hoist jig system (brackets/measuring)
- Scissor-type hoist jig system with pulling equipment (mechanical or electronic measuring)
- Drive-on bench systems with tower ram pulling (mechanical or laser measuring)
- Multifunctional repair systems with pulling and measurement systems.

Pulling equipment: general description

Pulling equipment available for major accident damage can range from a single mobile pulling unit to multi-pulling units; these may be arranged around a bracket jig system or a measuring bridge system; they may be set up on mobile benches, centre-post or four-post lifts; and they may incorporate brackets or measuring systems or a combination of the two. The choice of systems is varied and wide and depends on the cost involved in the initial outlay, the requirements of the individual workshop and the workload of the shop (based on the number of vehicles repaired each week).

Alignment and repair systems

Jig brackets: general development
Vehicle design and manufacture have progressed considerably over the years. The development of crash repair systems has

likewise mirrored these changes. When cars were chassis built, the chassis tended to be relatively flat, so that the repair was mainly two dimensional. The change to lighter monocoque construction introduced one other dimension which demanded further consideration. First, the heights of the critical points on the vehicle's underside were controlled relative to an imaginary datum line, and so height became of vital importance. Second, accuracy became a far more critical consideration than in the past as suspension systems became less adjustable and more complex.

Jig brackets were introduced to serve as simple go/no-go gauges, providing a continuous visual indication of the extent of damage. During repairs brackets were found to yield further benefits. They provided progressive anchorage, meaning that once a critical point had been repaired and locked in place with the jig bracket, it would not be disturbed by subsequent repairs. In addition, during building, the jig brackets act as a welding jig to ensure that new parts are correctly installed and that the positions of important steering and suspension pick-up points are not disturbed by welding heat.

Concurrent with these developments, vehicle manufacturers were concentrating on particular aspects of car design. They were attempting to improve body weight by the introduction of high-strength steels and vehicle safety by incorporating strong passenger compartments with front and back and crumple zones to absorb impact on collision. These two developments mean that cars are more susceptible to damage in the crumple zone areas at the front and rear ends as a direct result of the built-in safety characteristic of deformation of structures. Consequently more cars now need alignment and repair requiring the use of jig equipment. In particular there has been a growth in the number of cars sustaining damage which, while not major, still requires the use of proper diagnostic and repair systems to affect a repair. The repairer must therefore choose the repair system most appropriate to the degree of damage present. Consequently there has been a growth in the development of measuring systems which can be used in a wide variety of repair situations with speed and accuracy. With dual systems, which combine both brackets and measurement, the repairer can carry out fast and accurate repairs whatever the degree of damage. However, with a body alignment system, whether jig brackets or a measuring system, it is vital that the completion of repair should be followed by a full suspension and wheel alignment check to make sure that the car is completely roadworthy.

Since the early 1950s there has been considerable progress surrounding the design and use of jig brackets. On the technical side, developments mean that the jig brackets are now available for a wide range of vehicles and are lighter, easier to use and more readily available. For ease of use and to prevent possible errors, integral brackets fit to a regular-spaced hole pattern on top of five integral beams which are common in all integral bracket sets. The front integral beams run longitudinally, as do chassis legs, thus allowing first improved access to the engine compartment and second the production of stronger, simpler brackets. All integral sets incorporate a number of check brackets which are fitted without stripping out mechanical components. Each integral bracket carries a plate which shows a reference number, which side of the jig it fits to and which edge points forward.

Body scanner

Alongside the jig and bracket system has been the development of the electronic measuring system. Key parts of the vehicle body are identified and can be measured using a laser scanner. This helps to ensure that the repaired vehicle is perfectly straight and does not drive in a sideways motion, referred to as crab fashion.

Figure 14.2 Genesis electronic system components (*Chief Automotive Ltd*)

1 Computer	5 Workstation	9 Target attachments
2 Display monitor	6 Body scanner	10 Upper body bar
3 Keyboard	7 Scanner tray	11 Lower body bar
4 Printer	8 Targets	12 Scales

The body scanner automatically reads all reference points simultaneously from strategically placed targets positioned on the vehicle underbody. These readings are then compared with published data and are displayed in colour on the monitor screen. Upper-body and strut tower positions are similarly located to provide height, width and length specifications relevant to the lower body. Genesis first displays visual guides, showing where to place each target. It then graphically illustrates how existing positions compare with vehicle specifications and provides direction and distance information for planning the repair. The entire system stays in place while repairs are being made.

Setting up the equipment, hanging the targets in position, entering the vehicle's information, and positioning the body scanner take a qualified bodyshop technician approximately 10 minutes. Calibration of the vehicle is then carried out automatically by Genesis in a few seconds.

To check progress at any moment the computer can provide either a visual display or a printout verification of the vehicle's structural condition. This can be of great benefit

Figure 14.3   Computer measuring screen (*Chief Automotive Ltd*)

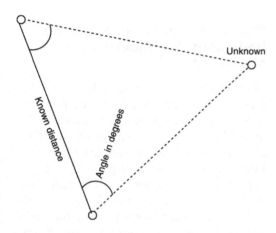

Figure 14.4 Triangulation (laser measurement) (*Chief Automotive Ltd*)

for insurance companies for quick approval of additional repairs and also for customers requiring verification of repairs.

Choice of systems

The design concept of monocoque construction with its crumple zones which absorb energy through deformation of panel structures and the increasing use of high-strength steels have made essential the use of repair and alignment systems even for minor damage repairs. This has been encouraged by the research carried out at the Motor Insurance Repair Research Centre Thatcham and by the promotion by the VBRA and the AA of improved standards throughout the vehicle body trade.

Owing to the wide range available of repair and alignment systems, it is increasingly difficult to select any one system, as they each have many differing advantages.

The following are some of the important criteria to be considered in the selection of suitable equipment.

- Suitability for the type of repair work normally handled
- Suitability for the volume of work handled
- Availability of initial training and further training as required
- Adequate service and backup by the manufacturer
- Availability of accurate data information which is constantly reviewed and updated

- Expected lifespan of the equipment and the possibility of it becoming outdated
- Mobility, portability and ease of storage
- Firm securing of vehicle during repair
- Good system of alignment that can be used without too much dismantling of mechanical components (dual systems are sometimes preferable)
- Multi-pulling facilities for simultaneous repair over 360° and enough pulling power to cope with HSS
- Adjustable working height positions
- Adequate holding facilities to allow the welding in of new panel assemblies

Alignment of the modern integral body

Misalignment and vehicle control

Any serious impact on any part of the all-welded body of the mono-constructed vehicle will cause distortion throughout the structure, which will not necessarily be adjacent to the damaged area. Since the front and rear suspension units are connected, either directly or through the medium of a front or front and rear sub-frame, distortion of the body shell means a disturbance of the relationship of the running wheels one with another. When this occurs the result will be crabwise running in which the car seems to travel in a sideways manner, commonly called dog-tracking or crab fashion. It is often necessary for the driver of such a vehicle to struggle constantly to maintain a straight course, and this becomes greater as the speed increases.

Misalignment of car underbody can also cause tyre wear and may affect the brakes and steering control, which will not be cured however much checking and adjustment is carried out. These faults can render a car outside the uniform standards required by law in respect of the fitness of the car for use on the road. A car with a damaged or weakened underbody can be a menace to safety on the road. Aside from

the effects it has on steering control and tyre wear, it can also place undue stress on other mechanical parts of the vehicle. The alignment of the engine with the clutch and transmission may be affected. This could cause a manual transmission to jump out of gear and might result in permanent clutch failure. Any radical change in the angle of transmission to the rear axle may cause excessive wear in the universal joints, noise in the rear axle or axle failure. Moreover, whenever the underbody has been damaged by collision there is a possibility of a broken hydraulic brake line which could lead to complete brake failure. Electrical connections also may become broken. There are a number of modern cars where the rear track varies substantially from the front to the back, and there are several which have up to 25 mm variation in track; these points are taken into consideration when checking for distortion.

Checking a mono-constructed body shell for underframe alignment

In order to make an alignment check of a mono-constructed underbody, the first step is to assess the extent of the damage, the probable amount of force involved which caused it and the direction of that force. This information will act as a guide by indicating where to look for possible distortion. With this knowledge a visual check should be made of doors, bonnets, boot lids, roof and centre pillar positions for any distortion which may be visible. Inconsistent gaps around these panels show that the panel assembly has been moved during the collision.

The underbody should be checked next. This can only be done by jacking the car up and inserting safety stands or by using lifting equipment in the form of a hoist, which will allow a more thorough examination of the underside of the vehicle. The sub-frames and cross members can now be examined for kinks or buckles with the aid of a portable light, but where sighting is difficult or impossible the members are examined by touch. This is not conclusive evidence of distortion, but it will generally be sufficient to convince the repairer whether a more precise check should be made by using the drop-line method, the gunsight gauge method or a jig alignment system.

Drop-line method of checking underbody alignment

For many years in the body repair trade the drop-line method was the basic means of checking the alignment of both composite and mono-constructed vehicles. When a car has been damaged the relationship of the vital points, such as rear suspension spring hangers, front hub centres and front suspension centres, must be checked on a flat floor surface with the vehicle jacked clear of the floor, all four road wheels removed, and safety axle stands in position. With a plumb bob and length of cord, plot onto the workshop floor a series of points common to both right- and left-hand underbody members. At least eight such points should be selected, four on each side, which could include front and rear spring shackle bolts, bumper bar bracket bolts or any factory-formed holes, rivets, bolts, or intersection of cross members common to both side members. On a composite designed body the body bolts provide ideal locating points. Variation in wheelbase can be checked by plotting the centres of each of the four wheels. The plumb-bob line should be held against the centre of each point selected and the position where it strikes the floor should be marked with a pencil cross. The pencil is more easily seen if the floor is previously chalked in the approximate area where the plumb bob will fall.

After all the points have been plotted and the wheels replaced, the car can then be rolled away, leaving the chalked pattern on the floor. These points are joined diagonally

Figure 14.5 Drop-line check on front and rear sub-frames of a Classic Mini

A – A	660.40 mm (26.0 inch)	Front sub-frame front mounting set screws
B – B	412.75 mm (16.25 inch)	Front sub-frame rear mounting set screws
C – C	1282.70 mm (50.50 inch)	Rear sub-frame front mounting set screws
D – D	977.90 mm (38.50 inch)	Rear sub-frame rear mounting block set screws

All dimensions taken at centre line of set screw or set screw hole.

by using a chalked length of cord which is held tight and then flicked to the floor, leaving a chalked straight line between each pair of points. These lines will indicate if any variation has occurred in the diagonal measurements. The next stage is to establish a chalked centre line through the diagonals. This is done by bisecting the lines joining the front and rear pairs of points and using the chalked cord to mark the centre line. If the underbody is in perfect alignment, this centre line should then pass where each diagonal intersects the centre line. When the centre line does not pass within 3 mm to these intersecting points, measurements must be taken to establish which member is distorted. A further check can be made by joining the sets of points transversely. These transverse lines will be at right angles to the centre line if the underbody is in correct alignment.

Owing to the inaccuracy of this system and the increase in the complexity of design structure of vehicles and their critical suspension geometry, the drop-line method has been superseded by the use of alignment jigs, using either a bracket alignment check system or a universal measuring system.

Tech note
The Classic Mini was the first car to pioneer the use of sub-frames both front and rear. These were fitted simply by the use of nuts and bolts, which meant that it could easily become out of alignment.

Gunsight gauge method of checking underbody alignment

Another method of checking underbody alignment for distortion is that of using gunsight frame gauges. These gauges consist of two sliding bars which can be adjusted by either pulling out or pushing in to make them fit correctly across a chassis frame or underbody. At the middle of the gauge is a sighting pin which always remains central irrespective of the distance to which the gauge is extended and which is used to establish the datum line (the vehicle's centre line). There are adjustable hanging rods on both ends so that the gauge may be raised or lowered. These hanging rods are also used to attach the gauge to the underbody or frame at symmetrical points on it, which may be factory-formed holes, mounting points or any part of the underframe which has identical positions on both sides of the vehicle. Never hang gauges on movable parts such as control arms, springs, torsion bars or any mechanical moving parts, as false readings will be obtained. A minimum of three gauges is essential, but the more gauges used the easier it will be to determine exactly where the distortion is located, as sighting along four or five gauges is quicker than relocating three gauges back and forth along the length of the vehicle.

When setting up the gauges to check underbody alignment, the in-line condition will be when all sighting pins are in line and all gauge bars are parallel. If one gauge must be lowered to clear some obstruction on the underbody, all

Figure 14.6 (a) Gunsight gauge frame (b) diamond detector gauge (*Blackhawk Automotive Ltd*)

other gauges must be lowered a corresponding distance to maintain a level sighting. The condition in which the gauge bars are not parallel indicates that the underbody has moved up or down at some point along its length. The condition in which the sighting pins are not in line while the gauge bars are still parallel indicates that the underbody members have moved to one side only (sway). A twisted underbody is recognized by the gauge condition in which sighting pins are out of line in addition to the bars not being parallel.

actual location is found by visual inspection. The main advantages gained by using this system of alignment check are the speed at which it can be set up to diagnose distortion and the convenience it offers in making regular checks during repair procedure.

Chief precision alignment gauge system

The principle of the self-centring siting pin is used by Chief in their alignment gauge system.

These gauges are hung underneath the car from specific data reference points, giving an instant alignment check according to the siting on the pins.

Underbody alignment using jig fixtures

Using an alignment and repair system is the only method of guaranteeing a comprehensive and accurate check on the alignment of the underbody of a vehicle. The monocoque body is produced on a production line using a series of quality-controlled jigs and fixtures for holding all panels in the correct position for spot welding. To the production engineer this is the only way of consistently reproducing a basic shell fitted with all its mechanical parts in the correct places. When it comes to repairing a body which has been damaged or distorted in an accident it is only logical that accurate alignment jigs should be used to ensure a safe, roadworthy repair and one which has the appearance of a new vehicle. The introduction of systems of body jigs was a major step forward in body repair techniques. These systems were designed and developed to cope with the ever-increasing call on the accident repair trade. Every vehicle presented a different problem, but, equally, a body shell had certain points which always could be relied upon to be accurately controlled during factory production, and it was around these points that all manufacturers of alignment equipment designed their products. This approach and the enthusiastic

Figure 14.7 Frame gauges showing distortion (*Blackhawk Automotive Ltd*)

As well as these basic checks an additional gauge is available, known as a diamond detector gauge, which is specially made to fit on the top or bottom of the self-centring gauges. This provides extra sighting pins to help to detect sideways movement (sway) and diamond damage (when the underbody has been forced into a diamond shape), which may not be evident when using self-centring gauges only. The only indication of a diamond condition would be that the two pins of the diamond detector gauge would be out of line with all other self-centring pins. Having determined the type of damage, the

Figure 14.8 Manufacturer's alignment points

No.	Standard measurement point	Hole-Size shape. mm (in)
1	Centre of front bumper position hole – RH	0–9 mm (0.354 in)
1a	Centre of front bumper position hole – LH	0–9 mm (0.354 in)
2	Centre of front side member positioning hole – LH	0–25 mm (0.984 in)
2a	Centre of front side member positioning hole – RH	0–25 mm (0.984 in)
3	Centre of suspension crossmember mounting hole – LH	0–16 mm (0.629 in)
3a	Centre of suspension crossmember mounting hole – RH	0–16 mm (0.629 in)
4	Centre of front floor side member positioning hole – LH	0–15 mm (0.59 in)
4a	Centre of front floor side member positioning hole – RH	0–15 mm (0.59 in)
5	Rear portion of rear seat crossmember positioning hole – LH	– 25 × 38 mm (0.984 × 1.496 in)
5a	Rear portion of rear seat crossmember positioning hole – RH	– 25 × 38 mm (0.984 × 1.496 in)
6	Centre of rear floor side member drain hole – LH	0–10 mm (0.393 in)
6a	Centre of rear floor side member drain hole – RH	0–10 mm (0.393 in)
7	Rear portion of rear floor side member extension positioning hole – LH	– 18 × 67 mm (0.708 × 2.637 in)

co-operation of the vehicle manufacturers at all stages ensured that when a car was mounted on the equipment, the damage of distortion showed up in an unmistakable fashion. The products now available to the body repairer include the following:

- Alignment systems that incorporate bracket mounting
- Alignment systems that incorporate either a mechanical or a laser light measuring system
- Dual equipment, which incorporates both measurement and bracket systems
- Cassette-type multifunctional alignment and repair systems

- Alignment systems that incorporate either a mechanical or an electronic measuring system

The manufacturers are continuing their research to develop appropriate body repair equipment as car design and methods of assembly are influenced by technological progress.

Vehicle body alignment: upper structure

Damage to the inner construction of a car body might be quite severe although not obvious. In some cases the damage to the

Figure 14.9   Manufacturer's transverse alignment points

outer panels of the body can be repaired, and the original damage will appear to have been corrected although it has not. Failure to check the correction of the internal damage by alignment methods can result in incorrectly fitting doors, bonnets and boot lids.

The alignment methods are based on comparative measurements in which one distance is compared, by means of a trammel, tramtrack gauge or measuring equipment incorporated in an alignment measuring system, to another which should be the same. Comparative measurements are universally used and are regarded as the easiest and quickest method of measuring collision work. These measurements are supplied by most manufacturers on data sheets or in body alignment manuals. Body measurements are made in the same way in both composite and monoconstructed vehicles. Checking body measurements is done by a system of diagonal comparisons, commonly called X checking. The location and extent of the damage is determined by measuring the body. Further measuring during the repair will indicate when alignment is restored. In this system of measuring the body is divided into two basic sections: the front section, which is the area from the centre pillar forwards, and the centre/rear section, which is the area from the centre pillar backwards to the rear boot compartment. In addition to these sections, there are special sets of measurements for

the alignment of the bonnet lid, the boot lid, and the front and rear doors. Some alignment systems incorporate upper body alignment ensuring equipment.

Door alignment

The doors of a car body are said to be misaligned when they do not fit the contour of the body or when they do not fit in the door opening correctly. Doors must provide a good seal against dust, water and air, and in order to do this they must match the contour of the body at all points. When closed, doors must fit correctly into the opening provided for them, because if the door is incorrectly positioned it will not close properly, and the gap around the door will not be uniform. In some cases in which the door appears to be misaligned the door itself is not the cause of the fault, and so it is advisable to X check the body at the forward and centre positions to determine whether the failure to match the body contour is due to distortion of the body or distortion of the door itself. In some cases lack of uniform clearance at the front of the door might be due to mis-positioning of the front wing.

Door alignment is checked by making a visual close inspection of the door and its relation to the surrounding body panels. First open and close the door and observe any movement at the edge of the door at the lock pillar; if there is any up-and-down

Figure 14.10 Front diagonal alignment check

No.	Standard measurement point	Hole-Size shape. mm (in)
26	Front pillar positioning notch (Inner side) – LH	–
26a	Front pillar positioning notch (Inner side) – RH	–
39	Centre of rear door striker mounting hole (Upper section) – LH	0–10 mm (0.393 in)
39a	Centre of rear door striker mounting hole (Upper section) – RH	0–10 mm (0.393 in)
48	Centre of hood opener cable routing hole – LH	0–5.3 mm (0.208 in)
48a	Centre of hood opener cable routing hole – RH	0–5.3 mm (0.208 in)
49	Centre of front door striker mounting hole (Upper section) – LH	0–12 mm (0.472 in)
49a	Centre of front door striker mounting hole (Upper section) – RH	0–12 mm (0.472 in)

movement, the door is out of alignment but is being corrected by the dovetail of the lock as the door closes. Look for signs of rubbing or scraping by the bottom edges of the door on the sill panels or centre pillar. Scrub marks are evidence of a condition known as door sag. In cases in which the door is severely damaged and a good deal of straightening is necessary, it is essential to check the door before any attempt is made to reinstall and to align the door with the body. Measurements should be taken on the damaged door, working from points that can be easily established on an undamaged door, and the readings compared to determine whether the damaged door requires further attention. Always be sure to take measurements from the same points on both doors.

Bonnet alignment

Bonnet alignment can be easily checked by visual inspection. When the bonnet is closed the gap all around it should be uniform; if it is not, misalignment exists. Manufacturers' specifications concerning this gap are

Figure 14.11   Side body check

No.	Standard measurement point	Hole-Size shape. mm (in)
25	Front pillar positioning notch (Upper section)	–
26	Front pillar positioning notch (Lower section)	–
27	Notch of roof side rail	–
28	Notch of front pillar outer and side sill outer	–
29	Centre pillar positioning notch (Lower section)	–
30	Centre pillar positioning notch (Upper section)	–
31	Notch of rear pillar outer and side sill outer	–
32	Notch of centre pillar outer and side sill outer	–
33	Rear pillar positioning notch (Upper section)	–
34	Rear pillar positioning notch (Lower section)	–

available; however, a good general rule is that this clearance should be between 3 mm and 6 mm wide. Correct bonnet alignment with the wings and front grille preserves the streamlined contours of the front end of a car. A uniform gap should exist between the sides of the bonnet and the wings. At the rear end of the bonnet the body is recessed to form the scuttle panel into which the bonnet fits to give a continuous flowing line. If the bonnet is too high or too low a recognizable step occurs at this point. If the bonnet is too far forward a large gap will exist between the rear of the bonnet and the raised portion of the scuttle panel; however, if the bonnet is too far back this gap will be too small, and there is a possibility that when the bonnet is opened it will catch the scuttle panel and chip the paint on either or both of these panels. When a condition of misalignment exists that affects the bonnet opening, it will be necessary to X check the opening before deciding whether to move either the wings or the grille panel. In some cases this can be done by adjustment; in other cases, where these assemblies are fully welded, hydraulic equipment must be used. The bonnet hinges on most cars are also constructed to give adjustment; therefore the bonnet can be moved forwards, backwards and sideways.

Boot lid alignment

Correct boot lid alignment exists when the boot lid matches the contour of the body surrounding panels. Misalignment of the boot lid is checked visually by an inspection of the gap all round the boot lid between it and the body. It is often necessary to check more closely for boot lid alignment, because an incorrectly aligned boot lid could allow the entrance into the luggage compartment of water and dust which might cause damage to anything being carried there. A simple check is to chalk the edges of the body flanges which contact the weatherstrip on the boot lid, then close and reopen the lid, thus transferring the chalk to the weatherstrip at each point where contact is made. If this chalk line is visible around the entire weatherstrip, the boot lid sealing is perfect. Where the chalk line does not appear, the boot lid is not sealing properly and realignment will be necessary. Realignment may be possible by adjustment of the latch or hinges, but in some cases it may be necessary to bend or twist the boot lid back to its correct alignment. When the boot lid opening is at fault it must be carefully X checked, and it can only be realigned by the use of hydraulic equipment.

Major repair techniques

In major repair work the method of repair is to analyse the crash, establish the order in which the collision damage occurred and reverse the order when correcting the damage. Distortion of the vehicle's underframe in the case of mono-constructed vehicles, or of the chassis in composite-constructed vehicles, must be rectified first before any other part of the repair is attempted. Any vehicle body under major repair will never hold its correct alignment unless underframe damage has been completely repaired first. This essential stage in the repair of the vehicle is carried out by first inspecting the damage and checking its alignment by one of the recognized methods to locate the exact position and extent of the damage.

Correction is carried out by using alignment and repair equipment in conjunction with hydraulic pulling and pushing equipment to realign any underframe members which do not align with the specified locating points. Any members which are beyond economical repair can be cut out and replaced with new sections by using a combination of power tools and MIG welding equipment. The use of this hydraulic equipment together with the alignment and repair system will ensure precise accuracy of the complete underbody.

The next stage of the repair is the realigning and reshaping of the body shell structure and is normally known as roughing out the damage. After alignment checks have been made, hydraulic equipment is used to push and pull the body shell back to its correct shape. It is advisable to rough out all damaged sections, even though some of these may eventually need replacing with new panels. The reason for this is that the successful fitting of a replacement section will depend on the correct alignment of the surrounding areas, which can best be achieved by restoring them to as near the true shape as possible before the removal or cutting away of the sections beyond repair. It has been found that the use of multiple hydraulic pulling equipment used simultaneously is the most effective method when dealing with the majority of major collision work, together with standard hydraulic body jack equipment as a secondary pushing/pulling facility to aid the repair of the damage. The main advantage is the fact that the pulling is done externally from any angle while the car is anchored securely. In cases in which there is a tear in the damaged section, this must be welded up before any pulling or pushing is attempted; otherwise the tear will hinder the rectification and possibly become greater on the application of force. An important factor is that heat can be used to relieve stress in correcting damaged sections that have been badly creased or buckled as a result of impact. In this case the stress resulting from the buckle helps to retain the metal in its damaged state, and even after it has been restored to its correct shape by the application of pressure from hydraulic equipment, the section will tend to return to its damaged position if pressure is released. Therefore it is of considerable importance that before the pressure is taken off, these buckles or creases are hammered out with the use of heat and hand tools. When using heat for relieving stresses, it is important not to overheat the section beyond dull red and, wherever possible, never to heat any point more than once, as excess or repeated heating on one spot causes surface oxidation and annealing, both of which weaken the member or panel area. Heat can only be used on low-carbon steel panels and not on high-strength steels, because heat weakens the latter by disturbing the heat treatment which originally

strengthened them. This is particularly so in the case of frame members, where part of the strength of the member is derived from internal stresses set up during the forming operations in the manufacturing processes.

When the roughing-out stage is complete and any major body misalignment has been corrected, then the panel assemblies which are beyond repair must be cut out in such a way that the remaining connecting flanges are left intact. These are then straightened with the use of hand tools to facilitate the locating of the new panel section, which after positioning must be checked for alignment with its surrounding panels. When the alignment of the new piece is correct it is welded into position; since the vehicle is still in the repair equipment and correct alignment is maintained, check that all doors, the bonnet and the boot lid fit and operate in the appropriate body shell openings. Minor damage such as small dints and scrapes in the original panels should now be repaired using hand tools and the conventional techniques of planishing or filling.

On completion of all repairs to structural and panel damage, the vehicle can be removed from the alignment and repair system. All dismantled mechanical and body trim parts can be reassembled, including any necessary replacements to windscreen, rear window and door windows. It is imperative that after repairing all major accident work the vehicle's steering geometry is checked with optical alignment equipment and a road test carried out to assess its roadworthiness and handling capabilities. When the repairer is satisfied with the vehicle's performance and that all repairs have been completed, the vehicle will need refinishing to return it to its original factory finished condition.

Although every vehicle receiving major collision damage must be individually treated, the following is a basic approach to repair procedure:

- Inspect and check underframe or chassis for alignment using conventional equipment.

- Rectify any misalignment of underframe or chassis using hydraulic pulling and pushing equipment, hydraulic body jack, and alignment and repair system.
- Realign body shell, correcting damage with a combination of pulling equipment and body jack, and the use of heat and hand tools (not HSS). Also cut out and weld in any replaced buckled panels and reinforcing members which are beyond repair.
- Repair doors; fit glass and body trim to doors and check for ease of operation and alignment of the door in the opening in the body shell. Also check alignment of bonnet and boot lid and ease of operation.
- Smooth and finalize minor damage to body panels with hand tools.
- Replace all mechanical, trim and body parts.
- Replace windscreen and rear windows if necessary.
- Make an optical check on the vehicle's steering geometry.
- Road test the vehicle.

- Refinish to original factory condition.
- Valet car's interior and exterior.
- Final bodyshop check for quality control of completed work prior to returning vehicle to customer.

Repairs to monocoque bodies by pulling

The method of correcting damage by pulling from the outside is not new, but the application of this principle has developed into two main methods. These are direct and vector pulling.

Direct pulling

This is achieved by using systems which are designed around a base beam, sometimes mounted on castors for easy movement, with an upright beam pivoted at one end. A hydraulic push ram is mounted across the angle between the two beams. When the ram

Figure 14.12 Blackhawk general dozer set-up

Figure 14.13   Blackhawk front pull set-up

is extended, the upright beam is pushed in an arc backwards, thus putting tension onto the chain attached to the damage. This produces a direct pull between the repair and the beam.

Vector pulling

This is achieved by attaching a chain to the damaged area, then passing it over a hydraulic ram set as near to 45° as possible, and anchoring it to either a rack, a tie-down system or a special clamp. Pressure is applied to the chain by extending the ram; this puts tension in the chain, which in turn creates a pulling force on the damage. As pressure is increased the angle will continuously change in response to the pressure applied.

The principle involved in a pulling repair is to apply force to the car body that is opposite to the force that originally damaged it. With a monocoque body, the engine, suspension and steering rack all contribute to the

final strength and shape of the vehicle, so it is important not to remove them unless absolutely necessary for the type of repair being carried out.

Before beginning actual pulling, establish exactly the nature and type of damage sustained. This can only be achieved by the use of alignment and repair equipment; visual inspection is not satisfactory and can lead to inaccuracies. However, the vehicle must still be considered as a unit, and the pulling force applied must act on the whole of the body and not just on the damaged parts.

The development of pulling systems has meant that in dealing with major body repair work it is possible to get a direct upward or downward pull to the front, rear or either side of the vehicle using single or multiple pulling equipment.

When repairing mono-constructed bodies, basic body repair methods and a knowledge of panel beating are absolutely essential. The combination of methods always necessary in repairs is as follows: heavy external pulling, use of body jack, use of heat (only on low-carbon steels and not on high-strength steels), use of power tools and cutting out buckled sections or reinforcing members. Before any pulling commences, the angle of the pull is already known, but difficulties can arise in setting up the precise angle with the equipment. This problem can be solved by using the principle of parallel lines. First examine the damaged parts carefully and take accurate measurements to see in which direction and by how much the damaged parts have moved. This means that both lateral and vertical movement must be determined to make sure the correct reversal of this damage can be achieved. When this is known, the first pulling must always start as near to the major impact as practicable.

Heavy external pulls are a must for all severe mono-constructed body damage because the impact has stressed the body metal. It is important when pulling damaged sections back to their original shape not to apply all the force at one place at the same time. Spread the restoring forces by using multi-pulls, which will allow the pulling forces to be activated from several points on the damaged section simultaneously. This method of pulling is more effective than single pulls on severely damaged vehicles. The body jack can also be used to assist by applying additional forces at the same time so they help each other in working the panel back into place.

With all mono-constructed bodies, special attention must be given to certain load-bearing or structural members such as frame side members, inner wheel arch panels, cowl areas, cross members and the floor pan that form a part of the cross members or box sections. These must be brought back to the equivalent of their original condition, which means returned to their original shape without wrinkles or excess thinning of the metal.

Figure 14.14   Celette Sevenne body jig

Figure 14.15   Simultaneous multipulls on two vehicles (vector pulling) (*Blackhawk Automotive Ltd*)

Figure 14.17   Celette Griffon body jig system

Figure 14.16   Simultaneous multipulling using direct pulling (pulling towers) (*Kroll (UK) Ltd*)

The first step in making any hook-up is to anchor the pulling equipment into anchor pots, a rack system, or an alignment and repair system. Next, a pulling attachment such as a sheet metal clamp or pull hook is secured to the damaged part. Work the damage out of the metal with a hammer and dolly while the metal is under tension from the pulling equipment. Use heat sparingly as required to remove severe wrinkles. Where sheet metal damage is widespread, several hook-ups may be required to bring the damage out gradually to prevent tearing and further damage.

With the vector method of pulling, parallel pulling can be carried out in the following way. First connect the chain to the damaged part of the car, then hold the chain tightly in the direction you wish to pull and adjust the pulling unit or ram to the measured angle. It is not necessary to have the chain vertically in line with the direction of pull, but it is important the ram should not be set at an angle lower than 45°, that the chain links are set in line without any twist and that the

chain between the ram and chain anchor point should always lean towards the car. As soon as tension is applied to the chain, safety wires must be fitted in case the clamp or chain tears away from the main structure of the car.

Rectification of various damage types

Although every accident is different and must be treated individually, it is possible to generalize and divide accident damage into several types. Each type can be pulled and pushed straight in a number of ways.

For instance, if a vehicle has been in a head-on collision, and has suffered a straightforward **crumple**, secure the undamaged part of the underbody or chassis leg and pull out the crumple section in the direct line of damage. If the vehicle was struck by another, or if it has hit a low obstacle, it will have **sagged**. To repair this, secure one end of the vehicle, provide a support at the point of the bend, and pull down the raised end. With a **twist**, hold and support the undamaged part and apply a downward force to the front while allowing the other end to pivot up. A **diamond** can be difficult. Hold one side of the vehicle and pull along the line of the diagonal. The tendency will be for the whole of the vehicle to move, so hold the opposite corner to the pull. To repair severe **side damage**, push out and stretch the crushed side. It will also help to pull from the outside; this is because the passenger compartment is very strong, and repairing damage to it requires considerable force.

Repairs to underbody damage of monocoque vehicles

The underbody is the foundation on which the vehicle is built. It is of the utmost importance, therefore, first to determine the extent of the collision damage to the underbody and then to correct it. The repairer must be able to determine what, if any, damage exists before attempting a repair. It is also necessary to determine whether any underbody misalignment exists which requires repair before attempting to correct body alignment and damage. The majority of body repair shops have alignment and repair equipment capable of handling this type of repair. Although at present the majority of British manufacturers are producing motor vehicle bodies designed and built on mono or unitary construction principles, there are still some special sports models which have independent chassis construction, in addition to the available European and American vehicles which are based on composite construction. Consequently body repair establishments must make provisions to deal with any such vehicles which may be brought in for repair.

The main idea of straightening underbody damage is to exert force in the reverse direction from that of the collision impact. Alignment and repair equipment permits several pulling set-ups to be made at one time. This type of equipment allows the vehicle to be aligned using either brackets or a measuring system. Repairs are carried out using a combination of pulling equipment, body jacks, chains, clamps, wood blocks and welding equipment. The severity of the damage decides the method of repair and techniques to be used. In the case of a vehicle which has received extreme damage and in which the underbody has twisted badly, the method of repair is that the vehicle must be mounted onto the repair system and the exact amount of distortion established by using either brackets or a measuring system. Whilst the vehicle is still mounted on the system, the underbody must be pulled back into alignment first and then checked against the manufacturer's data, especially the critical suspension and engine mounting positions.

First rough out the damage but do not remove any panels unless it is necessary. Body damage and alignment should, wherever possible, be corrected at the same time as any underbody damage. Work can be made easier by stripping off any body panels or mechanical parts which might interfere with accessibility to the damaged area. As side and cross members are usually formed from low-carbon steel, heat may be used to relieve stresses in badly creased areas. However, when these members are made from high-strength steel, heat must not be used because of its weakening effect; the metal must be repaired in a cold state. For low-carbon steel, heat may be applied with an oxy-acetylene welding torch. Only a very little heat is required to bring the colour of the heated member to a dull red, at which time it is at a proper temperature to be worked. Start to heat the buckled portion or damaged area well out near the edge. Pan the flame over the entire buckled area so that it is heated uniformly. Best results can be obtained by moving the flame in a circular motion until the entire area is heated. Heating well out towards the edges of the damaged area first will remove the chill from the surrounding area, thus preventing the damaged area from cooling too rapidly.

Pulling equipment can be used for heavy external pulling by anchoring it to the underbody members using special bolt-on clamps or chain attachments and pulling the damaged section back into line. Where it is difficult to obtain an anchor point owing to inaccessibility, a substantial metal tab is welded onto the member; this can be used for pulling and later removed. In some cases a combination of pulling and pushing is needed when correcting damage, and this requires a heavy-duty body jack to be used in conjunction with pulling equipment. Wood blocks can be used to spread the load to prevent distortion. As the repair proceeds with a combination of pulling, pushing and heating, continuous checks should be made for alignment, as it is sometimes necessary to overcorrect to allow for spring-back in the metal structure. A final alignment check should be made after the tension is released.

Wheel alignment (steering geometry)

Over the years there has been increased awareness of the importance of wheel alignment, and particularly four-wheel alignment, as an integral part of the major crash repair procedure. Straightening a vehicle involves two separate alignment operations. First, the body must be aligned, making sure that suspension mounting points are in the correct position. Once the vehicle is jigged, in many cases a mixture of old and new suspension components is fitted. Secondly, a full geometry check must be undertaken to satisfy the repairer that the car is within the manufacturer's tolerances after the repair.

Wheel alignment has an important role to play in the overall safety of the modern vehicle, and proper testing equipment in this area is now vital. Much of the new technology caters for what is now regarded as four-wheel or total alignment. Modern vehicles have become far more sophisticated. They may have active rear suspension, which can also be computer controlled. These set-ups have an element of steering built in to complement work done by the front wheels; the system is commonly known as four wheel steering. Rear wheel alignment with a fractional variance from the manufacturer's specification can be critical to the car's handling and steering. It is the norm for cars to veer to left or right: suspension systems are designed and set up by the manufacturer so that the car will travel straight, regardless of road crown.

The main purpose of wheel alignment is to allow the wheels to roll without scuffing, dragging or slipping on the road. The proper alignment of suspension and steering systems centres around the accuracy of six control angles: **camber**, **caster**, **steering angle inclination (SAI)**, **scrub radius**, **toe** (in and out) and **turning radius**.

King-pin inclination (KPI)

Figure 14.18 Steering angles from the front of the car

Caster angle

Figure 14.19 Steering angles from the side of the car

Inspection and test drive

Thorough visual inspection and a test drive on a straight, flat road should show whether the car needs to go for wheel alignment. The basic checks are:

1 Measure ride height.
2 Does the vehicle appear level, or is there any sagging?
3 Are the tyres properly inflated and of correct size and specification?
4 Is there any uneven wear on the tyres?
5 Low-speed wobble may show up a separated tyre tread.
6 High speed may reveal steering wear and odd noises, indicating loose or damaged parts.

Pre-alignment check

This breaks down into two areas:

1 Tyre inspection
2 Steering component inspection.

Tyre inspection, front and rear, is as follows:

1 Check for any unusual wear patterns: camber, toe wear, inflation wear, cupping.
2 Check for any physical problems: ply separation, dry rot.
3 Check tyre size and type: same size side to side, same brand and tread pattern. Do not mix radial-ply and bias-ply tyres.
4 Tyre pressure and wheels: check pressure and set to specifications; check for bent or egg-shaped rims.

Steering component inspection is as follows:

- Tie rod assemblies
- Idler arm and bushings
- Centre link and joints
- Manual steering gear: check for excessive play and leaks, check U-joint at steering shaft
- Power-assisted steering (run engine): check for excessive play; inspect all hoses and seals; check fluid level and belt condition; check power assisted balance

- Ball-joints and wheel bearings
- All bushings, front and rear
- Coil springs, torsion bars and shock absorbers
- Calibrate alignment equipment

Wheel alignment procedure

Two-wheel alignment is still used in many workshops as it takes less workshop time, is easier and requires less expensive equipment. However, its efficiency relies on the car having two parallel axles and front and rear wheels perfectly in line with each other.

Four-wheel alignment owes much to computer technology, which can instantly check and define the alignment of a vehicle. As full wheel alignment starts from a centre-line reference on the rear axle and then checks all steering geometry angles on all four wheels, a simultaneous four-wheel check will speed up the entire operation. Computer-based alignment systems use computer power to carry out the calculations so that you can be certain of centralizing the steering wheel and alignment of the vehicle. This ensures that the steering wheel spokes are horizontal when the car is travelling straight ahead.

Replacement of damaged panels

Damaged panels can be removed from vehicles by using:

- A hacksaw and metal cutting snips
- A hammer and thin narrow-bladed chisel
- Either a conventional drill, a Zipcut tool or a spot-welding removal tool (air operated) to drill out each spot weld
- An air chisel with appropriate cutting tools
- A power saw having either a rotary vibrating blade or a straight reciprocating blade
- A plasma arc cutting torch
- A combination of any of these methods.

A severely damaged inner panel such as a wheel arch can be cut out using the oxy-acetylene torch. However, this method is not recommended because of the very ragged cut edge left by the torch. A better method would be the use of the plasma cutting torch, which leaves a fine cut edge for panel replacement. With both of these methods it is important to be aware of the fire risks.

Lapped and spot-welded joints of panels must be drilled first and carefully separated with a narrow-bladed cold chisel. The positions of the spot welds are indicated by small circular discolorations of the metal. These are centre popped, then drilled 4.8 mm or 5.6 mm, sufficiently far to break the joint; it is not always necessary to go right through both parts of the panel, as in most cases if the top panel is drilled the joint will break. Alternatively a cobalt drill specially designed for cutting out spot welds can be used, either in a conventional air or electric drilling machine or fitted to an air attachment designed specifically as a spot-welding removal tool. Another alternative is the Zipcut tool, which again will fit into a conventional drilling machine and which cuts round the spot weld, leaving a hole in the top panel surface. In some cases a fine-toothed hacksaw blade, fitted in a pad-saw handle, and a pair of snips or shears can be used to remove damaged sections of panels. A tool for removing damaged panels is the power chisel set, which comprises a compressed-air gun having interchangeable chisel heads. This sheet metal cutter is capable of speedy and accurate removal of panels or parts of panels that have to be replaced. Power saws can also be used in the removal of damaged sections: either the rotary saw with a vibrating blade or the straight reciprocating blade. Both of these tools give an excellent fine cut edge suitable for instant re-welding without any dressing.

The bodywork having been straightened and aligned as much as possible, the new panel should be positioned and held in place by two or three clamps and all adjacent panels checked for alignment with the new panel. To ensure a perfect fit it is often necessary to trim the adjoining panels of ragged edges and to straighten out the locating flanges using hand tools. When making a close butt weld, exact alignment is ensured by positioning and clamping the new panel outside the old so that a scriber can be run down the edge of the old panel where the joint will come. Excess metal then being carefully trimmed and the edge dressed up with hammer and dolly, the new panel is ready to be welded into place. Always brush on weld-through anti-corrosive primer before welding. The panel should be aligned, clamped and checked. It should next be tack welded at several points; then the clamps should be removed and the panel rechecked for alignment. If this is not satisfactory the tack welds are easily broken. When alignment is correct, further tack welds may be placed between the first set. To avoid distortion only short sections between the tack welds should be welded at a time and the metal allowed to cool. In some cases it is helpful in confining the heat, and thus reducing the chance of distortion, to place wet cloth packing on either side and a little distance away from the weld; alternatively there is a foam type of material which will act as a heat barrier.

Panels that were originally spot welded but cannot be spot welded back can be MIG welded in the course of repair; the welds are made in the same place by welding through the holes in the adjacent panel which were left when the original spot welds were drilled out. This method is known as plug welding. Another method of securing panels is to MIG braze them in place, but one must remember that although MIG brazing will cause less distortion than MIG welding, it is not as strong. Therefore if the particular panel is a load-bearing panel, it must be welded in fully to obtain the maximum strength for the particular panel assembly. MIG welding has the advantage of reducing distortion to a minimum, owing to the effect of the inert gas shielding the spread of heat around the weld zone. This makes butt joints in panel assemblies more easy to accomplish, as less dressing up and planishing are required than in gas welding. Another advantage arises from the automatic feeding of filler material into the weld pool, which results in better penetration in all weld positions. As this equipment can be used both to weld and to spot weld, it is widely employed as a means of welding in new panel assemblies.

Partial panel replacement

The reason for replacing only part of a panel can vary according to the type of repair. Usually it is quicker and therefore less expensive than full replacement. If sectioning does not save on time or money, then the part is completely replaced: that is, when the re-panelling and straightening reaches the cost of the complete new panel. Some manufacturers design repair panels especially for these purposes. Part panel replacement has long been established as an effective way of repairing body damage on all cars. Depending on the type of damage involved, it used to be common practice to replace an entire sheet metal panel; by means of a part panel repair it is possible to replace only those panel sections which are actually damaged. The possibilities for sectional repairs to load-bearing and non–load-bearing parts have been taken into account during the initial phase of development by manufacturers. All instructions given regarding the locations of cut lines for sectional repairs to load-bearing parts must be strictly adhered to. Sectional repairs to vehicles damaged in an accident should be confined to body parts in which the damage has not produced any loss of strength sufficient to jeopardize safety. Only welding materials reaching the design weld strength should be used. Moreover, a body alignment and welding jig must always be used for repair operations such as sectional replacement of the side-member/apron-panel assembly. This ensures satisfactory repairs and adherence to the specified dimensions. Manufacturers have also carried out numerous strength tests, and extended trials (crash tests) have proved that part panel replacement is just as effective as replacing complete new panels. Provided repairs are expertly carried out, all repair procedures using part panels will result in the same standards of strength and operational safety being maintained as for a new vehicle.

To carry out this type of repair, the initial procedure is to make certain that the sections to be joined are from identical models. Then determine the most suitable point at which to make the joint: this will be influenced by the length of the weld, the amount of distortion likely to occur in making the weld, and the ease with which it can be dressed up. Remove the bulk of the unwanted section by cutting it away with a power saw, leaving an allowance of about 20 mm from the joint line, which can be trimmed to size when the pieces of the panels are separated. Where it is possible to fit the two sections on the vehicle, it is considered good practice to trim only one edge and to set the panels up with one section overlapping the other, as this enables a check to be made on the alignment of any surrounding or adjacent panels such as bonnet edges and door edges. A good butt joint is achieved by cutting through. When the alignment is correct, the two sections are tack welded along the length of the joint. The joint is then hammered level and fully welded. After this it is planished using normal techniques to achieve a smooth, undetectable joint. If this is impossible owing to inaccessibility, the weld should be hammered in using a shrinking dolly and the weld area filled.

Vehicles that have undergone an extended corrosion-prevention treatment to prevent body perforation by rust must be repaired using replacement panels coated by the new (cathodic) priming process. These panels are identified by corresponding stickers. Therefore remove as little as possible of the cathodic primer coating on replacement panels during sanding or other abrading operations, for example, on spot weld flanges or in areas where welds are to be made later. When repairs are being carried out in the underbody area, the zone in question must then be treated with underbody wax. The areas outside the side members and the wheel houses must also be treated with underbody protection material.

Repairing extensively damaged panels

Whenever possible it is best to repair severely damaged panels while they are still attached to the body shell. One advantage of leaving the panel in position is that it remains rigid whilst being processed. As already explained, it is difficult to hold a panel firmly when it is not attached to the body, and to beat a panel that moves with each blow is most unsatisfactory. Second, the shape is maintained while the panel is in position, and this allows the edges of the panel to be lined up with such parts as bonnets, doors and boot lids. In addition the use of the body jack while the panel is fixed to the body allows full use of the attachments designed for roughing out such parts.

Having decided whether the panel will be repaired on or off the job, the actual repair can then be roughed out, keeping in mind the fact that the shape needs reforming with as little stretching of the metal as possible. The original damaging force could have stretched the panel and, while reshaping, further stretching should be reduced to a minimum by the use of such methods as pushing out the damage with a body jack fitted with suitable rubber flex head, using a hardwood block between the body jack and panel or using hardwood blocks in conjunction with the hammer to eliminate the metal-to-metal contact. Where necessary, anneal creased sections with the oxy-acetylene welding torch prior to removing them. As the roughing out proceeds, weld any tears in the metal so as to gain rigid panel as soon as possible. Very stretched sections can be shrunk by the use of heat spots during the roughing out. A boxwood mallet, in conjunction with a suitable dolly block for the shape of the panel being repaired, should not be used as an initial smoothing or levelling step prior to the final planishing. The

actual smoothing with the planishing hammer and dolly should never be started until the correct shape is formed in the panel. At this stage the normal processes of planishing, dressing up, filing and picking up low spots should not be adopted until the job is completed. Either because certain parts of the repair are inaccessible or because of their over-stretched condition, it may be necessary to use solder filling or plastic filler to achieve a smooth finish.

Repair methods using heat barrier material

This is a method of repair using a specially formulated compound where heat would cause a problem of distortion or overheating of adjacent materials when welding.

The Cold Front heat sink chemical compound was originally developed for the American space programme. This heat barrier material, based on a chemical formula of magnesium aluminium silicate, prevents the conduction of heat through metal, thus eliminating the risk of distortion to panels or damage to areas immediately adjacent to any welding operation. When the material is applied to a metal panel on a vehicle body undergoing repair, it prevents the transfer of heat past the Cold Front barrier, giving complete protection to glass, rubber, plastic, paint and any material likely to be damaged by close proximity to the high temperatures which are generated during the processes of welding, brazing or soldering.

The method of application is as follows:

1 Apply generously, as using too little can allow heat damage. Apply to one side of the metal in light-gauge work (20 SWG) and both sides of the metal when working with heavy gauge (above 3 mm) for maximum protection.
2 When the mound of material begins to warm up, it becomes saturated with heat. Once saturated, it will allow heat to be conducted through the metal. If there is still work to be carried out, place more

Cold Front material on top of the existing mound to delay the conduction of the heat.

3 Always apply the material as close to the area of repair as possible.

4 Damage can be caused by radiant heat as well as conducted heat. The material acts as an effective heat shield against radiant heat when it is applied over the surface to be protected. Be sure not to leave any cavities or depressions when spreading.

5 It is safest to protect the repair area completely around the heat zone. However, if protection is only needed in one direction, the heat barrier should be spread five to six times as wide as the area to which the heat is being applied.

6 Metals have different conducting characteristics: copper, brass and aluminium conduct heat more rapidly than steel. Therefore the better the metal is as a conductor, the more heat barrier should be applied when working with that metal.

Cold Front heat barrier is odourless, non-toxic and harmless to skin. It will not stain surfaces, and any residue can be washed off with cold water once the repair is completed.

The ability of Cold Front to act as a total heat barrier is demonstrated by holding a strip of 20 SWG metal in the bare hand after applying a line of Cold Front halfway along the strip, then heating the portion of the strip

Figure 14.20 Application of Cold Front heat barrier material (*Gray–Campling Ltd*)

above the line to red heat. The part of the strip being held in the hand remains completely cold.

Door hanging

Door hanging requires particular care, as the fitting of doors is one of the most essential features of crash repair work. The most important aspect when dealing with the hanging of doors is that a door which is known to be the correct shape should never be altered to fit a door opening that is out of alignment. In addition, the resetting of door hinges must not be used as a method of rehanging doors without first checking for and correcting any misalignment.

Hinges used on the modern vehicle can take a good deal of force without altering their settings. In cases where hinges have been subjected to collision damage, the hinge transmits the damaging force to the pillar or door frame, causing twisting and misalignment. Sometimes hinges are so excessively damaged that they need replacing, although this is not the complete answer to correcting misalignment troubles arising from such damage. Where new hinges are necessary a close inspection and comparison should be made with the identical hinge mounting on the opposite side of the car, and any misalignment of this mounting position should be rectified before fitting the new hinge.

When it is considered that the hinge mounting points are correct, the doors should be hung in their respective positions and all screws or hinge fixing bolts placed loosely to allow the door to be realigned according to the car manufacturer's adjustments. The fixing bolts should be tightened if possible while the door is in its closed position. In most cases this is not possible, which means that the doors have to be held open whilst at the same time endeavouring to keep alignment correct while the bolts are tightened. If the door aperture is correct and the door is not twisted, with slight adjustment to the hinge position the door can be made to fit into its appropriate opening.

When the door will not fit into the body opening, the hinges should be tightened in the main adjustment position and the striker plate of the lock mechanism should be removed from the locking pillar to allow the door to swing freely. This gives a truer indication of the swing of the door as the opening is being corrected with the use of hydraulic equipment. The door should then be closed as far as possible, and the points stopping the door from entering the door aperture should be noted. Correction in most cases can be made by diagonal pushing with the body jack within the aperture. Measurements should be taken both before and after applying the pressure so that the effectiveness of the realignment of the door opening can be assessed. This checking and pushing must continue until the door is a good fit inside the door aperture.

Motor vehicle glazing systems

Since the introduction of legislation on the compulsory wearing of seatbelts for front-seat passengers in the early 1980s; both the type and the style of the glass fitted into windscreens have changed. Prior to 1980, toughened glass, which shatters into tiny, relatively harmless pieces when broken, and which can dangerously obscure a driver's vision, was used by many vehicle manufacturers. This glass was indirectly glazed onto the body and secured using a self-tensioning rubber weatherstrip fitted to the aperture. With the seatbelt legislation in force, however, safety standards were improved, bringing an increased use of laminated glass, which is the type that remains in place when broken and therefore allows the driver to continue the journey.

This resulted in a change in glazing methods, with laminated glass being increasingly secured by the process of direct glazing. This is a method of fitting a glass unit to a vehicle by chemically bonding it to the prepared aperture using specialized compound sealants. This method provides two definite design benefits. It gives an aerodynamic profile and also increases the strength of the windscreen, thus making it safer and more weather resistant. As well as being easier to fit on the production line, the majority of direct glazing is now done by robots. A direct-glazed windscreen can also be replaced to its original specification without affecting the structural integrity of the vehicle.

Indirect-glazed windscreen removal and replacement

The two basic types of safety glass used in the car windscreen are **laminated** and **toughened**. Laminated glass is of a sandwich construction and consists of two pieces of thin glass, one on either side of and firmly united to a piece of transparent reinforcing known as the interlayer. Although such a glass may crack, forming a spider's-web pattern, it holds together and the panel remains in one piece except under conditions of the most violent impact. *Toughened* glass is produced by the process of heating a solid piece of glass and then rapidly cooling it so that its liability to fracture is greatly reduced and its strength increased to about six times that of

Figure 14.21 Laminated glass

Typical fracture of laminated glass

Example of toughened glass after fracture

Figure 14.22 Characteristic breaking pattern of laminated and toughened glass untreated glass. If fracture should take place, the resulting fragments consist of very small, comparatively harmless particles.

In major accident damage, removal and replacement of the windscreen is an essential feature of the repair. Many windscreens and rear windows are held in place by a rubber weatherstrip which is usually of single-piece construction. Both glass and any outside trim mouldings, if used, are recessed into the weatherstrip. It also has a recess which fits over the body flanges around the opening for the glass. It is this portion of the weatherstrip that holds the entire assembly in place.

Indirect types of glazing systems can be identified for each specific model by referring to the body section of the relevant manufacturer's repair manual. The rubber weather seals used are either mastic sealed, dry glazed, self-sealing or push fit. The self-sealing types are held in place by a removable centre filler strip and can be either mastic sealed or dry glazed. All the systems are used with either a rubber weatherstrip moulding with a sealing compound or a rubber weatherstrip moulding without a sealing compound.

The rubber weatherstrip has two rebates, one for the glass and one to seat over the metal flange of the body aperture. It is the contact area of the aperture that requires careful attention whenever the glazing is disturbed or removed. In all cases it is essential

(a)

(b)

(d)

(c)

Figure 14.23 Indirect glazing system: (a) mastic-sealed, (b) dry glazed, (c) self-sealing, (d) push fit that the aperture is correctly painted prior to refitting any of the weather seals.

In order to remove a windscreen which is fitted with self-tensioning weatherstrip, the following procedure must be carried out. First cover the bonnet to protect the paintwork. Then, working from the inside of the car, remove any instrument panel mouldings or trim which may interfere with the removal of the windscreen. Also, working from the outside of the vehicle, remove any windscreen chrome mouldings which may be fitted into the weatherstrip. Again working from the inside and starting from one of the top corners, pry the rubber weatherstrip off the body flange with a screwdriver. At the same time press firmly against the glass with your hand adjacent to the portion of the rubber lip being removed, or bump the glass with the palm of the hand; this should force the entire assembly over the body flange. The windscreen can then be removed from the body opening.

To replace the windscreen, first clean the glass channel in the rubber weatherstrip to be sure that all traces of sealer and any broken glass are removed. Place a bead of new sealing compound all around the weatherstrip in the glass channel. Work the glass into the glass channel of the weatherstrip and be sure it is properly seated all the way round. Lay the windscreen on a bench, suitably covered to prevent scratching, so that the curved ends are pointing upwards, and in this position the pull cords used to seat the rubber lip over the body flange can be inserted into the metal channel on the weatherstrip. One or two cords may be used. Where two cords are normally used they can be pulled in opposite directions, which will make the fitting of the windscreen easier. Commence with one cord from the bottom right-hand side of the windscreen, leaving about one foot spare for pulling. Start the second cord from the bottom left-hand corner and proceed to the right, following round up to the top left-hand corner. Apply sealer all around the metal lip on the body aperture. Place the entire unit in the windscreen opening with the cord ends hanging inside the vehicle. An assistant can help

by pressing against the glass from the outside while each cord is pulled out of the groove. As the cord is pulled out, the rubber lip is seated firmly over the body flange. A rubber mallet can be used to tap the unit around the outside so as to seat the entire windscreen assembly firmly in place. Any surplus sealer can be cleaned off with white spirit.

The procedure for removing a windscreen fitted with self-sealing weatherstrip is first to locate the joint in the filler strip which runs right through the centre of the weatherstrip. The strip is then carefully eased up and pulled out slowly around the entire assembly. This relieves the tension on the glass, which can then be lifted out of the weatherstrip by easing one corner first. To replace this type of windscreen the metal lip on the body aperture is sealed with a sealer, and the appropriate channel in the weatherstrip is then placed over this lip and seated all the way round the body aperture. The windscreen is fitted into the glass channel on the weatherstrip with the aid of a special tool which enables the glass channel lip to be lifted, thus allowing the glass to slip easily into position. A small brush is used to apply a solution of soap and water to

Figure 14.24 Using pull cords to fit windscreens

the filler channel. This assists the filler strip installation. A specially designed tool is used to insert the filler strip into its channel.

Direct-glazed windscreen removal and replacement

The method of direct glazing of windscreens, also known as bonding, is very popular with motor manufacturers. This process involves the bonding of glass into the aperture. Some of the advantages claimed for direct glazing are as follows:

1   Increased rigidity and strength caused by integration of the glass into the body of the vehicle, reducing the need for supporting members and improving visibility with enlarged glass areas.
2   Weight reduction.
3   Improved aerodynamics by the deletion of rubber surrounds, aiding fuel economy.
4   A better seal against the weather, eliminating the need for resealing.
5   Increased protection against car and contents thefts.

Various bonding materials are used in direct glazing. Some require heating to induce a chemical reaction to create adhesion, whilst others will cure at room temperature.

Polyurethane and silicone materials are usually supplied as a pumpable tape. They are cold cured, and the material is dispensed onto the glass through a specially formed nozzle out on the end of a cartridge. This can be done with the aid of a hand-operated or compressed-air cartridge gun. These materials are highly viscous in their uncured state, enabling a high degree of manoeuvrability within the glass aperture to ensure a good seal.

The sealant itself is either one- or two-component polyurethane. The latter contains an accelerator compatible with the adhesive; this is applied evenly to the perimeter of the windscreen.

Once this is achieved, the glass is placed into the aperture using glass suckers and aligned before securing. When a one-component polyurethane is used the car can be driven away in

Figure 14.25 Direct glazing with moisture-cured urethane

4 to 6 hours: if a two-component polyurethane is used it takes only 30 minutes to cure, allowing the vehicle to be back on the road within an hour of the repair starting.

To remove the glass from the vehicle, the bonding material has to be cut. This can be achieved by using a piano wire, a special cutter or a hot knife, as follows:

• *Piano wire* – Before using this method of removal there are certain safety precautions that should be observed. Gloves should be worn to protect the hands, and safety glasses or goggles should be used for eye protection in the event of the wire or glass breaking. When the finishers or mouldings have been removed to expose the bonded area of the glass, the wire has to be fed through the bonding. This is done by piercing a hole through the bonding and feeding the wire through with the aid of pliers. Handles can be fixed to the ends of the wire to allow a pulling action.
• *Pneumatic or electric cutter (oscillating)* – This is an air- or electric-powered tool to which special shaped blades to suit specific vehicle models are fitted. The tool removes windscreens that have been bonded with polyurethane adhesives.

• *Hot knife* – The hot knife can be used by one person cutting from the outside of the vehicle. It can cut round the average bonded windscreen and clean off the excess remaining adhesive while minimizing the risk of damaging the car bodywork. Before use, all trim around the windscreen, both inside and outside, should be removed. The cutting medium is provided by a heated blade, which is placed under the edge of the glass and pulled around the perimeter, melting the bonding to release the windscreen. It is used in conjunction with an air supply, which constantly blows onto the cutting area of the blade. This prevents the heat dissipating along the blade and gives a constant temperature as well as eliminating smoke and fumes emitted from the cutting operation. Overheating can result in toxic fumes being given off, and a charcoal filter mask should be worn as a safety precaution. Before glazing can be replaced, the bonding surfaces must be prepared correctly. The residue of the original sealer left in the aperture has to be trimmed to ensure there is a smooth layer for the new sealer to adhere to and to allow the glass to seat in the correct position. Various sealants for windscreen replacement are available.

The surfaces are then prepared with the materials included in the individual manufacturer's glass replacement kit, that is sealer application gun with bonding material, suction cups and leak test equipment, following the instructions provided. There is little variation in the sequence of operations, which is typically as follows:

1 Using a sharp knife, trim and level off sealer remaining in the vehicle windscreen aperture. Remove loose sealer. Areas where bare metal is exposed must be treated with etch primer.

Figure 14.26 Removing direct glazing using a piano wire

Figure 14.27 Electric windscreen cutter (*Fein/George Marshall (Power Tools) Ltd*)

Figure 14.28 Welwyn hot knife (*Welwyn Tool Co. Ltd*)

2 If the original windscreen is to be refitted, level off the remnants of the old sealer. Take care not to damage the black ceramic edging.

3 Apply the cleaning solvent provided in the repair kit to the windscreen aperture and the inboard edges of the screen, then wipe both aperture and screen with a clean cloth. **Warning:** When working with solvents and primers, use PVC gloves or apply barrier cream to hands.

4 Take the spacer blocks from the repair kit, remove the adhesive backing, and fit them in the original position on the windscreen aperture. Ensure that the differently shaped blocks are located correctly.

5 Fit the suction cups to the outer face of the windscreen and place the screen in the vehicle. Centralize the screen in the body aperture. Take strips of masking tape and attach them across the join between the body and the screen. These strips afford a guide when finally fitting the screen after sealant has been applied. Cut the masking tape strips at the glass edges and remove the screen.

6 Shake the tin containing the glass primer for at least 30 seconds. Then, using the felt swab, apply the primer evenly along the screen perimeter. When dry, wipe with a clean cloth. Repeat the process, this time by cutting the applicator head and using it to apply a coat of primer 17 mm wide (0.7 in) to the windscreen aperture.

7 Immerse a cartridge of sealer in water at a temperature of 60°C for approximately 30 minutes. Pierce the sealer cartridge in the centre of its threaded end and fit the pre-cut nozzle. Remove the lid from the other end, shake out any crystals present and install the cartridge in the applicator gun.

8 Apply the sealer in a smooth continuous bead around the edge of the windscreen. The sealer bead should be 12 mm (0.5 in) high and 7 mm (0.25 in) wide at its base.

9 Lift the screen using the suction cups and offer it up to the vehicle, top edge leading. Carefully align the masking tape strips, then lower the screen into position. Remove the suction cups. Testing for leaks, using either leak testing equipment or spraying with water, may be carried out immediately.
Caution: If a leak or leaks are detected, apply sealer to the area and retest. Do not remove the screen and attempt to spread the sealer already applied.

10 If leaks are not present, fit the windscreen finishers. Heat the windscreen finishers to 45°C and, commencing at one end of the screen lower corners, press into position around its entire perimeter. Remove the fascia's protective covering. Trim the lower spacer blocks. Fit retaining clips as necessary.

11 Fit the windscreen wiper finisher panel and air intake mouldings and the wiper arms and blades.
Caution: Do not slam the vehicle's doors with the windows fully closed until the screen is fully cured.
Warning: The integrity of the vehicle's safety features can be impaired if incorrect windscreen replacement bonding materials and fitting instructions are used. The manufacturer's instructions should be adhered to at all times.

Water leak detection

The general principles for searching for the actual location of where water leaks could potentially occur on the vehicle are not specific to any particular model and can apply to all vehicles.

First start by obtaining as much information as possible from the customer; such as when, where and how the leak occurs, and also whether the water appears to be clean or dirty. If these facts are not known, considerable time could be spent checking the wrong areas of the vehicle.

Tools and equipment

The following tools and equipment are recommended:

- Garden spray (hand-pump pressure type)
- Wet/dry vacuum cleaner
- Torch
- Mirror
- Seal lipping tool
- Trim panel removing tool
- Small wooden or plastic wedges
- Dry compressed-air supply
- Hot air blower
- Sealant applicator
- Ultrasonic leak detector.

Locating the leak

Locating the source of the water leak involves a logical approach together with a combination of skill and experience. For the purpose of locating the leak the vehicle should be considered in three specific areas: the front interior space, the rear passenger space and the boot space. From the information provided by the customer the body repairer should be able to determine which area is the right one on which to concentrate.

Having identified the area of the leak, the repairer must find the actual point where the water is entering the vehicle. An ordinary garden spray, of the type which can be pressurized and adjusted to deliver water in the form of a very fine spray or a small, powerful jet,

has been found to be very effective in helping to locate most leaks. Using a mirror and torch will help you to see into any dark corners.

Testing

The sequence of testing is particularly important. Always start at the lowest point and work upwards. The reason for this is to avoid masking a leak in one place while testing in another. If for example testing was to commence around the windscreen, water cascading down could leak into the car via the heater plenum or a bulkhead grommet. However, it could be wrongly assumed that the windscreen seal was at fault.

The visible examination of door aperture seals, grommets and weatherstrips for damage, deterioration or misalignment, as well as the actual shut of the door against the seals, are important parts of identifying an area where water can pass through.

Leak detection using ultrasonic equipment

When the vehicle is in motion, the body shape may produce eddy currents and turbulence which can force air and water through the smallest orifice. When the vehicle is stationary, it can be difficult to reproduce these conditions to a realistic level.

The ultrasonic equipment works on the principle of a transmitter creating ultrasonic waves which penetrate the smallest orifice in the vehicle body and are then picked up by a receiver fitted with a suitable probe, which can in turn pinpoint the exact leak point. The transmitter is placed in a base plate which automatically switches it on via a reed switch in the base of the transmitting body.

Ultrasonic waves will then penetrate out through the unsealed area, including the front windscreen, wiper spindle and washer jet fitment, roof seams, bulkhead seams and grommets, A-post area, door seals, front wheel arch seams and heater air intakes, to the receiver.

An important feature of this equipment is the incorporation of a sensitivity control which enables the operator to check a wide range of fitments. For example, a boot seal is a very light touch seal, whereas a urethane screen fitment is at the other end of the scale setting of the equipment and would need a high setting to determine a leak.

To ensure satisfactory leak location and testing, it is recommended not to use a leak detector on a wet vehicle, which should be blown dry prior to the test. As with any equipment it is important to read the instructions to enable the equipment to be used to its full potential.

Sealing

Having located the point of entry of the water, it is then necessary to carry out satisfactory rectification. Door aperture seals and weatherstrips should be renewed if damaged or suffering from deterioration. Alternatively the seals can be adjusted by carefully setting the mounting flange after making sure the fit of the door is correct.

Leaks through body seams should be sealed from the outside wherever possible to ensure water is excluded from the seam. First dry out with compressed air or a hot air blower.

When leaks occur between a screen glass and the weatherstrip, or in the case of direct glazing between the glass and the body, avoid removing the glass if possible. Use recommended sealers to seal between the glass and the weatherstrip or the body. If the vehicle is wet then it should be blown dry prior to sealing. It is difficult to seal a wet vehicle because, if a fault is found, it cannot be rectified quickly before moving to another area.

Windscreen repair: Glas-Weld Pro-Vac

This system of repair involves extracting all the air from the damaged glass and replacing the void with a resin which, when cured with ultraviolet light, is optically clear and is stronger than the glass that has been replaced. The system is widely accepted, and repair to Glas-Weld standard would not cause the vehicle to fail the statutory Department of Transport test for windscreens: 80 per cent of

stone-damaged windscreens can be repaired by this method.

When glass breaks, an air gap opens up. Sometimes crushed glass may also be present. There are also flat surfaces present inside the glass on either side of the air gap. These three things result in the visibility of the damaged area being affected. The damage will eventually spread further as the temperature changes and the vehicle flexes in normal use.

The system works as follows:

1 All foreign materials and crushed glass are carefully removed from the centre of the damaged area to open up an airway into the break. Certain types of damage may need to be precision drilled.
2 The patent Pro-Vac injector is filled with the appropriate resin, depending on local temperature and humidity, before being mounted on the screen. The injector is then threaded through the stand until the outer seal makes airtight contact with the screen.
3 The Pro-Vac injector is capable of creating a total vacuum within the damaged area, which is essential for top-quality repairs. Using alternating vacuum and pressure cycles, all of the air in the break is withdrawn, and the void is filled with resin. Once the damage is optically clear, curing can commence.
4 Using a special ultraviolet lamp, the resin is cured. Once the curing process is complete the repair area is not only optically clear but also structurally sound; in fact it is stronger than the glass it replaces.

Automobile plastics welding

The car manufacturing industry, in its endeavours to reduce weight and improve fuel economy, forecasts a greater use of plastics in future designs. Considering the extensive use of plastics on current models, coupled with an even greater anticipated usage, it is therefore important that engineers and repairers are knowledgeable in the various types of plastic that are being used and are aware that some plastics can be repaired. Many body components are manufactured in

Figure 14.29   Before and after repair (*Glas-Weld Systems (UK) Ltd*)

Figure 14.30   Damaged windscreen (laminated safety glass) (*Glas-Weld Systems (UK) Ltd*)

Figure 14.31   Cleaning the damaged area (*Glas-Weld Systems (UK) Ltd*)

a variety of plastics. Bumpers, grilles, light surrounds and even complete body panels have enabled designers to enhance vehicle aerodynamic styling and cosmetic appeal while retaining impact resistance and eliminating corrosion from these areas.

Plastic offers the same structural strength as steel in a body component because of its greater elasticity. Minor impacts that could deform steel beyond repair can be absorbed by plastic. Where damage is incurred it is capable of repair by welding. Cracks, splits, warping and even the loss of material from plastic components can be

remedied with the aid of the Leister Triac hot-air welding equipment. Where a steel component with equivalent damage would be replaced at some cost, the repair of the plastic part can save time and expense, particularly when winter

Figure 14.32 Fitting injector to the windscreen (*Glas-Weld Systems (UK) Ltd*)

accident periods make great demands of the vehicle repairer's parts stock. Welding plastics does not produce fumes when the correct procedure is followed. In a short time a plastic component can be restored to an 'as new' condition without the need for fillers or special treatments. The combination of welding and recommended paint procedures will show no trace of a repair that should last the life of the vehicle.

Identification of the material

There are two basic types of plastic used in cars:

- thermosetting
- thermoplastic.

Thermosetting plastics are cured by heat during manufacture. They cannot be welded; any more heat applied to them will break down their molecular bonding. The way to repair them is by using adhesive. **Thermoplastic** materials can melt without breaking down, so they can be manipulated like steel. With the right amount of heat they can be bent, softened and welded. The skill is in the selection of the appropriate temperature. The majority of plastics employed in vehicle manufacturing are thermoplastics. There are different types of thermoplastics, each having a specified temperature for welding operations.

Experienced repairers of plastic may recognize broad types of plastic by the degree of hardness or softness of the component surface. A quick test for identifying to which group a plastic belongs can be made by cutting a very thin sliver off the component. The thermoplastic curls right back on itself, but the thermoset stays straight.

In their body repair manuals, some vehicle manufacturers give details of plastic parts, while others mark all plastic parts with code letters. However, when no information or coding is available, identification can be carried out in two ways. The first is the method of combustion testing, where a small strip of material is held in a butane flame and the flame colour is noted. The other method, which works on the majority of the most common plastic materials, is to use an organic solvent test kit. These kits are produced by an automotive paint manufacturer, and they work by a process of solvent application to the material surface. A normally concealed area should be chosen and degreased, and any paint applied to the surface should be removed from the test area with the aid of an abrasive. Identification is confirmed by reaction between the plastic and test solvent in accordance with the test kit manufacturer's instructions. Gloves and face masks should always be worn when using solvents.

The usual plastic identification codes are as follows:

- ABS – acrylonitrile butadiene styrene
- ABS/PC – polymer alloy of ABS
- PA – polyamide (nylon)
- PBT – polybutylene terephthalate (POCAN)
- PC – polycarbonate
- PE – polyethylene
- PP – polypropylene
- PP/EPDM – polypropylene/ethylenediene rubber
- PUR – polyurethane (not all PUR is weldable)
- PVC – polyvinyl chloride
- GRP/SMC – glass-fibre–reinforced plastics (not weldable)

Equipment for plastic welding

The Welwyn Tool Company supply the Leister Triac welding guns. These blow hot air out of their nozzles, and sophisticated electronics make sure that the temperature of this air is exactly right. A potentiometer is set to control the power, a photo-electric cell monitors the colour of the heating element and, as a failsafe, a thermocouple double checks the output. The guns can maintain a critical temperature setting to ±5°C, independent of any mains voltage fluctuations. It is important to use a welding rod of the same material as the component: hence the importance of correct identification. The welding rods come in various profiles, and there is a speed-weld nozzle available to suit each rod.

Repair method

Suppose a bumper has suffered a long, straight crack which makes it particularly suitable for repair by welding. When carrying out the repair process, thorough preparation of the surfaces is important. When working on a painted bumper it is necessary to sand off any paint in the vicinity of the area to be welded; a mask for facial protection is then necessary. To minimize the risk of damage spreading, the ends of the crack are drilled out with a 2 mm drill. The edges of the crack are then shaped into a V with a burring bit which has a cutting edge on both its circumference and its face. This will allow the welding rod, which in this case is supplied in a V shape, to fit snugly into the crack. Care should be taken not to penetrate more than two-thirds into the depth of the material. The extent of the damage should now be clearly visible.

In any welding operation it is important that the two surfaces to be joined should remain in alignment, so they should be tacked together using a tacking nozzle. The section of the temperature chart relating to this nozzle must be consulted to determine the correct gun setting; the temperature can then be adjusted on the potentiometer. After allowing the gun to heat up for at least 2 minutes, the tacking operation can be carried out. The nozzle is held at an angle of

Figure 14.34 Rotary control on rear of hot air tool for temperature setting (*Welwyn Tool Co. Ltd*)

Figure 14.33 Leister Triac electric welding tool and accessories (*Welwyn Tool Co. Ltd*)

Figure 14.35 Split bumper (*Welwyn Tool Co. Ltd*)

Figure 14.36 Rotary burring tool cutting a V-groove (*Welwyn Tool Co. Ltd*)

about 20° and is run along the crack without applying any pressure.

The speed-weld nozzle is fixed to the gun, and the temperature is again adjusted prior to carrying out the weld. The end of the welding rod is trimmed into a point to ensure a smooth start to the weld. The welding rod is inserted into the nozzle, and as soon as the edge shows signs of melting the weld can be carried out.

Repair or replacement

The repair of all collision damage can be divided into three categories:

1   Repairing by realigning and reshaping the damaged section which can be economically repaired
2   Repairing by removing sections beyond economical repair and replacing them with new sections
3   Repairing by using a combination of realigning and reshaping together with the replacement of appropriate sections.

Before deciding whether to repair or replace, consideration must be given to the extent and type of damage, whether it is possible to repair a particular panel or not (this fact is influenced by the equipment which is available), and also whether it will be cheaper to repair or replace any one section. A good

body repair worker would probably be able to repair almost any part of a car damaged in a collision; however, the labour and time cost involved in attempting to repair extensively damaged panels would be an uneconomical proposition. Only experience in body repair work will give the ability to estimate the time required to do a particular job. Therefore the final decision depends on what is possible to repair and what is economical to repair. As the construction of the modern vehicle advances, body panel assemblies are becoming very complicated, and therefore their cost is constantly increasing.

The cost of replacement parts differs between one manufacturer and another, and even between models. In most cases the following parts are usually replaced rather than repaired:

1   All chrome exterior trim. Owing to the intricate design and nature of the materials from which they are made, trims do not lend themselves to re-straightening, and the cost of re-plating is greater than the cost of a new part.
2   Interior trim which is difficult to clean and repair, especially tears and gashes.
3   Certain mechanical components which it would be ill advised to repair and reuse owing to their critical alignment and strength factors.

On mono-constructed vehicles, the correct positioning of new panels, which may incorporate reinforcing members, becomes increasingly important and calls for expensive jigging to ensure the accurate alignment of panels, which will have to be fitted with major mechanical components and suspension units. The highest proportion of any repair bill is made up of the cost of labour involved in stripping out to gain access to the damaged area and rebuilding after replacement is effected. In certain cases, therefore, it can be more economical to repair a damaged panel than to replace it because of the time and labour cost involved for the amount of work over and above that required to repair the part. Consequently in any body repair workshop the dividing line between repairing and replacing is determined not only by what it is possible to repair and what it is economical to repair but also by the availability of the most up-to-date equipment to deal with all types of major collision damage.

Costing

The determining factor when deciding whether to repair or replace lies in the comparison between the cost of repairing the damaged part as against the cost of a new replacement part. When the owner of a car is personally paying the cost of repairs the repairer will need to come to an agreement with him concerning an acceptable standard of repair for the amount of money the owner is able to pay. When it is an older car the extra cost involved in striving for perfection in the repair is not always justified, since these costs could be more than the current value of the vehicle under repair. In cases of this nature a compromise must be arrived at between the repairer and owner as to the quality of repair and cost of repair.

Where an insurance company pays for repairs to the damage of a new or nearly new car, the quality of the work must be such that when the repair is finished there is no indication of the damage having occurred. When an insurance company is paying for repairs to an older car, and the condition of the car (in many cases due to corrosion) affects the quality of the repair, the possibility of achieving perfection of appearance is limited, and a mutual agreement must be reached between the owner and the insurance company on a satisfactory quality of repair. Time is always an important factor in the repair of collision damage; however, the majority of car owners are not willing to sacrifice quality of repair for speed of repair.

Estimating

The ability to estimate can only be gained through practical experience in the field of body repair work. An estimate must be competitive to be acceptable to both the private car owner and the insurance company, for the majority of collision jobs are paid for by an insurance company. In estimating, the financial gain for the bodyshop will depend on the estimator's skill in assessing the damage, his knowledge and experience in repair techniques and the capabilities of the equipment available.

Small body repair establishments have no separate estimator and therefore it is usual for the man who prepares the estimate to be responsible for carrying out the repair work as well. The larger establishments employ one member of staff who is responsible for all estimating, while the actual repair work is carried out by the tradesmen on the shop floor. Consequently there must be no lack of coordination and understanding between these two and the jobs they perform. The decisions of the estimator are all important, as they instigate the organization of work on the shop floor. Moreover, no matter how skilled the tradesmen are, with bad estimating the financial profits can be drastically affected.

In collision work the estimate is considered as a firm commitment to do the work involved for the amount of money shown and should be detailed so that insurance companies or private owners can determine from the estimate exactly what it is proposed to do to the damaged vehicle. The estimate must include cost of parts and labour costs at the recognized retail rate of the bodyshop. In the preparation of an estimate, the crash damage should be itemized into the number of damaged panel sections and assemblies, and therefore it is important to have a knowledge of vehicle body construction. The outer panels which make up the body shell should be referred to by the manufacturer's recognized names, such as:

- Near-side front door (NSF door)
- Near-side rear door (NSR door)
- Off-side front door (OSF door)
- Off-side rear door (OSR door)
- Roof panel

- Boot lid
- Bonnet
- Near-side centre pillar (NS centre pillar)
- Radiator grille
- Near-side front wing (NSF wing)
- Off-side front wing (OSF wing)
- Near-side sill panel (NS sill panel)
- Off-side sill panel (OS sill panel)
- Front bumper bar assembly with valance
- Rear bumper bar assembly with valance.

In addition this can be further broken down to all internal panel structures according to the position of the damage.

It is essential to determine the exact amount of stripping necessary for either the repair or the removal and replacement of a damaged section. To calculate the actual cost of a job it is necessary to establish a set rate per hour for all repairs done in the repair shop, and therefore an accurate total repair time can be calculated on the time taken to repair each section of the body. This then provides the actual cost estimate for repairing each section or part of the job. The retail rate or set rate per hour is the amount of money that the repairer charges the customer for labour and is made up of wages paid to the tradesmen; the cost of overheads, which should include such items as supervision, depreciation of equipment, rent, heat, light, electrical power, advertising, telephone accounts, cleaning, office staff, stationery and postage and workshop materials; and also a reasonable amount of profit.

Computer estimating

Computers are as indispensable to the bodyshop as the spray booth and chassis jig. Offering benefits in all aspects of bodyshop operation, there are computers and software packages dedicated to quick and easy repair estimates, stock control, invoicing and management accounts, paint and materials, parts lists, mixing and colour matching, job cards, workshop loading, booking in, job efficiency reports, invoice production, reports and letters. Costs have tumbled, and the benefits

can be reaped by all bodyshops, regardless of size.

The key to success in any bodyshop is the production of an accurate, well-presented estimate. An estimate has to be a clear, concise statement of what you consider to be a fair return for the care and work you carry out on a crash-damaged vehicle. By utilizing the memory capacity of the computer you will have readily available all necessary details to help you draw up an estimate quickly and accurately.

Your computer system will have the facility to prepare not only a detailed document for the customer but also a comprehensive breakdown for the insurance company. The documents produced will be tailor made to your own letterhead stationery. The insurance estimate can often be sent direct by fax to the company involved, thereby further reducing the time gap between estimating and authorization by that company. Apart from the major bonus of speed, a computerized estimating system will greatly enhance your company's image as a professional organization. You will have instant access to the progress of any job: most systems include this as standard. The more time and money you spend at the outset, the more control you will have over staff movement and job completion.

Write-offs

Write-off classifications
Although all vehicles can be repaired, the question is often whether the repair is **economically viable**. That is, is it going to cost more to repair than the vehicle is worth. This decision is also effected by the excess to be paid by the insured party, especially on older, lower-value vehicles, which may have other faults too.

When a vehicle is written off the insurance company are required to inform the **DVLA**, the purpose of this is to reduce the number of **ringers** – cars made from the parts of two or more damaged vehicles. When the DVLA have been informed they will not issue any documents on that vehicle, and it will be marked as such on the DVLA database. If the vehicle is repaired, as it may be if it is **Category C** or **Category D**, it must be subjected to a **Vehicle Identity Check (VIC)**. The VIC only confirms the originality of the vehicle; it does not judge the standard of the repair or the roadworthiness of the vehicle. At this point a separate **MOT** may be needed.

The write-off categories are:

- *Category A* – the vehicle is so badly damaged that it must be completely destroyed; this usually means **crushing** for scrap
- *Category B* – the vehicle is severally damaged, but parts may be removed and sold, that is used as spare parts, before the remains are crushed for scrap
- *Category C* – the cost of the repair is greater than the value of the vehicle
- *Category D* – other reasons, these mainly relate to the outstanding finance being greater than the value of the vehicle and **stolen recovered** vehicles which may have only cosmetic damage

Problems around write-off

The writing-off of a vehicle is often an emotional time for the owner. As a body repairer you need to be aware of this, and to be able to offer some sound information. Particularly common problems are:

- If it is a new car, the pay-out from the insurance company may be much less than the customer paid only a short while ago.
- It may be a cherished classic that the owner wanted to keep forever.
- The car may be a ringer.
- The vehicle may have previous unseen damage.
- The car may be greatly under-valued or over-valued.
- There may be another serious fault on the vehicle that affects its value.

Tech note

It is important to understand that when the vehicle is written off, the driver suddenly has no vehicle, and this may put them in a situation in which they cannot do their work to make a living or be with family for an important reason.

Skills and questions

1. Define the terms 'primary damage' and 'secondary damage'.
2. Explain the theory of collapsing metal panels when they are involved in collision.
3. Define the terms 'visible damage' and 'concealed damage'.
4. Define the term 'direction of damage'.
5. Describe what is meant by direct damage to metal panels.
6. With the aid of a sketch, explain the difference between a single- and a double-crowned panel surface.
7. Use a sketch to explain the principle of the Mitek system.
8. Describe, with the aid of a sketch, the principle of the Korek system.
9. State two advantages of Dozer equipment when used in repair.
10. Describe how the vector principle is applied in repair work.
11. With the aid of a sketch, describe pressure areas and tension areas on a damaged panel section.
12. Explain the basic use of the pulling tower system in the repair of vehicle bodies.
13. Draw a sketch to show how a Dozer can be used to repair a front end damaged by collision.
14. What are the comparable advantages of portable and fixed body jigs?
15. Explain the term 'bracketless jig systems'.
16. What are the advantages of using bracketless jig systems?
17. State the procedure to be followed in order to mount a vehicle on a bracket-type jig system.

18 State the advantages of a dual alignment and repair system.

19 Explain how the measurement system works on the Dataliner equipment.

20 How can misalignment of a car body affect the control of the vehicle when it is being driven on the road?

21 Use a simple sketch to show how the drop-line method of alignment is carried out.

22 Use a simple sketch to show how gunsight gauges are used in repair.

23 State the factors which determine whether damaged panels should be repaired or replaced.

24 List, in logical sequence, a basic approach to the repair procedure adopted in major collision damage.

25 What is meant by repairing using partial panel replacement?

26 List four methods of panel removal and the tools used.

27 Why is a vehicle road tested following major repair work?

28 Describe a method of repair by replacement of a front wing which has received major accident damage.

29 What is the importance of care of customers' vehicles when received for repair in the workshop?

30 Describe the difference between a laminated glass windscreen and a toughened glass windscreen.

31 With the aid of a sketch, describe the fitting of an indirect-glazed windscreen.

32 Describe the removal of a direct-glazed windscreen and the type of equipment that would be used.

33 Describe in detail how to check for door aperture alignment.

34 Explain the method of repairing and aligning a damaged body shell when using a jig and hydraulic pulling equipment.

35 Explain why a careful study of accident damage is necessary before any stripping or repair is carried out.

36 State the advantages of having a body jig system incorporated in a lift system.

37 Why is it essential to check the steering geometry of a vehicle after major repair work has been carried out?

38 Describe the four stages of the procedure for dealing with repairs carried out under an insurance claim.

39 How are material costs established on an estimate?

40 State the importance of the clearance certificate.

41 A suspension frame member is to be replaced by welding; explain the correct method of achieving its alignment beforehand.

42 A vehicle is to have the rear boot floor, wheel arch and quarter panel assemblies removed. State the repair procedure to reinstate these panel assemblies.

43 With the aid of a sketch, show the position of a pull-dozer to realign a centre pillar after side impact.

44 Explain the repair procedure necessary for part panel replacement on the quarter panel assembly of a four-door saloon.

45 With the aid of labelled sketches, show the difference between direct glazing and indirect glazing.

46 Outline the procedure for mounting a vehicle on a jig using the MZ bracket system.

47 Show, with the aid of a sketch, how the laser on Dataliner equipment can achieve a 90° angle.

48 State the advantages of the use of split measuring bridges on a jig measuring system.

49 Explain what is meant by the term 'universal jig bracket'.

50 State the safety precautions necessary when carrying out pulling in repair.

51 Describe a computer-based estimating system.

52 Describe how to use the Glassmatrix estimating system.

15

Bodyshop planning

Initial planning

Pre-planning

Setting up a bodyshop capable of carrying out repairs that will meet customers', insurance companies' and motor manufacturers' quality standards requires significant investment in time and money. Therefore to set up or redevelop a bodyshop, professional advice is essential.

There are a number of suppliers, manufacturers and independent consultants offering a range of planning, design and consultancy services, from simple equipment layouts to sophisticated three-dimensional computerized blueprints. The design and planning services provide everything for the re-equipping of new bodyshops in existing premises, extensions to original bodyshops or the total development of greenfield sites. Preplanning is essential, whether for the building of a new bodyshop or the remodelling of existing premises, because it is necessary to be certain that the finished bodyshop will meet all the operational requirements before construction is started. It is therefore important to conduct as much market research as possible before finalizing your plans. This research should include asking bodyshop staff what improvements they would like to see in the new bodyshop, visiting other bodyshops, talking to other managers to collect advice and recommendations about

good practice and possible problems to be avoided and employing consultants to help you through the process.

Consultants

There are a number of specialist consultants in the bodyshop field, they can offer a variety of advice and assistance packages or be just used for special parts of the project which you have limited experience or knowledge in. It is worth doing a skills analysis of yourself and your prospective partners before you start on the actual project. At that point you can then decide if one or more consultants is needed. Typical consultant roles are:

- *Bodyshop consultants* – usually specializing in advising on systems and procedures to be used; these will impact the layout of the bodyshop and the choice of equipment
- *Management consultants* – providing advice on a number of strategic and operational issues
- *Financial consultants* – providing financial advice. Usually they will help generate loans and grants. Often they provide this service free to the user as they get commission from the finance houses who lend the money.
- *Safety consultants* – providing advice, signage and documentation for health and safety issues

The Repair of Vehicle Bodies, Andrew Livesey, 9780815378709, 2019, Taylor & Francis

- *Fire consultants* – providing advice on fire protection systems, they usually give free advice and designs if you buy their products
- *Environmental consultants* – providing guidance and possibly selling environmental equipment and systems, such as the safe and legal disposal of scrap and other waste
- *Energy consultants* – of particular interest to the bodyshop is the safe and economical use of electricity and gas. They will generally give new businesses advice on the options available free if you are intending buying either equipment or energy from them.

Choosing a site

The location of any building or bodyshop is a crucial factor in its profitability: no matter how efficient may be the management and staff, customers will not be encouraged to bring their vehicles for repair if access to the site is difficult.

A new bodyshop located on an industrial estate usually presents few problems. It also usually means a lack of sensitive residential neighbours. Access is generally good, and the essential services of gas, electricity, water, drains and telephone are already available. On sites lacking these services it could prove costly to have them installed.

A site that is relatively cheap may have distinct disadvantages such as a narrow road, steep hills, a particularly bad surface, sharp corners, one way in and one way out only; all these affect access to the premises. Ideally the customer should be able to see the site before he or she is on top of it. Your outside sign should be simple, easy to read and brightly coloured; otherwise you run the risk of customers missing it, driving past your bodyshop and ending up frustrated even before walking through the door. The premises should be easy to drive in and out of, and there should be adequate space for vehicles to turn safely. These requirements apply equally to suppliers delivering bodyshop materials, especially with large commercial vehicles. Parking facilities should be adequate and made attractive by either landscaping or the addition of planters. A prospective customer will not be impressed if he has to park halfway down the street to walk to your bodyshop. The choice of site will of course depend to a large extent on the amount of capital available. The operator who buys a leasehold site pays proportionally less but may not be able to make the necessary profit within a set number of years. Freehold land is more expensive but a much better investment. Another option is renting an existing bodyshop. Alternatively, converting a purpose-built building might be considered, although this could be costly and may not ultimately fulfil the necessary criteria for an efficient bodyshop, although it is probably the most popular option.

The nature and volume of work to be undertaken will largely depend on the type of building and the total area needed. The site area should allow separate parking for vehicles waiting for repair and those waiting for collection, in addition to turning space. This reduces the risk of accidental damage on site and ensures a more efficient workflow since the bodyshop personnel know exactly where to collect the next vehicle requiring attention or repair. Also there should be a secure outside compound for seriously damaged vehicles awaiting examination by the insurance assessor (third-party claims and total write-offs).

To purchase a site without regard to possible future expansion could be a mistake. The increase in vehicle population ensures that a body repair business that is well situated and planned is almost certain to grow. However, when there is a lack of room for expansion, vehicles would have to be shunted between sites, thereby affecting the cost of repair. Suitable land adjacent to the premises would be an asset and a good investment for future development. The major factor in site selection is planning permission, without which no building may be erected.

Legal requirements for the planning stage

When building a new bodyshop, the following factors are important: the best location possible, high profile, easy access and sufficient space. Having found the location, make absolutely sure that it will be permissible to build exactly what is required. Check local building regulations, and ask to see proposals for the future of the selected site and surrounding area at the town hall. This advice also applies when extending an existing bodyshop, when it is essential to ascertain that extension building plans do not contravene building regulations.

Planning permission to erect commercial buildings must be obtained before building work commences; failure to do so may result in heavy fines and the possible destruction of any unauthorized building. The first step therefore is to visit the local council office and arrange an appointment with the planning officer, who will want details of dimensions of proposed buildings and construction materials. The planning authority must be satisfied that the proposed building will meet the standards set by the Building Regulations, the Town and Country Planning Acts, the Fire Precaution Regulations, the Storage of Flammable Materials Regulations, the Storage and Use of Liquids and Gases Regulations, the Control of Substances Hazardous to Health (COSHH) Regulations and the Environment Protection Act (EPA).

The supply of essential services (electricity, gas, water, drainage, telephone) is also dealt with at this stage. If the site is a long way from existing services, determine in advance whether the cost of connection would be worthwhile. On-site valeting (vehicle washing) would require special drainage to cope with excess dirt and grease. All such factors should be taken into account in addition to any relevant legal requirements.

Planning officers are trained to help to implement the following of correct procedures in the planning, erection and completion of all buildings. This is a complex business, often involving not only planning officials but also building and sanitary inspectors as well as the environmental health department officer; therefore it is usually advisable to employ an architect. However, before any plans are prepared for building works, advice should be taken on bodyshop planning. The architect can only build or alter a building around the business if both parties are sure what shape the business will take; therefore bodyshop planning must come first.

Planning the workshop layout

The layout and design of the bodyshop will depend upon the type of work to be undertaken: minor repair, major crash repair, complete refinishing, PDI paint rectification, special customizing, the repair of light commercial vehicles and any combination of these. However, efficiency is paramount in any workshop, and for this reason the location of tools, equipment, parts and personnel must be carefully organized. In such an environment, speedy access to tools and equipment is very important. Also important is a basic understanding of work study, which should help to ensure the smooth running of the workshop.

Planning plays an essential role in the day-to-day running of a bodyshop and has a dramatic impact on the overall profitability. In the first instance it is necessary to determine the number of operatives at the bodyshop, the type of work to be carried out within the bodyshop, whether the bodyshop is part of a main dealership or is independent, and any special considerations. Upon this information can be determined the number, types and mix of bays required in the bodyshop.

Allowing two to three bays per productive unit gives the total number of bays required. From this it is possible to determine the mix of bays, such as paint shop or panel shop; add to this an estimating bay and valeting bays, and the plan takes shape. A good plan gives a flow of work (flow-line system) and eliminates any possibility of bottleneck problems. This should allow a vehicle to move from the stripping area to the repair area, on to the preparation area, next to the paint area, then into the refit area, then to the valeting area, and finally into the inspection area. The routing of vehicles is probably the most important feature of bodyshop layout. Workbays should be arranged so that the movement between them is minimized and major crash repairs and non-runners do not cause bottle-necks. The optimum layout should allow vehicles to move easily from one area to the next in a logical sequence. The plan should also aim to provide flexibility between body and paint bays to allow for trouble-free fluctuations of the repair sequence. Safety in the bodyshop has become a priority, with an ever-increasing need to ensure a safe and healthy working environment. To comply with HASAWA, COSHH and EPA regulations it is essential to incorporate such items as purpose-built mixing rooms, dust extraction, fume extraction, extracted gun cleaning tanks and safe solvent handling procedures.

Basic principles and dimensions for bodyshops

Entrances

The minimum opening width for entrance and exit doors should be 4.5 metres for cars and 5.0 metres for commercials. The minimum height for cars should be 4 metres, but this must be increased if commercial work is to be undertaken. Many bodyshops favour a combined entrance and exit for ease of control and to allow easier traffic movement, but often site conditions dictate the entry and egress situation.

Roof supports

Wherever possible the working area should be clear with no roof supports (single spans of 30 metres), and this point should always be taken into consideration when working on a new workshop layout. Interior roof heights of 5.5 metres not only give the height for a mezzanine floor but are more acceptable to others if the building is subsequently sold.

Workshop floor space

Using all available space in the interests of efficiency does not mean in any circumstances that the premises should be so filled with vehicles that considerable time is taken up with moving and re-parking them. It has been established that a large percentage of non-productive time can be attributed to this factor. The ideal is to ensure that every vehicle is placed so that it can be removed from its location, without obstructing or interfering with its neighbour, to any other bay or booth.

- Panel bays should be 6 metres × 3.6 metres.
- Paint preparation bays should be 6 metres × 3 metres.
- A passenger vehicle undergoing repair on a fixed jig requires a floor space of approximately 4 metres × 7 metres. With a mobile jig, repair can be accommodated in the 6 metres × 3.6 metres of floor space of a panel bay.
- Jigs, lifts and other panel bay movable equipment should be a minimum of 600 mm from any fixed obstruction to allow adequate safe working distances.
- Gangways between bays should ideally be 4.5 metres.
- Turning circles should be based on a medium/large saloon car.
- Walkways should be a minimum of 1 metre.

Planning the areas of the workshop

Reception area – Reception facilities for customers could include toilets, seating accommodation, tea and coffee facilities, TV.

Administration offices – This could include manager's office, foreman's office, accounts office with secretarial staff and if possible office space for visiting insurance engineers.

(a) A specialist import facility for high volume minor repairs

(b) Existing building with access from one side

(c) Small workshop complex

Figure 15.1 Bodyshop layout (*ICI Autocolor, Ernest W. Godfrey, Pickles Godfrey Design Partnership*)

Parts department for trade and retail – This could include parts manager's office, parts store area, and parts department staff.

Staff facilities – These could include toilets, washroom and showers, restroom, canteen, first aid room.

Fire-fighting equipment – This must be approved by the local fire officer.

Working areas should be tailor-made to suit the needs of the particular bodyshop. Typical requirements are as follows:

Vehicle damage assessment area – This is where the vehicle is first assessed for the degree of damage in order that an estimate may be written for customers and insurance companies.

(d) Greenfield site development

Figure 15.1b (*continued*)

Stripping area – This is where damaged vehicles are, prior to repair, stripped of body panels, body trim and electrical and mechanical assemblies.

General body repair area – This is where minor repairs, general repair work, and work not requiring a jig can be carried out.

Specialist repair area or jig area – This is designed for major accident damage where the vehicle must be repaired on a jig system.

Preparation area – This is for the pretreatment of the repair in the form of sanding, priming and masking before painting.

Refinishing area – This is where the vehicles are repainted and baked to original factory condition.

Refit area – This is where interior and exterior trim, as well as mechanical and electrical assemblies, are refitted after the vehicle has been painted.

Valeting and wax injection area – This is designed for the cleaning of the vehicle's interior and exterior back to showroom condition as well as for replacing the corrosion-prevention treatments of undersealing and injection wax treatment.

Final inspection area – This is where the vehicle can be checked for any minor defects after the repair process is completed.

Reception area

The reception area is very important. Here customers get their first impressions by meeting the receptionist, who represents the company. Reception staff play a very important part by instilling confidence into customers through their courteous approach and knowledge of vehicle repair. The area should be clearly signed, and the customer should walk into a pleasant, clean and tidy reception office. Natural colours are probably the best for decorating the walls, and wall space should be used to promote the bodyshop by displaying framed certificates of the staff's qualifications, association memberships and insurance company approvals. The area should be large enough to accommodate customers and the reception engineer and to hold customers' records, bodyshop loading display boards, telephone and workshop intercom systems. Lighting should be efficient and floor covering attractive but practical and easy to clean. Furniture should be serviceable and comfortable without being elaborate and sufficient to seat at least six people. The area must be sited near an entrance door, and ideally provide customers' toilets and a drinks vending machine. A window installed between the work area and the reception area allows customers to view tasks being carried out. Professional technicians at work on vehicles are a good selling point, as customers are always interested in seeing work being carried out. However, if customers can see into the bodyshop, it must be kept tidy at all times.

Administration offices

The bodyshop manager's office should also be adjacent to the reception office for easy access to liaise with the customers. He or

Figure 15.1c Very large-size bodyshop layout

Figure 15.1d Small-size bodyshop layout

through the bodyshop should be carefully planned with control cards from the raising of the job to its completion and final costing. This process is now being aided by the use of computerized technology. The foreman's office should be situated overlooking the main bodyshop areas so that he has instant access to staff as they carry out their individual repair tasks.

Parts department

Within the vehicle repair industry a considerable number of parts are received, issued and used every working day. Main dealerships sell body parts to outside trade and retail outlets. The receipt and issue of parts, the stock held and the costs are therefore very important to any bodyshop. The parts department is a very valuable item in the profitability and efficient running of the business. Its function is to satisfy the needs of all the departments within the organization, and also both trade and retail customers when necessary. As with the repair shop and paint shop, the parts department must be efficient and well organized. It should be adjacent to and with direct access to the reception office in order to control the flow of paperwork between the two departments and also to enable the ordering and receipt of the parts necessary to complete the repairs.

Assessment area

When considering the assessment area in the workshop plan, this again could be on public view and so should be designed to impress. It should be kept clean, tidy and uncluttered at all times. It should also be efficient, with adequate facilities for detailed assessment of damage and an accurate method of estimating repair times needed and parts required. This is where computerized estimating should be located, and customers will always be impressed by a speedy estimate on the screen in front of them. When a customer calls to have damaged panels repaired or paintwork rectified, the works manager

she needs to be easily accessible to the secretarial staff employed by the company for the day-to-day clerical administration of the company. The progress of a vehicle's path or foreman usually completes the estimate for work to be carried out. A separate estimating area away from the main workshop area is ideal, as vehicles will need to be thoroughly inspected all round and underneath. A vehicle lift and high-quality lighting are required, especially at ground level, together with wheel alignment systems which give a quick check on suspension and alignment of the wheels.

Repair areas

The bodyshop should make the best possible use of floor space to provide maximum productivity while still taking into account health and safety regulations. The best floor plan is probably to drive the car in at one door, move it through the bodyshop in stages and then drive it out at another door, unless strip and refit are carried out by the same person.

When building a new workshop it is possible to plan this, but if remodelling an existing operation this may not be possible. The rule of the bodyshop is flexibility. A number of large repair organizations have adopted flow-line systems: these can only be successful if allowance is made for several lines accommodating various speeds of work but must still retain the facility to move a vehicle easily out of the line in the event of unexpected hold-ups. Medium establishments use a series of accessible working bays from which work can be fed from the stripping area to the repair area to the preparation area and on to the paint shop.

To use floor space to its best advantage, it is important to decide exactly what work will be carried out, how many employees will be needed, and what equipment will be needed for all the jobs anticipated. It is important that the workforce does not have to wait to use the same piece of equipment; therefore the location of equipment should be care-fully planned so that a job goes smoothly from stage to stage without wasting time. The relevance of shop floor planning is to remember that the bodyshop is a place of movement and not a static collection of equipment in a working area. The highest labour hour drain is caused by the unnecessary movement of men and work within a badly planned workshop. As far as possible, no job should be accepted into the workshop until the parts have been obtained to complete the repair without holding up the flow of the work. Stripped body parts should not be left to clutter the floor but should be labelled for identification and stored in the main store department or in portable storage bins which can be moved with the vehicle.

When handling major crash repairs, the computer-linked jig systems can give fast and accurate readouts before, during and after repair.

Preparation and refinishing area

When planning your bodyshop, productivity, health, safety and environment regulations must all be taken into consideration.

Start with the preparation area. There must be space to pressure wash vehicles to get cleaner paintwork before they enter the paint shop.

The dust created by sanding in the preparation stage must be prevented from being drawn into the paint area. The best solution is an efficient extraction system or extraction tools. Avoid walls or curtains, as they lead to rigid space allocation and are inflexible because they prevent bodyshops reacting to variable workloads. If possible, design the paint shop so that the vehicle moves forward in a relatively straight line or an easy loop, and try to keep it inside the workshop to avoid the problem of coming into contact with dirty, dusty air.

Ideally the refinishing area should form a self-contained area separate from the mechanical and repair side of the workshop and the preparation area. The main objective is to minimize the amount of handling of the vehicles by arranging for a flow system from the vehicle's entry to its exit from the spray booth, thus preventing any dust entering the spray booth. Consequently spray booths and ovens can be used to their full capacity throughout the working day.

In the paint shop, health and safety regulations will dictate many decisions and must be taken into account in your plans. The ramifications of the EPA and COSHH legislation are far reaching for all bodyshops. In effect, any bodyshop currently operating without a properly constructed, fully operational and well-maintained spray booth is operating outside the law and risks summary closure.

A spray booth is in most cases the single biggest purchase decision made by the bodyshop owner. Most bodyshops are now fitting combibooths. A combibooth is a combined spray booth and low-bake oven with an electrically controlled operational cycle which provides the correct temperature, filtration and lighting requirements for the spraying and baking of paint. The purpose of the combibooth is to provide an efficient, clean area, free of humidity and within the temperature band 20 to 25°C. The booth must be sealed off to prevent overspray escaping into the workshop and must be force ventilated to remove solvent fumes and microscopic paint particles whilst drawing in clean and dust-free air. Once spraying is completed, the combibooth can be put into the bake cycle, in which the internal booth temperature is raised to promote rapid curing of the paint. One important point to consider is the choice of the spray booth. This must meet all the requirements of the EPA in relation to the throughput of air and the placement of the exhaust stack. This point is particularly important in residential areas. The booth must operate under negative pressure and, to comply, must be fitted with a pressure gauge. As a further precaution, emissions from spray booths should be tested at least once a year to monitor particulate matter emitted. The EPA stipulates that concentration of total particulate matter in final discharge to air from the booth should not exceed 10 mg/m³.

When constructing the paint storeroom, remember to provide space for the mixing system and for the safe storage of paint. Again, follow all the necessary regulations.

Valeting

The vehicle should now be back at the valeting bay for both interior and exterior valeting, to emerge as good as new with the minimum wasted time. Careful valeting work can transform the final appearance of the vehicle, thus creating customer satisfaction by returning the car looking like it did when new. If the customer's first impressions on collecting the vehicle exceed expectations, he will be more than willing to recommend the company's quality of service.

First aid

First aid equipment should be supplied at set points throughout the workshop, and staff should be encouraged to attend first aid classes. The Health and Safety at Work Act requires that where more than 50 personnel are employed, one qualified first aid person must be appointed.

Fire-fighting equipment

Consult the local fire officer regarding fire points and types of fire appliances, exits and fire doors. Organize staff fire drill and training procedures, and make sure that all fire extinguishers are maintained.

Bodyshop heating

The body repair shop can be a difficult building to heat because of the very nature of the work that is carried out. First, it needs to be a reasonable height so that vehicle bodies can be elevated and moved about easily; and second, doors are constantly being opened and shut to allow for the movement of vehicles in and out of the workshop.

Careful consideration of bodyshop heating requirements will not only result in energy consumption savings but also make the bodyshop more comfortable. Whatever type of heating systems are used, care should be taken to ensure that heaters are placed where they will not interfere with work in progress, will not take up unnecessary floor space and will provide employees with the maximum benefit.

Most of the conventional methods of space heating can be found in bodyshops, ranging from portable gas-fired units through wet systems to radiant systems.

Radiant heating

Infrared (IR) heating is the transfer of energy by IR electromagnetic radiation; this is the portion of the electromagnetic spectrum between visible light and the top section of the radio-radar wavebands. Any object will emit IR energy; the higher the temperature, the more IR radiation is emitted. Therefore when two objects are adjacent to one another and one is hotter than the other, there will always be an exchange of heat from the hotter object to the cooler one. Radiant heaters operate by the transfer of heat to solid objects rather than by heating the air itself; therefore high-ceiling buildings can be heated with a high degree of efficiency. By installing radiant heaters above a cold concrete floor, there will be a transfer of IR radiant energy downwards towards the cold concrete, which will absorb energy until it becomes warm. Other objects at floor level, such as vehicles, benches and machinery, as well as the walls, will become warm either by receiving direct IR energy or by the transfer of energy from the floor. All of these items become secondary heat sources, transmitting energy into the surroundings of the workshop. Ceiling-mounted radiant heating units are ideal for bodyshop use in that they do not take up usable space and encroach into the working area; however, they are not suitable where the ceiling height is less than 3.5 metres. Also radiant heat absorbed by its surroundings does not easily escape through service doors: therefore heat loss is reduced and a saving is made on energy costs. Furthermore, heating the floor keeps it dry, and the fact that vehicle bodies become heated reduces problems associated with condensation as the cold vehicle bodies are brought into a heated environment.

Electric heating

Electric spot heating allows comfortable heat to be concentrated where and when it is wanted. Linear quartz heaters can be used for the spot heating of specific work areas, for zone heating of a particular area of the bodyshop or for full-coverage heating. For spot heating, the heaters should be positioned about 3 to 5 metres apart, at a height of approximately 2.5 to 3.5 metres from the floor, depending upon the heater rating. Heaters should be rated at 1.5 kW for sheltered locations, rising to 3 kW or even 4.5 kW for more adverse conditions. For zonal heating, the heating requirement should be calculated on the basis of 150 to 300 W/m^2. The actual value selected will depend upon the state of the building, internal draughts and so forth. If the zone is more than about 15 metres wide, then additional rows of heaters may be required. For full coverage, the heaters should be positioned around the perimeter of the building at spaces of about 4 to 6 metres, using load densities of 100 to 200 W/m^2 based upon heat loss calculations. If the building is more than about 17 metres wide, additional heaters can be installed.

When installing these heaters, the general principle is for the heater to be angled down at about 45° when wall mounted and up to 90° when mounted overhead.

As with all forms of heating, care should be taken with these heaters if there is a risk of combustible dusts or flammable gas hazards. The lamp/filter assembly is the hottest part of the heater, with surface temperatures of 750 to 800°C, broadly similar to the surface temperature of the lamp in a tungsten halogen floodlamp. In case of concern over this point, consultations should be held with the factory inspector.

Gas heating

Radiant gas heaters come in two forms: tubular and ceramic plaque.

The radiant tube heater typically incorporates single or twin emitter tubes through which hot gases are drawn by a vacuum fan to produce the IR heat. An automatic burner fires into the mouth of the emitter tube, providing rapid warm-up without any need for preheating. The burner operates in a totally enclosed combustion chamber, with the safety controls housed separately to protect them from dust and contamination. If it is required to exhaust the products of combustion to the atmosphere, the individual heaters can be linked into a manifold system with one exhaust fan. Typical capital costs are about half those of an equivalent central boiler system and about the same as a ducted warm-air system. The heaters can either be wall mounted and inclined or suspended at heights from 3 to 12 metres. The units are able to cope with high air exchanges, and they can also reduce the spread of airborne dust and fumes.

Ceramic plaque heaters are silent in operation and are marketed as being ideal for spot heating and for the heating of lofty uninsulated buildings which suffer high air movement. They incorporate a totally enclosed burner with a skirt to protect the ceramic plaque from cold air washing over the hot surface.

Conventional fixed heating systems

Conventional industrial central heating systems are to be found in many bodyshops. Central heating relies on a central boiler feeding a number of convector radiators. These radiators warm the surrounding air, therefore raising the ambient temperature of the whole environment and anything contained in it. The amount of energy required is in direct relation to the volume of air contained within the area to be heated. As soon as a door or window is opened, warm air is pushed out by the heavier cold air streaming in from outside. This makes such heating extremely unsuitable for bodyshops, where doors are constantly being opened to allow vehicles to be moved in and out. A further drawback is that the convector radiators need to be mounted on the walls of the workshop, thus taking up valuable space.

Portable heaters

Portable heaters come in a variety of sizes and types, from electric fan heaters and propane gas to forced-air space heaters using gas, oil or paraffin, and even mobile IR quartz heaters. Whilst mobile forced heaters are relatively expensive to run, they are cheap to buy and flexible in use. They generate considerable air movement, which can create dust problems. It is important to ascertain that the design of the unit being considered is suitable for the area to be heated and will not cause any safety problems.

Bodyshop lighting

Lighting in the bodyshop plays a critical role in the final finish of a vehicle body. Effective lighting can highlight faults which otherwise may not be seen; this not only reduces job times but also minimizes rectification work.

Effective lighting is not necessarily a question of increasing the level of lighting: this can be counter-productive, as it produces what is known as disability glare, which can effectively blind the operative, preventing him from seeing surface faults. Both the level and type of lighting can be crucial to the vehicle painter's colour-matching ability.

Planning and specifying bodyshop lighting requirements are an extremely complex and technical subject because many factors have to be considered in conjunction with calculations. Therefore it is recommended that consultant specialist bodyshop lighting companies should be contacted when either designing a bodyshop or re-equipping an existing workshop with new lights.

There are four important areas of a bodyshop, each of which has different lighting requirements.

Stripping and straightening areas

The stripping and straightening areas require only walk-around vertical lighting with relatively low illumination. A suitable lighting level for this type of area would be between 300 and 500 lux.

Preparation area

The lighting in the preparation area is of critical importance because the quality of preparation fixes the final finished paint level. It needs to be of much higher illuminance and multidirectional to enable the most minor faults to be picked up. It is also important that glare and shadow are kept to a minimum and that side lighting is of a good level. The lighting level required for the preparation area is between 750 and 1000 lux on both vertical and horizontal surfaces. Mounted standard fluorescent fittings at a 45° angle are unsuitable as the disability glare, and uneven light levels can effectively prevent even major faults from being seen. Specialist bodyshop lighting units that give glare-free vertical and horizontal lights are available. They improve the productivity of the operatives and also reduce the amount of remedial work necessary.

It is important that light readings are taken with a vehicle *in situ*. The light meter should be placed in a variety of positions, not just for straight-down light readings, in order to ensure that adequate light is available on the sides and lower portions of the vehicle.

By use of the correct lighting, any preparation faults will be found straight away and hence will not go through to the final finish, where they would be both difficult and time consuming to rectify.

Paint mixing room area

The paint mixing room usually contains only one or two light fittings, but if any colour comparison is to be done in this room the correct tubes must be used. The light levels for this area should be between 800 and 1100 lux.

Spray booth and painting area

The lighting requirements in the spray booth are similar to those of the preparation area but with an even greater necessity for avoiding glare. This is because the surfaces of the vehicle, once painted, will be of a very high gloss and hence reflective, exacerbating any disability glare and masking faults in the paintwork. This problem necessitates a much more sophisticated lighting system than the rest of the bodyshop.

The key factor should be avoidance of cross-booth glare. This is where the operative, standing on one side of the booth, has his vision impaired by the glare emanating from the lighting on the opposite side of the booth.

Most spray booths are designed with integral lighting systems. Hence it is as well, when choosing a booth, to carry out a dummy run with a car inside the booth. You should then stand in a variety of positions as if you were the paint operative: if, in any one of these positions, you are receiving direct light from one or more of the lighting tubes installed in the booth, you will be suffering from some degree of glare, which will probably be disabling and will certainly be tiring.

The light intensity inside the spray booth should be in the region of 900 to 1200 lux and approximately the same at the sill level and the gutter line.

Lighting is only one of the many factors that will be taken into account when choosing a spray booth. However, if a booth is excellent in all other respects, the lighting can be modified by a specialist company without causing too many problems. The same applies where improved lighting is required in an existing booth. Lighting is certainly an area worthy of meticulous attention by the bodyshop operator.

Light sources in bodyshops

There are several general types of lamp found in bodyshops:

- *Tungsten bulbs* (general lighting service [GLS] lamps) – These are still used for hand lamps and toilets, but are steadily being replaced by low-energy compact fluorescent lamps.

- *High-pressure sodium* (SON) – These give a warm golden light (SON DL) or a pleasant, less yellow light (white SON). Both are good for mechanical areas and forecourts, but their colour rendering makes them totally unsuitable for bodyshops.
- *Mercury lamps* (MBF) – These give white light similar to fluorescent, although not good enough for colour matching. They are suitable for general areas.
- *Metal halide* (MBIF) – These lamps are now both long life and suitable for colour matching.
- *Fluorescent tubes* (MCF) – These are the industry standard and provide an excellent form of low-cost lighting. There are many different types of fluorescent tube on the market, and it is important that the correct type of tube is fitted in each area of the bodyshop. Spray booths should be fitted with colour-matching tubes, and it is advisable that this type of tube is used in the preparation area too. It is vital that all tubes fitted in a spray booth should be the same colour. If one tube in the spray booth should fail and a direct matching replacement is not available immediately, it is actually better to run with a tube missing than to temporarily substitute a non-matching tube. In areas where critical colour matching is regularly performed, it is definitely worthwhile to change lamps on a regular basis about once a year and so to avoid the problems of lamp failures and the colour variations which occur with age.

Essential equipment for the bodyshop

In the utilization of equipment there are two important factors:

Return on investment

This is simply the profit made from the amount invested in the workshop and equipment.

Return on assets

This is the assessment of how the workshop and equipment are being used. This relates to net profit achieved against an asset, which would be the work being produced in the workshop. This base is divided between fixed assets and current assets. Fixed assets are money in buildings and equipment. Current assets are money tied up in the work in progress and in stocks held within the stores department. To achieve the necessary return it is therefore absolutely essential to use the site and the expensive equipment required for vehicle repair to the fullest extent.

Essential equipment guide

Before purchasing and installing expensive equipment an analysis of any past work and a forecast of the future are essential to forecast the probable availability of the work.

The equipment requirements can be categorized as follows:

- Specialist equipment: essential and desirable
- General workshop equipment: essential
- Hand tools and expendable items: essential.

Specialist equipment needed for stripping, repairing and painting

If full and efficient service is to be given to the customer, certain equipment is required in the workshop. All equipment used in the workshop must meet all current legislation.

- Alignment and repair jigs for measuring and straightening
- Pulling equipment for reshaping damaged body sections
- Hydraulic body jack equipment for panel repair
- Welding equipment: oxy-acetylene, MIG/MAG welding, TIG welding, spot welding (double and single sided)
- Plasma arc cutting equipment
- Dust and fume extraction systems (portable and static)

- Wheel alignment equipment (four-wheel alignment)
- Combibooth
- Panel booth
- Mixing room with mixing system
- Infrared driers
- Wall-mounted breathing air filters
- Air regulators
- Respirators and air-fed visors
- Spray guns (standard and/or HVLP guns)
- Gun cleaning tank, extracted
- Combined paper baler and can crusher
- Pedal waste bin for solvent contamination waste
- Plunger cans to dispel solvent
- Screw compressor (suitable for the total capacity of air requirement of the workshop).

General workshop equipment
- Hoist (wheel-free type)
- Headlight focusing equipment
- Axle stands and trolley jacks
- Folding crane (1 tonne)
- Battery charger (portable)
- Parts trolleys
- Masking machine
- Filler dispensers
- Solvent dispensers
- Panel stands
- Impact wrenches
- Torque loading spanners for wheel nuts
- Waste oil dispensers
- Wet and dry vacuum
- Valeting machine
- Polisher
- Wax injection equipment
- Vehicle moving skates
- Benches (mobile).

Hand tools and power tools
- Panel tools and wallboards
- Air tools, electric tools: grinders, chisels, drills, saws
- Windscreen cutting tool
- Spottle spot welding cutter
- Spot welding dresser

- Random orbit sanders and dust extraction
- Block sanders and dust extraction

Miscellaneous and expendable items
- Welding goggles approved to BS
- Welding headscreens approved to BS
- Fire blankets approved to BS
- Welding curtain/screen approved to BS
- Gloves: disposable, plastic, leather, rubber, canvas
- First aid equipment
- Fire extinguishers
- Storage cabinet/lockers.

Dust and fume extraction (extraction and arrestment systems)

Polluted air is often invisible to the naked eye. However, the effect it can have on the health of a workforce and the overall efficiency of an organization can be dramatic. The most effective way to purify air is to capture airborne pollutants at source and, depending on individual applications, to either recirculate fresh and preheated air or vent the pollutants away from the working environment to a safe collection point. The systems should actually be extraction and arrestment systems and should extract the pollutant materials and collect them in a safe and manageable form.

In the context of bodyshops, the main problems are fillers and paint dust from the rubbing down and flatting processes, paint and solvent fumes from the wiping down and painting processes and welding fumes.

Filler and paint dust generated in the preparation area is best collected as soon as it is produced by using off-the-tool extraction. This can either be by portable units serving one or two operators or by fixed systems with a central extraction unit serving a number of fixed extraction points located in the workshop.

Specially designed extraction equipment can be tailor made to an individual bodyshop for the removal of dust and fumes.

Gases, powders and chemical vapours are all types of hazardous elements to which fume extraction can be applied. The range of self-supporting arms, combined with the versatility to mount the system on ceilings, floors, benches and walls, make the access to applications unlimited. Furthermore, the easily manoeuvred suction hoods create extraction right at the source of the problem. The fans are designed to draw the polluted air through the extraction arms, dispersing the fumes via the assembly ducting. Alternatively, to recirculate the purified air, an electrostatic unit can be employed to eliminate harmful particles and utilize existing preheated air.

A range of vehicle exhaust extraction systems is available: a choice can be made from a simple drop system through to the rail system, which allows vehicles to be driven whilst maintaining at-source extraction with both advanced infrared remote controlled and electrically motor driven reels. At-source extraction of welding fumes is far more energy efficient than using central ventilation systems. Harmful fumes can be captured and disposed of in a safe and simple way regardless of the welding environment. Irritation, fever, poisoning and fibrosis are a few of the effects that can be minimized with the extraction of fumes from the welding operations. Many welding processes create noxious and harmful fumes, which can be eliminated with portable welding smoke eliminators. These provide complete extraction where confined areas pose a problem, especially in body repair workshops. There is a wide range of lightweight smoke eliminators which take up very little floor space and can be carried from job to job. The filters in these portable extraction systems can be changed very quickly, and some have an alarm light fitted to warn the user that the filter needs replacing.

It is a specific requirement of the COSHH regulations that equipment is maintained in efficient working order. The physical system must be regarded as part of a broader health and safety housekeeping policy aimed at keeping the whole area dust and fume free.

Bodyshops and legal requirements

Health and safety legislation has made the vehicle body repair industry increasingly aware of the need to provide adequate facilities for employees, both as a legal duty and to improve the working environment. Within this framework of a safe working environment the employer must also promote efficient work methods, which together should result in improved productivity.

Tech note
In previous editions of this book the author has produced a definitive list of the acts, regulations and rules relating to the vehicle repair industry. The explosion of rules and regulations, especially from the impact of European law changes, means that this would now fill a library.

The following is a sample of the list of statutory regulations which the vehicle repair bodyshop MUST COMPLY with.

Statutory legislation

- Abrasive Wheel Regulations
- Classification, Packaging and Labelling of Dangerous Substances Regulations
- Control of Pollution (Special Wastes) Regulations
- Control of Pollution Act
- Control of Substances Hazardous to Health (COSHH) Regulations
- Electricity at Work Regulation
- Environment Protection Act
- Factory Act
- Fire Precautions (Factories, Offices, Shops and Railway Premises) Order
- Fire Precautions Act
- Health and Safety at Work Act
- Highly Flammable Liquids and Liquefied Petroleum Gases Regulations

- Management of Health and Safety at Work Regulations
- Manual Handling Operation Regulations
- Motor Vehicle Construction and Use Regulations
- Personal Protection Equipment at Work (PPE) Regulations
- Petroleum (Consolidation) Act
- Petroleum (Mixtures) Orders
- Protection of Eyes Regulations
- Provision and Use of Work Equipment Regulations
- Road Traffic Act (MOT)
- Weights and Measures Act
- Workplace (Health, Safety and Welfare) Regulations

Tech note

Because of the complexity of health and safety and environmental issues, anybody starting up a new bodyshop or taking over an existing business is advised to employ a consultant to provide the specialist advice and guidance which is needed.

Health and safety legislation

The Health and Safety at Work Act 1974 places a duty upon employers to ensure, so far as is reasonably practicable, safe working conditions and the absence of risks to health in connection with the use, handling, storage and transport of articles and substances.

The act places a statutory duty on employers to have a declared safety policy for a business in which more than five persons are employed. The act does not specify what you should do; it merely provides the framework in which you should operate, together with establishing the Health and Safety Executive and the Health and Safety Commission. The act regulates all working methods. The importance of this act cannot be over-emphasized; no working methods may be employed that can be seen to be a health or safety hazard to employees.

First a company safety policy must be established and a safety committee formed. The committee should consist of members with specialized knowledge of the risks of a particular area, that is bodyshop, paint shop, parts department, offices. The chairman of the committee should be a senior member of the company.

Safety committee's main objectives

1 To ensure that the company's premises outside and inside are safe and healthy
2 To set up an administrative system to maintain a safe working environment
3 To set up some kind of information system on all matters relating to health and safety
4 To ensure that employees receive adequate training in health and safety matters

---

Tech note

Regular inspections of all equipment as outlined in the safety practice code are now a legal requirement. Failure to meet this requirement can result in imprisonment or unlimited fines or both.

Employer's duties
To provide and maintain:

1 Safe equipment
2 Safe place of work with safe access and exit
3 Safe systems of work
4 Healthy working environment
5 Welfare arrangements (washing facilities, eating facilities, first aid)
6 Proper training with time off for this purpose.

Employee's duties

1 To take reasonable care for his/her own safety and the safety of others who may be affected by his/her acts
2 To cooperate with his/her employer or any person on whom a duty or requirement is imposed

The company should establish, in conjunction with the safety committee, its own code of practice on safety procedures within its establishment, and all personnel should receive a copy.

Objectives of the act
1 To repeal, replace or modify existing regulations
2 To maintain or improve standards of health and welfare of people at work
3 To protect people or other workers against any risk to health or safety arising out of the activities of people at work
4 To control the keeping and use of dangerous substances including explosive and highly flammable materials (paints, solvents, resins)
5 To control the emissions of toxic substances into the atmosphere (dust, fumes, smoke, gases)

Control of Substances Hazardous to Health (COSHH) Regulations 1988

Since 1974 a number of regulations have been introduced which describe in detail the requirements for specific safe working practices. One of the most recent is the Control of Substances Hazardous to Health (COSHH) Regulations 1988. COSHH is a major piece of health and safety legislation. It tightens the general obligations of an employer under Section 2 of the Health and Safety at Work Act by specifying comprehensive rules on how substances should be controlled. COSHH applies to all substances in all forms, including gases, vapours, solids, dusts, liquids and even micro-organisms. It also covers mixtures and preparations.

A substance may be a 'hazard to health' if the substance itself is harmful or if:

1 There are hazards from impurities.
2 Dust or fumes are generated during use.
3 It is dangerous when used in combination with other materials.

COSHH attempts to follow the principle of good occupational hygiene practice:

1 Assess the hazard (Regulation 6).
2 Control it (Regulations 7, 8, 12).
3 Maintain the control (Regulations 9, 10, 11).

Assessment: Regulation 6
An employer shall not carry on any work which is liable to expose any employees to any substance hazardous to health unless he has made a suitable and sufficient assessment of the risk created and the steps needed to achieve and maintain control.

Control of exposure: Regulation 7
Exposure must be prevented or adequately controlled. 'Adequately controlled' means that repeated exposure must not cause damage to health. Maximum exposure limits must not be exceeded, or employers may be prosecuted.

Use of controls: Regulation 8
Employers should take all reasonable steps to ensure that control measures are properly used, and employees shall make full and proper use of the controls provided.

Maintenance of controls: Regulation 9
Controls must be maintained in efficient working order and good repair. Thorough examinations and tests must be made at suitable intervals. Records of tests and repairs must be kept for five years.

Exposed monitoring: Regulation 10
Air sampling is required to monitor exposure every 12 months if:

1 It is needed to maintain control.
2 It is needed to protect health.
3 Listed carcinogens are in use.

Health surveillance: Regulation 11
Health surveillance is required at least every 12 months if:

1 Valid health surveillance techniques exist.
2 Specified substances are in use.

Surveillance may include biological monitoring, clinical examinations and review of health records. Records must be kept for 30 years.

Information, instruction and training: Regulation 12
The employees must be told:

1 The risk to health
2 The results of the tests
3 The control measures to be used.

They must be given appropriate instruction and training. Any person carrying out work under the regulations must be competent to do so (Figure 15.2).

Carrying out a COSHH assessment
Gather information

1 What is used, handled, stored
2 Intermediates, by-products, wastes.

HEALTH SURVEILLANCE RECORD			
EMPLOYEE'S NAME	TYPE OF SURVEILLANCE	RESULT	DATE CARRIED OUT
MR JOHN SMITH	RESPIRATORY FUNCTION SCREENING.	WITHIN NORMAL LIMITS.	30-1-93
MR DEREK BROWN	CHEST X-RAY	WITHIN NORMAL LIMITS.	6-3-93
MR WILLIAM BLACK	AUDIOMETRIC HEARING TEST.	CATEGORY 5 (NORMAL)	20-4-93

Figure 15.2 Health surveillance record (*Akzo Coatings PLC*)

TRAINING RECORD		
EMPLOYEE'S NAME	TYPE OF TRAINING	DATE CARRIED OUT
MR JOHN SMITH	SAFETY TRAINING ON THE USE OF AIR-FED MASKS	1-3-93
MR DEREK BROWN	TRAINING IN THE USE OF PORTABLE FIRE-FIGHTING EQUIPMENT FOR THE WORKSHOP.	10-4-93
MR WILLIAM BLACK	TRAINING IN THE SAFE USAGE OF MIG WELDING EQUIPMENT.	5-5-93

Figure 15.3 Training record (*Akzo Coatings PLC*)

Identify hazards (a hazard is the potential to cause harm)

1 Nature of hazard
2 Route of exposure
3 Possible interactions.

Evaluate the risk (risk is the likelihood that a substance will cause harm in the actual circumstance of use)

1 How is the substance used?
2 How is it controlled?
3 What is the level of exposure?

Decide on controls

1 Type of controls (substitution, local exhaust ventilation (LEV))
2 Maintenance and testing procedures
3 Air monitoring
4 Health surveillance
5 Record keeping.

Record assessment – A record must be kept of all assessments other than in the most simple cases that can be easily repeated.

Review – Assessments should be reviewed, for example, when:

1 There are any significant changes to the process, for example, plant changes, volume of production.
2 New substances are introduced.
3 Ill health occurs.
4 New information becomes available on hazards or risks.

Environment Protection Act 1990

The Environment Protection Act 1990, like the Health and Safety at Work Act, allows the Secretary of State to introduce regulations to control, amongst other things, the release of substances harmful to air, water and land. The legislation is wide in scope, with the long-term aim of minimizing environment pollution. The paint application industry is among the third-largest group of environment polluters, and it is because of this that the EPA introduces many refinishing industry-specific controls. The object of the act is to reduce and eventually eliminate the use of volatile organic compounds (VOCs) in surface coatings, using the best available techniques not entailing excessive costs (BATNEECs). VOCs are substances which react with nitrous oxides in sunlight to create low-level ozone. This photochemical air pollution causes damage to vegetation and can also cause serious breathing difficulties in humans and animals. All users will have to use BATNEECs to prevent, minimize and render harmless any release of VOCs. If there is alleged contravention, the user will have to prove that no better techniques were available.

Bodyshops purchasing more than 2 tonnes of solvent per year will be required to register under the classification of Part B 'The Respraying of Motor Vehicles' of the act. Any bodyshop currently operating below these levels is not required to register.

Two separate pollution control regimes are established in Part 1 of the Act to control industrial processes falling into the categories A and B. Category A is integrated pollution control (IPC) operated by HMIP in England and Wales and HMIPI in Scotland. Category B is local authority air pollution control.

Environmental Protection (prescribed processes and substances) Regulations 1991

These identify the respraying of road vehicles as a category B process (local authority air pollution control) if the process may result in the release into the air of any particular matter or of any volatile organic compound (solvents) where the process is likely to involve the use of 2 tonnes or more of organic solvents in any 12-month period. Cleaning agents, fillers, stoppers, primers, gunwash and many other products in the refinishing of motor vehicles all contribute to this figure. The figure should represent the amount consumed in the process, that is, lost through evaporation, spillage and transfer.

Table 15.1 Guide to safe working practices: to be filled in by the bodyshop. The example gives a typical process carried out in most paint shops every day. The initial assessment has identified there is a serious risk to people's health from the way the hazardous substance is being used. Immediate steps must be introduced to control and reduce the risks. Failure to comply with the approved code of practice by an employer may lead to prosecution.

Activity	Hazard	Risk	Precautions		Products used	Products contain	Remarks
			Personal	General			
Wet sanding of fillers and stoppers	Wet sludges of unknown composition	Skin contact with unknown materials Eye contact with splashes of unknown materials	Gloves Goggles or visor	Carry out in a properly ventilated area			
Priming		Remarks:	Conclusions:	Conclusions:			
Mixing (hardener addition and or thinning)	Solvents and, depending on the particular product, isocyanates, acids or polyamides may be present	Inhalation of solvent vapours Skin contact with solvents Eye contact with vapours, solvents or paint splashes	Respirator suitable for solvent vapours where extraction inadequate Gloves Goggles or visor	Carry out in a well ventilated area or with local exhaust ventilation to minimize vapour build-up. NB air-fed respirators are not required when mixing isocyanate containing products at ambient temperatures under conditions of good ventilation	Washfiller 580	Zinc chromate <5% MEL Butyl acetate 10–25% OES Plus blends of additional solvents	Product substitution not possible. Issue staff with respirator and eye protection Arrange for extraction to be installed in mixing room
		Remarks: High risk Carried out three times daily	Conclusions: Strong smell of solvent from acid hardener No respirator or eye protection used whilst mixing	Conclusions: No LEV in mixing area	Washfiller phosphoric hardener	Ethanol 15–30% OES Methyl isobutyl ketone 30–45% OES+ Plus blend of additional solvents including phosphoric acid	Reduce mixing from three times a day to once daily

Application						
Spray mist containing solvent and, depending on the particular product, isocyanates, zinc or strontium chromates, polyamides or acids	Inhalation of solvent vapours or spray mists	Respirator suitable for solvent vapours and particulates Air-fed respirators *must* be worn (i) whenever isocyanates are present or (ii) when zinc or strontium chromate containing primers are used and the ventilation in the spray enclosure is inadequate for the particular job	All spraying *must* be carried out in a properly ventilated spray enclosure	As above	As above	Issue replacement charcoal respirator, set up inspection and maintenance records. No spraying to be done outside spray booth
	Skin contact with solvent vapour or spray mists	Gloves				
	Eye contact with solvent vapour or spray mists	Goggles or visor				
	Remarks: High risk Process carried out daily	Conclusions: Cartridge mask used but in poor condition	Conclusions: Spraying in open workshop Poor extraction			

COSHH ASSESSMENT RECORD				RECORD NO.

PROCESS IDENTIFICATION ....

....

PROCESS LOCATIONS ....

FREQUENCY OF USE ....

NUMBER OF OPERATORS ....

SUBSTANCES/PRODUCTS USED	SUPPLIER	HAZARDS	OES	MEL

CONTROLS

PRECAUTIONS CURRENTLY USED ....

HOW EFFECTIVE ARE THEY ....

LIKELY HAZARD ....

– INHALATION ....

– SKIN CONTAMINATION ....

– EYE CONTACT ....

ACTION REQUIRED ....

....

NEW CONTROLS INTRODUCED ....

....

....

IS A MAINTENANCE SCHEDULE REQUIRED	YES	NO

....

HEALTH SURVEILLANCE REQUIRED	YES	NO

....

CARRIED OUT BY ....

SIGNATURE .... DATE ....

Figure 15.4   COSHH assessment record (*Akzo Coatings PLC*)

If the process involves the use of 2 tonnes of organic solvents in any 12-month period then the bodyshop operator must seek authorization from his or her local authority to carry out the process. If the operator of the bodyshop calculates the solvent usage to be less than 2 tonnes, then the best advice would be to monitor solvent usage in order to demonstrate the fact to the local authority.

First check with the local authority, who will require information about your operation such as:

1   Who and where you are
2   What you do and how you do it
3   What is released to air, how much and where from
4   What you do to prevent, control and monitor such releases
5   Evidence or proposals that the objectives of BATNEEC will be met.

Upgrading of existing bodyshops to comply with the act

The engineering controls that the EPA imposes upon equipment are very significant, particularly regarding spray booths if emission exceeds the 2 tonnes VOCs limit. For spray booths already working there is a requirement to submit an upgrading programme to the local authority within 12 months of the initial registration.

The requirements for upgrading are as follows:

1   Filtration systems have to ensure that the concentration of paint particulate matter in the final discharge to atmosphere from the spray booth does not exceed 10 mg/m^3.
2   Manufacturers or companies upgrading booths will be required to provide a guarantee confirming that the equipment conforms to the emissions limit.
3   The vent velocity of the extract duct must achieve a minimum of 15 m/s for dry filter systems.

4   The vent velocity of the extract duct must achieve a maximum of 9 m/s where a wet method is used.
5   Pressure control systems must be provided which will shut down the spray booth if it is over-pressurized and activate an audio alarm.
6   Exhaust duct openings must not be fitted with plates, caps or cowls that could act as restrictors or deflectors.
7   Exhaust ducting (the chimney) must be a minimum of 8 metres above ground level, but it also has to be a minimum of 3 metres higher than the roof height of any nearby building which is within a distance of five times the uncorrected chimney height.
8   Preventive measures must be taken for fugitive emissions of odour, fumes and particulate matter from mechanical operations like welding, grinding or sanding. All these activities should take place in a building.
9   Shot blasting emissions must not exceed 50 mg/m^3.
10 Paint mixing and equipment cleaning are to take place in an adequately ventilated area.
11 An automatic totally enclosed machine for cleaning spray guns and equipment must be provided.
12 Spray gun testing must be into an extracted area.

Waste management: EPA Part 11

Section 34 of the EPA makes it a criminal offence to 'treat, keep or dispose of controlled waste in a manner likely to cause pollution of the environment or to harm human health'. In order to meet this stated aim, the Part 11 provisions impose a 'duty of care' on everybody involved in the chain of waste management. This means that if you are involved in the creation of waste, then you are responsible for its safe and proper disposal. Typical relevant bodyshop waste includes dirty solvent, paint residues, empty cans and dirty rags.

Those subject to the duty of care must try to achieve the following four things:

1 To prevent any other person committing the offence of disposing of 'controlled waste', or treating it, or storing it without a waste management licence; or breaking the conditions of a licence; or in a manner likely to cause pollution or harm to health
2 To prevent the escape of waste which is, or at any time has been, under their control (this has implications for waste storage facilities and waste containment)
3 To ensure that if waste is transferred it goes only to an 'authorized person' or to a person for 'authorized transport purposes'
4 When waste is transferred, to make sure that there is also transferred a written description of the waste, which is good enough to enable each person receiving it to avoid committing any offence under 1 and to comply with the duty at 2 to prevent the escape of waste.

Under the code of practice, the waste must be:

1 Identified (paint, solvent, paper and tape, dust, loaded extract filter media, scrap metal, tyres and batteries)
2 Categorized
3 Kept in appropriate containers (external skip, internal container, paper baler, can crusher, solvent/paint 'closed loop' system, parts for recycling)
4 Collected and disposed of by a registered operator (transfer note signed by the disposer and collector, and a written description of the waste)
5 Documented at all stages.

Management of Health and Safety at Work Regulations 1992

These regulations set out broad general duties which apply to almost all work activities in Great Britain. They are aimed mainly at improving health and safety management and can be seen as a way of making more explicit what is required of employers under the HSW Act. The main provisions are designed to encourage a more systematic and better-organized approach to deal with health and safety.

The regulations require that employers should:

1 Assess the risk to health and safety of employees and of anyone else who may be affected by the company's work activity. This is so that the necessary preventive and protective measures can be identified. Employers with five or more employees will have to record the significant findings of the assessment.
2 Make arrangements for putting into practice the health and safety measures that follow from the assessment.
3 Provide appropriate health surveillance for employees where the risk assessment shows it to be necessary.
4 Appoint competent people either from inside the organization or from outside to help devise and apply measures needed to comply with duties under health and safety law.
5 Set up emergency procedures.
6 Provide employees with information they can understand about health and safety matters.
7 Cooperate with other employers sharing the work site.
8 Make sure that employees have adequate health and safety training and are capable enough at their jobs to avoid risk.
9 Provide temporary workers with some particular health and safety information to meet special needs.

The regulations will also:

1 Place duties on the employees to follow health and safety instructions and report danger.
2 Extend the current law which requires the employer to consult employees' safety representatives and provide facilities for them.

Provision and Use of Work Equipment Regulations 1992

These regulations will place general duties on employers and list minimum requirements for work equipment to deal with selected hazards whatever the industry.

In general the regulations will make explicit what is already somewhere in the law or is good practice. If equipment has been well chosen and well maintained there should be little need to do more than follow the guidance on the regulations. Some older equipment may need to be upgraded to meet the minimum requirements.

Work equipment will be broadly defined to include everything from a hand tool through machines of all kinds to a complete plant. Use of the equipment will include starting, stopping, repairing, modifying, installing, dismantling, programming, setting, transporting, maintaining, servicing and cleaning.

The general duties require the employer to:

1 Make sure that equipment is suitable for the use that will be made of it.
2 Take into account the working conditions and hazards in the workplace when selecting equipment.
3 Ensure that equipment is used only for operations for which and under conditions for which it is suitable.
4 Ensure the equipment is maintained in an efficient state, in efficient working order and in good repair.
5 Give adequate information, instruction and training.
6 Provide equipment that conforms with EC product safety directives (see what follows).

Specific requirements will cover:

1 Guarding of dangerous parts of machinery (replacing the current law on this)
2 Protection against specific hazards, that is, falling/ejected articles and substances, rupture/disintegration of work equipment parts, equipment catching fire or overheating, unintended or premature discharge of articles and substance, explosion
3 Working equipment parts and substances at high or very low temperatures
4 Control systems and control devices
5 Isolation of equipment from the source of energy
6 Stability of equipment
7 Lighting
8 Maintaining operations
9 Warnings and markings.

The Regulations implement an EC directive aimed at the protection of workers. There are other directives setting out conditions which much new equipment (especially machinery) will have to satisfy before it can be sold in the EC member states. They will be implemented in the UK by regulations made by the Department of Trade and Industry.

Personal Protective Equipment (PPE) at Work Regulations 1992

These Regulations set out in legislation sound principles for selecting, providing, maintaining and using PPE. They do not replace recently introduced law dealing with PPE, for example COSHH or noise at work regulations.

PPE is defined as all equipment designed to be worn or held to protect against a risk to health or safety. This includes most types of protective clothing and equipment such as eye, foot and head protection, safety harnesses, life jackets and high-visibility clothing. PPE should be relied upon only as a last resort, but where risks are not adequately controlled by other means the employer will have a duty to ensure that suitable PPE is provided free of charge for employees exposed to these risks. The regulations say what is meant by suitable PPE, a key point in making sure that it effectively protects the wearer. PPE will only be suitable if it is appropriate for the risks and the working conditions, takes account of workers' needs and fits properly, gives adequate protection and is compatible with any other item of PPE worn.

Employers also have duties to:

1 Assess the risks and PPE intended for issue to ensure that it is suitable.
2 Maintain, clean and replace PPE.
3 Provide storage for PPE when it is not being used.
4 Ensure that PPE is properly used.
5 Give training, information and instruction to employees on its use and how to look after it.

PPE is also subject to a separate EC directive on design, certification and testing and will be marked by the manufacturer with a CE mark.

Manual Handling Operations Regulations 1992

These will apply to any manual handling operations which may cause injury at work. These operations will be identified by the risk assessment carried out under the Management of Health and Safety at Work Regulations 1992. They will include not only lifting of loads but also lowering, pushing, pulling, carrying or moving them, whether by hand or other bodily force.

Employers will have to take three key steps:

1 Avoid hazardous manual handling operations wherever reasonably practicable. Consider whether the load must be moved at all and, if it must, whether it can be moved mechanically.
2 Assess adequately any hazardous operations that cannot be avoided. An ergonomic assessment should look at more than just the weight of the load. Employers should consider the shape and size of the load, the way the task is carried out, the handler's posture, the working environment (cramped or hot), the individual's capacity, and the strength required.

Unless the assessment is very simple, a written record will be needed.

3 Reduce the risk of injury as far as reasonably practicable. A good assessment will not only show whether there is a problem but will also point to where the problem lies.

Workplace (Health, Safety and Welfare) Regulations 1992

The regulations will cover many aspects of health, safety and welfare in the workplace. Some of them are not explicitly mentioned in the current law, though they are implied in the general duties of the Health and Safety at Work Act.

The regulations will set out general requirements in four broad areas.

Working environment
1 Temperature in indoor workplaces
2 Ventilation
3 Lighting including emergency lighting
4 Room dimensions and space
5 Suitability of workstations and seating

Safety
1 Safe passage of pedestrians and vehicles (traffic routes must be wide enough and marked where necessary, and there must be enough of them)
2 Windows and skylights (safe opening, closing and cleaning)
3 Transparent and translucent doors and partitions (use of safety material and marking)
4 Doors, gates, escalators (safety devices)
5 Floors (construction and maintenance, obstructions and slipping and tripping hazards)
6 Falling a distance into a dangerous substance
7 Falling objects

Facilities
1 Toilets
2 Washing, eating and changing facilities

3 Clothing storage
4 Drinking water
5 Rest area (and arrangements to protect people from the discomfort of tobacco smoke)
6 Rest facilities for pregnant women and nursing mothers

Housekeeping
1 Maintenance of workplace, equipment and facilities
2 Cleanliness
3 Removal of waste materials

Risk assessment

For every task which is carried out in the workshop a risk assessment should be carried out. Risk assessment is about identifying hazards and the severity of the hazard, deciding on the level of likelihood of the occurrence and setting control to deal with them. For instance a can of solvent with a secure top in a cupboard is relatively safe. The same can of solvent left without a top and placed on the floor is likely to get kicked over, or the vapour could cause a fire. That is, it now has a potential to cause a major injury and is very likely to happen.

UNISON's 25 steps to risk assessment
UNISON have produced a step-by-step approach to risk assessment. The 25 steps are:

1 Set up a programme of risk assessments.
2 Consult safety representatives about the appointment of competent persons.
3 Appoint competent people.
4 Decide on methods and approach.
5 Identify any other specific health and safety legislation which applies.
6 Collect information.
7 Consult safety representatives and employees about work and perceived hazards.
8 Observe what happens in practice.
9 Identify hazards.
10 Identify the harm that could arise from hazards.
11 Identify those at risk.
12 Identify how they may be harmed.
13 Evaluate the likelihood of harm occurring.
14 Evaluate the likely severity of the harm.
15 Evaluate the likely numbers who could be harmed.
16 Identify the control measures already in place, including information, instruction and training.
17 Evaluate the effectiveness of the control measures.
18 Decide what more needs to be done to eliminate or control risks, in accordance with the accepted priorities of risk prevention and control measures.
19 Record the assessments.
20 Provide safety representatives with copies of the assessments and supporting information.
21 Draw up an action plan and prioritize risks to be tackled.
22 Draw up a timetable for completion of action (action plan or development plan).
23 Allocate financial and staff resources for carrying out the action plan.
24 Implement measures.
25 Monitor the effectiveness of control measures and review the risk assessments at agreed regular intervals and whenever changes require it.

Risk grid
One way to assess risks is to give both the severity of the hazard and the likelihood of occurrence, a factor between 1 and 5. Multiply the factors to get a severity of risk. This will give you a matrix from 1 to 25. That is, 1 is low risk; 25 means somebody is likely to be killed.

Possible hazards
Hazards come in different forms. The following lists are not exhaustive but give a good framework to start from.

Table 15.2 Risk grid

Likelihood of occurrence	Hazard security index	Death	Major injury	Significant injury	Minor injury	Trivial injury
	Very likely	25	20	15	10	5
	Likely	20	16	12	8	4
	Quite possible	15	12	9	6	3
	Possible	10	8	6	4	2
	Not likely	5	4	3	2	1

Physical hazards

- Asbestos
- Awkward posture
- Chemicals
- Display screen equipment
- Drugs
- Electricity
- Fire
- Lead
- Machinery
- Manual handling
- Noise
- Paint
- Petrol/diesel
- Radiation
- Slips, trips and falls
- Solvents
- Transport
- Vibration

Biological hazards

- Animal allergens
- Hepatitis
- HIV
- Legionnaires disease
- MSRA
- Plant allergens
- Tuberculosis
- Weil's disease

Psycho-social hazards

- Boredom
- Bullying
- Isolation
- Lack of control
- Lack of support

- Long hours
- Monotonous work
- Shift work
- Stress
- Violence
- Work overload

Using this information the Risk Assessment Form can be completed. Such a form should be completed for each discrete task in the Bodyshop.

Trending developments

As mentioned in other chapters, body repair skills and vehicles are changing, but the biggest changes are taking place in the business model. That is the way in which the industry conducts its business and how the repairs are carried out. A case study of the setting up of a major body repair facility is discussed in Chapter 18. This is a traditional body repair shop which uses the latest technology and can carry out almost all repairs on site. The trend is the development of single or limited-range services. The use of body repair pods for specialist services is discussed in Chapter 19. Two of the business models which are trending are franchise working and outsourcing.

Franchise – This is where a business is run using a well-known product or brand which is nationally advertised and quality controlled. You may know that many of the high street fast food outlets are franchises. The quality-controlled burgers or pizzas are delivered to the outlet, cooked by the owner and team and sold in branded packaging. Variations of this in the vehicle repair industry include smart repairs, windscreen replacement and some spare parts outlets. These are starting to be developed in the vehicle body repair industry using small units on industrial estates and repair pods. The company advertises on a national basis, with a call centre and/or on-line booking system, the customer books in and pre-pays for the service – the customer is then directed to the most convenient repair facility. This allows small businesses to concentrate on doing the job, not having to worry about advertising, making booking or collecting cash.

Outsourcing – This works in two ways. Often companies, such as large dealerships and vehicle importers, need body repairs done on an irregular basis. Body repairers either working at the dealership or in their own premises in effect provide a subcontractor service. The other use of the term is that as a body repairer, you may use an outsource company to provide human resource services or invoicing and the collection of payments.

> **Safety note**
> Remember, if in doubt, don't do it no matter what the reward.

Skills and questions

1 Explain the role of a bodyshop planning consultant in the planning of a new workshop.
2 Explain the importance of the choice of site in locating a new workshop.
3 State the legal requirements necessary when building a new workshop.
4 Define the following abbreviations: COSHH, HASAWA, EPA.
5 Explain the importance of entrances, roof supports and floor space when planning a workshop.
6 List the necessary areas to be considered for inclusion in a new workshop.
7 Give reasons for the difference in floor areas between a panel bay and a bay with a static jig.

Brooklandsgreen Motor Company

BODYSHOP RISK ASSESSMENT

Support Team:			SAFETY MANAGER:		DATE:	
ACTIVITY	HAZARDS	RISK FACTOR High/Med/ Low	EXISTING OR PLANNED CONTROLS	THOSE AT RISK	IF NOT CONTROLLED, NOTE ACTION REQUIRED	

Figure 15.5 Risk assessment form

8 Discuss the importance of the reception area as part of the workshop.

9 Why should a new workshop have a separate vehicle assessment area?

10 What are the four types of heating system that could be used in workshops?

11 Explain why bodyshop lighting plays a crucial role in the repair and final finish of a vehicle.

12 Make a list of the essential equipment needed for the stripping, repairing and painting areas.

13 Make a list of hand tools, power tools and safety equipment needed in a workshop.

14 Explain the importance of dust and fume extraction systems.

15 Name five of the main health and safety legislation acts which affect the workshop.

16 Discuss volatile organic compounds (VOCs) and their effect on the environment.

17 Which act applies to waste management?

18 Explain the importance of the Personal Protective Equipment (PPE) at Work Regulations 1992.

19 Explain what is meant by the Manual Handling Operation Regulations 1992.

20 Explain the term BS EN ISO 9001–2000 and its relevance to bodyshops.

Reinforced composite materials

Introduction

Composite materials came into the automotive arena to fulfil the needs of small
post-war car manufacturers. As metal
was rationed for 12 years after the end
of World War 2 small companies sought
alternatives. The automotive market was
developing at the same time as the small
boat market; both developed along the
same lines using a glass fibre and a resin
lay-up procedure (**GRP**) which is still
used today by many kit car manufacturers.
The use of **composites – a product made
up of more than one material which
is bonded together to provide special
properties** – became more specialized
as **carbon based technical materials**
became available

Depending on the materials used in composite construction, the following properties of composites may influence their
choice:

- Components can be produced on a one-
  off basis with minimum tooling.
- Compound curvature can be produced
  with constant material thickness.
- Composites provide lightness for a given
  strength.
- Composites are resistant to corrosion.
- Different finishes are available.
- There is a style cache in the use of car-
  bon fibre.

Basic principles of reinforced composite materials

The basic principle involved in reinforced
plastic production is the combination of
polyester resin and reinforcing fibres to
form a solid structure . Glass-reinforced
plastics are essentially a family of structural
materials which utilize a very wide range of
thermoplastic and thermosetting resins. The
incorporation of glass fibres in the resins
changes them from relatively low-strength,
brittle materials into strong and resilient
structural materials. In many ways glass
fibre–reinforced plastic can be compared to
concrete, with the glass fibres performing
the same function as the steel reinforcement
and the resin matrix acting as the concrete.
Glass fibres have high strength and high
modulus, and the resin has low strength and
low modulus. Despite this the resin has the
important task of transferring the stress from
fibre to fibre, so enabling the glass fibre to
develop its full strength.

Polyester resins are supplied as viscous
liquids which solidify when the actuating
agents, in the form of a catalyst and accelerator, are added. The proportions of this
mixture, together with the existing workshop conditions, dictate whether it is cured
at room temperature or at higher temperatures and also the length of time needed for
curing. In common practice pre-accelerated
resins are used, requiring only the addition

of a catalyst to affect the cure at room temperature. Glass reinforcements are supplied in a number of forms, including chopped-strand mats, needled mats, bidirectional materials such as woven rovings and glass fabrics and rovings which are used for chopping into random lengths or as high-strength directional reinforcement. Other materials needed are the releasing agent, filler and pigment concentrates for the colouring of glass fibre–reinforced plastic.

Among the methods of production, the most used method is that of contact moulding, or the wet laying-up technique as it is sometimes called. The mould itself can be made of any material which will remain rigid during the application of the resin and glass fibre, which will not be attacked by the chemicals involved and which will also allow easy removal after the resin has set hard. Those in common use are wood, plaster, sheet metal and glass fibre itself, or a combination of

Table 16.1 Advantages of reinforced composite plastics (*Owens-Corning Fiberglas*)

Compared with metals	Compared with laminated thermoplastics	Compared with injection moulded thermoplastics	Compared with injection moulded thermosets	Compared with wood
1 Higher strength weight ratio	1 Greater scope in mouldable shapes	1 Far higher strength	1 Far higher strength	1 Much higher strength
2 Easier and cheaper manufacture of complex shapes	2 Higher strength	2 Improved dimensional stability	2 Ability to be formed into thin flat sections or panels	2 Greatly increased strength/weight ratio
3 Good corrosion resistance	3 Comparable or better electrical properties	3 Higher temperature resistance		3 Improved dimensional stability
4 Ability to incorporate self-colours				4 Better weathering properties
				5 Higher water resistance
				6 Ease of fabricating complete structures

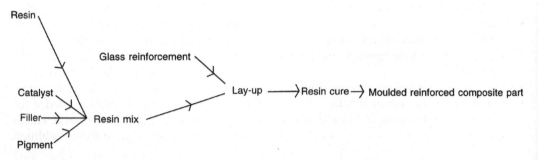

Figure 16.1 Flow chart showing the principles of reinforced composite materials

these materials. The quality of the surface of the completed moulding will depend entirely upon the surface finish of the mould from which it is made. When the mould is ready the releasing agent is applied, followed by a thin coat of resin to form a gel coat. To this a fine surfacing tissue of fibre glass is often applied. Further resin is applied, usually by brush, and carefully cut-out pieces of mat or woven cloth are laid in position. The use of split washer rollers removes the air and compresses the glass fibres into the resin. Layers of resin and glass fibres are added until the required thickness is achieved. Curing takes place at room temperature, but heat can be applied to speed up the curing time. Once the catalyst has caused the resin to set hard, the moulding can be taken from the mould.

Manufacture of reinforced composite materials

When glass is drawn into fine filaments its strength greatly increases over that of bulk glass. Glass fibre is one of the strongest of all materials. The ultimate tensile strength of a single glass filament (diameter 9–15 micrometres) is about 3,447,000 kN/m². It is made from readily available raw materials and is non-combustible and chemically resistant. Glass fibre is therefore the ideal reinforcing material for plastics. In Great Britain the type of glass which is principally used for glass fibre manufacture is E glass, which contains less than 1 per cent alkali borosilicate glass. E glass is essential for electrical applications and it is desirable to use this material where good weathering and water resistance properties are required. Therefore it is greatly used in the manufacture of composite vehicle body shells, both for private and for commercial vehicles.

Basically the glass is manufactured from sand or silica, and the process by which it is made proceeds through the following stages:

1   Initially the raw materials, including sand, china clay and limestone, are mixed together as powders in the desired proportions.

2   The 'glass powder', or frit as it is termed, is then fed into a continuous melt furnace or tank.

3   The molten glass flows out of the furnace through a forehearth to a series of fiberizing units usually referred to as bushings, each containing several hundreds of fine holes. As the glass flows out of the bushings under gravity it is attenuated at high speed.

After fiberizing, the filaments are coated with a chemical treatment usually referred to as a forming size. The filaments are then drawn together to form a strand which is wound on a removable sleeve on a high-speed winding head. The basic packages are usually referred to as cakes and form the basic glass fibre which, after drying, is processed into the various reinforcement products. Most reinforcement materials are manufactured from continuous

Figure 16.2   Manufacture of glass fibre. The story starts with molten glass, heated in a modern furnace. As it is pulled through tiny holes, that liquid mass is transformed into fibres smaller in diameter than a human hair. These fine fibres are then assembled into textile yarn, or into reinforcement products, including glass fibre roving, mat, chopped strands, glass web and milled fibres material (*Owens-Corning Fiberglas*).

filaments ranging in fibre diameter from 5 to 13 micrometres. The fibres are made into strands by the use of size. In the case of strands which are subsequently twisted into weaving yarns, the size lubricates the filaments as well as acting as an adhesive. These textile sizes are generally removed by heat or solvents and replaced by a chemical finish before being used with polyester resins. For strands which are not processed into yarns it is usual to apply sizes which are compatible with moulding resins.

Glass reinforcements are supplied in a number of forms, including chopped-strand mats, needled mats, bidirectional materials such as woven rovings and glass fabrics, and rovings which are used for chopping into random lengths or as high-strength directional reinforcements.

Types of reinforcing material

Woven fabrics

Glass fibre fabrics are available in a wide range of weaves and weights. Lightweight fabrics produce laminates with higher tensile strength and modulus than heavy fabrics of a similar weave. The type of weave will also influence the strength (due, in part, to the amount of crimp in the fabric), and usually satin weave fabrics, which have little crimp, give stronger laminates than plain weaves which have a higher crimp. Satin weaves also drape more easily and are quicker to impregnate. Besides fabrics made from twisted yarns, it is now the practice to use woven fabrics manufactured from rovings. These fabrics are cheaper to produce and can be much heavier in weight.

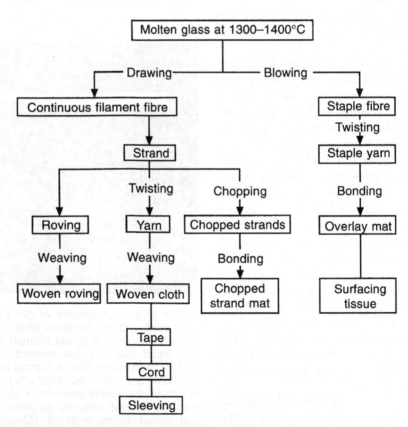

Figure 16.3 Derivation of glass fibre reinforcement

Figure 16.4 Manufacture of rovings. Rovings can be supplied to suit a variety of processes, including projection moulding, continuous laminating, filament winding, pultrusion and as reinforcement for sheet moulding compound. These rovings consist of continuous glass strands, gathered together without any mechanical twist and wound to form a tubular, cylindrical package (*Owens-Corning Fiberglas*).

Figure 16.5 Manufacture of mats. Glass fibre mats are used as resin reinforcement in contact and compression moulded applications. Chopped-strand mats, made from fine chopped-glass strands bonded with a powder of emulsion binder, are used in both areas (*Owens-Corning Fiberglas*).

Figure 16.6 Manufacture of chopped strands. Chopped strands are widely used to reinforce thermoplastic compound (GRTP), polyester bulk moulding compounds (BMC) and in the manufacture of wet-laid glass webs. Chopped strands consist of continuous glass strands chopped to a desired length and are available with a wide variety of surface treatments to ensure compatibility with most resin systems. They are generally solvent and heat resistant and offer excellent flow properties (*Owens-Corning Fiberglas*).

Figure 16.7 Manufacture of yarn (*Owens-Corning Fiberglas*)

Chopped-strand mat

Chopped-strand glass mat (CSM) is the most widely used form of reinforcement. It is suitable for moulding the most complex forms. The strength of laminates made from chopped-strand mat is less than that with woven fabrics, since the glass content which can be achieved is considerably lower. The laminates have similar strengths in all directions because the fibres are random in orientation. Chopped-strand mat consists of randomly distributed strands of glass about 50 mm long which are bonded together with a variety of adhesives. The type of binder or adhesive will produce differing moulding characteristics and will tend to make one mat more suitable than another for specific applications.

Needle mat

This is mechanically bound together, and the need for an adhesive binder is eliminated. This mat has a high resin pick-up owing to its bulk and cannot be used satisfactorily in moulding methods in which no pressure is applied. It is used for press moulding and various low-pressure techniques such as pressure injection, vacuum and pressure bag.

Rovings

These are formed by grouping untwisted strands together and winding them on a 'cheese'. They are used for chopping applications to replace mats either in contact moulding (spray-up) or translucent sheet manufacture of press moulding (pre-form). Special grades of roving are available for each of these different chopping applications. Rovings are also used for weaving, for filament winding and for pultrusion processes. Special forms are available to suit these processes.

Chopped strands

These consist of rovings prechopped into strands of 6 mm, 13 mm, 25 mm or 50 mm lengths. This material is used for dough moulding compounds and in casting resins to prevent cracking.

Staple fibres

These are occasionally used to improve the finish of mouldings. Two types are normally available, a compact form for contact moulding and a soft, bulky form for press moulding. These materials are frequently used to reinforce gel coats. The weathering properties of translucent sheeting are considerably improved by the use of surfacing tissue.

Figure 16.8 Woven fabric (*Scott Bader Co. Ltd*)

Figure 16.9 Rovings (*Owens-Corning Fiberglas*)

Figure 16.10 Chopped strands (*Owens-Corning Fiberglas*)

Figure 16.11 Staple fibres (surface mat) (*Owens-Corning Fiberglas*)

Application of these materials

Probably chopped-strand mat is most commonly used for the average moulding. It is available in several different thicknesses and specified by weight: 300, 450 and 600 g/m². The 450 g/m² is the most frequently used and is often supplemented with the 300 g/m². The 600 g/m² density is rather too bulky for many purposes and may not drape as easily, although all forms become very pliable when wetted with the resin.

The woven glass fibre cloths are generally of two kinds, made from continuous filaments or from staple fibres. Obviously,

most fabricators use the woven variety of glass fibre for those structures that are going to be the most highly stressed. For example, a moulded glass fibre seat pan and squab unit would be made with woven material as reinforcement, but a detachable hard top for a sports car body would more probably be made with chopped-strand mat as a basis. However, it is quite customary to combine cloth and mat, not only to obtain adequate thickness but because if the sandwich of resin, mat and cloth is arranged so that the mat is nearest to the surface of the final product, the appearance will be better.

The top layer of resin is comparatively thin, and the weave of cloth can show up underneath it, especially if some areas have to be buffed subsequently. Chopped fibres do not show up so prominently, but some fabricators compromise by using the thinnest possible cloth (surfacing tissue as it is known) nearest the surface, on top of the chopped-strand mat. When moulding, these orders are of course reversed, the tissue going onto the gel coat on the inside of the mould, followed by the mat and resin lay-up.

It is important to note that if glass cloths or woven mat are used, it is possible to lay up the materials so that the reinforcement is in the direction of the greatest stresses, thus giving extra strength to the entire fabrication. In plain-weave cloths, each warp and weft thread passes over one yarn and under the next. In twill weaves, the weft yarns pass over one warp and under more than one warp yarn; in 2 × 1 twill, the weft yarns pass over one warp yarn and under two warp yarns. Satin weaves may be of multi-shaft types, when each warp and weft yarn goes under one and over several yarns. Unidirectional cloth is one in which the strength is higher in one direction than the other, and balanced cloth is a type with the warp and weft strength about equal. Although relatively expensive, the woven forms have many excellent qualities, including high dimensional stability, high tensile and impact strength, good heat, weather and chemical resistance, low moisture absorption, resistance to fire and good thermo-electrical properties. A number of different weaves and weights is available, and thickness may range from 0.05 mm to 9.14 mm, with weights from 30 g/m² to 1 kg/m², although the grades mostly used in the automotive field probably have weights of about 60 g/m² and will be of plain, twill or satin weave.

Carbon fibre

This is another reinforcing material. Carbon fibres possess a very high modulus of elasticity and have been used successfully in conjunction with epoxy resin to produce low-density composites possessing high strength.

Resins used in reinforced composite materials

The first man-made plastics were produced in this country in 1862 by Alexander Parkes and were the forerunner of celluloid. Since then a large variety of plastics has been developed commercially, particularly in the last 25 years. They extend over a wide range of properties. Phenol formaldehyde is a hard thermoset material; polystyrene is a hard, brittle thermoplastic; polythene and plasticized polyvinyl chloride (PVC) are soft, tough thermoplastic materials; and so on. Plastics also exist in various physical forms. They can be bulk solid materials, rigid or flexible foams, or in the form of sheet or film. All plastics have one important common property. They are composed of macro-molecules, which are large chain-like molecules consisting of many simple repeating units. The chemist calls these molecular chains polymers. Not all polymers are used for making plastic mouldings. Man-made polymers are called synthetic resins until they have been moulded in some way, when they are called plastics.

Most synthetic resins are made from oil. The resin is an essential component of glass fibre–reinforced plastic. The most widely used is unsaturated polyester resin, which can be cured to a solid state either by catalyst and heat or by catalyst and accelerators at room temperature. The ability of polyester resin to cure at room temperature into a hard

Table 16.2   Advantages and disadvantages of glass fibre reinforcement

	Advantages	Uses	Disadvantages
Rovings	Unidirectional strength	Spray-up	Limited use in hand lay-up
		Local longitudinal strength	
		Mechanical bond for bulkheads	
		Making tubes	
Woven rovings	Easy to handle	To increase longitudinal and transverse strength	Poor interlaminar adhesion
	Bidirectional strength		
	High glass content		Traps air, causes voids
	High impact resistance		High cost
Chopped-strand mat	Multidirectional strength	Various	
	Low cost	General-purpose reinforcement	
	Good interlaminar bond		
	Can be moulded into complex shape		
Woven cloth	High strength	Sheathing	Very high cost difficult to 'wet out'
	Smooth finish	As a fire barrier	
Surface tissue	Fine texture	Reinforcing gelcoat	Low strength

material is one of the main reasons for the growth of the reinforced plastics industry. It was this which led to development of the room-temperature contact moulding methods which permit production of extremely large integral units.

Polyester resins are formulated by the reaction of organic acids and alcohols, which produces a class of material called esters. When the acids are polybasic and the alcohols are polyhydric they can react to form a very complex ester which is generally known as polyester. These are usually called alkyds, and they have long been important in surface coating formulations because of their toughness, chemical resistance and endurance. If the acid or alcohol used contains an unsaturated carbon bond, the polyester formed can react further with other unsaturated materials such as styrene or diallyl phthalate. The result of this reaction is to interconnect the different polyester units to form the three-dimensional cross-linked structure that is characteristic of thermosetting resins. The available polyesters are solutions of these alkyds in the cross-linking monomers. The curing of the resin is the reaction of the monomer and the alkyd to form the cross-linked structure. An unsaturated polyester resin is one which is capable of being cured from a liquid to a solid state when subjected to the right conditions. It is usually referred to as polyester.

Catalysts and accelerators

In order to mould or laminate a polyester resin, the resin must be cured. This is the name given to the overall process of gelation and hardening, which is achieved either by the use of a catalyst and heating or at normal room temperature by using a catalyst and an accelerator. Catalysts for polyester resins are usually organic peroxides. Pure catalysts are chemically unstable and liable to decompose with explosive violence. They are supplied, therefore, as a paste or liquid dispersion in a plasticizer, or as a powder in an inert filler. Many chemical compounds act as accelerators, making it possible for the resin-containing catalyst to be cured without the use of heat. Some accelerators have only limited or specific uses, such as quaternary ammonium compounds, vanadium, tin or zirconium salts. By far the most important of all accelerators are those based on a cobalt soap or those based on a tertiary amine. It is essential to choose the correct

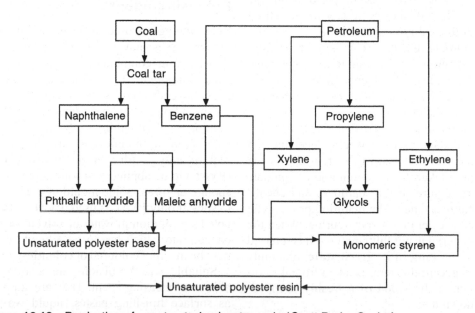

Figure 16.12   Production of unsaturated polyester resin (*Scott Bader Co. Ltd*)

type of catalyst and accelerator, as well as to use the correct amount, if the optimum properties of the final cured resin or laminate are to be obtained.

Pre-accelerated resins

Many resins are supplied with an in-built accelerator system controlled to give the most suitable gelling and hardening characteristics for the fabricator. Pre-accelerated resins need only the addition of a catalyst to start the curing reaction at room temperature. Resins of this type are ideal for production runs under controlled workshop conditions.

The cure of a polyester resin will begin as soon as a suitable catalyst is added. The speed of the reactions will depend on the resin and the activity of the catalyst. Without the addition of an accelerator, heat or ultraviolet radiation, the resin will take a considerable time to cure. In order to speed up this reaction at room temperature it is usual to add an accelerator. The quantity of accelerator added will control the time of gelation and the rate of hardening.

There are three distinct phases in the curing reaction:

*   *Gel time* – This is the period from the addition of the accelerator to the setting of the resin to a soft gel.
*   *Hardening time* – This is the time from the setting of the resin to the point when the resin is hard enough to allow the moulding or laminate to be withdrawn from the mould.
*   *Maturing time* – This may be hours, several days or even weeks depending on the resin and curing system and is the time taken for the moulding or laminate to acquire its full hardness and chemical resistance. The maturing process can be accelerated by post-curing. When the material is not fully matured it is referred to as being in its **green state**, or simply as **green**; this term is taken from the colour of the wood of a freshly chopped-down tree.

Fillers and pigments

Fillers are used in polyester resins to impart particular properties. They will give opacity to castings and laminates, produce dense gel coats, and impart specific mechanical, electrical and fire-resisting properties. A particular property may often be improved by the selection of a suitable filler. Powdered mineral fillers usually increase compressive strength; fibrous fillers improve tensile and impact strength. Moulding properties can also be modified by the use of fillers; for example, shrinkage of the moulding during cure can be considerably reduced. There is no doubt, also, that the wet lay-up process on vertical surfaces would be virtually impossible if thixotropic fillers were not available.

Polyester resins can be coloured to any shade by the addition of selected pigments and pigment pastes, the main requirement being to ensure thorough dispersion of colouring matter throughout the resin to avoid patchy mouldings.

Both pigments and fillers can increase the cure time of the resin by dilution effect, and the adjusted catalyst and promoter are added to compensate.

Releasing agents

Releasing agents used in the normal moulding processes may be either water-soluble film-forming compounds or some type of wax compound. The choice of releasing agent depends on the size and complexity of the moulding and on the surface finish of the mould. Small mouldings of simple shape, taken from a suitable GRP mould, should require only a film of polyvinyl alcohol (PVAL) to be applied as a solution by cloth, sponge or spray. Some mouldings are likely to stick if only PVAL is used. PVAL is available as a solution in water or solvent or as a concentrate which has to be diluted, and it may be in either coloured or colourless form.

Suitable wax emulsions are also available as a releasing agent. They are supplied as surface finishing pastes, liquid wax or wax polishes. The recommended method of application can vary depending upon the material to be finished. Hand apply with a pad of damp, good quality mutton cloth or equivalent, in straight even strokes. Buff lightly to a shine with a clean, dry, good quality mutton cloth. Machine at 1800 rpm using a G-mop foam finishing head. Soak this head in clean water before use and keep damp during compounding. Apply the wax to the surface. After compounding, remove residue and buff lightly to a shine with a clean, dry, good-quality mutton cloth.

Wax polishes should be applied in small quantities since they contain a high percentage of wax solids. Application with a pad of clean, soft cloth should be limited to an area of approximately 1 square metre. Polishing should be carried out immediately, before the wax is allowed to dry. This can be done either by hand or by machine with the aid of a wool mop polishing bonnet.

Frekote is a semi-permanent, multi-release, gloss-finish, non-wax polymeric mould release system specially designed for high-gloss polyester mouldings. It will give a semi-permanent release interface when correctly applied to moulds from ambient up to 135°C. This multi-release interface chemically bonds to the mould's surface and forms on it a micro-thin layer of a chemically resistant coating. It does not build up on the mould and will give a high-gloss finish to all polyester resins, cultured marble and onyx. It can be used on moulds made from polyester, epoxy, metal or composite. Care should be taken to avoid contact with the skin, and the wearing of suitable clothing, especially gloves, is highly recommended. These products must be used in a well-ventilated area.

Adhesives used with GRP

Since polyester resin is highly adhesive, it is the logical choice for bonding most materials to GRP surfaces.

Suitable alternatives include the Sika Technique, which is a heavy-duty, polyurethane-based joining compound. It cures to a flexible rubber which bonds firmly to wood, metal, glass and GRP. It is ideal for such jobs as bonding glass to GRP or bonding GRP and metal, as is often required on vehicles with GRP bodywork. It is not affected by vibration and is totally waterproof. The Araldite range includes a number of industrial adhesives which are highly recommended for use with GRP. Most high-strength impact adhesives (superglues) can be used on GRP laminates.

Most other adhesives will be incapable of bonding strongly to GRP and should not be used when maximum adhesion is essential.

Core materials

Core materials, usually polyurethane, are used in sandwich construction, that is basically a laminate consisting of a foam sheet between two or more glass fibre layers (Figure 16.13). This gives the laminate considerable added rigidity without greatly increasing weight. Foam materials are available which can be bent and folded to follow curved surfaces such as vehicle bodies. Foam sheet can be glued or stapled together, then laminated over to produce a strong box structure, without requiring a mould. Typical formers and

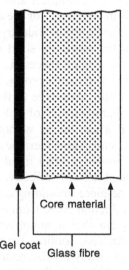

Figure 16.13 Typical sandwich construction (*Scott Bader Co. Ltd*)

core materials are paper rope, polyurethane rigid foam sheet, scoreboard contoured foam sheet, Termanto PVC rigid foam sheet, Term PVC contoured foam sheet, and Termino PVC contoured foam sheet.

Formers

A former is anything which provides shape or form to a GRP laminate. They are often used as a basis for stiffening ribs or box sections. A popular material for formers is a paper rope, made of paper wound on flexible wire cord. This is laid on the GRP surface and is laminated over to produce reinforcing ribs, which give added stiffness with little extra weight. The former itself provides none of the extra stiffness; this results entirely from the box section of the laminate rib. Wood, metal or plastic tubing and folded cardboard can all be used successfully as formers. Another popular material is polyurethane foam sheet, which can be cut and shaped to any required form.

Moulding techniques for reinforced composite laminates

Contact moulding

This is the oldest, simplest and most popular fabrication technique for the automotive, reinforced-plastic body industry. It is normally used for relatively short runs, but it has also been adapted successfully for series production. It is the only production method which takes full advantage of the two most important characteristics of polyester resin, namely that it can be set without heat and without pressure. A considerable industry has been built around contact moulding, which has facilitated the cost-effective production of large one-piece mouldings, particularly for low production runs. Contact moulding advantages are that a minimum of equipment is required, tooling is inexpensive, there are practically no size restrictions, and design changes are easily made. Disadvantages are that the labour context is high and

the quality of the moulding depends on the skill of the operators. The lay-up and curing times are comparatively slow, and only one good surface finish is achieved.

The contact moulding process is carried out in the following manner. A master pattern or model is made, representing in all its dimensions the finished product. This could be, for example, a full-size wood model of the type of body shell or cab shell required, or it might equally well be a steel or aluminium panel-beaten structure of composite type or even a plaster model reinforced with wire mesh. From this is made a master mould, which must be female or concave for the most part, and this would in all probability be made in reinforced plastics similar to those used for the final product. It is important to differentiate, as a matter of terminology, between 'mould' and 'moulding', one being the production tool and the other the product itself.

Figure 16.14 Typical formers: (a) metal tube, (b) paper rope, (c) cardboard tube, (d) foam strip, (e) wood, (f) folded cardboard (*Scott Bader Co. Ltd*)

An important aspect of the process is that the surface of the mould will inevitably be faithfully reproduced in the moulding, and accordingly if the mould is bumpy or rough so will be the final article. It does not usually matter if the unseen or partly hidden side of the moulding is rough (indeed, it usually is), but for the displayed surface to be unsightly is not normally tolerable. This is why a female mould is usually used. If the original pattern was very smooth, so will be the inside of the mould and, therefore, the outside of the moulding. This is why so much trouble is taken over the pattern. If in wood, it is tooled and sanded to perfection, and if in metal, it is panel beaten with the greatest possible skill, then ground and polished if necessary. It is of the greatest importance that separation of the moulding from the mould be easy. A special compound or a polish such as carnauba wax and/or silicone lubricant can be used; plastics film is occasionally used as a separating membrane, as it will not adhere strongly to either the mould or the moulding. For contact moulding, the equipment is relatively simple and inexpensive. Contact moulding can be further subdivided into hand lay-up and spray lay-up.

The application of the release agent to the mould is followed by brush or spray application of a gel coat of resin. There has been a constant improvement and development in polyester resins, and among other things this has led to the introduction of successful semi-flexible gel coats, which are of particular interest to the motor industry. The gel coat is a continuous skin on the working face of a moulding. It is almost pure resin, and its object is to give a good finish as well as to protect the working surface from corrosion and other damage. It also hides the fibre pattern of the reinforcement. The gel coat can be colour impregnated or otherwise specially formulated, for example, for extra abrasion or impact resistance. It should be as even in thickness as possible, as thicker areas are very susceptible to accidental damage, while thin patches can lower the resistance of the structure to moisture and to chemical attack.

Even in hand lay-up the spray method may be used for this stage and for the application of the separating agent, so that there is a small element of mechanization. A fine surfacing tissue may be applied to the gel coat while wet, or it may simply be allowed to gel as it is. Further resin is sprayed on or brushed on, and mat or woven cloth, which has been carefully cut to patterns, is laid in position. Consolidation and air removal are then effected by manual means. It is customary to use rollers made up of split washers for this operation, which is an extremely important one if consistency and strength of the moulding are to be obtained. More mat or cloth is added in order to build up the requisite thickness of reinforced plastics, and the moulding is then allowed to set or cure. Curing normally takes place at room temperature, but sometimes under a certain degree of heat if the process is to be accelerated. It should be remembered that curing is itself a heat-producing process. Contact moulding can also be carried out by simple mechanical means, but the general principle is always that of bringing the materials into contact with the mould, without the use of any dies.

The following is a summary of the contact moulding process:

1 The master mould must be spotlessly clean.
2 A release agent is applied to the entire surface of the mould face.
3 Gel-coat covering is applied by brush or spray.
4 Catalyst is added to the resin and the catalysed resin is smoothed over the gel-coated mould.
5 Glass fibre mat, pre-cut to exact size, is laid on the mould, and a further small quantity of resin is poured onto the mat. With brushes and hand roller, the resin is drawn through the mat. A second layer of mat is applied and the drawing-through process is repeated until the required thickness is achieved. It is critical that all air bubbles be removed by the brushing and rolling.

6 The mould is allowed to cure naturally or heat is used to speed up the curing process.
7 After curing, the moulds are broken and the completed sections are removed (Figure 16.19).
8 They are then trimmed and ready for use.

Figure 16.15 Application of release agent to mould surface (*Scott Bader Co. Ltd*)

Figure 16.16 Application of gel coat covering (*Scott Bader Co. Ltd*)

Figure 16.17 Impregnation of glass mat with resin (*Scott Bader Co. Ltd*)

Figure 16.18 Rolling impregnated mat (*Scott Bader Co. Ltd*)

Figure 16.19 Final mould and moulding (*Scott Bader Co. Ltd*)

Spray-up technique

A development from the basic manual contact process, which is employed with increasing frequency in the automotive body industry, is known as spray-up. In this method, rovings are automatically fed through a chopping unit, and the resultant chopped strands are blown or carried by the sprayed resin stream onto the mould. The glass-and-resin mix applied in this way is consolidated, and the air pockets or bubbles are removed by manual rolling, as in simple hand lay-up.

There are several commercial spraying systems available, in which the glass fibre and resin are deposited simultaneously on the mould face. They consist of two principal types. In the **twin-pot system** a twin-nozzle spray gun is used, and in order to prevent gelation in the gun the resin is divided into two parts, one of which is catalysed and the other accelerated. The two streams of resin spray converge near the surface of the mould simultaneously with a stream of glass fibre ejected by a glass rovings chopper. In the other type, the **catalyst injector system**, accelerated resin is sprayed from a single-nozzle gun, but liquid catalyst is metered into the resin in the gun itself. A glass rovings chopper delivers the reinforcement to the mould surface as in the former system.

Although much of the manual labour of the hand lay-up is eliminated by using the spray process, thorough rolling is still necessary, not only to consolidate the deposited glass resin mixture but also to ensure that the accelerated and catalysed portions of the resin are adequately mixed. Considerable skill is needed to control the thickness of the laminate when using the resin glass spray gun. Spraying reduces labour costs, especially when the volume of production is large enough to keep the equipment in constant use.

Hot press moulding

This process involves the use of chopped-strand glass mat, pre-impregnated with polyester resin, which is then in general principle formed in presses in a similar manner to that used for forming steel sheet. In this case, however, the dies, which are pre-heated, have to remain closed for the curing cycle of the pre-impregnated mat, which may involve a period from 15 to 30 seconds. Hot press moulding using matched dies provides a good finish on both surfaces. Further, this method enables high glass content and uniform dimensional properties and appearances to be achieved at lower cost than by other methods for runs greater than 1000 units. Such a process reduces the labour content of producing panels, but much increases the initial tooling charge.

Resin Glass Resin
      fibre

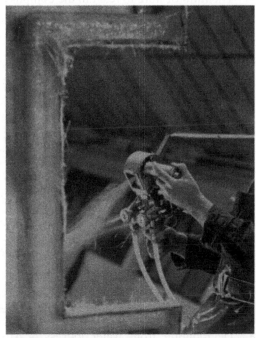

Figure 16.20 The spray-up technique (*BP Chemicals (UK) Ltd*)

Figure 16.21 Hot press moulding

Production composite moulding processes

There are many fibreglass reinforced plastic moulding processes available to the designer. Each process has its own characteristics, as well as its own limitations as to part size, shape, production rate, compatible reinforcements, and suitable resin systems. The most common moulding processes are as follows (Figure 16.22): hand lay-up, spray-up, resin transfer moulding, compression moulding, injection moulding thermoplastics, injection moulding thermosets, pultrusion and reinforced reaction injection moulding.

The three techniques used in the production of body panels are as follows:

- *Cold press mouldings* – This is used in the manufacture of the boot lid. The boot lid is formed by cold pressing a mineral reinforced resin-coated glass fibre mat in a gel-coated mould, forming a component which is very stiff in relation to its weight.
- *Reinforced reaction injection moulding* – The RRIM technique is used for all the vertical body panels such as the front and rear wings, front grille and bumper assembly and rear panel and bumper assembly. RRIM polyurethane has the properties of good recovery from deformation, outstanding resistance to wear, impact and abrasion and a fast cycle time in manufacture. The use of this material for all exposed corners of the car helps to reduce minor body damage repair.
- *Pressure-assisted resin injection moulding* – The bonnet is a pressure assisted moulding of sandwich construction with polyester resin exterior on either side of a rigid urethane core. The underside of the moulding is an intumescent fire barrier which is a major safety factor for an engine compartment cover.

Composite theory

In its most basic form a composite material is one which is composed of two elements working together to produce material properties that are different to the properties of those elements on their own. In practice, most composites consist of a bulk material called the matrix and a reinforcement material of some kind which increases the strength and stiffness of the matrix.

Polymer matrix composite (PMC) is the type of composites used in modern vehicle bodywork. This type of composite is also known as fibre-reinforced polymers (or plastics) (FRP). The matrix is a polymer-based resin and the reinforcement material is a fibrous material such as glass, carbon or aramid. Frequently, a combination of reinforcement materials is used.

The reinforcement materials have high tensile strength but are easily chaffed and will break if folded. The polymer matrix holds the fibres in place so that they are in their strongest position and protects them from damage.

The properties of the composite are thus determined by:

- The properties of the fibre
- The properties of the resin
- The ratio of fibre to resin in the composite–fibre volume fraction (FVF)
- The geometry and orientation of the fibres in the composite.

Resin

The choice of resins depends on a number of characteristics, namely:

- Adhesive properties – in relation to the type of fibres being used, and if metal inserts are to be used such as for body fitting
- Mechanical properties – particularly tensile strength and stiffness
- Micro-cracking resistance – Stress and age hardening causes the material to crack; the micro-cracks reduce the material strength and eventually lead to failure.
- Fatigue resistance – Composites tend to give better fatigue resistance than most metals.
- Degradation from water ingress – All laminates permit very low quantities of water to pass through in a vapour form. If the laminate is wet for a long period,

the water solution inside the laminate will draw in more water through the osmosis process.

- Curing properties – The curing process alters the properties of the material. Generally oven curing at between 80°C and 180°C will increase the tensile strength by up to 30%.
- Cost – The different materials cost different prices

The main types of resins are: polyesters, vinylesters, epoxies, phenolics, cyanate esters, silicones, polyurethanes, bismaleides (BMI) and polyamides. The first three are the ones mainly used for vehicle bodywork, as they are reasonably priced. Cyanates, BMI and polyamides cost about 10 times the price of the others.

Reinforcing fibres

The mechanical properties of the composite material are usually dominated by the contribution of the reinforcing fibres. The four main factors which govern this contribution are:

1 The basic mechanical properties of the fibre
2 The surface interaction of the fibre and the resin, called the interface
3 The amount of fibre in the composite – FVF
4 The orientation of the fibres.

The three main reinforcing fibres used in vehicles are glass, carbon and aramid. In addition, the following are used for non-body

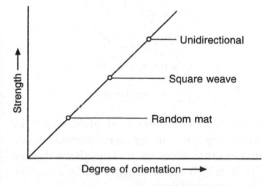

Figure 16.22   Orientation of fibre reinforcement

purposes: polyester, polyethylene, quartz, boron, ceramic and natural fibres such as jute and sisal.

- *Aramid* – Aramid fibre is a man-made organic polymer, an aromatic polyamide, produced by spinning fibre from a liquid chemical blend. The bright golden-yellow fibres have high strength and low density, giving a high specific strength. Aramid has good impact resistance. Aramid is better known by its Dupont trade name Kevlar.
- *Carbon* – Carbon fibre is produced by the controlled oxidation, carbonization and graphitization of carbon-rich organic materials – referred to as precursors – which are in fibre form. The most common precursor is polyacrylonitrile (PAN); pitch and cellulose are also used.

Fabric types and constructions

To weave any material warp threads and weft threads are needed. The warp is the longest part, the part which is wound around the roller or folded. The weft is the shorter part, running at 90° across the warp. The spun warp threads are wound onto a giant roller, and the weft threads are woven into the warp threads in different patterns for different purposes.

Plain

(Courtesy of SP Systems)

Each weft thread passes alternately under and over each warp fibre. The resultant weave is symmetrical with good stability and reasonable porosity. However, because of the high level of crimp of the fibres the mechanical properties are low compared to other weave patterns; this also means that it does not drape very well.

Twill

(Courtesy of SPSystems)

One or more weft fibres alternate by passing over or under two or more warp fibres. This gives the visual appearance of a diagonal rib. The threads therefore have less crimp, giving higher mechanical properties than the plain weave. The fabric looks smoother and drapes better.

Satin

(Courtesy of SPSystems)

Satin weaves are fundamentally twill weaves modified to produce fewer intersections of warp and weft. The harness number used in the designation, typically 4, 5 and 8, is the total number of fibres crossed and passed under before the fibre repeats the pattern. A crowsfoot weave is a form of satin weave with a different stagger in the repeat pattern. Satin weaves are very flat and have low crimp which means that they have excellent mechanical properties and drape well. Because of the pattern shape one face of the material will have sections of fibre running open in one direction, giving an asymmetric imbalance. Care must be taken when laying up several layers to ensure the pattern is set to minimize stress levels.

Basket

(Courtesy of SPSystems)

Basket weave is fundamentally the same as plain weave except that two or more warp fibres alternately interlace with two or more warp fibres. An arrangement of two wefts crossing two warps is designated 2 × 2 basket. The arrangement of fibres need not be symmetrical. Basket weave is flatter and through less crimp is stronger than plain weave. Basket weave is less stable and is best used with thicker fibres, called high tex, to avoid excessive crimping.

Leno

(Courtesy of SPSystems)

Leno weave is used to improve the stability in open fabrics which have a low fibre count. It is a form of plain weave in which the adjacent weft fibres are twisted around consecutive warp fibres. Fabrics in leno weave are usually used in conjunction with other weave-style fabrics, as leno weave alone is unlikely to produce sufficient strength because of its openness.

Mock leno

(Courtesy of SPSystems)

Mock leno is a version of plain weave in which the fibres under–over interlace every two or more fibres apart. This give a thicker fabric which has a rougher surface texture and very good porosity.

Table 16.3 Properties of fibreglass composites and alternative materials

Material	ASTM test method	Glass fibre by weight %	Specific gravity D792	Density lb/in³	Tensile strength 10³psi D638	Tensile modulus 10⁶psi D638	Elongation % D638	Flexural strength 10³psi D790	Flexural modulus 10⁶psi D790	Compressive strength 10³psi D695	Impact strength Izod ft lb/in notched at 73°F D256	Hardness Rockwell (except where noted) D785	Flammability UL–94
Glass fibre reinforced thermosets	Polyester SMC (compression)	30.0	1.85	0.066	12.00	1.70	1.0	26.00	1.60	24.00	16.00	Barcol 68	5V
	Polyester SMC (compression)	20.0	1.78	0.064	5.30	1.70	.4	16.00	1.40	23.00	8.20	Barcol 68	5V
	Polyester SMC (compression)	50.0	2.00	0.072	23.00	2.27	1.7	45.00	2.00	32.00	19.40	Barcol 68	5V
	Polyester BMC (compression)	22.0	1.82	0.065	6.00	1.75	.5	12.80	1.58	20.00	4.26	Barcol 68	5V
	Polyester BMC (injection)	22.0	1.82	0.065	4.86	1.53	.5	12.65	1.44	–	2.89	Barcol 68	VO
	Epoxy (filament wound)	80.0	2.08	0.061	80.00	4.00	1.6	100.00	5.00	45.00	45.00	M98	VO
	Polyester (pultruded)	55.0	1.69	0.060	30.00	2.50	–	30.00	1.60	30.00	25.00	Barcol 50	VO
	Polyurethane, milled fibres	13.0	1.07	0.038	2.80	–	140.0	–	0.037–0.053	–	–	SD*65–75	VO
	Polyurethane flaked glass	23.0	1.17	0.042	4.41	–	38.9	–	0.15	–	2.10	–	VO
	Polyester (spray-up/lay-up)	30.0	1.37	0.049	12.50	1.00	1.3	28.00	0.75	22.00	13.00–15.00	Barcol 50	VO
	Polyester, woven roving (lay-up)	50.0	1.64	0.059	37.00	2.25	1.6	46.00	2.25	27.00	33.00	Barcol 50	VO

(Continued)

Table 16.3 (Continued)

Material	ASTM test method	Glass fibre by weight %	Specific gravity D792	Density lb/in³	Tensile strength 10³psi D638	Tensile modulus 10⁶psi D638	Elongation % D638	Flexural strength 10³psi D790	Flexural modulus 10⁶psi D790	Compressive strength 10³psi D695	Impact strength Izod ft lb/in notched at 73°F D256	Hardness Rockwell (except where noted) D785	Flammability UL-94
Glass fibre reinforced thermoplastics	Acetal	25.0	1.61	0.058	18.50	1.25	3.0	28.00	1.10	17.00	1.80	M79	HB
	Nylon 6	30.0	1.37	0.049	24.00	1.05	3.0	29.00	1.11	24.00	2.20	R121	HB
	Nylon 6/6	30.0	1.48	0.053	23.00	1.20	1.9	35.00	0.80	26.50	2.20	M95	HB
	Polycarbonate	10.0	1.26	0.045	12.00	.75	9.0	16.00	0.60	14.00	2.00	M80	V-1
	Polypropylene	20.0	1.04	0.037	6.50	.54	3.0	8.30	0.52	25.00	1.10	R103	HB
	Polyphenylene sulphide	40.0	1.64	0.059	22.00	2.05	3.0	37.00	1.90	21.00	1.50	R123	V-O/5V
	Acrylonitrile butadiene styrene (ABS)	20.0	1.22	0.044	11.00	.90	2.0	15.50	0.87	14.00	1.20	R107	HB
	Polyphenylene oxide (PPO)	20.0	1.21	0.043	14.50	.92	5.0	18.50	0.75	17.60	1.80	R107	HB
	Polystyrene acrylonitrile (SAN)	20.0	1.22	0.044	14.50	1.25	1.8	19.00	1.10	17.50	1.10	R122	HB
	Polyester (PBT)	30.0	1.52	0.054	19.00	1.20	4.0	28.00	1.17	18.00	1.80	R118	HB
	Polyester (PET)	30.0	1.56	0.056	21.00	1.30	6.6	32.00	1.25	25.00	1.80	R120	HB
Unreinforced thermoplastics	Acetal	–	1.41	0.051	8.80	.41	40.0	13.00	0.38	16.00	1.00	M78–M80	HB
	Nylon 6	–	1.12	0.040	11.80	.38	30.0	15.70	0.39	13.00	0.60	R119	HB
	Nylon 6/6	–	1.13	0.041	11.50	.40	60.0	17.00	0.42	15.00	0.80	R120, M83	V-2
	Polycarbonate	–	1.20	0.043	9.50	.34	110.0	13.50	.34	12.50	16.00	M70	V-2
	Polypropylene	–	.89	0.032	5.00	.10	200.0	5.00	0.13–0.20	3.50	1–20	R50–96	HB

	Material											
	Polyphenylene sulphide	–	1.30	0.045	9.50	.48	1.0	14.00	0.55	16.00	0.50	R123	V-O
	Acrylonitrile butadiene styrene (ABS)	–	1.03	0.037	6.00	.30	5.0	11.00	0.35–0.40	10.00	3–6	R107–115	HB
	Polyphenylene oxide (PPO)	–	1.10	0.039	7.80	.38	50.0	12.80	0.33–0.40	12.00	5.00	R115	V-1
	Polystyrene acrylonitrile (SAN)	–	1.05	0.038	9.50	.40	.5	14.00	0.55	14.00	0.30–0.45	M80–85	HB
	Polyester (PBT)	–	1.31	0.047	8.20	.28	50.0	12.00	0.33–0.40	8.60	.80	M68–78	HB
	Polyester (PET)	–	1.34	0.048	8.50	.40	50.0	14.00	0.35–0.45	11.00	0.25–0.65	M94–101	HB
Metals	ASTM A-606 HSLA steel (cold rolled)	–	7.75	0.280	65.00	30.00	22.0	–	–	65.00	–	B80	–
	SAE 1008 low-carbon steel (cold rolled)	–	7.86	0.280	48.00	30.00	37.0	–	–	48.00	–	B34–52	–
	AISI 304 stainless steel	–	8.03	0.290	80.00	28.00	40.0	–	–	80.00	–	B88	–
	TA 2036 aluminium (wrought)	–	2.74	0.099	49.00	10.20	23.0	–	–	49.00	–	R80	–
	ASTM B85 aluminium (die cast)	–	2.82	0.102	48.00	10.30	2.5	–	–	48.00	–	Brinell 85	–
	ASTM AZ91B magnesium (die cast)	–	1.83	0.066	33.00	65.00	3.0	–	–	33.00	–	Brinell 85	–
	ASTM AG40A zinc (die cast)	–	6.59	0.238	41.00	10.90	10.0	–	–	41.00	–	Brinell 82	–

*Shore D

Table 16.4 Compatibility of materials and processes for fibreglass composites

Process	Thermosets					Thermoplastics									
	Polyester	Polyester SMC	Polyester BMC	Epoxy	Polyurethane	Acetal	Nylon 6	Nylon 6/6	Polycarbonate	Polypropylene	Polyphenylene sulphide	ABS	Polyphenylene oxide	Polystyrene	Polyester PBT
Injection moulding	•		•	•	•	•	•	•	•	•	•	•	•	•	•
Hand lay-up	•			•											
Spray-up	•			•											
Compression moulding	•	•	•	•							•				
Preform moulding	•			•											
Filament winding	•			•											
Pultrusion	•			•											
Resin transfer moulding														•	•
Reinforced reaction injection moulding	•			•	•		•								

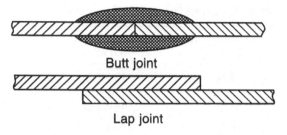

Butt joint

Lap joint

Figure 16.23 Bonded joints

Figure 16.24 Mercedes McLaren SLR body-in-white

Figure 16.25 Carbon fibre composite front nose of Mercedes McLaren SLR

Table 16.5 Resin comparison (*Courtesy of SPSystems*)

Material	Advantages	Disadvantages
Polyester	Easy to use; lowest cost resin	Only moderate mechanical properties; high styrene emissions on open moulds; high cure shrinkage; limited range of working times
Vinylester	Very high chemical/environmental resistance; higher mechanical properties than polyesters	Postcure generally required for high mechanical properties; high styrene content; double cost of polyester; high cure shrinkage rate
Epoxy	High mechanical and thermal properties; high water resistance; long working times available; temperature resistance up to 220°C; low cure shrinkage	Critical in mixing; corrosive; dangerous to handle; triple cost of polyester

Pre-impregnated material (pre-preg)

Woven material is available pre-impregnated with resin. It is referred to as pre-preg. This means that the material has exactly the right amount of resin applied to it. The resin is fully coating the material so that there are no dry spots which could lead to component failure. Pre-preg is therefore quicker to use, and the resin density is accurate.

Pre-preg has a limited shelf life which is compounded by the fact that it must be stored at –18°C. A deep-freeze cabinet is therefore

Table 16.6 Properties of reinforcement fibres and other materials (*Courtesy of SPSystems*)

Material type	Tensile strength (MPa)	Tensile modulus (GPa)	Density (g/cc)	Specific modulus
Carbon HS	3500	160–270	1.8	90–150
Carbon IM	5300	270–325	1.8	150–180
Carbon HM	3500	325–440	1.8	180–240
Carbon UHM	2000	440+	2.0	200+
Aramid LM	3600	60	1.45	40
Aramid HM	3100	120	1.45	80
Aramid UHM	3400	180	1.47	120
Glass – E glass	2400	69	2.5	27
Glass – S2 glass	3450	86	2.5	34
Glass – quartz	3700	69	2.5	31
Aluminium alloy (7020)	400	1069	2.7	26
Titanium	950	110	4.5	24
Mild steel (55 Grade)	450	205	7.8	26
Stainless steel (A5–80)	800	196	7.8	25
HSS steel (17/4 H900)	1241	197	7.8	25

Figure 16.26 Front nose attached to main tub of Mercedes McLaren SLR with aluminium frame

needed for storage. The pre-preg cannot be unrolled nor cut when it is in the frozen state, so it must be removed from the freezer and brought up to normal room temperature. It is only possible to freeze and defrost the pre-preg a limited number of times, so the material must be managed carefully. The usual way to do this is by means of a control card. The dates and times of defrosting are recorded, as is the amount of material taken off the roll. That way the life of the roll and the amount of material left can be seen without removing the roll from the freezer.

Curing

The resin, whether it is by wet lay-up or pre-preg, needs time and heat to dry it out and make it hard. When the hardener is added to the resin it will generate heat chemically. Be careful – this heat can cause fire and other damage. However, at normal temperature, 20°C, it will take about 5 days for the resin to become fully hard. During this time period the component should not be moved nor should any stress be applied. To speed up the hardening process and to add extra strength to the component it is normal to use an oven. The oven may be a simple box with a heating element or an autoclave, which is a cylindrical shaped oven that can be pressurized or evacuated inside. The normal procedure is to place the newly made component in the oven or autoclave, then rack up the temperature gently, over a period of about 30 minutes. Maintain the temperature typically at 150°C for about 5 hours, then gradually lower the temperature, again over about a 30-minute period. The best way to do this is with a computer control system.

Core materials

Engineering theory tells us in most cases that the stiffness of a panel is proportional to the cube of its thickness. That is, the further apart that we can keep the outer fibres the stiffer the panel will be. Putting a low-density core between two layers of composite material will add stiffness with minimum weight and at reasonable cost.

Foam

A variety of materials are used. One of the most common is foam. Foam can be made from a variety of synthetic polymers. Densities of foam can vary between 30 and 300 kg/m^3, and thicknesses available are from 5 to 50 mm.

Honeycomb

Honeycombs are made from a variety of materials, including extruded thermoplastic–ABS, polycarbonate, polypropylene and polyethylene-bonded paper, aluminium alloy and, for fire-resistant parts, Nomex. Nomex is a paper-like material based on Kevlar fibres.

Heat

A point to be noted is that most carbon fibre materials are affected by heat. Thermal expansion can lead to micro-cracking.

A carbon fibre panel which is painted black will absorb a lot of heat if left in the sun for a long period. This can cause the panel to expand, which could lead to micro-cracks in the panel and cracks in the paint work. This will then allow in moisture, which will cause further deterioration of the panel.

Repair of reinforced composite bodies

The repair of reinforced composite bodies and component panels is not difficult, and unless major damage has been sustained it can quite readily be carried out by a competent body repair shop. If the correct materials and tools are used and operations are carried out in the correct sequence, repair is usually considerably easier than to a metal body damaged in an accident of similar severity. The equipment required is simple, the only two items not generally found in the average workshop being a split-washer roller and a graduated measuring cylinder. Power tools can speed preparation, but hand tools will be just as effective. Much of the success of the repair depends on the correct preparation of the resin mix. If polyester resin is supplied from a manufacturer, the correct proportions of resin, catalyst and accelerator will be stipulated. In calculating the amount of resin required for the repair, first weigh the cut patches of chopped-strand mat and activate three times this weight of resin. If a

Table 16.7 Synthetic polymers and their characteristics for use in foam making

Foam material	Abbreviation	Characteristics
Polyvinyl chloride	PVC	Good resistance to water; available in plain sheet or grid scored
Polystyrene	PS	Very light; low mechanical strength
Polyurethane	PU	Good for thermal and acoustic insulation; can be used at up to 150°C
Polymethyl methacrylamide	Acrylic	Very strong, but expensive
Polyetherimide	PEI	Good fire resistance, used for interior trim on public carrying vehicles
Styreneacrylonitrile	SAN	High impact strength

large repair is being attempted, do not prepare more than half a pound (0.2 kg) of resin at a time to avoid the mix curing before it can be used.

The first essential when considering the repair of a reinforced composite moulding is to ensure that the area to be repaired is clean and dry, including the rough edge of any torn portion. In many cases a vehicle with minor damage will be driven back from the site of the accident and road moisture and dirt is deposited on the damaged area. The dirt must be washed off and the area dried using some convenient form of heating, but care must be taken that the moulding is not further damaged by too high a temperature. The area of the damage must then be checked and marked for cutting. It is usually found that any break in a moulding is surrounded by an area of bruising where the resin is crushed, and this must be removed. It is unusual for this to extend more than about 8 cm from a break, and the easiest way of checking the exact limits is to shine a powerful lamp through the laminate. The bruising will then show up as a dark or light patch depending on the colour and content of the laminate, and it can be marked out for removal.

After all damaged areas have been removed, the cut edges should be feathered on the non-weathering side and then the surface roughened for about 2 inches back from the cut. A single layer of chopped-strand mat and polyester resin is then laminated into the roughened area. While the area is still wet, a sheet of cellophane large enough to cover this material and the hole is applied and pressed into the uncured resin. The cellophane is then supported by sheet metal, hardboard or card which can be fastened to the moulding with adhesive tape and struts fixed to hold it to contour. The hollow is then almost filled by applying polyester resin and chopped-strand mat, continuing to build it up until it is nearly flush with the original surface. The laminate is allowed to cure and then finished off flush and smooth with a standard two-compound gap filler.

Accident damage assessment

All damage to the body is covered by the following categories:

- Highly stressed
- Moderately stressed
- Lightly stressed.

On that definition depends the original construction and therefore the repair method to be employed.

As a general rule there should be a bonded joint wherever two panels touch, or wherever they close on important points. It is usually possible to check these bonds both visually and physically for fractures and breaks. Ascertain the cause of damage and the direction of impact and examine all panels or bonds which may have been affected.

A front-end impact, for example, may easily cause the bonds at the bulkhead to split without the defect being easily visible. To facilitate a closer examination, it may be necessary to remove parts and mechanical components to determine the extent of the damage. When determining the replacement underbody sections and panels to be ordered, make sure the new underbody section and panels will be attached to firm composite materials: avoid badly crazed areas and badly burnt areas.

Fire damage is the most difficult to assess, but generally only the obviously burnt or charred sections will need to be replaced or reinforced.

After a moderate or severe accident, checks should be carried out on the A- and B-posts, chassis datum points and suspension and wheel alignment.

Accident repair

For repair purposes, accidents may be defined as

- Front section
- Centre section
- Rear section.

This definition determines the original construction and therefore the repair method employed.

Resin and gel mix

The panels on GRP vehicles are manufactured using different types of materials. Therefore, when carrying out repair, the type of resin and gel mixes used will depend on the original materials used in manufacture.

Superficial defect repairs

Pinholes or air voids – Either drill or rout out so as to leave a larger hole with near-vertical walls or enlarge by gouging or picking out. Then fill the holes with polyester stopper or filler.

Surface crazing – There are various causes of surface crazing, but most results from sharp impacts or accidental damage. During an accident some panels may flex sufficiently to cause the surface to craze without immediate apparent damage to the painted surface. The crazing may not work its way through the paint surface for some weeks, so it may be necessary when assessing accident damage to carefully examine all panels, particularly near cracks or split bonds. In case of doubt it may be possible to promote the appearance of the crazing by applying gentle heat.

Surface crazing itself generally stops at the first layer of glass fibre and is consequently not structurally serious. However, extensive crazing near damaged areas should be taken as an indication of over-stressing and the panel should be reinforced or replaced. It is not possible to remedy crazing by simply resurfacing with a further layer of resin.

Replacement sections and panels

Where the repair of a damaged vehicle calls for replacement sections and panels, they may be obtained in most cases direct from the manufacturer of the vehicle under repair.

Where severe damage has been sustained, the damaged section or panel may be cut away and a replacement section or panel grafted in. Before cutting away any damaged section or before ordering replacements, the repairer should ascertain the proposed method of repair and the positioning of joint lines. Determine a method of correct positioning of replacement panels and sections such as B-posts, A-posts or any prominent feature from which measurements can be made, and scribe these clearly onto the section which is to be used. Use a marker to define the lines on which it is proposed to cut the section or panel. Study these lines to see that any damaged section or panel removed will allow adequate pick-up points on the replacement section to ensure it can be accurately positioned and that the proposed outline transverses longitudinal, lateral and horizontal definition points to assist easy lining up of the new section or panel in all three planes.

Common faults in moulded laminates

Many complaints concerning the appearance and performance of these mouldings stem from the basic cause that the resin is under-cured. There are, however, several problems in the form of visible flaws or other defects, the remedies of which will become apparent from the analysis of the causes.

Wrinkling

This is caused by solvent attack on the gel coat by the monomer in the laminating resin due to the fact that the gel coat is under-cured. Wrinkling can be avoided by ensuring that the resin formulation is correct and the gel coat is not too thin, and by controlling temperature and humidity and keeping the work away from moving air, especially warm air. If the workshop is equipped with hot-air blowers, these should be directed away from the moulds.

Figure 16.27 Wrinkling (*Scott Bader Co. Ltd*)

Figure 16.28 Pinholing (*Scott Bader Co. Ltd*)

Pinholing

Surface pinholing is caused by small air bubbles which are trapped in the gel coat before gelation. It occurs when the resin is too viscous or has a high filler content or when the gel coat resin wets the release agent imperfectly.

Figure 16.29 Poor adhesion of the gel coat (*Scott Bader Co. Ltd*)

Poor adhesion of the gel coat resin

Unless the adhesion of the gel coat to the backing laminate is very poor, this defect will be noticed only when structure is being handled and pieces of gel coat flake off. Areas of poor adhesion can be detected sometimes by the presence of a blister or by local undulations in the surface when it is viewed obliquely. Poor gel coat adhesion can be caused by inadequate consolidation of the laminate, by contamination of the gel coat before the glass fibre is laid up or, more generally, by the gel coat being left to cure for too long.

Figure 16.30 Fibre pattern (*Scott Bader Co. Ltd*)

coat is too thin or when the reinforcement has been laid up and rolled before the gel coat has hardened sufficiently or when the moulding is removed too early from the mould.

Fibre pattern

The pattern of the composite reinforcement is sometimes visible through the gel coat or prominently noticeable on its surface. This usually occurs when the gel

Fish eyes

On a very highly polished mould, particularly when silicone-modified waxes are used, the gel coat is almost non-existent.

Figure 16.31 Severe blisters (*Scott Bader Co. Ltd*)

This shows up as patches of pale colour usually up to 6 mm in diameter. It can also occur in long, straight lines following the strokes of the brush during application. This fault is rarely experienced when a PVAL film is correctly applied.

Blisters

The presence of blisters indicates that there is delamination within the moulding and that the air or solvent has been entrapped. Blisters which extend over a considerable area may also indicate that the resin is under-cured, and this type of blister may not form until some months after moulding. Blisters can also occur if the moulding is subjected to an excessive amount of radiant heat during cure. A possible cause of this defect is the use of MEKP rather than cyclohexanone peroxide paste. If, on the other hand, the blister is below the surface, the cause is likely to be imperfect wetting of the glass fibre by the resin during impregnation. This would be due to the fact that insufficient time had been allowed for the mat to absorb the resin before rolling. Blisters of this kind can usually be detected by inspection as soon as the moulding has been removed from the mould.

Crazing

Crazing can occur immediately after manufacture, or it may take some months to develop. It appears as fine-hair cracks in the surface of the resin. Often the only initial evidence of crazing is that the resin has lost its surface gloss. Crazing is generally associated with resin-rich areas and is caused by the use of an unsuitable resin or resin formulation in the gel coat. The addition of extra styrene to the gel-coat resin is a common cause. Alternatively the gel-coat resin may be too hard with respect to its thickness. In other words, the thicker the gel coat the more resilient the resin needs to be. Crazing which appears after some months of exposure to the weather or chemical attack is caused by either under-cure, the use of too much filler or the use of a resin which has been made too flexible.

Star cracking

This is the result of having an over-thick gel coat and occurs when the laminate has received a reverse impact. Gel coats should never be more than 0.5 mm thick.

Internal dry patches

These can be caused by attempting to impregnate more than one layer of mat at a time. The presence of internal dry patches can be readily confirmed by tapping the surface with a coin.

Figure 16.32 Crazing (*Scott Bader Co. Ltd*)

Figure 16.33 Internal dry patch (*Scott Bader Co. Ltd*)

Figure 16.34 Star cracks (*Scott Bader Co. Ltd*)

Leaching

This is a serious fault. Leaching occurs after exposure of the laminate to the weather, and is characterized by a loss of resin from the laminate, leaving the glass fibres exposed to attack by moisture. Leaching indicates either that the resin used has not been adequately cured or that it is an unsatisfactory resin for that particular application.

Cracking, shrinking and discolouring

The identification of this fault is visual. Deep cracks appear in the resin, and the colour changes from a green to a mauve purple, and it is hot to the touch. The cause of this

fault can be large build-up of resin due to drainage or to excess application; both lead to extreme exotherm. Alternatively, incorrect amounts of catalyst or accelerator may be used, usually an excess. The effect of the build-up of exotherm is to cause excessive shrinking of the moulding and internal stress. The extreme exotherm can also damage and distort the mould itself. This can be prevented by avoiding large amounts of resin (but where this is necessary the build-up should be gradual), using the correct amount of ingredients, being aware of any increase in workshop temperatures and noting any variations in the correct percentages of catalyst or accelerator used.

Low-rigidity laminate

The first identification is by touch; the laminate will feel spongy and more flexible than usual. Alternatively, check by applying a Barcol hardness tester. The common causes of this fault are low resin content or under-cure of the resin. To prevent this fault, ensure the correct ratio of resin to glass and eliminate any draughts. Large areas should be made to gel quicker to cut down styrene loss, and correct proportions of catalyst and accelerator must be used.

Safety precautions

The handling of polyester resin, glass fibre and ancillary materials such as catalysts presents several hazards which can be reduced to a minimum if the correct precautions are taken. Most glass fibre materials and resins are perfectly safe to use provided the potential hazards are recognized and reasonable precautions are adopted. Normally you will have no problems if you follow these rules:

1 Do not let any materials come into contact with the skin, eyes or mouth.
2 Do not inhale mist or vapours, and always work in a well-ventilated workshop.
3 Do not smoke or use naked flames in the workshop.

Storage precautions

Liquid polyester resins are flammable but not highly flammable, most of them having a flashpoint of 31°C. Resins and accelerators should preferably be kept in a brick-built store conforming to the normal fire regulations for a paint store. The storage life of polyester resin is about 6 to 12 months provided the resin is kept below 20°C in the dark (in metal drums). Storage at higher temperatures, even for only a few days, will considerably reduce the shelf life.

Catalysts are organic peroxides and present a special fire hazard. They should be stored in a separate area, preferably in a well-ventilated fire-resisting compartment. If kept reasonably cool they will not burn or explode. In case of fire in the vicinity, they can be kept safe by drenching the containers with water.

Operating precautions

Most polyester resins contain monomeric styrene, which is a good grease solvent and may cause irritation to the skin. The most effective method of protecting hands is the use of a barrier cream or rubber gloves, and this is strongly recommended. Resin can be removed from the hands with proprietary resin-removing creams or with acetone followed immediately by a wash in warm soapy water. In sufficient concentration styrene vapour is irritating to the eyes and respiratory passages, and therefore workshops should be well ventilated. When resin is sprayed a gauze mask should be worn to protect the mouth and nose. This also applies to trimming operations, during which resin and glass dust can cause irritation.

Catalysts are extremely irritating to the skin and can cause burns if not washed off immediately with plenty of warm water. Particular care must be taken with liquid catalysts to avoid splashing, spilling or contact with the eyes. Protective goggles should be worn as a necessary precaution.

Workshop conditions

The building should not be damp, and it should be adequately heated and ventilated. Good head room is desirable, and sufficient space should be allowed for all operations. The floor area should be divided into sections as follows: preparation of reinforcement, mixing of resins, moulding, trimming and finishing. Resin and curing agents should be stored away from the working area in a cool place, observing the necessary precautions for flammable liquids and keeping in mind the special hazards associated with organic peroxides. Glass fibre should be stored and tailored under dry conditions and separately from the moulding area. The temperature of the building should be controlled between 15°C and 25°C. Ventilation should be good by normal standards, but draughts and fluctuations in temperature must be avoided, so doors and windows should not be used for ventilation control. Dust extraction in the trimming section should be of the down-draught type. Cleanliness is important both for health of the operators and for preventing contamination of resin and reinforcement. As far as possible, health and comfort depend in the first place on planned extraction, and in the second place on workshop education in the nature of the materials used. Almost all the offence comes from the styrene vapour and glass filaments, both of which advertise their presence before the concentration reaches a danger level. As far as is known the only real source of physical harm is the dust produced in grinding, but all the materials and by-products contribute to discomfort, and sensible evasive action is essential.

Spillage and disposal

Most of the following products are covered by the terms of the Deposit of Poisonous Wastes Act:

*   *Polyester resin* – Absorb spillages in dry sand and dispose by landfill or controlled incineration.

- *Furane resin* – Extinguish all naked lights, open doors and windows. Absorb spillage into sand or chalk. Pack into drums, seal and store prior to collection by specialized chemical disposal company.
- *Catalyst* – Absorb into vermiculite, remove to landfill or controlled incineration. Wash down remaining traces with copious water.
- *Accelerator* – Absorb into dry sand and dispose by landfill or controlled incineration.
- *Release agents* – Wash down with water.
- *Mould cleaner* – Absorb into sand or earth, remove to landfill or controlled incineration. Flush contaminated area with water.

Fire hazards

Many resins and associated products either are flammable or contain flammable additives. Styrene, catalyst and acetone are particularly dangerous. Do not smoke or use naked lights, oil burners or similar heating devices in the working area. If a fire does start, do not attempt to put it out with water unless it is a catalyst. Dry-powder extinguishers can be used on accelerator, mould cleaner, acetone, resins and release agents.

Fires can be started if catalysed but uncured resins are thrown away. The waste resin will continue to cure, and the heat generated by the curing process can ignite other waste materials. Therefore unwanted resin should be left in a safe place until it is fully cured; it can then be discarded without risk of fire.

Skills and questions

Working with reinforced composite products is a specialized job. Some colleges, private training providers and manufactures put on special courses for working with these materials. If you are looking to work in this field, you would be well advised to enrol on a specialist course to get the necessary hands-on training.

1 Explain the following abbreviations: RRIM, VARI, PVAL.
2 State three advantages of reinforced composite materials when used in vehicle body construction.
3 List the main physical properties of glass fibre composite materials.
4 Explain the advantages and disadvantages of reinforced composite materials when used as an alternative to low-carbon steel.
5 Explain briefly the function of (a) the releasing agent, (b) the gel coat, (c) the catalyst.
6 Describe the process of contact moulding.
7 Explain the purpose of pre-accelerated resins.
8 With the aid of a sketch, show the lay-up of a laminate in the mould.
9 Describe the type of tools that would be used to trim the edges of reinforced laminates.
10 Describe the sequence of repair to a damaged glass fibre composite body panel.
11 Describe how a patch mould can be used during the repair to damage of a GRP laminate.
12 Describe the likely damage that would occur to a panel made from GRP and which had been subjected to a heavy blow.
13 Name the types of materials that could be used to reinforce polyester resin.
14 Name three reinforcing materials that could be added to a GRP moulding to give strength to the laminate.
15 Explain why a GRP panel may offer better resistance to minor damage when compared with a low-carbon steel panel.
16 In the automobile industry, why is GRP limited in use to a small specialist sector?
17 Describe, with the aid of a sketch, a test to show that GRP is more elastic and less ductile than aluminium sheet.
18 State the reasons why GRP has not replaced low-carbon steel as the material used to manufacture vehicle bodies.
19 State the reasons why certain manufacturers of sports cars prefer to make their vehicle bodies from GRP.
20 Explain how to repair a deep scratch in a GRP body panel.

17

Automotive finishing and re-finishing

The method by which the priming coat on vehicle body shells is applied in the factory is known as **electro-deposition**. A large dip tank containing about 2500 litres of a water-borne paint is included in the production line. An overhead conveyor carries the body shells from the pre-cleaning area to the dip tank. The paint is charged with electricity and the shell is earthed through the conveyor. The thinner of the paint, being water, acts as an electrolyte; the paint solids, i.e. pigment and binder, are ionized and are attracted to the earthed car body. An even coating of paint is thus applied, even on thin metal edges. The thickness of the coating can be varied according to the electrical potential introduced. When the car body moves out of the tank, surplus paint drains out of it and the shell is then rinsed off under sprinklers, which does not affect the electrodeposited coating. The car body is then dried off and baked.

Glossary of terms used in spray painting

In order to be able to appreciate more fully the descriptions of processes and practices in the paint shop, you may like to learn some of the trade terms and names of equipment.

- *Air delivery* – The actual volume of compressed air delivered by the compressor after making allowances for losses due to friction. It is measured in litres per second.
- *Air duster* – A tool which, when fitted to an air line, is useful for blowing water from recesses and for drying a surface quickly prior to painting.
- *Air pressure* – The pressure of air which has been mechanically compressed. It is measured in bars or pounds per square inch (psi).
- *Air receiver* – A reservoir or storage tank to contain compressed air.
- *Atomization* – The breaking up of paint or other materials into very fine particles. Good atomization is essential in spray painting.
- *Double-header coating* – This results from the practice of spraying one coat immediately after another without allowing a flash-off period.
- *Dry coating* – Several thin coats of paint can be applied fairly rapidly if they are sprayed 'dry'. This is done by increasing the ratio of atomizing air to paint at the spray head of the gun. Dry coating is particularly useful when carrying out local repairs to paintwork.
- *Evaporation* – The solvents or thinner added to paints to reduce the viscosity evaporate during the drying process, and consequently some of the liquid content of the paint vanishes. When this happens the paint film becomes thinner and projections on the substrate come through it.

The Repair of Vehicle Bodies, Andrew Livesey, 9780815378709, 2019, Taylor & Francis

Figure 17.1 Electro-deposition of priming coat

- *Feather edging* – The rubbing down of a damaged area of paintwork until there is no perceptible edge between the paint and the substrate.
- *Feathering the gun* – To ease the pressure on the gun trigger whilst spraying, thereby reducing the volume of paint passing through the fluid tip. This is done mainly when spraying local areas to achieve a feather edge.
- *Flash off* – To allow the greater part of the more volatile solvents in a sprayed coat of lacquer or enamel to evaporate before proceeding with the application of another coat or with stoving.
- *Fluid cup* – A container for the paint attached to the spray gun in conventional spray painting, or a separate item in pressure-feed systems connected to the spray gun by means of a fluid hose.
- *Fluid nozzle* – The orifice in the fluid tip.
- *Ground coats* – The paint coats between the primer and finishing coats. Ground coats are usually of a similar colour to the enamel.
- *Guide coat* – A thin coating applied as evenly as possible over a surface to be rubbed down. Following rubbing down, no trace of the guide coat should remain, so that complete flattening of the surface is achieved. The guide coat should obviously be of a contrasting colour to that over which it is applied.
- *Hold-out* – The degree of imperviousness of a dried paint film. Some filler coatings in particular are porous and so

they tend to absorb the binder or medium of finishing coats, thus reducing their effectiveness as a glossy finish.
- *Matt finish*– A surface finish which has no glossy effect.
- *Shrinkage* – This refers to the manner in which some paints decrease in size not only vertically but also horizontally. As with sinkage, this is caused by solvent evaporation. Nitrocellulose materials are particularly prone to this phenomenon, which can affect the adhesion to the substrate.
- *Sinkage* – This trade term can have two interpretations. When paint is applied over a particularly porous surface it will sink into it, and if this paint is a finishing material the gloss will be impaired. The other explanation of the term brings in solvent.
- *Tacking off* – To wipe over a surface with a specially treated cloth, which is slightly sticky, to remove dust. Tacking off is essential before applying finishing coats.
- *Viscosity* – The degree of resistance to flow of a liquid. More simply, it refers to the thickness or otherwise of a fluid such as paint.

Basic composition of paint

- *Pigments* – Fine solid particles which do not dissolve in the binder. They give colour and/or body to the paint. Some pigments possess good anti-corrosive properties and are used in paints designed to give protection to the substrate. Extenders are cheaper than pigment, but when used in the correct proportions they carry out many useful functions such as improvement of adhesion and ease of sanding.

- *Binder* – Reacts to form a film and binds the pigments together and to the surface. The binder is often referred to as the medium of the paint.
- *Thinner* – Some of the liquid of the paint is often withheld from the paint container and supplied separately as a thinner. The user adds thinners to adjust the viscosity to suit his requirements.
- *Additives* – Small quantities of substances which are added to carry out special jobs. Wax in varnish creates a matt finish, and silicones in metallic paint give a hammer finish.

Types of paint

Cellulose synthetic

This dries by the evaporation of the solvent. The main advantage of this material is rapid air drying. However, there are a number of disadvantages. The coating dries rapidly only when thin films are applied; otherwise drying is delayed by solvent retention. The high proportion of solvent used (60 per cent in most cases) results in shrinkage, which causes the film to adhere poorly to the substrate. The absence of chemical change means that the dried film does not increase in chemical resistance and is readily softened by the original solvent.

Oil paints

The drying of an oil paint depends on the ability of certain drying oils to dry by a reaction that involves atmospheric oxygen, a process which is confined to relatively thin films.

Synthetic paints

These are mixtures of drying oils and synthetic resins. The most obvious limitation of a paint based solely on a drying oil is slow drying. To improve this property and to give tougher films and improve the gloss, a resin is added to the oil, and they are cooked together for a period so that they chemically combine. The varnishes produced can be divided into two main classes based on their oil-to-resin content: long oil and short oil.

Stoving paints

These are also mixtures of oils and resins that require exposure to an elevated temperature to produce a cure (dried film). The time of exposure is mainly dependent on the temperature: 60 minutes at a temperature of not less than 138°C and 10 minutes at a temperature of not less than 205°C.

Figure 17.2   Composition of paint

Blacking paints

Chassis black is a cheap black paint generally based on bitumen. It has good adhesive qualities on bare metal and is a good rust inhibitor.

Tyre black is also a cheap black paint, being of low viscosity. Several proprietary brands are available.

Two-pack paints

These are probably the most widely used paints in the vehicle refinishing trade, with more than 80 per cent of refinishers preferring them. They present special health hazards, and the user should be equipped with an air-fed mask and face visor to prevent inhalation of the vapours when spraying. A canister mask of the CC type can be used as an alternative, but these can prove to be expensive, as the canister is only useful for 15 minutes' continuous use and should then be discarded. Precautions should also be taken to prevent the mixed material or spray vapours making contact with the skin, as this can cause dermatitis.

These paints consist of a base material and a catalyst or activator. When they are mixed together, a chemical reaction takes place which results in complete polymerization. Two-pack (or 2K) paints have a limited pot life after mixing, but when curing is complete they can equal stoving paints in hardness and durability. They are characterized by high solids content and low solvent content, which results in high build and good scratch filling with the minimum number of coats, thus resulting in savings on labour time and overspray wastage.

The gloss from the gun is good, and no burnishing or polishing is needed unless dirt is present in the finish. Should this be the case, the finish can be wet flatted with P1200 paper, using soap as a lubricant, and burnishing can be carried out using fine rubbing compound and a polishing machine of 6000 rev/min. A clean, dry lambs-wool mop is recommended for best results.

Materials included in this group are acrylic and polyurethane primer undercoats and finishes, including base-coat-and-clear finishes, epoxy resin primers and finishes and polyester spraying fillers.

It is common practice in most refinishing paint shops to force dry these materials using low-bake ovens on large areas for 30 minutes at 60°C or using infrared lamps on small areas.

Low-bake finishes

These are modified stoving paints which can be completely cured at a temperature between 66 and 93°C. The material was formulated for the refinisher to enable him to match more closely the original finish of the car manufacturers. These are now losing favour to two-pack materials.

Materials used in refinishing

Primers and dual-purpose primers

Primer
The priming coat is the first coat of paint on any surface. Its functions are to gain maximum adhesion to the substrate, to provide a sound base for subsequent coatings and, on metals, to act as a corrosion inhibitor. The priming paint should be selected to suit the type of surface onto which it is to be applied.

Etch primer
The best types of etch primer are two-pack materials, the base and activator being supplied separately. They are mixed 10 minutes before use but have a limited pot life (about 6 hours), although long-life etch primers are now available. Brushing activators are available for those shops where spray painting is not practical. Etch primers have a fairly high water absorption characteristic and should be coated with surfacer or filler after the appropriate drying time to avoid moisture absorption from the surrounding

atmosphere. Special thinners are provided for etch primers and should be kept for this purpose only.

The pigment for this type of paint is zinc chromate, which makes it an ideal primer for aluminium, although it can be used also with good effect on most other metals. The activator contains phosphoric acid which etches the surface, thereby ensuring good adhesion. An extremely thin coating gives best results. Once the base and activator have been mixed together they should on no account be returned to the tin of the base.

Primer surfaces

A primer surfacer does the work of both a primer and a surfacer.

Self-etching primer surfacers are available which eliminate the need for using an etching primer and then coating over with surfacer or filler. Self-etching primer surfacers have gained in popularity, the main advantage being improved adhesion to metal substrates when compared with standard primer surfacers.

Primer filler

A primer filler is similar in function to a primer and a filler.

Polyester primer filler is a two-pack paint, first used extensively on the European continent. When introduced into Britain it was viewed with a certain amount of distrust, as the claims made for it (primer, stopper, surfacer rolled into one) appeared too good to be true. There was good cause for this distrust, as one of the difficulties encountered was that blistering occurred when it was used with nitro-cellulose materials. Another factor against it was that it required 4 hours to cure. However, modifications have been made to it which have resolved the problem of blistering and reduced the curing time to 2 hours at an ambient temperature of 25°C, which can be still further reduced to 15 minutes when force dried.

This paint is particularly suitable for use on rough surfaces where heavy coatings are required to level up the surface with the minimum of effort. Several coatings can be applied wet-on-wet as it dries by catalyst action. The amount of sinkage after drying is virtually nil. Polyester primer/filler possesses exceptionally good build qualities and when dry can be very easily smoothed down with abrasive paper used either wet or dry. Facemasks should be worn when dry rubbing is carried out.

Best results are obtained if the paint is applied with a gravity-fed spray gun. When spraying local repairs the paint can be applied at the supplied viscosity, but for a large area a small quantity of the appropriate thinner may be added to obtain better atomization and to provide a smoother coating. Spray guns must be thoroughly cleaned immediately after being used with polyester primer filler. As the paint solvent is extremely volatile, the cap must be screwed down tightly on the tin when not in use. Polyester primer filler can be used with most car refinishing processes and is very useful when used with low-bake enamels.

Fillers and levellers

Surfacer

A surfacer is applied over the primer. Its function is to build up the coating thickness whilst filling up minor defects such as scratches.

Filler

Filler is a heavy-bodied material used for levelling defects which are too deep to be filled economically with surfacer. Fillers are manufactured to suit the method of application: there are spraying, brushing and knifing types. Fillers are available as both single-pack and two-pack, cellulose and synthetic types.

Stopper

Stopper is a putty-like substance used for filling up defects too deep for satisfactory levelling with either filler or surfacer.

Though deep indentations are normally filled up by the body repair worker, the painter is sometimes required to carry out levelling work. A two-pack stopper, usually based on polyester resin and a catalyst, is used for this purpose. It dries rapidly in heavy layers, unlike cellulose or oil-based stoppers, which must be applied in thin layers with a drying period between applications. Polyester stopper is intended for use on bare metal or over high-baked primers. It cannot be used as an inter-coat stopper over etching primers or between enamel coats. It can, however, be coated over with most standard paint systems used in refinishing, including cellulose synthetic, coach finish and low-bake synthetics. The normal curing time of a polyester stopper is about 30 minutes at 20°C.

Sealers

There are three types of sealers used by the refinisher:

- standard
- isolators
- bleed inhibitors.

Standard sealers have a low pigment and high binder content. They are supplied ready for use and are applied over the final coat of surfacer or spray filler to provide hold-out of the finishing material and promote higher gloss. They also reduce the risk of crazing when applying acrylic lacquer-type finish over cellulose-based undercoats.

Isolating sealers are more heavily pigmented and are used to avoid a reaction between different types of paint systems, for example when applying a lacquer-type paint over a synthetic enamel.

Bleed-inhibiting sealers contain carbon black pigment, which is able to absorb floating colour. These sealers are recommended when carrying out a colour change over a colour which is suspected of being a bleeder, that is, some of the pigment or dyestuff will float into the new coating and discolour it. Several reds, particularly those containing organic pigments, are prone to this behaviour. Whichever sealer is used, it can only do its job if it is a continuous film. For this reason it must not be flatted down, through a light denib, carefully done, is permissible.

Finish

Finish is the term used to describe the finishing colour coats. They have a comparatively low pigment content as opposed to surfacers and fillers. The high percentage of the vehicle or binder provides the glossy effect.

Abrasive papers

The type of abrasive paper mainly used by the spray painter is known as wet-or-dry paper. The abrasives used are silicon carbide and aluminium oxide, and these are attached to a treated paper backing by means of waterproof resin glue.

Wet-or-dry paper is available in various grades ranging from 80D (coarse) to P1200 (very fine): low numbers denote coarse grades and high numbers identify the finer grades. This type of abrasive paper is normally used in conjunction with water to avoid a build-up of paint particles, which would affect the abrading effectiveness of the paper.

In addition to wet-or-dry paper, the spray painter should have in his stock emery paper and production papers. These are much coarser than wet-or-dry papers and are used for rougher work such as removing rust or mill scale. These two papers are most effective when used with power tools in the form of circular discs, being attached to the pad with specially formulated disc adhesives. As they are normally used dry, the particles are attached to the paper or cloth backing with a good-quality hide glue.

Abrasive papers fall into two categories, based on the amount of space between the particles. If these are widely spaced the paper is referred to as open coated and is used to remove paint or rust, which both tend to fill up the spaces on finer grades.

When the particles are tightly packed the paper is known as close coated and is used to rub down smoother surfaces.

Masking

Masking tape

This is a paper tape, one side of which is coated with an adhesive of a non-drying composition. It is supplied in rolls in a variety of widths, but the most widely used are those measuring 20 mm and 25 mm. Wider tapes are considered to be uneconomical, though narrower tapes such as 13 mm do have limited uses. Where two-tone work is carried out, a finer edge can be achieved by the use of gummed paper than is possible with masking tape, but the time spent on removing it makes it less popular.

Masking machines are also available in which a roll of masking paper and various widths of masking tape are mounted. As the paper is pulled out of the machine, a strip of the tape is automatically attached to one of the edges.

Masking paper

Brown paper (kraft paper) is an ideal masking material, though newspaper is very widely used for this purpose. It should be

Figure 17.4 Method of fixing masking tape to paper

borne in mind that newspaper is extremely porous and subject to solvent penetration, and it is therefore recommended that at least two layers be used over existing paintwork. It is common practice in some shops to oil or grease the newspaper, but this cannot be recommended because of the risk of grease contamination to those areas to be sprayed.

Burnishing

Burnishing compound

There are both burnishing pastes and polishing liquids. These are generally emulsions of mineral oils and water with the addition of an emulsifying agent. They also contain mild abrasives to 'cut down' the final coat of enamel and promote a good lustre. Burnishing compounds are also used during the carrying out of local paint repairs to remove overspray. Standard compounds and creams or liquids contain ammonia to keep them fresh in the tin, but ammonia can cause staining of clear coatings. Special ammonia-free compounds and liquids are available for the burnishing of clear-over-base finishes.

Mutton cloth

In days gone by, butchers and slaughterers wrapped joints of meat in this material – hence its name. Mutton cloth is available in different **grades: coarse, medium and fine**. The coarser grades are used with rubbing compound to remove overspray when carrying out localized paint repairs. The fine grades are used with fine compounds and creams for final burnishing to promote a deep gloss.

Paper roll

Masking tape

Figure 17.3 Masking machine

Solvents

The paint shop should be well stocked with the appropriate solvents for the types of paint to be used. The importance of using the correct thinner for a particular type or make of paint cannot be over-stressed, as many painting defects can be traced to the incorrect use of solvents. Solvents can prove to be an expensive item, and in consequence a cheaper form of cleaning solvent should be stocked for the purposes of cleaning spray guns and equipment.

Solvents fall into two categories, high boilers and low boilers. Those which require a high temperature to bring them up to their boiling point tend to be slow in evaporating and consequently slow in drying. Low boilers, on the other hand, evaporate and dry quickly. Low boilers are best used with primers, surfacers and fillers where it is required to build up the coating thickness fairly rapidly. High boilers can be used in finishing coats to promote better flow, helping to eliminate an orange-peel defect. A good-quality solvent, however, will contain both high and low boilers in well balanced proportions.

Spray painting equipment

The items of equipment in a spray painting shop are basically as follows: air compressing unit, air line, air filter, pressure regulator, air hose and finally spray gun.

Air compressor

When selecting an air compressor for a particular workshop, one should calculate the volume of compressed air that will be required to operate the various tools throughout the workshop, such as rubbing-down tools and spray guns. The size of the compressor chosen should be capable of giving a higher free air delivery than is required.

In a workshop where more than one activity is carried out, such as panel beating and spray painting, it is advisable to install a stationary two-stage compressor. This should be bolted to the floor at least 300 mm from any wall and sited where it can receive an ample supply of clean, dry air. A two-stage compressor consists of an electric or fuel oil motor, the compression unit itself and a storage tank. In addition to these basic components there can also be various refinements in the way of safety devices and operating switches.

Taking a piston-type compressor as an example, the operational sequence is as follows. The motor drives the compressing pistons, which are situated within two cylinders set above the storage tank. Air at normal atmospheric pressure is drawn into the first cylinder via the air intake, to which is attached a filter. The air is then compressed to an intermediate pressure by the action of the piston moving upwards and reducing the volume of the cylinder. When a sufficient pressure has been attained, the air is pumped through a chamber (or intercooler) into the second smaller cylinder, where it is compressed even further by piston action and finally discharged into the storage tank for subsequent use.

The compressed air is stored in the receiver (or storage tank) for a cooling-off period, during which moisture and vapours within the air will condense and collect on the floor of the tank. The very act of compressing air will generate heat, and so the cooling period is essential. In a further effort to keep the air cool, the compression cylinders have external fins to provide a greater surface area for

Figure 17.5 Single-acting two-stage compressor

1. Compressor
2. Motor
3. Air receiver or storage tank
4. First and second stage valves
5. Pressure switch automatically controls volume of air in tank
6. Centrifugal pressure release on large models relieves motor starting against load
7. Safety valve
8. Starter
9. Drain cock
10. Safety guard
11. Air intake
12. Intercooler
13. Pressure gauge
14. Release cock

Figure 17.6 Principal parts of a stationary two-stage compressor

the heat generated within them to dissipate into the atmosphere. In addition to storing the compressed air and providing a cooling-off period, the air receiver acts as a buffer between the compressor and the spray gun, blanketing out pulsations so the compressed air can be drawn from it at a steady even pressure. It is important that the size of the air receiver be in direct proportion to the size of the compressor. Contrary to popular belief, air volume is more important to the spray painter than is air pressure. A typical suction-fed spray gun will require 3 to 4.5 litres of free air delivered (FAD) per second to allow it to operate satisfactorily. The gun must have this volume of air from the air receiver regardless of the pressure. In a typical two-stage piston-type compressor, the free air delivered will only be 70 to 75 per cent volumetric efficiency. Free air delivery is in consequence the prime factor to consider when purchasing a compressor.

Air lines

The compressed air is drawn from the air receiver and led to the air transformer through a galvanized tube known as an air line. The three main requirements of an air line installation are:

1 Low pressure drop between the compressor plant and the points of air consumption
2 Minimum of air leakage
3 High degree of contamination filtering throughout the system.

Pressure drop in an air line is caused by the frictional action of the air molecules on the inside surface of the pipe, as well as pressure build-up at angles. The greater the distance between the compressor and consumption points, the greater will be the pressure drop. For this reason the air line must be of sufficient internal diameter to carry a satisfactory volume of compressed air into the spray room. For the average refinishing shop an internal diameter of 25 mm will suffice. Pressure drops can be increased where angles are introduced into the system, and in consequence the number of angles should be kept to a minimum. Also the more couplings there are in the installation, the greater is the risk of air leakage, with a consequential drop in pressure.

It is inevitable that some moisture vapour will be drawn into the air line from the air receiver, and for this reason the air line should slope down towards the air receiver to enable condensed moisture to run into it rather than to the air transformer. The service line leading to the transformer should

Air from compressor

Drain cocks

Figure 17.7 Air line installation

be tapped from the top of the main air line and should preferably be in the form of a U-bend rather than elbow joints with right-angled bends on them.

Air transformer

This consists of two units: a condenser or filter and a pressure regulator. The *condenser* allows the compressed air to expand into a chamber, thus assisting cooling, and then removes moisture from it by means of a removable filter. A drain cock is situated at the base of the chamber to drain off accumulated impurities periodically. The *regulator* is a reducing valve with which to reduce the air pressure from the compressor to that required for spraying. The air transformer should be fitted with a pressure gauge, giving an accurate reading of the pressure of air passing through the regulating valve. A small lightweight filter can be fitted to the handle of the spray gun as an additional safeguard in conditions where exceptional humidity exists.

Air hose

The compressed air is led from the transformer to the spray gun by means of an air hose. This consists of a rubber tube covered with cotton braid enclosed within a rubber covering; the three layers are vulcanized into one. Multi-braid hoses are available for high-pressure work. At each end of the hose are couplings for attachment to the transformer and the spray gun.

Though the interior wall of the air hose is smooth, it will still create a certain amount of resistance to the flow of air, particularly when long lengths are used. Hoses with an interior diameter of 6 mm should never be used in lengths exceeding 4 m because of the high pressure drop encountered. For example, if the air transformer is delivering air at a pressure

of 3.5 bar (50 psi) the pressure drop over 4 m would be 0.7 bar (10 psi). However, using an air hose of diameter 8 mm over the same length and at the same supplied pressure, the drop would only be 0.2 bar (3.5 psi). From this it can be seen that a hose of 8 mm diameter will give best results on lengths exceeding 4 m. When working in a spray booth, the spray painter rarely, if ever, requires an air hose greater than 6 m in length.

Types of spray gun

Of all the tools and techniques used in paint shops, the spray gun and spray painting have provided the most satisfactory method of applying paint. Unless there is a complete change in the design and construction of the motor vehicle, the spray gun will be used for many years to come. The spray gun, like all tools, is only effective in the hands of a skilled operator, and therefore a painter should know as much as possible about what has become the main tool of the trade.

A spray gun is a precision instrument which uses compressed air to atomize the fluid paint and break it up into small particles. The air and paint enter the gun through separate passages, mix and are then ejected at the front of the gun.

Spray guns can be divided into groups:

1 By methods of paint supply, such as suction feed, gravity feed or pressure feed
2 Those with detached or attached paint containers
3 Internal-mix or external-mix types
4 Bleeder or non-bleeder types.

The most widely used spray gun is the suction-feed, external-mix, non-bleeder model.

Paint supply methods

Suction-feed gun
With this type, a stream of compressed air creates a vacuum at the fluid tip which allows atmospheric pressure within the fluid

Figure 17.8 Single-braid air hose

cup to force the paint up the fluid tube to the fluid tip and air cap. The paint container (fluid cup) is limited to one litre (1000 cm³) capacity to enable the gun to be handled without fatigue. The suction-feed spray gun is easily identified, as the fluid tip protrudes slightly beyond the air cap.

Gravity-feed gun

The fluid cup is mounted above the spray head and paint is fed to the gun by the force of gravity. The fluid cup is usually limited to 0.5 litre capacity, which makes the gravity-feed gun unsuitable for the spraying of large areas. However, it is very useful for painting local repairs where heavy-bodied fillers are applied and where rapid colour changes are necessary.

Figure 17.10   Typical suction-feed spray gun

1	Air cap	6	Air valve
2	Fluid tip	7	Trigger
3	Fluid needle	8	Fluid-packing nut
4	Fluid-control	9	Gun body
	screw	10	Air inlet
5	Spreader control	11	Fluid inlet

Figure 17.9   Gravity-feed spray gun

Pressure-feed gun

This type of gun sprays paint that has been forced from the paint container by compressed air. The air cap of this gun is not designed to create a vacuum, and the fluid tip is flush with the front of the air cap. Pressure-feed guns are used where a large quantity of a particular paint is to be sprayed or where the material is too viscous or heavy to be siphoned from the fluid container as in suction-feed guns.

The fluid container or pressure vessel is connected to the gun by means of a reinforced fluid hose and normally ranges in size from 2 to 25 litres. The smaller-pressure vessels can be carried in the operator's free hand, but

Figure 17.11 Gravity-feed spray gun (*DeVilbiss Automotive Refinishing Products*)

Figure 17.13 Typical pressure-feed tank

ones from 10 litres capacity upward can be mounted on wheels for easy portability.

Mix methods

Internal-mix gun
This gun mixes air and paint inside the air cap and is used with low air pressure to apply slow-drying materials.

External-mix gun
This is the most widely used type of gun and can be used to spray most types of paint. It is the best type of gun for spraying quick-drying materials. The air and paint mix beyond the air cap, and perfect atomization can be achieved.

Figure 17.12 Pressure-feed spray head

Figure 17.14 Internal-mix spray head

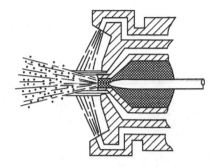

Figure 17.15 External-mix suction-feed spray head

Bleed methods

Bleeder gun

This is designed without an air valve. Air continually passes through the gun, thus preventing a build-up of pressure in the air hose. The gun is usually used with small compressors having a limited output and having no unload or pressure switch. The trigger on the gun controls the flow of paint only.

Non-bleeder guns

These are equipped with an air valve. The trigger controls both the flow of compressed air passing through the gun and the flow of paint.

High-volume low-pressure (HVLP) spray guns

Legislation which is part of the Environmental Protection Act requires that spray painters must reduce the amount of paint vapours being released into the atmosphere as a result of their working activities.

In conventional air-atomized spray painting, about 35 per cent of the mixed paint actually reaches the job surface. The remaining 65 per cent is either extracted to the atmosphere or collects on the workshop floor and walls. Of this waste material, 30 per cent is solid whilst 70 per cent is classified as volatile organic compounds (VOCs).

It is with these VOCs that the legislation is concerned.

In an effort to conform to the EPA, spray gun manufacturers have developed spray guns which atomize the paint at low air pressure, that is 0.6 bar (10 psi) as opposed to the usual 4 to 5 bar (60–75 psi). This reduced air pressure results in greatly reduced overspray, and as a further bonus, 65 per cent of the mixed paint reaches the surface. Savings on paint wastage are obvious.

Though atomization is achieved with much lower pressure, these spray guns still require large volumes of compressed air to operate them. A typical air pressure at the air transformer may be 4 bar (60 psi), but this is reduced in the gun body by means of an air restrictor, which reduces the air velocity at the gun outlet in the ratio of about 6:1. Other manufacturers like DeVilbiss do not have restrictors in the guns and only need 25 psi at the inlet to give 10 psi at the cap.

Figure 17.16 High-volume low-pressure (HVLP) cup gun (*DeVilbiss Automotive Refinishing Products*)

A special air cap is provided with the gun. This has a pressure gauge attached to it which enables the spray painter to adjust the outlet pressure. When the pressure has been adjusted to 10 psi, the air cap is removed and a more conventional type fitted. The fluid cup is pressurized and can be either attached to the gun or remote from it. Gravity types are also available.

Figure 17.19 HVLP spray gun using a remote fluid cup (*DeVilbiss Automotive Refinishing Products*)

Figure 17.17 Sectional view of an HVLP cup gun (*DeVilbiss Automotive Refinishing Products*)

Figure 17.18 Air cap with pressure gauge attached (*DeVilbiss Automotive Refinishing Products*)

Figure 17.20 HVLP gravity-fed spray gun (*DeVilbiss Automotive Refinishing Products*)

Basic parts of a standard spray gun

Air cap – This is the nozzle at the front of the gun that directs compressed air into the stream of paint, thus atomizing it and forming the spray pattern. Air caps are designed to give a wide variety of spray patterns and sizes and in addition ensure perfect atomization over a wide range of paint viscosities. The choice of an air cap depends on the following:

1   Volume of compressed air available
2   Type of paint feed system being used
3   Type and volume of paint to be sprayed
4   Size of the fluid tip (air caps are usually designed to operate with a particular fluid tip)
5   Nature and size of the surface to be painted.

Fluid tip – This is situated behind the air cap and meters out the paint. The volume of paint passing through the fluid tip and into the stream of compressed air is governed by the diameter of the orifice in the fluid tip. The choice of fluid tip depends, in the main, on the type of material to be sprayed. Heavy, coarse or fibrous materials require large nozzle sizes to prevent clogging, whilst thin materials, which are applied at low pressures, require small nozzle sizes to prevent an excessive flow of paint. The fluid tip provides a seating for the fluid needle.

Fluid needle – This seats in the fluid tip, its function being to start and stop the flow of paint. For the gun to operate efficiently, the fluid tip and fluid needle should be selected as a pair and should be of the same size.

Fluid control screw – This is an adjustment control which limits the length of travel of the fluid needle, governing the flow of paint from the fluid tip.

Trigger – The function of the trigger is to operate the air valve and also the fluid needle.

Air valve – This is situated in the handle of the spray gun (or gun body) directly behind the trigger by which it is operated. Its function is to control the passage of air through the gun.

Gun body – This can be regarded as a basic frame onto which the spray painter will mount a suitable set-up to suit his requirements.

Spreader control – This is of great importance in controlling the volume of air passing to the horn holes of the air cap. If air is cut off from the horn holes a narrow jet of paint giving a spot pattern is ejected, but when air is allowed to pass through the horn holes a fan spread is obtained, the width of the fan varying according to the volume of air.

Spray gun maintenance and cleaning

Cleaning a suction-feed gun

First, loosen the gun from the paint container, allowing the fluid tube to remain in the container, and unscrew the air cap a few turns. Holding a piece of cloth over the air cap, pull the trigger to divert air down the fluid tube and drive the paint back into the container. Next, empty paint from the container and rinse it out with the appropriate solvent and an old paintbrush. Pour a small quantity of the solvent into the container and spray it through the gun to flush out the fluid passages. Remove the air cap and immerse it in clean solvent, then dry it out by blowing with compressed air.

If the holes in the air cap become blocked with dried paint, a stiff brush moistened with solvent will usually remove the obstruction. If not, a toothpick or sharpened matchstick can be used to clean out the hole. On no account must wire or a nail be used, for this can distort the holes and permanently damage the air cap, resulting in a distorted spray pattern.

Figure 17.21 Basic parts of the HVLP spray guns (*DeVilbiss Automotive Refinishing Products*)

1	Nylon air cap retaining ring	23	Mini QD female valve and male stem
2	Front air cap baffle (nylon)	24	Adaptor to fit regulator to QD valve
3	Kit of two baffle seals	25	Cup pressure regulator
4	Rear baffle assembly	26	Kit of two regulator elbow connectors
5	Spreader/fan adjustment ring	27	Kit of five regulator-to-cup tubes
6	Fluid needle packing set	28	Cup pressure tube connector
7	Fluid needle packing gland	29	QD stem
8	Kit of five fluid needle springs	30	QD valve
9	Kit of five gun body bushing gaskets	31	Complete cup lid assembly three in BSP
10	Fluid needle body bushing	32	Fluid tube three in BSP female fitting
11	Fluid needle adjusting screw	33	1-quart KR cup yoke
12	Complete air valve assembly	34	Kit of five washers
13	Air valve stem	35	Cam lever
14	Kit of three air valve (trigger) springs	36	KR 1-quart pressure cup lid
15	Kit of five trigger pivot screws	37	Kit of three KR cup lid gaskets
16	Kit of five female trigger pivot studs	38	Washer
17	Chrome-plated trigger	39	Fluid tube retaining nut
18	Male/male air connector one in BSP	40	Kit of five drip-free diaphragms
19	1-pint nylon gravity cup assembly	41	1-quart (1.14-litre) PTFE-lined pressure cup
20	Gravity cup lid assembly	42	KR cup and lid assembly (PTFE lined)
21	Kit of five drip-check lids	43	Fluid inlet connector three in BSP
22	Kit of five O-rings for gravity cup	44	Fluid inlet locknut

Figure 17.22   Air cap, fluid tip and fluid needle (*DeVilbiss Automotive Refinishing Products*)

Cleaning a gravity-feed gun

Remove the cup lid, empty out surplus paint and replace it with a small quantity of solvent. Replace the lid, unscrew the air cap a few turns and, holding a piece of cloth over the air cap, pull the trigger. Air will be diverted into the fluid passage, causing a boiling action and flushing it clean. Spray solvent through the gun and clean the air cap. Finally, with a solvent-soaked rag wipe the outside of the paint container clean.

Cleaning a pressure-feed set-up

Shut off the air supply to the pressure tank, release pressure in it and loosen the lid. Unscrew the air cap a few turns, hold a piece of cloth over it and pull the trigger. The pressure will force the paint from the gun and fluid hose back into the tank. Empty surplus paint from the tank and pour in a small quantity of solvent, replace the lid firmly and pressurize the tank at about 0.3 bar (5 psi). The pressure will force the solvent to the gun. Hold a piece of cloth over the air cap and pull the trigger, and the higher atomizing pressure will force the solvent back into the pressure tank. If this is repeated about a dozen times, the purging action will clean out the fluid passages. Disconnect the fluid

hose and blow it out with compressed air. Steep and clean the air cap and finally dry out the tank with a piece of cloth.

Whichever type of gun is used, it must not be immersed in solvent, as this causes the lubrication oil to be washed away and will cause paint leakage from the gun. In addition to this, the air passage could become blocked with pigment sludge.

Lubrication

After the gun has been cleaned, a drop of oil should be applied to the fluid needle packing, air valve packing and trigger fulcrum screw.

A spray gun will function efficiently provided that it is clean and well maintained, but neglect will eventually cause the gun to malfunction.

Spray gun motion study

The spray gun is not a difficult tool to master, but a study of the following text and the accompanying diagrams will be invaluable to the inexperienced sprayer. Any person who is using a spray gun for the first time should obviously spray a few practice panels such as disused car doors, wings, bonnets and so on to get the feel of the gun before undertaking actual work.

Spraying flat surfaces

The gun should be held at right angles to the work and at a distance of 150 to 200 mm. Should the gun be held too close to the surface this will result in too much paint being applied, causing runs and sags. Holding the gun too far from the surface creates excessive overspray and a sandy finish. The relationship of gun distance and stroke speed is easily understood, and with a little practice the sprayer is able to adjust his speed of movement to suit the distance between the gun and the surface. The gun distance should be kept as constant as possible, and arcing of the gun must be avoided to obtain

Figure 17.23 Basic parts of a standard spray gun (*DeVilbiss Automotive Refinishing Products*)

1	Retaining ring for air cap	15	Gun body bushing
2	Air cap	16	Fluid needle adjusting screw
3	Air cap and retaining ring	17	Air valve assembly
4	Corrosion-resistant fluid tip and gasket	18	Air valve
5	Baffle	19	Kit of three springs
6	Kit of five seals	20	Kit of five screws
7	Kit of five JGA-7 fluid needle packings	21	Kit of five trigger bearing studs
8	Fluid needle packing nut	22	Locknut
9	Valve assembly	23	Connector three in BSP
10	Kit of five circlips	24	Trigger
11	Kit of five O-rings	25	Air-flow valve
12	Fluid needle	26	Retaining ring
13	Spring	27	Connector one in BSP
14	Kit of five gaskets	28	Baffle

an even coating thickness. The correct gun action is acquired by keeping the wrist flexible. Do not tilt the gun; hold it perpendicular to the surface. Tilting will give an uneven spray pattern, resulting in lines across the work.

When spraying a panel, the technique of 'triggering' the gun must be mastered. The stroke is started off the panel and the trigger is pulled when the gun reaches the edge of the panel. The trigger is released at the other edge of the panel, but the stroke is carried on

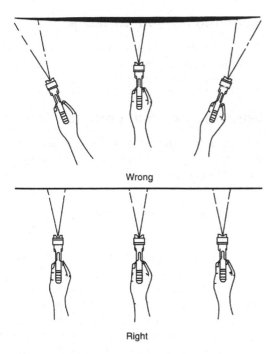

Wrong

Right

Figure 17.24   Maintaining spray distance

Too close   Too distant

Figure 17.25   Spray distance

Figure 17.26   Tilting spray gun

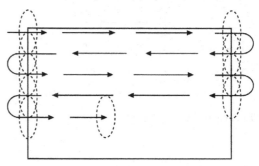

Figure 17.27   Panel spraying method

for a short distance before reversing for the second stroke. This triggering action must be practised and perfected to avoid a build-up of paint at the panel edges and to reduce paint wastage due to overspray. edge of the panel, and from then on the aiming point is the bottom of each previous stroke. This gives the 50 per cent overlap necessary to obtain a wet coating. An alternative method is shown in Figure 17.30. In this method the ends of the panel are first sprayed with single vertical strokes, the panel then being completed with horizontal strokes. This technique reduces

overspray and ensures complete coverage of the surface.

Long panels such as those encountered on furniture vans require a different approach. A certain amount of arcing is permitted to avoid a build-up of paint where the strokes overlap, and the triggering of the gun is very important. The length of each horizontal stroke is 450 to 900 mm approximately, or whatever the sprayer can manage comfortably. the panel being sprayed in separate sections, each section overlapping the previous one by about 100 mm.

When spraying level surfaces such as car roofs and bonnets, always start on the near side and work to the far side to re-dissolve any overspray. A certain amount of gun tilting is usually unavoidable when reaching across a car roof, and overspray is thus created.

Figure 17.28 Alternative panel spraying method

Spraying curved surfaces

As previously stated, the gun should be kept at right angles to the surface and as near a constant distance from it as possible.

Spraying external corners

The point to watch is to half trigger the gun to avoid applying too much paint as you go round the corners.

Spraying internal corners

Spraying directly into a corner gives an uneven coating but is satisfactory for most work. When an even coating is necessary, such as with metallic finishes, it is better to spray each face separately, starting with a vertical stroke at the edge of the panel. The vertical stroke should be followed with short horizontal strokes in order to avoid over-spraying or double-coating the adjoining surface.

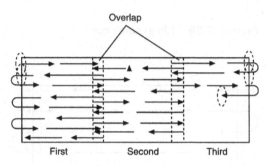

Figure 17.29 Spraying large panels

Figure 17.30 Spraying curved surfaces

Figure 17.31 Spraying external corners

Figure 17.32 Spraying vertical corners

Figure 17.33 Spraying internal corners

Figure 17.34 Spraying internal corners: method for better finish

Spraying sequence

An automobile should be sprayed in sections, spraying one section at a time before moving on to the next one. Start at the top

Figure 17.35 Suggested sequence of spraying a car

and work downwards, going around the car in an anti-clockwise direction.

The painter must decide on his approach before commencing to spray and then work methodically round the car, finishing at a point where an overlap is least noticeable. As the bonnet of the car is the panel which attracts most attention by the customer, most spray painters prefer to spray this last to avoid the risk of overspray falling on it.

Spraying defects

No matter how excellent the spraying equipment is, sooner or later some small trouble shows itself which, if it were allowed to develop, would mar the work done. However, this trouble can usually be very quickly rectified if the operator knows where to look for its source. The following sections contain the causes and remedies of all the troubles most commonly encountered in spraying.

Fluttering spray

Sometimes the gun will give a fluttering or jerky spray, caused by an air leakage into the paint supply line. This may be due to the following:

1 Insufficient paint in the cup or pressure feed tank so that the end of the fluid tube is uncovered
2 Tilting the cup of a suction-feed gun at an excessive angle so that the fluid tube does not dip below the surface of the paint
3 Some obstruction in the fluid passageway which must be removed
4 Fluid tube loose or cracked or resting on the bottom of the paint container
5 A loose fluid tip on the spray gun
6 Too heavy a material for suction feed
7 A clogged air vent in the cup lid
8 Loose nut coupling the suction feed cup or fluid hose to the spray gun
9 Loose fluid needle packing nut or dry packing.

Figure 17.36 Identification of parts of gun responsible for fluttering spray – numbers in diagram correspond with those in the text

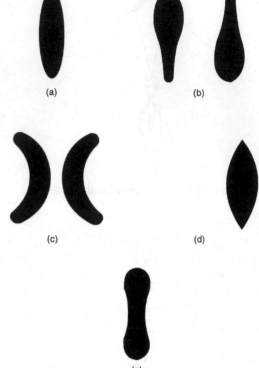

Figure 17.37 Spray patterns

Faulty spray patterns

(a) The normal spray pattern produced by a correctly adjusted spray
(b) Top- or bottom-heavy pattern caused by horn holes in air cap partially blocked, obstruction on top or bottom of fluid tip or dirt on air cap seat or fluid tip seat
(c) Heavy right- or left-side pattern caused by right or left side horn hole in air cap partially clogged or dirt on right or left side of fluid tip
(d) Heavy centre pattern caused by: too low a setting of the spreader adjustment valve on the gun, atomizing air pressure which is too low, paint which is too thick, pressure feed at too high a fluid pressure, a flow of paint which exceeds the normal capacity of the air cap or the wrong size fluid tip for the paint being sprayed
(e) Split spray pattern caused by the atomizing air and fluid flow not being properly balanced.

Spray fog

If there is an excessive mist or spray fog, it is caused by:

1 Too-thin paint viscosity
2 Over-atomization, due to using too high an atomizing air pressure for the volume of paint flowing
3 Improper use of the gun, such as making incorrect strokes or holding the gun too far from the surface.

Paint leakage from gun

Paint leakage from the front of the spray gun is caused by the fluid needle not seating properly. This is due to the following (numbers correspond to those on the figure):

1 Worn or damaged fluid tip or needle
2 Lumps of dried paint or dirt lodged in the fluid tip

Figure 17.38   Fluid needle assembly

3   Fluid needle packing nut screwed up too tightly
4   Broken fluid needle spring
5   Wrong size needle for the fluid tip.

Faulty packing

Paint leakage from the fluid needle packing nut is caused by a loose packing nut or dry fluid needle packing. The packing can be lubricated with a drop or two of light oil, but fitting new packing is strongly advised. Tighten the packing nut with the fingers only to prevent leakage but not so tight as to bind the needle.

Air leakage from gun

Compressed air leakage from the front of the gun is caused by the following (numbers correspond to those on figure):

1   Dirt on the air valve or air valve seating
2   Worn or damaged air valve or air valve seating
3   Broken air valve spring
4   Sticking valve stem due to lack of lubrication

Figure 17.39   Air valve assembly

5   Bent valve stem
6   Air valve packing nut screwed too tightly
7   Air valve gasket damaged.

Oil in air line

If the air compressor pumps oil into the air line, it can have the following causes:

1   The strainer on the air intake is clogged with dirt.
2   The intake valve is clogged.
3   There is too much oil in the crank case.
4   The piston rings are worn.

Compressor overheating

An overheated air compressor is caused by:

1   No oil in the crankcase
2   Oil which is too heavy
3   Valves which are sticking or dirty and covered with carbon
4   Insufficient air circulating round an air-cooled compressor due to it being placed too close to a wall or in a confined space
5   Cylinder block and head being coated with a thick deposit of paint or dirt
6   Air inlet strainer clogged.

Sanding and polishing machines

Sanding and polishing by hand can prove to be both laborious and expensive. Unfortunately there are many parts of vehicle surfaces where there is no alternative but to carry out these processes without the aid of power tools. However, surfaces such as the roof, bonnet, boot lid and parts of the doors and wings of cars can be rubbed down or polished more economically and efficiently with power tools. Damaged areas of paintwork can be rubbed down very quickly with these machines, but final feather edging is best done by hand. Two types of machine are favoured by the refinishing painter, the rotary sander and the orbital sander. Both are obtainable as either compressed air-operated or electrically driven machines. In addition, orbital sanders are available which operate with compressed air and can be connected to a water supply so that the worked surface is continually washed with clean water whilst rubbing down takes place.

Sanding processes

In order to produce smooth, glossy finishes the substrates and undercoats must be levelled down without leaving deep scratches. Non-sand primer surfaces have virtually eliminated many of the problems associated with scratch swelling, but they must be applied over satisfactorily prepared surfaces. Where repair work has been carried out which includes the use of polyester fillers and/or surfaces, the repaired area must be carefully sanded prior to applying the non-sand coatings.

The wet sanding of high-build undercoats and fillers has long been the accepted method of levelling these materials in order to avoid the creation of excessive dust. In addition, this method of sanding helps to clean the surface but also presents certain problems. Water penetration behind window and windscreen rubbers, door checks and so on can lead to lengthy drying-out times. All undercoats are slightly porous and, if the moisture from the sanding process is not completely dried out, problems of micro-blistering and faulty adhesion may result. The two main objections to dry sanding have always been that the abrasive paper clogs up and the process creates excessive dust in the workshop.

However, clogging up of the abrasive paper is no longer a big problem since the introduction of coated abrasives. These abrasive papers have a coating over the grit particles of a material which is based on zinc stearate. This coating allows the sanded-off residue to be shaken from the paper to produce a fairly clean sanding surface, thus creating more mileage from each sheet or disc of the abrasive paper. Coated abrasive papers are available in production paper grades (e.g. P40 to P120) and also in finer grades of lubricoat papers (e.g. P150 to P600).

It should be noted that, following sanding down with these coated abrasives, the surface should be cleaned with a proprietary spirit cleaner to remove particles of the stearate coating from the surface before applying paint. Failure to do so may result in the appearance of 'fish eyes'.

Dry sanding of painted surfaces and surfaces to be painted has increased in popularity as a result of the development of sanding tools which incorporate dust-extraction methods. These dust extractors can be in the form of either a vacuum bag attached to the sanding tool or a large vacuum dust-collecting unit remote from the tool and to which two or more sanders can be connected. An added bonus with the remote machines is that they can also be used as a vacuum cleaner for the workshop. The sanding tools have a pad with eight or more holes in them to which are attached sanding discs with similar holes. In operation, the dust created by sanding is drawn through the holes and deposited in the collecting unit or bag.

Sanding machines available include types which are dual acting: that is, they can be set to either rotary or orbital (eccentric) actions. The rotary action is in the region of 420 rev/min, whilst the eccentric action moves at 12,000 strokes/min. The machines can be either electric or driven by compressed air. The rotary action is suitable for removing old paint films and surface rust, and when fitted with a polishing head, the machine can be used for final polishing of paintwork. The eccentric action is suitable for feather edging, flatting surfaces prior to painting, and levelling surfaces and fillers. The sanding discs can be either self-adhesive (synthetic resin adhesive) or of a type which has a looped velvet reverse side which is simply pressed onto a special pad. These pads themselves are available as hard-soft and super-soft types depending on the type of work for which they are required.

Polishing machines

There are many types of machine suitable for polishing, and they all operate with a common circular movement. To be suitable for polishing, however, the machine must not rotate with too many revolutions per minute, as scorching of the film will easily result. The pad is covered with a lambs-wool disc or foam pad, which must be kept clean at all times. Remember that polishing is the final process; should grit be picked up on the disc, scratches will result, which can ruin the finish on a vehicle. Polishing machines are best used with liquid polishes and should be used with a sweeping movement.

Figure 17.40 Air-driven polisher with foam pad (*Desoutter Automotive Ltd*)

Tech note

Even with slower-revolving polishing machines; polishing in one spot for too long can cause scorching and blistering of the paint film.

Preparation of a motor vehicle for repainting

Two methods of preparing a vehicle for repainting are open to the painter: (a) completely strip the existing finish down to the metal substrate, or (b) prepare the surface by rubbing down with abrasives. The main point to consider when deciding which process to use is the condition of the existing finish allied to the extent of damage in the case of a local repair. Which process to choose can only be decided upon by a close inspection of the vehicle. On a motor car that has been involved in a collision and received damage to its front wing, but not serious enough to warrant replacement, the panel beater will carry out his work first, bringing the wing as near as possible to its original shapes. The vehicle then becomes the responsibility of the painter, who will carry out a close inspection of the condition of the paintwork. Should the paint show signs of poor adhesion to the substrate, it is advisable to strip

it off completely using paint remover. However, if the damaged area is not extensive and the paintwork on the rest of the wing is in sound condition, the work can be carried out, without the use of paint remover, by rubbing down the area.

Preparation by using paint remover

If the existing paintwork is in poor condition, or if the damaged area is so extensive that the remaining paintwork on the damaged panel is not worth rubbing down, the existing finish can be removed with paint remover. The first task is to remove any flashes or chrome strips that can easily be removed. If the work is being carried out in a shop where several vehicles are being refinished, it is advisable to store these fittings in boxes, which should be labelled with details of the car such as customer's name, registration number and type of car. On a complete refinishing job it is also advisable to label each part to indicate from which side of the vehicle it was removed. Should it prove to be too difficult or impractical to remove any parts, these should be masked completely. Though it is not always necessary to mask the surrounding areas when stripping, it is advisable to do so, because if any paint remover accidentally came into contact with an adjoining door or bonnet, damage to these areas would occur. However, if masking paper is applied it should be of a fairly stout nature and laid in two or three thicknesses to ensure that any paint remover accidentally deposited on it does not immediately penetrate to the paintwork beneath.

Paint removers were formerly of a highly flammable nature and constituted a fire risk, but nowadays non-flammable types are available and are widely used. As well as removing fire risks they also evaporate at a much slower speed; hence they remain wet for a longer period and so have better penetration properties. Though they are quite easy to use, some precautions should be taken when working with these paint removers. They can be a cause of dermatitis by penetrating the pores of the skin, dissolving the natural oils beneath and so leaving the pores open to attack by bacteria. A simple precaution is to apply a barrier cream to the hands before commencing paint stripping and to rinse off with water any remover which comes into contact with the skin. Strong rubber gloves are very useful when using liquid paint removers. The operator should not smoke when using paint remover, even though it is not flammable, as it gives off a strong vapour which, on coming into contact with a naked light, becomes toxic in nature. This toxic vapour if inhaled into the lungs will cause dizziness, nausea and even vomiting. Adequate ventilation should be provided in any shop where paint remover is used, for the same reasons.

A fairly liberal coat of stripper should be applied to the surface and allowed sufficient time to penetrate through the various layers of paint. The softened paint is then removed with a stripping knife and the surface washed down with a generous quantity of water. Both the laying on and the washing off of the stripper can be done with old or cheap paintbrushes. To assist the removal of paint at awkward places, a wire brush is an invaluable tool. Very often a layer of scum or a fine stain of primer is left on the surface, but this can be quite easily removed with steel wool and water. It is imperative that all traces of paint remover

be thoroughly rinsed off; otherwise it will cause damage to any subsequent paint system applied to the job.

Preparation by using abrasive paper

Before the rubbing down of a local repair commences, the whole panel should be thoroughly washed down with a detergent to remove all traces of wax polish. Nowadays nearly all wax polishes contain silicones, which can create 'pinholing' in any paint applied over them. Water-miscible cleaning solutions, which are normally used for degreasing, are suitable for wax removal. Thorough rinsing with water and drying should follow. The grade of wet or dry paper chosen to rub down the damaged area depends very greatly on the degree of damage, but normally 180 grade is suitable. The paper should be folded and torn so that one-quarter of the length can be removed (Figure 17.49). This is then attached to a rubber rubbing block and the damaged area abraded, using a liberal quantity of water to keep the work clean so that continual inspection can take place. The rubbing down is normally carried out with a forward and backward motion until the paint is removed and a feather edge achieved. Should paint remain in recessed areas, it can be removed with a wire brush, coarse steel wool or abrasive paper without the rubbing block.

As an alternative to rubbing down by hand, power tools can be used to speed up the process. These can be either electric or driven by compressed air. Those which are electrically driven are used dry, but compressed-air tools can be used either wet or

Figure 17.41 Method of fixing abrasive paper into rubbing block

dry. The latter types are generally favoured for lighter work where damage to the paintwork is not extensive, but the electric tools are most suitable where severe rusting of the metal has occurred.

Treatment with phosphating liquid

Following the removal of the paint by either of the methods described, the bare metal should be treated with a proprietary phosphating liquid. These liquids (there are several brands available) have an acid content, mainly phosphoric, which etches the metal and completes the de-rusting of it. In addition to etching the metal, phosphating solutions deposit a layer of iron or zinc phosphates onto the metal which inhibits corrosion. These liquids should be used according to the manufacturer's instructions but are generally diluted in the ratio 1:1 with water, applied with an old paintbrush and washed off with water after about 15 minutes. The liquid, being of acid content, should not be allowed to come into contact with the skin, eyes or clothing of the operator, and rubber gloves should be worn to protect the hands. In the event of its accidentally doing so, the affected part should be rinsed thoroughly in running water. Following drying of the bare metal with an air duster and wash leather, the area is now ready for repainting.

Finishing and refinishing processes

There are four main car refinishing paints at the refinisher's disposal: cellulose synthetic (half-hour enamel) paint, acrylic resin paint, low-bake synthetic paint and two-pack paints. Cellulose synthetic and low-bake synthetics are used as repair materials over the high-bake synthetics applied by the car manufacturers; in addition, of course, they can be used for complete resprays. Acrylics are best used on repair work over an original high-bake acrylic finish but can also be used as a refinishing material on complete resprays.

To determine whether a paint is suitable for repairing a particular job, a flat area should be chosen and a wet coat of the paint sprayed onto it. If wrinkling or lifting occurs this is proof that the solvent is too strong and will probably lift the existing finish.

The following paint systems are typical of those carried out in refinishing shops. They do not follow any one particular paint manufacturer's specifications but are intended as a guide to the use of the various paint types. When using a particular brand of paint, the operator should always follow the maker's instructions as to viscosities, drying times and temperatures. The processes outlined are based on the assumption that the vehicle has been prepared for refinishing, followed by the necessary masking up.

Coach finish

This is the traditional material, used in coach painting workshops for many years, but it has lost favour to more modern materials mainly because of its lengthy drying time. However, in paint shops where spraying equipment is not available, this material still has a use. It is essential that the workshop is kept clean; otherwise dirt in the finish is a certainty owing to the prolonged period of paint film wetness.

The finishing material can be sprayed either hot or cold and can also be brushed. It is suitable for application on all vehicle construction materials including wood. Undercoats may also be sprayed hot when thinned 6:1 with white spirit. The paint system is of the simplest type. For example, for metal surfaces the procedure is as follows:

1   Prepare the surface.
2   Carry out any necessary masking.
3   Wipe the surface down with a proprietary spirit wipe or a mixture of methylated spirit and water.
4   Apply one coat of self-etching primer, either spraying or brushing type. Allow to dry and then apply ground coats as soon as possible.

5   Apply two coats of ground coat:
   (a) If cold sprayed, allow 2 hours between coats.
   (b) If hot sprayed, heat to 60°C and allow 1 hour between coats.
   (c) When brushed, allow to dry for 6 hours between coats.
6   Should stopping up be necessary, this should be done after the first ground coat, which should, of course, be allowed to through-dry completely. A synthetic resin stopper should be used and, when hard-dry, wet flatted with P360 wet-or-dry paper.
7   Should sanding down of the ground coats be necessary, allow to dry for at least 6 hours and dry-sand using P400 lubricoat paper. Dust off and tack off. If sanding down is not necessary, the finishing material can be applied following the flash-off times listed in point 5.
8   Apply finishing enamel:
   (a) When brushed, no thinning is required.
   (b) When hot sprayed, heat to 60°C without thinner.
   (c) When cold sprayed, thin 6:1 with white spirit or preferably with a thinner supplied by the paint manufacturer.
   Apply two coats as follows:
   (a) Brushed: allow at least 6 hours between coats, the time depending on workshop temperature.
   (b) Hot sprayed: one light coat; allow 30 minutes and apply a double-header coating.
   (c) Cold sprayed: one light coat; allow 60 to 90 minutes before applying a double-header coat, but take care not to apply too wet a coating; otherwise runs will result.
   Spraying pressures should be in the order of 3 to 4 bars (45–60 psi). Spray gun fluid tips should be of the smaller types (say 1.25 mm diameter) to avoid flooding the job. Spray gun distance from job is best about 200 mm.
9   Remove masking whilst the paint is at the tacky stage. This will allow the edges to settle down into place.

10 After overnight hardening, transfers, lining and lettering may be applied.

Cellulose synthetic (half-hour enamel): complete respray from bare metal

The process summary is as follows:

1   Spray a thin coat of etching primer, mixed 1:1 with activator, thinned to 20 to 25 seconds viscosity.
2   Allow to dry for 15 minutes.
3   Spray cellulose primer surfacer, thinned 1:1 to 19 to 22 seconds viscosity. Apply three full coats, allowing 10 to 15 minutes between coats.
4   Allow to dry for 1 hour.
5   Wet flat with P400 or P600 wet-or-dry paper.
6   Apply cellulose stopper where necessary, allowing 15 to 20 minutes between layers.
7   Allow to dry for 1 to 1.5 hours.
8   Wet flat stopper with 320 wet-or-dry paper.
9   Spray cellulose primer surfacer to stopped-up areas, and flat with P600-grade paper.
10 Blow off vehicle with air gun and tack off.
11 Spray finishing material thinned 1:1 to a viscosity of 21 to 23 seconds. Apply one coat and allow to dry for 15 to 30 minutes. Apply second coat.
12 Allow overnight drying.
13 Wet flat with P800-grade paper, dry with air gun, tack off.
14 Spray double-header coat, thinned as before.

Alterations to this standard method can be made where necessary. For example, an alternative to using cellulose primer surfacer at stage 3 would be to use a synthetic resin primer surfacer which has better build qualities and may be preferable on rough surfaces. It is thinned 4:1, giving a viscosity of 26 to 30 seconds. Spray two or three coats, allowing a flash-off period of 30 minutes between coats, and the surface will appear completely matt. Allow the final coat to dry for 4 to 6 hours before dry scuffing or to dry overnight before wet flatting with P600- or P400-grade paper. Cellulose stopper cannot be applied between coats of this material, and when used after the third coat must be oversprayed with cellulose primer filler. Cellulose finishing paints can be applied directly over synthetic resin primer surfacers, but if drying conditions are less than ideal a certain amount of lifting of the surfacer could be experienced. This is caused by the strong solvent used in cellulose finishes penetrating the still-soft synthetic primer surfacer and acting as a paint remover. Therefore when adverse drying conditions prevail, such as low temperature, dampness or high humidity, it is advisable to spray a coat of sealer or cellulose surfacer over the job before applying the cellulose-type finish.

At stage 11 in the process, an alternative method is to apply one coat of the finishing paint, allow to dry for 15 to 30 minutes, then spray a double-header coating. This is a quicker method than that mentioned in the process summary, but it does not produce the same standard of finish.

It should be noted that with the exception of tacking off and the actual application of the paint, all the process stages should be done outside the spray booth. Cleanliness within the booth or spray room is essential if high-class finishes are to be obtained. Rubbing down, even though done with water, will leave a scum on the workshop floor, which when dried out will leave behind a powdery residue. When disturbed this will cause air contamination. Burnishing and polishing cloths also cause contamination of the atmosphere, and these processes should be carried out in a part of the workshop away from the spray booth. These polishing cloths must not be put down on the floor or dirty work benches, for if grit should be picked up on them, damage to the new paint finish will occur. As to the correct air pressure to use when spraying, this will vary according to the type of spray gun being used, the paint viscosity and the type of paint. For cellulose synthetic paints and air pressure of about 4 bars (60 psi) is normal, though this can be adjusted by the spray painter to suit his own requirements. The golden rule is to use the lowest air pressure that will give satisfactory atomization. Stage 6 in the process summary refers to the application of cellulose stopper. It should be borne in mind that this material dries by the evaporation of the solvent content, and if it is applied in heavy layers the inside of the coating will remain wet for a considerable period. In addition to this, when drying does eventually take place the stopper will contract and sink. A much more satisfactory repair will be obtained if the material is applied in thin layers with a suitable drying period between applications. Plastic spreaders are to be preferred to metal knives when laying on the stopper, as these cause less damage to surrounding areas of paintwork. The lid must always be firmly replaced on the tin or tube of the stopper when not in use; otherwise the stopper will harden in the container and cannot be satisfactorily re-softened.

A coat of sealer could be added to the process, where costs permit, between stages 10 and 11. This would reduce absorption of the finishing coats by the surfacer and promote a high gloss.

Following the finishing coats, the vehicle may, if required, be burnished and polished.

Finally, it should be borne in mind that the use of a guide coat at each of the stages of sanding down is invaluable for locating surface defects.

Cellulose synthetic (half-hour enamel): refinishing over an existing finish

The majority of cars or vans that are refinished today are first patch repaired, followed by refinishing of the entire body. Only when the paintwork is in very bad condition is it considered necessary to strip it all off, and it is usually only essential to cut back to bare metal those parts where the coating is damaged or corrosion has set in. However, it is not always necessary to repaint the whole car or van. A typical example would be that of a comparatively new car with paintwork in excellent condition but with local damage such as would be sustained in a collision. In this case it would only be necessary to repaint the repaired areas.

A process summary would be:

1 Apply etch primer to bare metal parts only.
2 Spray one or two coats of cellulose primer surfacer, allowing 10 to 15 minutes between coats, onto the etch primer.
3 Apply cellulose stopper.
4 Rub down stopper with 360 paper.
5 Spray sufficient coats of primer surfacer or filler to bring the repaired area up to the level of the surrounding surface; use air pressure of 3 bars (45 psi).
6 Wet flat these areas with 360 paper.
7 Apply three full coats of primer surfacer to the entire body. Alternatively, synthetic resin primer could be used, providing more build to the system and allowing for easier flatting at stage 8.
8 Wet flat with P600-grade paper.
9 Apply cellulose stopper where necessary.
10 Rub down stopper with P400-grade paper and spray locally with cellulose surfacer. Wet flat these areas with P600 paper.
11 Dry off and tack off.
12 Spray finishing coats.

Cellulose synthetic (half-hour enamel): local repair

1 Rub down damaged area with P180-grade paper (wet) to obtain a feather edge. Wet flat surrounding paintwork with P400- to P600-grade paper. Alternatively, dry sand with a dual-acting sanding tool using P80- to P150-grade paper.

2 Treat exposed metal with phosphating solution.
3 Apply thin coat of etch primer to bare metal only; alternatively, spray one coat of primer surfacer directly onto bare metal, thinned 1:1.
4 Fill up defects with cellulose stopper in thin layers.
5 Wet flat stopper with 320- to 360-grade paper.
6 Spray sufficient coats of primer surfacer or filler (thinned 1:1) at 3 bars (45 psi) to bring repair up to level of surrounding surface.
7 Wet flat with P600-grade paper.
8 Burnish surrounding area with rubbing compound and a damp cloth to ensure a good colour match and better blending in.
9 Tack off.
10 Spray colour coats, thinned 1:1 to a viscosity of 19 to 23 seconds, at a pressure of 3 bars (45 psi). Spray several coats lightly until a good colour match is achieved. Allow to dry hard.
11 Wet flat with P800-grade paper, dry off and tack off.
12 Overspray the repair with a mixture of 75 per cent thinner, 25 per cent colour, carrying the spraying beyond the edge of the repair to obtain a soft blend.
13 Allow to dry hard (preferably overnight) and wet flat with P200-grade paper.
14 Burnish and polish.

Any masking off that may be required should be done between stages 2 and 3. A coat of sealer may be applied just prior to the colour coats to provide better hold-out, thus obtaining a smoother finish. Synthetic resin primers have not been included in this process, as they have a tendency to peel back from the edges when rubbing down takes place. Overspray from the spray gun can create unnecessary work when carrying out local paint repairs, but this can be restricted by using a narrower fan pattern than that used for spraying a whole panel. When this is done, a higher volume of paint will be applied, increasing the risk of runs, and so the fluid needle adjusting screw should be turned to the right until a satisfactory volume of paint issues from the fluid nozzle.

Occasionally the spray painter may be called upon to repaint a motor car on which the paintwork is in excellent condition, the customer simply desiring a change of colour. In this case he may only need to wet flat the surface with P600-grade paper using a weak solution of water miscible cleaning solution to remove any wax polish. Then, following thorough drying off and masking up, three coats of half-hour enamel will produce quite a good finish, which can be further improved by flatting, burnishing and polishing.

Acrylic lacquer

Though cellulose synthetic finishes are best left to dry in their own time to obtain best results, acrylic lacquer can be force dried without damage to the paint film, with a consequent speeding up of the process. This can be done with infrared lamps or in a heated booth (though not in excess of 50°C). The filler materials possess better build and flow-out than the cellulose-based materials, thus providing better surfaces for the finishing coats. The spraying viscosity of the finishing enamel is more critical than the half-hour enamel, and only the thinner recommended by the paint manufacturer must be used. Solvent evaporation from the wet paint film is governed to some extent by the workroom temperature, and should this be below 15°C a special quick-repair thinner should be used. This thinner evaporates very quickly but tends to produce a low gloss, which will require burnishing and polishing. However, should the ambient temperature be 15°C or above, a good hard glossy finish can be obtained straight from the gun which does not require polishing. As acrylic primer fillers have good adhesion properties, the use of etch primer, though recommended, is not essential provided that the metal substrate has been properly prepared and treated with a phosphating solution.

Acrylic lacquer: complete respray from pre-treated bare metal

1 Spray one coat of acrylic primer filler thinned to a viscosity of 21 to 23 seconds at 25°C. Allow 5 to 10 minutes to flash off.

2 Spray two coats of primer filler thinned 1:1, 26 to 29 seconds at 25°C. Allow 5 to 10 minutes between coats and 1 to 2 hours after second coat.

3 Apply cellulose stopper where necessary in thin layers, allowing 15 to 20 minutes between layers.

4 Wet flat stopper with P320- to P400-grade paper, dry off and tack off.

5 Spray stopper locally with primer surfacer, allow 5 to 10 minutes, then apply a full coat over the entire surface. Leave to dry for 1 to 2 hours.

6 Wet flat with P600-grade paper, rinse, dry off and tack off.

7 Apply one wet coating of acrylic sealer, allow to dry for 30 minutes, denib and dry. Do not wet flat. The lacquer coats should be applied within 2 hours. Tack off before spraying the enamel.

8 Spray lacquer coats as necessary using one light coat followed by a double header. Thin the enamel 2:3 to a viscosity of 16 to 19 seconds at 25°C. Use the appropriate thinner only. Should more than one coating of the lacquer be considered necessary, allow a flash-off time of 5 to 10 minutes between coats. The lacquer is touch dry after 15 minutes and can be safely handled after 1 to 2 hours, depending on room temperature. If necessary it can be burnished and polished after overnight drying, though this can be done after 4 to 6 hours if the enamel is force dried. The air pressure used when applying acrylic lacquer is between 3 bars (45 psi) and 4 bars (60 psi) depending on the make of spray gun employed.

Acrylic lacquer: complete respray over an existing finish

The method chosen here must obviously depend on the condition of the paint film. If the surface requires filling and stopping up, it should be wet flatted with 280-grade paper using a water-miscible solution or liquid detergent. After rinsing and drying off, the system described for bare metal can then be used. If filling and stopping is not considered necessary and the paint is sound, it can be wet flatted with P600-grade paper and a solution of liquid detergent. Following rinsing and drying off, tack off and apply the sealer and lacquer coats as described.

Acrylic lacquer: local repair

Acrylic lacquer can be used to repair high- and low-bake enamels but is not recommended for repairs to half-hour enamels or nitrocellulose-based air drying finishes. Nor is it suitable for use on wood or the repair of synthetic coach finishes. A typical system for a local or spot repair is as follows:

1 Degrease with a solution of liquid detergent.
2 Wet flat damage area with 180-grade paper.
3 Treat bare metal with metal conditioner or phosphating liquid, rinse and dry off.
4 Spot prime with acrylic primer filler.
5 Stop up with cellulose stopper where required.
6 Wet flat with 280-grade paper and rubbing-down block. Finish off with 320-grade paper, rinse and dry off.
7 Spray in with acrylic primer filler sufficient coats to level up the surface.
8 Wet flat with P600-grade paper, rinse, dry off and tack off.
9 Spray over the repair with acrylic sealer.
10 Denib and dry. Wet flat around the edge of repair with P800- to P1000-grade paper. Burnish surrounding panel.
11 Apply acrylic finish, thinned with quick-drying repair thinner, in light coats.

Finish off with a double-header coat to obtain a smooth finish.
12 When dry, wet flat with P1200-grade paper, burnish and polish.

Should the damage to the panel be too severe for satisfactory or economical levelling up with cellulose stopper, the two-pack polyester resin stopper could be used. This is applied to the bare metal prior to the paint system. It is best rubbed down and, after dusting off and tacking off, coated with primer filler.

The use of a sealer coat at stage 9 may be eliminated if the surface is carefully prepared. Sealers are supplied ready for use, and it is difficult to spray them without leaving an edge which is difficult to remove. Providing that the original finish is flatted 50 to 75 mm beyond the repair area, a satisfactory job can be produced by spraying the finishing coats onto the filler to slightly beyond its edge. Care should be taken not to overlap the colour onto un-flatted enamel.

Metallic finishes

Practically without exception, metallic colours being applied by car manufacturers in Britain are based on acrylic resins. However, they present problems not experienced when applying straight colours, which are caused by the metallic particle content. Without delving too deeply into the realms of paint technology, a metallic finishing paint could be described as a tinted, semi-transparent varnish containing finely ground particles such as aluminium, bronze and copper. Polished aluminium flakes, because of their silvery metal appearance, are the pigments most widely used. Because they are lacking in opacity, these paints are generally applied in several layers to achieve the desired effect of colour depth and an even distribution of the metal flakes. The best method is to apply a single coat followed by wet-on-wet (double-header) coats. The coverage may vary from colour to colour. Apply as many coats as may be necessary. Though car manufacturers favour acrylic-based metallic paints, they are, however, available in cellulose synthetic, slower-drying types of spraying synthetics and, most widely used nowadays, two-pack synthetics. They are not suitable for brush application. It is not necessary to outline a complete paint system for metallic finishes; this section deals with the application of the finishing coats. The actual spraying of these has a great influence on the finished appearance as regards colour. Spray gun technique is very important. The gun should be held at right angles to the surface and at a distance of 150 to 200 mm approximately. If the distance between the gun and the painted surface varies, dark and light patches will result. A 50 per cent overlap of gun stroke is essential to obtain an overall even colour and texture.

When too wet a coating is applied, the metallic flakes move freely within the wet film, and when solvent evaporation takes place they are generally in a fairly upright position. This tends to darken the final effect, as light does not reflect too well from the flakes in this position. Opacity is also reduced, and when sinkage takes place as a result of solvent evaporation, the particles tend to stick through the top skin of the paint film, causing the finish to have a seedy appearance.

Too dry an application will create a dusty effect, too pale a colour when viewed head on, and also because the metallic flakes have a tendency to lie parallel to the surface when a dry coating is applied. When repairing a metallic finish, two options are available to the refinisher. One is to extend the repair to a natural break in the car body lines and refinish the entire panel. Using this method there is always the danger of ending up with a mismatch when the masking is removed. A better method is to apply the colour coats

Position of metallic flakes in too wet coating

Position of metallic flakes in too dry coating

Figure 17.42   Coatings too wet and too dry

to the whole panel, remove the masking and then recoat but extend the colour further still, using a fade-out technique. Overspray thus created should be overcoated with a clear blend-in material to protect the overspray when burnishing is done. When using this second method with the base-coat-and-clear materials, it is best to coat the panel with the clear coating first and, whilst this is still wet, to apply the colour; this should then be overcoated with further coats of clear.

There are too many variables involved to make it possible to lay down a hard-and-fast system in obtaining a perfect match on repair work. No matter how much technical data is available, in the final analysis it becomes a system of trial and error, with the spray painter trying out various spray gun set-ups, varying the air pressure, speed of gun stroke, and distance from the surface. The spraying viscosities vary according to the type of material being used, but a rough guide is as follows: acrylic 16 to 19 seconds, cellulose synthetic 19 to 23 seconds, synthetic 23 to 28 seconds.

The damaged area should be levelled up as previously described and the whole of the surrounding panel wet flatted with P400-grade paper. Following drying and tacking off, the panel should be sprayed with primer surfacer or filler and flatted with P600-grade paper. After drying and tacking, the colour coats are applied. The function of this final coat of surfacer is to equalize solvent penetration from the finishing material. Surfacers and fillers are more porous than enamels, and when the surrounding panel is sprayed with the finishing colour only the repaired area can be detected by a slight variation in colour and texture. Where costs permit, a coat of sealer could be applied prior to the finishing colour. Light burnishing and polishing of metallic finishes can be done after overnight drying in the case of acrylic and cellulose-based materials. With the slower drying synthetics this process is best left for about 4 days to allow for complete solvent evaporation and to give the paint film time to harden off. When using two-pack materials, burnishing can be carried out after 16 hours at a workshop temperature of 15 to 20°C.

Quick air-drying synthetics

This type of material offers special advantages for low-cost re-sprays and rapid finishing of commercial vehicles. The main advantage is that it is suitable for application over all types of existing finishes, and is suitable for hot or cold spray application.

The rapid re-spray system is as follows:

1 Clean off all traces of traffic dirt, grease, wax and silicone polishes by using water-miscible cleaning solution.
2 Feather edge all damaged areas including parts damaged by stone chips (generally on sills). Treat bare metal with derusting liquid.
3 Spot prime any bare metal areas with cellulose-based primer surfacer and allow to dry for half to 1 hour. Stopping, if necessary, is best done before priming with two-pack polyester stopper.
4 Wet flat the whole vehicle with P600-grade paper, using liquid detergent or water-miscible solution as a lubricant and to remove wax. Rinse, dry and tack off.
5 Thin the enamel to a viscosity of 23 to 26 seconds (4:1). Spray one coat and allow to flash off for 10 minutes, then apply a double-header coating. The finish may also be applied hot, in which case it requires no thinning. The paint is heated to a temperature of 70 to 80°C, and one coat will be sufficient to give the required film thickness. The flash-off times between coats recommended by the paint manufacturer must be strictly observed. If too great a time lapse is allowed, then problems of lifting or wrinkling may be encountered.

Low-bake enamels

These are modified high-bake enamels rather similar in composition to those used by motor vehicle manufacturers. Though developed primarily for the refinishing trade, they are in fact used by vehicle manufacturers in the rectification of damaged or faulty-finished vehicles. One of the problems encountered by the refinisher has always been to obtain a close match to the original finish. When exposed to sunlight some colours have a tendency to fade, whilst a certain amount of discoloration takes place in others; some whites, for instance, tend to yellow with time. In addition to this, some pigments combine very well with a synthetic medium but not so well with cellulose-based vehicles. As the latter have always been the most widely used materials in refinishing, colour matching thus presented problems. These would have been less if refinishing shops were equipped with equipment similar to that of the vehicle makers, allowing them to use the same finishing materials. This, of course, would require a tremendous capital outlay, which would be well beyond the reach of the refinisher and would in all probability never be recovered.

In an effort to improve refinishing techniques, to reduce drying times and to obtain faster production, low-bake enamels (i.e. paints that cure at lower temperatures than the original high-bake finish) were developed. Being similar in structure to the high-bake material, these paints offer a closer match in colour and texture than do cellulose synthetic materials. Burnishing and polishing is seldom required, as these paints provide a good gloss from the gun.

Equipment

An obvious requirement for the drying of low-bake enamels is a stoving room, and various types are available. A stoving room can be acquired which can be connected to an existing spray booth, or where workshop space is limited, a combined spray booth and low-bake oven can be installed. The obvious limitation of a combined spray booth and low-bake oven is that no further paint spraying can be done during the stoving schedule, in addition to which valuable working time can be lost whilst waiting for the booth to return to a comfortable working temperature.

Where workshop space permits, a better proposition would be to install a low-bake oven which can be joined up to an existing spray booth with sliding shutter doors to seal the two areas. This can be further improved upon by having entrance/exit doors at each end of the unit which would provide a flow-line system of painting and stoving, thus reducing time wastage on vehicle movement. The type of installation must obviously be governed by the size and shape of the workshop. A further variation is one in which the vehicle being painted is moved sideways on rails and bogies.

Apart from the stoving room, the usual spray painting equipment used in refinishing shops is required.

Preparatory work

Deep indentations in the vehicle body can be filled with polyester stoppers which cure very quickly and have good adhesion properties when baked. Following rubbing down and cleaning off, the surface can be sprayed with synthetic resin primer surfacer if it is of a rough nature. As previously stated, these primers have exceptional filling properties, but when used with low-bake finishes the stoving schedule (1–1.5 hours at 82–93°C) can prove to be too long to be economical. Where the surface is not too rough, a better method would be to use cellulose primer surfacer thinned 1:1. Slight imperfections should be levelled with cellulose stopper, which must be over-coated with surfacer. The surfacer can then be stoved for 30 minutes at 70 to 80°C. A third material finding favour is the polyester primer filler, which obviates the need for stopper on small imperfections and provides a single coat build-up to the finishing stage. As previously stated, several coats can be applied wet on wet without flash-off periods, and force dried for 15 minutes at about 75°C. A further advantage is that polyester stopper can be used, if necessary, between coats of polyester primer filler. This material should be over-coated with cellulose primer surfacer before applying the finishing material. Whichever

of these materials is chosen, the final coating must be rubbed down wet, and following drying and tacking off, the colour coats can then be applied.

Spraying

The finishing material is thinned 7:2 and sprayed in one single coat followed by a double-header coat, with a 15 minute flash-off between coats. Another flash-off period of 15 minutes should be allowed before stoving.

Exterior view

As a spray-booth

As an oven

Figure 17.43 Spray booth combined with low-bake oven

Preparatory work

Deep indentations in the vehicle body can be filled with polyester stoppers, which cure very quickly

Preparation for stoving

It is current practice in the motor industry to repair areas damaged during assembly with low-bake enamels and then to stove the whole body fully trimmed. The temperature inside the vehicle does not rise sufficiently to damage the trim provided that the doors and windows are properly closed. Before stoving a vehicle refinished with low-bake enamel, check that doors and windows are closed, remove all exterior plastic fittings, increase tyre pressure by 0.5 bar (5 psi), check that the petrol tank is not too full and, if duotone work is being done, remove any masking. It is not necessary to remove the battery.

Stoving

The usual stoving schedule is 30 to 40 minutes at 80 to 100°C, following which the body should be allowed to cool for about 45 minutes, longer if necessary, before further masking up for duotone work is carried out. Certain parts of the car will be sheltered from the heat, and consequently paint applied to these parts will not be fully cured. A converter liquid is available from most manufacturers of low-bake paints which can be added to the enamel for brushing in door edges and insides of boot lids and bonnets, or these areas can be sprayed if the viscosity is reduced to 22 seconds with the appropriate thinner. Any areas touched in thus will dry quite hard at the normal paint shop temperature.

Finishing

Polishing of low-bake enamels is not normally necessary, as a good gloss from the gun is obtainable, but it should be borne in mind that these paints remain more open, that is wetter, than more conventional materials during the flash-off period prior to stoving. For this reason, a clean workshop is a necessity; otherwise dust may settle on the still wet film which will necessitate flatting and burnishing

Figure 17.44   Low-bake oven joined to spray booth

Figure 17.45   Oven joined to spray booth, with through doors for flow line

Figure 17.47   Short-wave infrared heater IRT 100 (*Infrarödteknik AB–Stanners*)

Figure 17.46   Joined oven and spray booth, with sideways flow

to obtain a smooth finish. The nibs should be removed with P800-grade wet-or-dry paper and the flatted area brought up to a high gloss with burnishing compound, followed with a good-quality car polish. Low-bake finishing can be carried out over both stoving and nitro-cellulose finishes providing they are in a sound condition. Red or maroon cellulose finishes which might bleed are best coated with a sealer before commencing the spraying of the surfacer coatings.

Short-wave infrared paint drying

Infrared drying lamps have been used in finishing shops for many years to force dry localized areas of paintwork and to accelerate the curing of two-pack polyester stoppers and fillers. A more recent development is the short-wave infrared heating module, which, instead of easily breakable lamps, has heating elements mounted into an aluminium cassette. One specialized piece of heating equipment has a reflector coated with a thin layer of gold to give maximum reflectivity and long life. In addition it is equipped with a fan for solvent vapour removal and a control

panel which guarantees efficient control of the drying process according to the type of paint being used. The unit is mounted on a stand with a flexible support lever, which makes it possible to locate the heating module in any position without locking devices. A variation on this type of heater is a vertical heater mounted on castors for force drying of doors, quarter panels and so on.

A further development is the infrared drying arch, designed to be mounted inside an existing spray booth. The arch consists of a number of heating modules mounted onto a steel frame in the form of an arch. This arch is stored at the rear end of the booth and is sheltered from overspray. Following the spraying process, it is moved quickly to the vehicle and then moves along the length of the vehicle at a slower speed. After the stoving of the vehicle, it then switches off automatically and moves back to the rear of the booth (see Figure 17.61). Being computer controlled, the arch is capable of providing the refinisher with the option of drying a complete vehicle or one or more panels at various positions on the vehicle body. It also caters for the various types of paint and colours.

Short-wave infrared units require no preheating, which makes for fairly high savings where energy costs are concerned. They also cool in seconds, which means the vehicle can be moved out of the booth almost immediately. When used to force dry two-pack materials, the drying times claimed are quite extraordinary: for example, a complete re-spray can be dried in 5 to 10 minutes and minor touch-ups in 1 to 2 minutes. Single panels finished in a material such as Bergers Standox 2K are baked in 4 to 5 minutes, whilst primers and fillers bake even faster. With drying times like this, the savings on energy costs and the gain in turnover of work are obvious.

> **Safety note**
> Radiant drying systems can easily damage your skin, and you will not notice it until too late. Always follow the correct safety procedures. Switch off the driers before entering, and allow time for them to cool.

Two-pack paint system

> **Safety note**
> Two-pack paints require very special PPE. Make sure that you are properly trained before you attempt to use it.

There are many variations available to the refinisher with these materials, and the paint manufacturers' literature must be referred to. It is inadvisable to mix one manufacturer's materials with another.

Although it is common practice in some workshops to use various primer surfacers under two-pack finishes, better build and inter-coat adhesion is obtained when two-pack undercoats are used. These undercoats are multipurpose materials which can be used as primers or non-sand surfacers. Selfetch 2K primer surfacers are available for use on bare, prepared substrates such as steel, aluminium and GRP and can even be applied over vehicle manufacturers' finishes. Transparent adhesion primers are available which can be applied onto manufacturers' finishes which have merely been cleaned with a cleaning spirit and a Scotchbrite pad. These primers can then be overcoated with the finishing material after 15 minutes as a wet-on-wet process. As previously stated, there are many variations of materials with these products, and so there is no basic system or process which can be described. A variety of thinners are available for use at various workshop temperatures. Various hardeners may be used depending on workshop temperature and humidity.

Panel cassette   Arch   Complete paint shop

Figure 17.48 Infrared cassette, arch and paint shop (*Infrarödteknik AB–Stanners*)

Two-pack paints are available as straight colours, metallic and base-coat-and-clear systems.

Tech note

It must be stressed that the time lapse between applying the base coat and the clear coatings recommended by the paint manufacturer must be strictly observed if inter-coat adhesion problems are to be avoided.

Water-based paint

Water-based paints use, as their name suggests, water instead of solvent. The use of water means that the emission of volatile organic compounds (VOCs) is eliminated. VOCs generate photochemical oxidants and suspended particulate matter (SPM); these are sources of air pollution which most countries are seeking ways to reduce.

Water-based paints are applied in the normal way, but care must be taken with the volume of paint delivered. The size of the fluid tip is therefore very important.

The use of water-based paint also allows the reduction of carbon dioxide (CO_2) by giving a faster drying time. Toyota have developed a system which does not need drying after the second coat, the third coat is applied straight away. This also saves painting time. This system reduces VOCs by 70 per cent and CO_2 by 15 per cent compared to conventional techniques. Water-based paint also offers easy clean-up and is therefore better in terms of waste management.

However, currently solvent clear lacquer coats are still needed, as the final finish needs to with stand the elements of the weather.

Safety note

Remember that you need the correct PPE for all types of paint spraying.

Burnishing, polishing and final detail work

Though brief reference to burnishing and polishing has been made earlier, these subjects are important enough to warrant fuller description. Even though most car refinishing paints are nowadays formulated to provide a good gloss from the gun, the final appearance of the vehicle can be further enhanced by careful burnishing and polishing. Burnishing helps to smooth out the surface whilst imparting a fuller gloss and revealing depth to the colour. Polishing, if carefully done, will improve the lustre still further and provide a protective coating over the paint film.

Burnishing

Where the final coat of enamel contains small particles of foreign matter, it is best flatted with P1200-grade wet-or-dry paper with soap as a lubricant. The soap must be rubbed into the paper and not into the painted surface. Care must be taken not to rub through the colour coats, especially where projections exist in the vehicle construction.

Following washing down, the car or van is now ready for burnishing. The mutton cloth (or any suitable soft cloth) is first wetted in clean water and wrung out. A small quantity of burnishing compound is applied to the cloth, and rubbing can commence, working in straight lines over a small area. A fairly firm pressure should be applied at first, but as the area shows signs of glossiness the pressure should be reduced. The friction caused by rubbing will generate heat, and when this takes place the cloth must be turned over and the burnishing continued until a smooth, glossy surface is achieved. Some of the colour will come away on the cloth, but provided that a sufficient coating thickness has been applied, this is unimportant. The cloths used for burnishing should be washed out periodically in clean water to remove the build-up of pigment which will hinder the action of the cloth.

(a)

Figure 17.49 Application of water-based paints is strictly controlled by the size of the fluid tip on the spray gun

Machines can be used for burnishing large areas of panel using liquid burnishing materials. These liquids, which contain milder abrasives, reduce the risk of swirls appearing on the surface, as would be the case should burnishing compounds be used with machines. Hand burnishing, however, will produce the better finish should costs permit. When using machines for burnishing, the precautions set out in 'Operating precautions' section should be observed. Following the burnishing process comes the final polishing.

Polishing

This is where the rotary-type polisher fitted with a lambs-wool pad comes into its own. These machines are best used with liquid car polishes but are also very effective when used with the more viscous conventional polishes. The polish is spread lightly over the surface with soft clean cloth, following which the finish is brought up to a high lustre with the machine. Heavy pressure is not required when polishing with a machine. Some parts of the vehicle body are obviously not suitable for machine polishing and must be done by hand. This is best done with soft yellow dusters rather than mutton cloth, which can leave striations on the surface and spoil the finished effect. The same care and precautions should be practised when polishing as when burnishing. The importance of working cleanly cannot be over-emphasized.

Final detail work

Attention to detail when completing a respray can make the difference between a professionally done job and an amateur one. The sequence of operations is usually as follows:

1 Remove overspray from windows with a razor-blade scraper and wash leather.
2 Clean chrome bumper bars with rubbing compound.

3   Clean and refit all parts which have been removed for the painting process.
4   Vacuum clean inside of car.
5   Paint tyres with tyre black. Rubber glazing strips can also be painted with this material.
6   Finally, carry out any lining work, using either lining tape, fine lining brushes or roller-type lining tools. The type of paint used is usually of a synthetic resin nature.

Rust-proofing

A vehicle body contains many recessed areas and box sections which are difficult, and in some cases impossible, to paint properly with a spray gun and conventional paints. The only time that they receive a coat of paint is in the electro-priming process carried out by the vehicle manufacturer. All primers are porous, and this material is no exception. Should moisture be allowed to penetrate the primer, it will eventually break down, and corrosion of the substrate will result. It is estimated that 90 per cent of corrosion on vehicle bodies is of the 'inside-out' type.

In an effort to prolong the life of a car body and provide an additional selling point, most manufacturers now carry out a rust-proofing process as part of the finished product or as a chargeable extra. A warranty is provided to the customer, but usually on condition that he or she has the vehicle checked for signs of corrosion at regular intervals and is prepared to pay for any rectification work that may be required. The material used to treat these cavity areas is based on wax dissolved in a solvent with rust inhibitors added. A compound of this type has good capillary attraction and can filtrate into joints, seams and otherwise inaccessible areas. Different types of wax coatings are available, some classed as penetrants and others as heavy-duty coatings. There are several firms worldwide who specialize in these types of materials and the equipment for applying them. Measures employed by car manufacturers in an effort to combat corrosion include the use of pre-coated steel, washing and spraying the assembled body and immersion in anti-rust primer, which is then baked. Spot-welded seams, which would be prime sites for corrosion, are sealed, and wheel arches, along with the underbody, are protected with anti-chip coatings. After a further baking period the car body is painted, and some of the critical cavities can be flooded with hot wax. However, a car is subjected to a great deal of stress throughout its life; it is scratched, scraped and banged, and panels may flex, seams move and joints vibrate. If untreated metal becomes exposed, rust will get in; therefore cars need to be treated regularly, and so rust-proofing becomes a continuous process.

The basic anti-corrosion tool is the spray gun with its four attachments. The rigid and flexible lances are best for all enclosed areas; the hook nozzle is designed for more accurate, directional work, while the fan spray tool is for coating open underbody surfaces.

Materials

Various anti-rust corrosion compounds include heavy-duty waxes and sealants for wings, wheel arches and underneath the vehicles, high-performance cavity waxes and penetrants for all enclosed areas, box sections, doors and pillars, special engine compartment waxes and waxes for special purposes. These compounds are waxes and inhibitors dissolved in a solvent. They are applied as liquids but solidify as the solvent evaporates. For ease of application make sure the materials and the vehicle to be treated are at room temperature (about 20°C). This is particularly important in winter.

Movement of vehicle in the paint shop

In any refinishing shop, the movement of vehicles is of primary importance. Much time and effort may be wasted in moving vehicles from one point to another as dictated by the finishing process. This is particularly true in those shops where expensive air-conditioned spray booths and/or spray booth ovens have been installed. To recoup the high capital outlay in such units it is essential that they should be used at full capacity for the whole of the working day. The simplest spraying/stoving unit, the combined spray booth and oven, is capable of handling six to seven vehicles per working day of 8 hours, providing it is kept in use over the lunch hour. Furthermore, with soaring building costs and high rateable values, coupled in some cases with restricted space for expansion, it is important that floor space be kept to a minimum. Whilst it is not possible to achieve a factory-type flow line for refinishing jobs that may vary tremendously in their requirements, for example, size, extent of repair and amount of work to be carried out, the desired continuous supply of vehicles to the spray booth may be achieved. Careful consideration of the problems of vehicle movement will enable this supply to be obtained with the minimum of time, labour and space.

There are several methods of moving vehicles within the refinish shop:

- *Under own power* – This method is the most economical of labour but requires the maximum of space. Also it gives rise to additional dirt, air pollution and fire hazards.
- *Manually* – Space requirements are as above, but demand on labour is heavier. Three men may be required to move an average car.
- *Mobile hydraulic jacks* – These permit some saving of space in that they allow a tight turning circle both front and rear.
- *Turntable* – This gives an excellent means of utilizing what would otherwise be a dead corner or may permit the spray

booth to be sited in an otherwise impossible position.

- *Rail and bogie system* – The vehicle is moved along railway-like tracks with bogies under each of its four wheels. This offers the greatest advantages of all sideways movement of the vehicles. Normally four sets of parallel rails are involved. The first pair constitutes the track on which the vehicles are first received into the paint shop and along which the preparatory work is usually carried out, that is, discing, flatting and stopping. The second pair is kept clear to allow movement of selected bodies to the spraying and/or stoving position. Vehicles are either driven or manhandled from one track to another. It is possible to have a layout utilizing only one track, but movements are more restricted than with two tracks. Two types of track are available: (a) those with sets of two rails and (b) those with sets of three rails. With type (a) bogies long enough to accommodate vehicles of differing wheel bases must be used. Type (b) permits the use of smaller, lighter bogies which are on selected pairs of rails from each set.

Figure 17.50 Suggested use of turntable

Figure 17.51 Lateral movement of vehicles using two sets of tracks

Figure 17.52 Two-track rails requiring long bogies

Figure 17.53 Smaller bogies can be used with three-track rails

Common spray painting defects

Blistering

Blisters are hemispherical projections on the painted surface, which will contain either air or water. Blisters can appear on surfaces which have been freshly painted or which have been painted for some time. Some of the causes of blistering are:

1 Moisture on the surface prior to painting
2 Excessive heat on article during service
3 Salts on surface prior to painting.

The third cause can lead to the defect known as 'contamination blistering', which appears mainly on those panels of a vehicle concealing the engine and which are subject to fluctuating temperatures. It can be avoided by using demineralized water for the final rinse prior to painting. Blistering can occur on old paint films where the metal substrate has corroded from the reverse side, for example, car wings, sills and lower door edges. Moisture penetrates through the metal to the paint film and causes it to swell at localized spots.

Blooming

This defect can be slight or severe. A slight case of blooming appears as a surface haze on a freshly dried paint film caused by moisture precipitation. It can be easily removed with a damp wash leather. If moisture is deposited onto a tacky paint film it can be absorbed by the binder or medium, which will swell and give a cloudy, milky appearance with loss of gloss. This type of bloom cannot normally be cured except by flatting and recoating. Where there is a high degree of humidity in the paint shop, it is best to use high-boiling (slow) thinner to avoid blooming.

Blushing

Blushing is similar in appearance to blooming but without loss of gloss. It is confined to wet paint films which are subjected to moisture precipitation, forming an emulsion with the binder. Blushing appears mainly with lacquers and can usually be cured by overspraying with a mist coat of anti-blush solvent.

Bridging

This occurs when the paint film contracts too much on drying so that it pulls away from recesses and corners. Too fast a thinner and too heavy a coating will produce this defect. Nitrocellulose lacquers particularly tend to exhibit this type of behaviour. Should the defect be discovered shortly after it has occurred, it can often be cured by slitting with a sharp knife or razor blade and over-spraying with a slow solvent or by adding slow solvent to any subsequent coats.

Cobwebbing

This is a defect of faulty atomization in which the paint leaves the spray gun in strings rather than small atomized particles. It is caused by incorrect thinning and is quite common in the spraying of acrylic enamels. When cobwebbing occurs, spraying should be stopped and solvent added to the paint. Though undesirable in the normal painting of motor vehicles, cobwebbing can be used to advantage to obtain decorative effects.

Dry spray

Though dry spraying can be an advantage to the spray painter, it is more often regarded as an undesirable defect, as it gives the work surface a sandy appearance. It is caused by:

1 Too high an air pressure
2 Using solvents which are too fast

3 Moving the gun too rapidly or too far from the job
4 Too high a paint viscosity
5 In hot spraying, too high a paint temperature or surrounding air too hot.

Excessive overspray

Overspray can give a dry spray appearance and is invariably the result of faulty spray gun manipulation. However, some articles are so shaped that oversprayed areas are difficult to avoid. The spray painter should attempt to keep these areas ahead of the gun so that with subsequent passes of the gun he can coat over the dry spray and dissolve it into the wet paint film. When this is not possible, overspray can be re-dissolved by mist coating the affected areas with a high-boiling (slow-drying) thinner or by flatting and burnishing.

Lifting

In this defect the existing paint on a surface wrinkles and then can be scraped or peeled off following the application of a fresh coating onto it. It is usually caused by the action of the solvent in the top coating attacking and softening the binder of the previous coat, such as when cellulose materials are applied over synthetic paints. When lifting occurs there is little alternative but to strip and recoat.

Orange peel

This is the most common of all spray painting defects. As the name implies, the dried paint film has the pebbly appearance of an orange skin and is caused by the failure of the wet paint film to flow out smoothly after application. This defect can be caused by any or all of the following reasons:

1 Paint of too high a viscosity (too thick)
2 Too fast a thinner (thinner evaporating too quickly)

3 Too low an air pressure, causing poor atomization
4 Too high a temperature in the spray room, causing the paint film to 'set up' before it can flow out
5 A sudden draught in the spray room (also causing the paint film to set up prematurely)
6 In hot spraying, too high a paint temperature.

Though orange peel is a defect of faulty flow-out, it can occasionally occur after the paint film has flowed out. This is the result of uneven solvent evaporation and the formation of vortices in the film, caused by improper solvent balance.

Pinholing

This is the sudden appearance of small vortices (pinholes) over the painted surface during drying. It is caused by using a solvent in the paint mixture which is too slow to evaporate and occurs when the topmost skin of the paint film has dried out. The slow solvent is trapped beneath the skin and ruptures it when escaping. Pinholing can also be caused by too high a room temperature and in hot spraying too high a fluid temperature.

Runs, sags, curtains

Each of these defects is the result of applying too heavy a coating thickness. Some of the factors which contribute to this are:

1 Paint which is too thin or too thick
2 Paint solvent which is too slow, retarding drying
3 Spray room temperature too low, also retarding drying
4 Spray gun held too close to the surface or spray gun movement too slow
5 Fluid nozzle too large for material being sprayed.

Tech note
Paint will not readily dry at temperatures less than 15°C. At this temperature the air in the UK and Europe is also very moist, so paint will remain wet and may become damaged by the moisture in the atmosphere.

Shelving

This term, though not widely used, is an extremely apt one to describe the defect in which the paint films tend to lie in separate layers or shelves, one on top of the other. This is caused by failure of the top coat to adhere properly to those previously applied so that one coat can be separated from the other by scratching with the fingernail or a knife. Shelving usually shows during sanding and can seldom be repaired. It is better to strip and refinish a surface rather than risk any subsequent coatings peeling off. The main causes of shelving are:

1 The solvent in the top coat is too weak.
2 The first coat is too dry before the second coat is applied.
3 There is grease or dirt between coats.
4 The top coat is too brittle.
5 There is poor natural adhesion in the top coat.

Colour mixing and matching

Colour mixing systems

Surveys have shown that, between them, 40 car manufacturers offer approximately 600 colours on current models. These are new colours only, and the total number of colours that the refinisher may have to match can run into thousands.

The development of a new car colour can be broken down into four stages:

1 Initiated by clothing and other fashions, new colours are reproduced in a range of paints. The car body stylist then selects a colour by trial and error to suit a new car model.
2 Paint manufacturers then reproduce this colour and carry out the usual tests for colour stability, durability and so on.
3 The vehicle manufacturer paints a number of car bodies with the colour to ensure that the paint is suitable for mass-production application.
4 Finally, the paint manufacturers are required to produce a coloured paint to suit the vehicle refinisher and to provide colour mixing formulae as well as basic colour hues and tints.

Figure 17.54 Volumetric colour mixing machine

There are two colour mixing systems available to the refinisher:

- *Gravimetric* – This system involves the use of a set of scales. The base colours, hues and tints are mixed according to their weight, the quantity of each being obtained from a given formula.
- *Volumetric* – The various quantities of each ingredient are added by volume. An adjustable measuring rod is the main tool involved.

In both systems the materials required to mix and match colours are base colours, full-strength tinters and reduced-strength tinters. The last are necessary in order to provide the refinisher with a measurable quantity. For example, instead of adding 0.5 grams of blue-black full-strength tinter, he can more easily add 5 grams of reduced-strength tinter.

A complete set of colour formulae is provided by the paint supplier, using manufacturers' paint codes, and is updated at regular intervals. Colour samples are also provided, which are added to as new colours are produced. There can be several slight variations of a colour as vehicles leave the production line, and these can provide the refinisher with real problems when trying to match a colour. In order to assist with solving these problems, the paint manufacturers produce samples of these colour variants and literature with hints on how to match them.

Colour mixing systems should not be installed in the workshop if at all possible. They constitute a fire hazard and can collect dust and overspray, which may affect the accuracy of the scales. A separate colour mixing room should be provided where possible which has plenty of natural light and also has 'daylight' fluorescent lighting.

It should be fully understood that the vehicle manufacturer uses high-bake materials, whereas the refinisher cannot. In order to match a colour, different pigments may have to be used in the refinishing material in order for them to be compatible with the binder. Consequently, a perfect match may be obtained in daylight conditions, but the colour may alter drastically when viewed under sodium or mercury streetlights. There is nothing that the refinisher can do about this, and it is advisable to explain this to the customer when only part of the car, such as a wing or door, is to be painted.

Hints on colour mixing and matching

Never try to match a colour by dabbing a sample of the colour onto the vehicle and making adjustments. Colours alter during the drying process. Always spray a sample card with the appropriate number of coats and allow it to dry before comparing it with the colour to be matched.

On a volumetric system, never use a dented tin or one with a concaved or convexed bottom to mix the colour in. On a gravimetric system, always remember to place the tin on the scales before zeroing them. When comparing the painted sample against the vehicle, daylight, preferably from a northern source, is best. Bright sunlight or artificial lighting can sometimes exaggerate or diminish a colour effect.

Vehicle cleaning – detailing

Detailing is being referred to as the new form of vehicle modification. It's in depth cleaning and paying attention to absolute details – hence the name. Detailing is very labour intensive and uses products which would not normally be used in general car cleaning and valeting.

Before handing the repaired vehicle back to the customer it is normal to clean the vehicle. The following discusses the different types or levels of cleaning that are often offered. That is, we have a hierarchy of car cleaning procedures. A body repairer might want to offer the customer a choice of levels of cleaning as an addition to repairing the accident damage or other work. This should generate additional income.

Basic types of car cleaning

Wash and Go – Five or 10 minutes either by hand-wash or drive through system. This simply removes the road dirt from the

outside of the car. In addition to the basic wash usually a few options are available. Typically, these are pressure washing to remove mud and grime from the wheels, wheel arches and the underside of the vehicle, drying by hand leathering to remove excess water and prevent streaking and vacuuming the inside of the vehicle. This is the basic procedure that most body repairers would include following the repair of a vehicle of any type.

Water re-cycling – Drive-through car wash facilities usually have a method of re-cycling the water and thus contribute to maintaining our environment. Hand-wash facilities must not allow dirty water to enter the surface water drains. They have a choice of re-cycling, draining into a tank to be removed in a sealed container or using the foul-water drains – using the foul-water drains needs written permission and will involve a cost. The penalties for non-compliance are quite heavy. Full details can be found in the Environmental Agency document Pollution Prevention Guidelines, available online.

Valeting – the next level up from the wash and go. This is washing, leathering and polishing the outside of the vehicle. It involves thoroughly cleaning the inside of the vehicle, including removing the floor mats, cleaning the seats and interior trim with cleaning products, adding de-odorizing products to make the interior smell fresh and cleaning the door-shuts, the parts that cannot be seen when the doors are closed. The final stage is treating the tyre side walls with tyre black. Valeting is the level of cleaning that you would expect to carry out when preparing a vehicle for the showroom, clean and sparkling as new. This process may take 1 or 2 hours.

Detailing

Detailing is about getting the vehicle to the absolute highest level of finish, cleaner and more impressive than when the vehicle was new. Detailing can take 100 hours or more. Such is the attention to detail.

Basic process – The following steps are offered as a general guide. The steps and the order of the steps will depend on the type of vehicle, the vehicle condition and the location of the vehicle. You will also need to check that the products you are using are safe for the particular job – that is, will not damage a particular finish or material. This information is usually indicated on the product. Also, it is wise to visit the manufacturer's website and obtain the relevant data sheet, putting a copy in your COSHH file for reference and adding a risk assessment for the process. Using cleaning products can be very hazardous, so the use of appropriate PPE is essential.

1 Use a spray cleaning product on the dirty parts of the vehicle. Look particularly at:
   - The sills and lower bodywork
   - Fuel filler cap area
   - Door-shuts and bonnet and boot closure areas. Often the higher-level areas are missed during normal cleaning, along with those which can only be seen from inside the vehicle. Be careful not to spray product on the vehicle's interior trim.
   - Also spray the wheels – however, you might choose to clean the wheels separately when they are removed to clean the inner sides.

   Leave the product to soak for a while as recommended by the manufacturer. Usually this is about 10 minutes.
2 Rinse the cleaning product off the vehicle using a hose or pressure washer as appropriate. Take care not to use excessive force or to allow any gritty deposit to scratch or damage the bodywork. The vehicle should look reasonably clean now.
3 Using soft water if possible and a high-quality car shampoo product, thoroughly clean the car. Soft sponges or car wash mitts are good for this.

4 Now rinse the car again with soft water and leather off and, if possible, blow dry so that the car is completely dry.
5 Rub the entire surface of the bodywork with a clay bar. This will remove the stubborn remains of tarmac, dead flies and other debris like paint overspray. Again, clean and dry the bodywork.
6 Now is a good opportunity to clean the wheels. Remove each wheel and, using special wheel cleaning product on both the inside and outside, thoroughly clean them. If they are damaged they could be re-furbished. This is also a good time to clean and may be paint touch-up, the hubs and brake callipers.
7 With the wheels replaced and covered with heavy-duty wheel covers to prevent damage. The body can be cut with cutting compound and then wax polished. That means going over the bodywork twice. When cutting it is advisable to check the paint thickness with a paint depth gauge to ensure that you don't remove the top coat. One way to apply the polish is with a machine polisher then clean off with soft cloths. Get into the door-shuts and other closure areas. Allow about 4 hours for this.

Risk assessment – At first appearance car cleaning may appear a relatively safe activity, but the following points must be considered when writing a risk assessment for this activity:

- Excessively hot water temperature or very low water temperature can cause scalding or frostbite.
- Continuously wet hands or other body parts can cause skin problems.
- Vibrations from polishing and other rotating parts can cause white finger.
- Inhaling particulates, dust or fumes from products can cause respiratory and other problems.
- Bending and stretching can cause back injury.

Cleaning PPE – Suggested PPE includes:

- Gloves – heavy-duty rubber gloves and light weight ones as appropriate
- Breathing mask
- Goggles and/or face mask
- Disposable overalls
- Waterproof boots.

COSHH – When using any product, you must obtain the data sheet for it. Generally these are available on the manufacturer's website. As well as having a copy in the COSHH file, the main points should be displayed clearly for all to see.

Interior – Interior cleaning is perhaps the most difficult because of the variety of materials used in vehicle interiors – almost all the different types of plastics, cloth, wool, leather and occasionally wood. There are also a lot of electronic devices in vehicle interiors – especially in seats – electronic seat adjustment controls, heated seats, seat belt warning and anti-dive electronics. So, cleaning product and water must be used with caution. When cleaning interiors it is a good time to change the heater pollen filter and use a low-pressure air line to blow out the heater vents. Products are available to clean the air conditioning and other ventilation systems. A pressurized canister releases product inside the vehicle whilst the engine is running and the heater is switched on. Ensure that the hand bake is firmly on, neutral or park engaged, and that normal exhaust gas precautions are taken, either outside or with an extractor. The vehicle must be watched at all times, as it is illegal to leave a vehicle running without supervision.

Leather seats – Often become worn on the squab on the side where the driver gets in and out. The first sign of this is the wearing away of colour. Colour-treatment product is available for this. It is usually dabbed on with a cloth and lightly rubbed so that it dries evenly. Leather saddle soap should be used to keep the leather soft and supple. Other wax-based products are also available.

Mechanical parts – Cleaning mechanical parts usually requires some form of de-greaser or steam cleaner. The oil-contaminated water must not be allowed to enter the drainage. For small-scale cleaning the use of drip trays is usually sufficient, the waste being collected with other waste products.

Soft water – Different areas of the country have different hardnesses of water. Water in chalky or lime stone areas tends to be hard. Hard water does not lather up or produce bubbles easily with shampoo. It is also more likely to leave stains, streaks or tide marks. There are a number of ways to soften water. One is to add softening chemicals; another is to pass the water through a water softener. Water softening using a de-ioniser is very cost effective. The de-ioniser is in effect a magnetic device that fits over the top of the water supply pipe. Warm water naturally dries more quickly, so warm, de-ionized water aids car cleaning.

Weather – Car cleaning should be carried out at a normal temperature – about 16°C to 24°C – so that drying time is not too fast or too slow. Avoid direct sunlight, as this leads to rapid drying, leaving a streaky finish.

Tyres – Tyre-black products are often made from what is called aliphatic hydrocarbon –a mixture of propane and butane. An advantage of this is that they dry quickly. You must avoid breathing in this vapour, as it is highly toxic. You should also avoid contact with your skin, as it can enter your body through your skin. The nature of these products and the associated vapours is that they can settle on the vehicle body. So apply the tyre black last, after the body has been sealed with the wax polish, keeping the spray in such a direction and small enough quantities so that it does not do any damage. Wipe away any excess with a soft cloth and dispose of this carefully, as it is both toxic and highly flammable. This procedure equally applies to most cleaning products.

Figure 17.55 Applying wheel cleaner with brush

Figure 17.56 Cleaner is left for about 10 minutes – it goes purple – seen here as darker – indicates dust from brakes.

Figure 17.57   Car covered in snow foam, an additional intermediate stage to remove stubborn road dirt. The snow form is sprayed on and allowed to run off naturally before a second wash and rinse.

Figure 17.58   Noodle mitt, especially useful for washing the lower part of the car

Figure 17.59 Microfibre towel, used to dry vehicle – especially horizontal surfaces

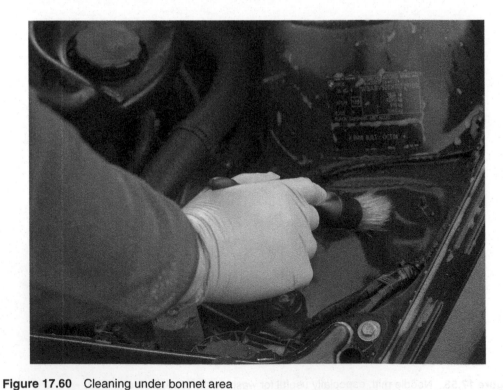

Figure 17.60 Cleaning under bonnet area

Figure 17.61 Dryer. These come in different sizes. This small one is good for areas around the mirror.

Figure 17.62 Clay bar. Keep folding it to ensure that the rubbing surface is clean.

Figure 17.63 This, with different heads and products, can be used for cutting the paint work, applying polish or other coats and polishing/buffing the final surface.

Safety note

Health and safety guidance and legislation is continuously changing, especially in the area of paint and other chemicals. In this book ,I have tried to bring most hazards to the attention of the reader, being aware that readers may be in different countries with different requirements. UK readers, and indeed those from any other country interested in reading about UK government regulations and advice – and it is only advice, as compliance does not guarantee safety or that you will be fully within the law – should go to www.hse.gov.uk and seek publications on the relevant area or topic. There you will find much free information which relates to current legislation.

Virtual-reality spraying

In this chapter, we have discussed paint and equipment and techniques. The difficulty for the student is actually perfecting the technique without wasting a lot of paint and causing a lot of mess. There is a virtual-reality (VR) spray system called SimSpray; see www.vrsim.com or www.simspray.net. It consists of a cabinet unit, spray gun, virtual-reality headset and a screen to give feedback with sound. SimSpray have a number of centres offering training using this setup in the United States and the UK. The student sprays various part of vehicles using the VR setup, and the system gives marks and guidance on how to perfect the process. No paint is used, and of course there are no health and safety hazards from paint.

Figure 17.64   Virtual-reality spraying trainer (*courtesy SimSpray.net*)

Skills and questions

Vehicle finishing is a high-level skill that requires regular practice. Every time you carry out a task, you should reflect on the activity to see how you can self-improve. The use of SimSpray will highlight your areas of skill and deficit.

1 State three of the advantages to be gained trom the method used by vehicle manufac-turers for applying the priming coat.
2 State two advantages to be gained by using a guide coat when sanding down a surface.
3 Explain two of the functions of a spray gun air cap.
4 Explain how a suction-fed (or syphon-fed) spray gun operates.
5 Determine the difference between thermo-plastic and thermosetting paint coatings.
6 What is the basic difference between those paints classed as lacquers and those classed as enamels?
7 What is the function of a ground coat?
8 Under what circumstances would a bleeder-type spray gun be selected in preference to a non-bleeder?

9 Why are bleeder-type colours a problem to the painter?

10 Explain the working principle of a single-acting, two-stage, piston-type compressor.

11 What is meant by the pot life of a two-pack paint?

12 Give two reasons why cellulose-based paints were replaced by stoving synthetics for mass-production finishing on new cars.

13 What type of vehicle finish is liable to become viscid when dry sanded with a sanding tool?

14 Reduced tinters have one-tenth of the tinting strength of a full-strength tinter. Why are they necessary for inclusion in colour mixing schemes?

15 What is the best type of priming paint to use on an aluminium surface?

16 What would be the most likely result of using two-pack polyester stopper as an inter-coat stopper in a lacquer-type paint system?

17 State the main disadvantage of a combined spray booth and low-bake oven unit.

18 What is the main health hazard associated with the spraying of two-pack paints?

19 Which method of heat transfer is used in (a) infrared and (b) low-bake heating units?

20 State two reasons why door handles should not be 'bandaged' when masking up.

21 In which way do thermoplastic vehicle finishes present problems to the vehicle refinisher when carrying out spot repairs?

22 What is an HVLP spray gun?

23 It is estimated that 80 to 90 per cent of corrosion on motor vehicles is classified as inside-out corrosion. What steps do vehicle manufacturers take to minimize this problem?

24 The spray pattern of a spray gun is top heavy. Describe a simple process to determine whether it is the air cap or fluid tip which is at fault.

25 State what is meant by 'VOCs'.

26 How do HVLP spray guns and VOCs relate to one another?

27 What is the main disadvantage of a combi-unit (spray booth and low-bake oven)?

28 What is the difference between baking and force-drying a paint film?

29 Describe a simple test to determine whether an existing paint finish is of the non-convertible type.

30 State three problems which can arise if cellulose stopper is applied too heavily.

18

Case study

This is a real case study of a real bodyshop in North London. This site has been developed from an existing wood-working business on an industrial estate. In this case study some of the important issues will be highlighted so that the potential bodyshop owner may benefit from the choices and concepts. This business is very efficient and environmentally very sound. It is unique in its combination of future proof-ness, customer friendliness, business efficiency and attitude towards the environment. The business is the MDS Accident Repair Centre.

General appearance

The images give an indication as to the general clean-cut approach to vehicle body repair by MDS.

Site selection

The site was selected for the following reasons:

- As an existing business was using it, there was no need for planning applications for change of use.

Figure 18.1 MDS Accident Repair Centre – exterior view. This blends in with the other buildings in the area, looking modern and neat. Other local businesses include the famous Elstree Film Studios.

The Repair of Vehicle Bodies, Andrew Livesey, 9780815378709, 2019, Taylor & Francis

Figure 18.2 Layout plan of workshop. To make full use of the space the paint booth in the corner position has been specially fitted with bi-folding doors so that access is possible with large vehicles. The paint booths also use a track-and-trolley system for moving the vehicles sideways.

Figure 18.3 Board room – The previous occupants had produced woodwork for the film industry, so the first floor offices were already well appointed. The board room is used for team meeting, training courses and meeting with insurance company engineers.

Figure 18.4   Live office. This office is fitted out with wi-fi, coffee machine and TV. It is available for customers to use when they are waiting for their vehicle to be repaired. Business customers use this, bringing their own laptops, which can log-on to the free wi-fi.

Figure 18.5   Reception area. This is bright and friendly, with a large-screen TV and coffee facilities.

- The building was high quality and fitted in with the proposed brand image.
- The location is near the main AI(M) and the M25, giving ease of access to a large percentage of the densely populated South East of England.

Pre-planning

To ensure that the business was successful, before the project was started the following pre-planning took place:

- An insurance company was approached and involved from the initial stage.

Figure 18.6 Valet bay. This is a portable bay which re-cycles all the water it uses.

Figure 18.7 A view of the bright, modern workshop showing the fronts of the paint booths

- Both bodyshop and business consultants were involved from the outset, bringing in a cost-saving IT system.
- Environmental and energy issues were tackled using well-respected suppliers.

Notable points

The following points should give the reader an idea of the level of detail carried out in setting up this business.

- Secure off-road parking for 60 vehicles, not visible from the road
- Each bay has independent trunking, giving three-phase power supply, air supply, dust extraction and radiant heating.
- The workshop has movement activated lighting of approximately 800 Lux.
- There are 32 CCTV cameras in the facility, which can be viewed remotely.

Figure 18.8   The workshop as seen from the customer viewing area

Figure 18.9   Customer viewing area as seen from workshop. Everything is open and visible to the customers – complete transparency in the repair process.

- An electronic lean management system is operated which tracks vehicle progress by the minute.
- The imaging system produces approximately 50 images of each repair, which is used as well as the paper file for each job, and the CCTV footage can be accessed if needed.

- Such is the quality control that insurance companies only carry out random samples – saving both parties costs.
- MDS deals with more than 50 vehicles each week, with a turn-round of 6 days. Each repair is given a 3-year warranty, subject to an annual inspection.

19

Mechanical, electrical and trim – MET

MET technicians are doing an increasingly large percentage of the repairs to damaged vehicles and the fitting of both new and replacement components. Many replacement body components are now supplied ready finished, so the job becomes to remove the damaged component and fit a new one. Of course, this is often quite a complex operation and entails lots of preparation and often removing several components to get at the damaged one. For instance, on some vehicles the complete front panel assembly must be removed to replace the headlamp. This entails working with electrical and mechanical components such as the radiator and air conditioning as well as the actual body panels.

In this chapter we can only deal with general principles. You need a workshop manual for the specific vehicle and of course lots of other information that is in the other chapters of this book.

Science

The MET technician needs to have a good understanding of science, as does anybody working in the vehicle repair industry. This section sets out to underpin your skills with relevant applied science, which should help you understand why cars get damaged and how some repair processes work.

SI System and common units

SI stands for System International, a system of measurement units which was developed following World War II. There are several different systems in use throughout the world, but for your examinations with UK-based examining bodies, SI only is used. It is worth noting that countries such as Germany and Japan use SI but with amendments and modifications. The Germans use DIN – Deutsch (Germany) Industrial Norm. The Americans use ANSI – American National Standards Institute – as well as SI. We'll also discuss some of the others, as they are used in America too. The Imperial System – so-called after the British Empire – is now known as the British Commonwealth, although it is very different now in how it operates.

For length the Imperial System of Measurement uses inches, feet, yards and miles. For mass it uses ounces, pounds and tons.

The following table is an approximate guide.

Decimals and zeros

It's very easy to make mistakes with decimal calculations and the use of zeros. As you will have seen from Table 19.2, the standard or basic units are often too big or too small in value. So there is a series of multiples and

Table 19.1 SI units

Quantity	Quantity Symbol	Unit	Unit Symbol
Length	l	metre	m
Mass	m	kilogram	kg
Time	t	second	s
Electric Current	I	ampere	A
Temperature	T	kelvin	K

Table 19.2 Imperial/SI approximations

Quantity	Imperial	SI
Length	1 inch (in.)	25 mm
Length	1 foot (ft.)	300 mm
Length	1 yard (yd.)	900 mm
Length	39 inches	1 metre
Length	1 mile	1.6 kilometres
Mass	1 ounce	25 grams
Mass	1 pound	454 grams
Mass	2.2 pounds	1 kilogram (kg)
Mass	1 ton	1000 kilograms

Table 19.3 Multiples and sub-multiples

Prefix	Symbol	Power	Number
giga	G	10^9	1,000,000,000
mega	M	10^6	1,000,000
kilo	k	10^3	1000
hecto	h	10^2	100
deca	da	10^1	10
deci	d	10^{-1}	0.1
centi	c	10^{-2}	0.01
milli	m	10^{-3}	0.001
micro	μ	10^{-6}	0.000 001

sub-multiples that is used to make calculations easier. For instance, *kilo* – meaning thousand – is added to *metre*, giving *kilometre* – in other words, 1000 metres. Going in the other direction, we use *milli*, meaning one thousandth. When we are talking about the very low voltages in vehicle electronics, we say *millivolts*.

Accurate measuring

Accurate measuring, called metrology, is used extensively in vehicle manufacture. Older car bodies were made with much larger tolerances than current ones. Tight, accurate tolerances on door-shuts, boot lids and screen apertures give a number of advantages, which are:

1 Improved vehicle quality and appearance, enabling the vehicle to be sold for the highest possible price
2 Reduction in wind noise, giving better passenger comfort
3 Improved aerodynamics, improving performance and fuel consumption
4 Enabling more economical use of materials, reducing vehicle costs

Four pieces of equipment which are used in vehicle manufacture are starting to find their way into vehicle body repair. These are currently scientific instruments that require skilled usage, but easy-to-use versions are now becoming available. Let's briefly discuss each one.

- *Co-ordinate measuring machine or CMM* – When we are making a panel to fit a vehicle the first important measurements will be the mounting points. All other measurements will be taken from these points, which we call co-ordinates, like the points on a map. A CMM measures distances from co-ordinates, even on the most irregular-shaped object such as a front wing. It does this to an accuracy of 1 micron (1μm, a millionth of a metre). This is necessary, as often the first panels are made by hand from the contours of clay models, called bucks. Therefore, the exact measurements may not be known.
- *Granite block* – This is what its name says, a gigantic block of granite on which a vehicle can be stood. This block of granite will have a mass greater than that of the vehicle and will be supported

Figure 19.1 CMM machine – Courtesy of Mitutoyo

on many hydraulic columns so that it can be kept perfectly level. Even the most level road surface will have a natural curvature, the curvature of the Earth. This granite block is made dead square. It enables dead accurate measurements to be taken of the vehicle, measurements which can be used during vehicle construction. Body construction is usually undertaken on jigs – sets of metal rails which support the floor assembly, with attachments to position the other panels before welding. The measurements from the granite block are used to inform the measurements for the production jig. For quality sampling, complete vehicles can be measured on the granite blocks. This is used extensively on racing cars, specifically to position the wings and fins, which control the aerodynamics of the car, again to an accuracy of 1 μm.

- *Lasers* are used for positioning panels during construction. Having accurate panel measurements – using the CMM and position measurements from the granite block. It is possible to transfer this information to the body jig to ensure that the panel is correctly positioned before fixing takes place, either welding or bonding. This is done by setting up one or more lasers on the outer part of the jig that will shine a light, or several spots will converge together, when the panel is accurately positioned. The panel can then be bonded or welded into place most accurately.

- *Scanners* – Measuring an irregular-shaped object, such as a grille panel, can be almost impossible without starting to mark the object, and then the measurements may not be transferable to metal to make another one. Scanning has two main uses. In manufacturing, the scanner can be used to take the profile from the clay buck, or first hand-made metal part, and convert it into a CADCAM file for the manufacture of a press die. The other use is to scan in the profile of vintage car parts to aid manufacturing new ones.

Capacity and volume

Most paint and other bodyshop consumables in the UK and Europe are sold in litres, parts of a litre or multiples of litres – for example, half-litre or maybe 5 litres. In the United States, it may be sold in pints or gallons.

A litre is defined as the volume of 1000 cubic centimetres, or 1000 cc. Water has a mass of 1 kg per litre at a temperature of 4°C. This is often referred to as density. Paint is lighter than water and has a density of about 0.9 kg per litre.

Sometimes the term *relative density* (rd) or *specific gravity* (sg) is used. Both terms mean the same thing: the density of the paint is compared to the density of water. So the paint, irrespective of its volume, if 1 litre weighs 0.9 kg, will have a rd of 0.9.

Still used by some companies and more so in the United States are pints and gallons. Beer is also frequently sold in pints too. Be aware that English pints and gallons are different to American ones.

English pints are made up of 20 fluid ounces, which means the volume of 20 ounces of water at 17°C. An English pint of water weighs a pound and a quarter. An English gallon is 8 English pints, weighing 10 pounds. It is defined as 4.54 litres.

American gallon is defined as 231 cubic inches, that is 3.78 litres. In water it weighs 8.34 lb, so it is considerably smaller.

The volume changes with temperature. The volumes of both paint and beer are legally measured in the UK at 16°C.

Temperature and heat

These two scientific terms are often mis-used, so let's get them cleared up so that you know what you are talking about when welding or brazing.

- *Temperature* – This is the hotness or coldness of an object. There are three temperature scales in use:

  Celsius (C) – also known as Centigrade because it has 100 degrees in it. It is related to the freezing and boiling point of water. Water freezes at 0° Celsius (0°C) and boils at 100°C.

  Fahrenheit (F) – Water freezes at 32°F and boils at 212°F.

  Kelvin (k) just uses the letter k. It is the absolute temperature scale. 0°C equals 273 k. 100°C equals 373 k. Absolute zero – lowest temperature achievable – is 0 k, which equals -273°C.

- *Heat* – This is the amount of energy used to raise the temperature of something. Heat is a form of energy. It takes 4200 joules of heat energy to raise the temperature of one kilogramme of water 1°C. Water is said to have a specific heat of 4200 J/kg C.

When you are using a gas torch to warm up something, for instance a piece of metal which you wish to bend, you will note that it takes time. Different metals take different times, and larger pieces take longer than smaller ones of the same metal. The longer time means that it is using more gas, which means more heat. As an example, typically 1 kg of propane will give 50 MJ.

Force and pressure

These two terms are also often confused or mis-used, so let's clarify them. It'll come in useful when you are pushing and bending or straightening something.

Force – We often use this in calculations but don't necessarily understand it. The unit of force is the newton (N), named after Sir Isaac Newton, who first discovered it. He noticed that if anything was dropped, it would go to the ground. This is due to the force of gravity (G). The farther an item drops the faster it goes. This is called acceleration due to gravity. The rate of acceleration on earth is typically 9.81 m/s/s. For terms of simple calculations, we often use 10 m/s/s as the equivalent of G.

$$\text{Force (N)} = \text{Mass (kg)} \times \text{Acceleration (G)}$$

Tech note

Mass is another name for weight when we are doing calculation with earth-bound objects.

So imagine an average person of mass 65 kg stood on the roof of a car.

Force of person's feet on car roof
= 65 kg × 10 (value of G)
= 650 N

That is a static force. If we have a dynamic force – say swinging a hammer – then to get an approximate value, we call it 2 sigma – in other words we double the total force. So for a large sledgehammer of 5.5 kg (12 lb) with a moderate swing, we get 2(5.5 × 10) = 110N.

Pressure is force divided by the cross-sectional area.

$$\text{Pressure (N/m}^2) = \text{Force (N)/Cross-sectional area (m}^2)$$

To avoid confusion with other units, the term pascal (Pa) is used for 1 N/m² pressure.

The pressure of 1 pascal is very low. Imagine an apple on your desktop. That is about 1 Pa. So we tend to use the term *bar*, which is equivalent to normal barometric pressure. One bar equals 101.3 KPa. The air compressor that provides air pressure for your power tools generates about 10 bar in pressure. Car tyres are typically inflated to between 2 bar and 3 bar.

Amps, volts, ohms, watts and Kirchhoff

Electricity is easy to understand if you get the basic terms clear. Although you can't see electricity, it behaves in a similar way to water. Providing that the plumbing in your workshop is connected to the mains supply, when you turn the tap on, water will flow out. Water comes out under pressure, a pressure from the main supply. The consumer standard for water pressure in the UK is enough pressure to fill a 4-5 litre bucket in 30 seconds. Typically this is 2 to 5 bar.

When we talk about electricity, the switch replaces the tap, the voltage (V) replaces the pressure, and the amps (I) replaces the bucket full. So the voltage should be high enough to force enough amps through to light up your lamp or power the starter motor. Items in a circuit provide a resistance (R). Like the tap, they can slow or stop the flow of electricity.

Safety note

Electricity can kill you
Please be aware that any voltage or amperage of electricity can kill you. It can also give you a non-lethal shock, which can make you flinch or jump and cause personal injury by hitting a hot or rotating part.

Ohm's Law is the relationship between amps (I) and volts (V), which will give a value for the resistance measured in ohms (R).

$$I = V/R$$

Watt (W) – This is a measure of power. Light bulbs usually have their power rating on them, and the same applies to electric motors and heating elements. It is normal to state the power and the voltage on these electrical items. With LED lights, the equivalent wattage is often given in two ways. For instance, some mains LED lights are rated as 9 W = 100 W. That is, they consume 9 W but give the equivalent light of a 100 W tungsten bulb – the older type of light bulb. They still work off a 230 V mains power supply.

$$\text{Watts (W)} = \text{Volts (V)} \times \text{Amps (I)}$$

Tech note
I = current in Amps
A is often used in a colloquial or less formal way, also meaning Amps.

If you have the wattage and the voltage, you can work out the current consumption in amps.

$$I = W/V$$

Knowing the current consumption is useful when you are fault finding or wiring up a new component. When fault finding, you can use an induction ammeter to see the actual current flowing. When fitting a component, it allows you to choose the correct cable size.

Kirchhoff – At any junction in an electrical circuit, the current flowing into the junction will equal the current flowing out. This gives you more information when testing a circuit and finding where the current is flowing to.

Safety note
- Before working on a vehicle's electrical system, always observe the following points:
- Always isolate the circuit concerned.

- Disconnect the battery where appropriate, checking that you have the ICE and other codes to re-set the system.
- Note any seat/window/electronic control/comfort settings so that you can re-set them to the customer's specification before handing back the vehicle.
- If SRS (air bags and set-belt tensioners) are involved, follow the manufacturer's instructions.
- On electric/hybrid or similar vehicles that use high-voltage or high-current systems, follow the required procedures before starting any work at all, whether it involves electrical circuits or not.

Friction and why cars roll over

Centre of gravity (CG) – Gravity attracts all bodies to the earth. It's worth remembering when you are jacking up a vehicle or raising one on a hoist. Front-wheel-drive cars have the bulk of their mass (weight) at the front, so their CG will be near the front. Mid-engined cars usually have evenly distributed mass so that they handle more smoothly – Ferraris, for example. When you are lifting a vehicle on a two-post hoist, you need to get the CG in the middle of the hoist. Most current cars have lifting pads on the sills to make accurate positioning of the car on the hoist fairly easy. Some large articulated vehicles have two levels inside them – lower and upper load-carrying areas. Car transporters often have three levels. It is important that the lower levels are more heavily loaded than the upper levels to maintain stability by reducing the height of the centre of gravity.

Centre of pressure (CP) – With large goods vehicles, especially articulated vehicles, the tall sides can cause instability on windy roads. This is because of the of the position of the centre of pressure not being in the middle of the vehicle. The wind pushes the vehicle sideways, and it is almost impossible to steer such a vehicle with high side winds.

Skidding happens when the friction between the tyres and the road is not sufficient to keep the car on a course. Friction is the ratio between the force acting downwards on the tyre (weight – W) and the force (F) needed to slide the tyre over the road, not rolling the tyre but making it skid. On a normal road with good tyre this ratio, μ (Greek letter mu), expressed as a decimal fraction, is about 0.8. If the road is covered in ice or wet leaves, the ratio can be as low as 0.01 – in other words, a vehicle can be pushed along with its brakes on.

$$\mu = \text{Force/Weight}$$

Rollover – This occurs when the sideways force on the centre of gravity exceeds the downward force caused by a vehicle's mass. This usually occurs when cornering fast. First the outer wheels lift. This moves the centre of gravity upwards, making the vehicle less stable until it eventually tips over.

Vehicles that are low and wide are least likely to roll over. Mid-engined layouts put the centre of gravity into the middle of the vehicle, which ensures that each wheel is equally loaded to the ground.

Some common laws of mechanical engineering

Newton's Laws are about force and acceleration they are numbered:

- *First Law* – A body – car, piece of metal, etc – will either stay where it is or continue moving uniformly unless another force is applied to it.
- *Second Law* – The force on a body is equal to its mass multiplied by its rate of acceleration. This is usually expressed in the form Force (F) = Mass (m) × Acceleration (a).
- *Third Law* – When a force is applied to a second body, the second body will

be exerting a force backwards. In other words, for each action, there is an equal and opposite reaction.

Hooke's Law is about how metal reacts to force. Metal stretches by an amount (X) proportionally to the force (F) applied to it until it gets to its elastic limit, when it breaks. The amount of stretch depends of the type of metal. The metal type is given a constant (k).

$$F = k\,X$$

Impact and momentum

Momentum – When a vehicle is travelling along a road, it processes momentum – Newton's First Law. That momentum (p) is the product of its mass (m) and velocity (v) – that is, speed multiplied with direction.

$$p = mv$$

The heavier the vehicle and the faster the speed, the greater the momentum. This is why we have speed limits.

Kinetic energy – We know that the moving vehicle possess momentum, and again applying Newton's Laws, this time the Second Law, we have to apply a force to stop.

Batteries

First of all, it's worth noting the dictionary definition of battery – 'similar object put together in a group'. General vehicle batteries consist of a number of cells, typically six cells each of 2 volts making up a battery of 12 volts, a 12-volt battery. Batteries work on the chemical reaction between two dissimilar (different) chemicals. The different chemicals give different voltages. The typical vehicle lead-acid battery actually has a cell voltage of 2.2 volts, so a 12-volt fully charged battery with six cells should read 13.2 volts. As most of the time batteries are not fully charged, we talk about the nominal (average or typical) voltage of 12 volts.

Battery rating

As well as being rated on their nominal voltage, batteries are rated in a number of other ways. Please be careful when comparing batteries and replacing batteries, as not even the ratings are equal. Different manufacturers use different criteria since there is no definitive set of standards. The typical headings are:

* *CCA* – cold cranking amperage, the amperage that a fully charged battery can deliver for 30 seconds
* *CA* – a little more vague than CCA; the amperage at which a battery can crank an engine
* *RC* – reserve capacity, how long a battery can deliver 25 amps
* *Ah* – ampere-hour, the rate at which a battery can deliver a constant amperage. Beware with this, as a battery rated 40 Ah might deliver 1 amp for 40 hours but not 40 amps for 1 hour.
* *Hot CA* – like **CA** but at 27°C.
* *PHCA* – pulse hot cranking amperes, the amperage available for 5-second bursts

Battery types

We have three distinct battery types: conventional batteries as used on cars and light commercial vehicles or LCVs, specialist high-capacity batteries for vehicles which have stop-start systems and batteries for

Figure 19.2 Typical battery

Figure 19.3 Battery installation in a VW Passat

electric and hybrid vehicles. Vehicles with stop-start systems frequently have two batteries, one for normal electrical running electrical/electronic equipment to maintain a constant supply voltage, the other for the stop-start system. Both batteries are usually 12 volts and lead-acid type, while the stop-start battery is of very high capacity in terms of Ah and CA.

Safety note
The sulphuric acid in lead-acid batteries is highly corrosive. Battery cases often get damaged in vehicle accidents, so look out for acid leaks and spillage.

The batteries used in electric vehicles and hybrids are very different. They use a much wider range of materials, but all are corrosive and harmful and operate at higher voltages. The batteries are put together in a battery pack, a special aluminium frame with an electronic control system. Battery packs are usually located centrally on the vehicle behind the crumple zone to prevent damage in an accident and to give even weight distribution. Replacement batteries are expensive, typically the price of a used small car.

The Toyota Prius a very popular car, uses 168 cells of nickel metal hydride. Each cell gives 1.2 volts for a total voltage of 201.6 volts. This weighs 53.3 kg, which is just less than the weight of an average person.

New battery types are being developed all the time. This is a list of the better-known ones:

- *CD* – cobalt dioxide
- *L-ion* – lithium ion
- *NCM* – nickel cobalt manganese
- *NCA* – nickel cobalt aluminium
- *MnO* – manganese oxide spinel
- *FePo* – iron phosphate

Figure 19.4 Toyota Prius battery pack

Figure 19.5 Gloves for electrical work – Courtesy of Salisbury

Safety note

Batteries in electric cars and hybrids are typically rated between 200 volts and 300 volts. Some vehicles use inverters to double this voltage at the motor – that's 600 volts. This voltage can kill you instantaneously.

Figure 19.6 Set of insulated tools (*courtesy 1000vtools.com*)

Battery disconnection

Battery disconnection procedures are different for different vehicles. This is offered as a general guide; however, you must read and observe instructions in the workshop manual for any particular vehicle before starting work.

- PPE – clean cotton non-conductive overalls, safety boots, electricians' gloves – 00 Rated for up to 500 volts, 0 rated for up to 1000 volts – safety goggles/glasses and, if appropriate underneath vehicle or in confined space, protective headgear
- Insulated tools which are clean and in good condition

- Ensure that the vehicle is turned OFF and gear selection is in 'P'.
- Check that you have access to any re-set codes, such as for the ICE and security system. Check any driver-configured settings, noting as appropriate.
- Some hybrids and electric vehicles have an off switch. This may be underneath the panels over the battery. Switch this off.
- Remove disconnect plug or, if there are separate terminals, remove the negative terminal first.

Class Color	Proof Test Voltage AC/DC	Max. Use Voltage AC/DC	Insulating Rubber Glove Label
00 Beige	2,500 / 10,000	500 / 750	10 ASTM D120 CLASS 00 TYPE I MAX USE VOLT 500V AC EN60903
0 Red	5,000 / 20,000	1,000 / 1,500	10 ASTM D120 CLASS 0 TYPE I MAX USE VOLT 1000V AC EN60903
1 White	10,000 / 40,000	7,500 / 11,250	10 ASTM D120 CLASS I TYPE I MAX USE VOLT 7500V AC EN60903
2 Yellow	20,000 / 50,000	17,000 / 25,500	10 ASTM D120 CLASS 2 TYPE I MAX USE VOLT 17000V AC EN60903
3 Green	30,000 / 60,000	26,500 / 39,750	10 ASTM D120 CLASS 3 TYPE I MAX USE VOLT 26500V AC EN60903
4 Orange	40,000 / 70,000	36,000 / 54,000	10 ASTM D120 CLASS 4 TYPE I MAX USE VOLT 36000V AC EN60903

ASTM Labeling Chart
Natural Rubber Electrical Insulating Gloves

Figure 19.7 Electrical glove labelling – ASTM

- Wait 15 minutes for any static electricity to drain.
- You can now proceed with your repairs.

Re-fitting will be in reverse order. Many cars have driver-configurable setting that can be disturbed when the battery is removed. These should be restored to the driver's setting. Typical driver configurable settings are:

- Window closure position
- Seat position settings
- Automatic lights operation, both interior and exterior
- Mirror control
- Security system and in-car entertainment.

Repair pods

The nature of vehicle repairs is changing, as we noted at the beginning of this chapter. The amount of time spent on each repair is much shorter now compared to only a few years ago. This has led to the creation of repair pods.

The repair pod is a combination of a micro-size body repair workshop and spray booth. The fittings inside the pod will depend on the target market – typically a tool set for general removing and refitting panels and components, a small paint mixing system and the necessary extraction system. Because the pods are portable, the planning regulations are very minimal, and the pods can readily be re-sited to suit business needs. Some repair pods have been located in supermarket carparks so that customers can do their shopping whilst the car is being repaired.

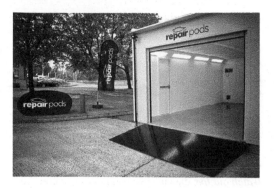

Figure 19.8 Moveable body repair pod (*courtesy Martin Penning, RepairPods.com*)

In the UK, there is now a trend, following on from the United States, of having several small businesses operating in and around supermarket locations. Typically these are shoe repairer, key cutting, dry cleaning or photo printers, so a repair pod fits in easily. The repair pod offers a simple way of setting up a new business. Some of these are operated on a franchise basis in which a central office takes booking and payments, allowing the body technician to do productive work.

Quick reference guide to vehicle systems

The MET technician is required to have a working knowledge of the vehicle systems. This knowledge is needed to be able to safely remove and re-fit the components without damaging them and to be able to check the operation of the systems when the vehicle is repaired. This next section does this very briefly, giving guidance on key points.

Component – **air inlet**

Watch out for: damaged air filter housing, damaged drive belts, the lubrication oil supply pipe from the engine

Component – **braking system**

Watch out for: brake fluid leaks and damaged brake lines. Brake fluid is corrosive to paint work.

Function: guide air into the engine; may have supercharger or turbocharger

Tech Tip: Also check the pollen filter after filing or spraying around the front of the car.

Function: stop the vehicle

Tech Tip: If any brake components are replaced, ensure that the brakes are working before moving the vehicle.

Component – **cooling system**

Watch out for: when hot the pressure is high – scalding. Anti-freeze is corrosive and can damage paint work.

Function: keeping the engine at a constant temperature, typically 80°C

Tech Tip: Always use new clips when replacing hoses; use pre-mixed anti-freeze to ensure that the ratio is correct.

Component – **drive line/final drive**

Watch out for: be aware of losing oil where shafts fit into housings. Check all casings for damage.

Function: transmit drive from the engine to the to the wheels

Tech Tip: Before disassembling, take photos to enable correct re-fitting. Check correct oil type and top up oil with car in level position.

Component – **electrical system**

Watch out for: damaged wiring looms and cracked component casings

Function: provide numerous electrical functions

Tech Tip: Check functionality of components and configurability of setting before stripping and then after re-assembly.

Component – **engine**

Watch out for: broken engine mounts, damaged drive belts, damaged sump or cam-box covers, oil leaks

Function: provides motive power for vehicle

Tech Tip: It is usually the heaviest component of the vehicle and may take a large percentage of frontal impact. Inspect very closely before carrying out any body repairs, as a damaged engine would usually mean a vehicle write-off.

Component – **exhaust**

Watch out for: cracked manifold, bent tail-pipe and especially fractured catalytic converter

Function: transmit burnt engine gases to the rear of the vehicle

Tech Tip: If possible, run the engine during inspection and listen to the exhaust. Look for leaks and tap the 'cat' gently to check for internal damage.

Component – **fuel supply system – petrol**

Watch out for: petrol leaks, damaged petrol tank, petrol leaks. Petrol is highly flammable.

Function: supply petrol to the engine

Tech Tip: Always drain petrol using a sealed and earthed petrol extractor pump into a sealed container.

Component – **fuel supply system – diesel**

Watch out for: very-high-pressure diesel fuel systems; some work at up to 2500 bars. Check for very fine cracks of components which will break under the high pressure.

Function: supply diesel fuel to the engine

Tech Tip: Diesel fuel is very hazardous Always use suitable gloves when working on diesel systems. Always drain into sealed containers.

Component – **gearbox – manual**

Watch out for: damaged casings and control levers or cables

Function: provide different gear ratios for varying road situations

Tech Tip: Check for free movement of gear lever and clutch operation.

Figure 19.9 Spark plug coil pack

Figure 19.10 Sectioned manual gearbox

Component – **gearbox – automatic**	*Function*: provide different gear ratios for varying road situations
Watch out for: damage to gearbox oil cooler and oil leaks	*Tech Tip*: Check steering wheel paddle controls and mode settings.
Component – **ignition system**	*Function*: provide high voltage to spark plugs
Watch out for: broken steering lock, and ECU damage	*Tech Tip*: Check the inertia switches for fuel and SRS before suspecting the ignition switch.

Component – **lubrication system**	*Function*: lubricates and cools the engine components
Watch out for: oil leaks and oil cooler damage	*Tech Tip*: Check oil only on level surface.
Component – **suspension and steering**	*Function*: provide smooth ride and control vehicle direction
Watch out for: bent or buckled wheels and tyre damage	*Tech Tip*: Carry out full suspension check if any suspension damage is suspected.

Figure 19.11   Engine management system ECU

Steering and suspension

> **Study note**
> Wheel alignment and steering geometry are also discussed in Chapter 14 – Major Accident Damage.

Toe-in is when the distance between the wheels is less at the front than the rear. Toe-out is the opposite.

Figure 19.12   Toe-in

Figure 19.13 Toe-out on turns – inner wheel turns through greater angle than the outer one

Ackermann principle is when the inner wheel turns through a greater angle than the outer one.

Front-end impact is very likely to damage steering or suspension components, which control the toe-in. The steering must be aligned correctly to ensure that the vehicle will steer safely.

Supertracker – This is piece of equipment – there are other makes and models – is used to check all four wheels at the same time. You can use it to check:

- Front wheel alignment – tracking
- Rear wheel alignment – tracking
- Using turn-tables check toe-out on turns
- Castor angle
- Camber angle
- KPI – also called swivel pin inclination
- That the front and rear wheels are aligned with each other.

When you are using this or similar equipment, you are advised to check the manufacturer's setting and tolerances and to note any specific points that may affect the reading, for example:

- Vehicles with torsion bar rear suspension may have a shorter wheel base on one side than the other to accommodate the long torsion bars.
- Vehicles may have wider track at the rear than at the front.
- Vehicles may have different wheels sizes and/or tyre sizes.

- Check tyre types – many owners now have two sets of tyres, summer tyres and mud and snow – M&S – for winter.

Advanced driver assistance systems – ADAS

This technology is developing very quickly and is outside the scope of this book, but the reader should be aware of it and seek

Figure 19.14 Supertracker (*courtesy of Supertracker.com*)

Figure 19.15 Supertracker (*courtesy of Supertracker.com*)

specific guidance from the particular manufacturer or from Thatcham Research when needed. The following section lists some of the ADAS used on current vehicles.

Thatcham Research

Originally called the Motor Insurers' Automotive Research Centre, it was set up in 1969 at Thatcham in Berkshire to look at economical and effective ways of repairing the then-new unitary-construction cars. Its work now involves crash testing and engineering consultancy for vehicle manufacturers.

- Head-up displays
- LED matrix head lamps
- Camera systems
- Automatic braking systems
- Automatic lighting systems
- Automatic door-locking systems
- Automatic mirror adjustment
- Automatic adjustable spoiler systems
- Cruise control
- Automatic screen wipers
- ABS
- Traction control
- Automatic parking systems
- Autonomous vehicles

Skills and questions

The role of the MET technician is constantly evolving and becoming more important in the repair of vehicles. As an MET technician, you need to be continually observant and looking at all new vehicles as they are released to see how the new systems work and how they are tested. Visiting trade shows and vehicle launches is a good way to improve your knowledge, and attending short courses run by specialist equipment manufacturers allows you to improve your technical and practical skills.

1 Explain the difference between heat and temperature.
2 How can you confirm that a lamp that you are replacing will give the correct amount of brightness?
3 What does an ECU do?
4 What are the differences between the following batteries: (a) starter battery in a stop-start car, (b) battery in a hybrid car, (c) ordinary battery in an older car?
5 State one of Newton's three laws, and give an example of how it might be useful in the body repair workshop.
6 What is a repair pod, and how may it be used in the repair of vehicle bodies?
7 What is Thatcham Research?
8 Name and state the function of three mechanical components.
9 You have been asked to remove a vehicle radiator. Using a bullet points, outline the procedure and clearly state any precautions to be taken.
10 State three of the safety precautions to be observed when removing the power battery from a hybrid car.

Appendix 1

Health and safety risk assessment and COSHH data sheets

These are examples of the information that should be read before doing this type of work and kept for easy reference in the workshop.

The format of these forms is quite comprehensive. The formats and indicative content may vary considerably between organizations and types of products and processes.

These examples should not be taken as definitive or as necessarily appropriate for any purpose. You are advised to carry out a risk assessment for every task that you do, and obtain data for every product that you use.

General Risk Assessment Form – for outdoor event – L'Eroica Britannia, Bakewell, June 2016

1 Organization – XXXXXXX				2 Assessor(s) XXXXXXX					
3 Description of Task – taster of brazing activity as used in vehicle restoration course									
Ref No.	What are the hazards?	Who might be harmed and how?	What are you already doing? List the control measures already in place	Likelihood × Severity = Risk (see Risk Table)			Risk Rating H, M, L	What further action, if any, is necessary? If so, what action is to be taken by whom and by when? Use additional sheet if required.	What is the risk rating now – H, M, L?
				L ×	S=	R			
01	Toxic fumes	Staff and students may be exposed to hazardous fumes, causing repertory damage.	This is an outdoor event – fumes controlled by clean flames. Activity clearly signposted. Gas securely stored when not in use.	3	3	9	Low	Viewing public will be at least 2 metres away from flames	Low
02	Naked flames/ high-intensity light	Staff and students at risk of burns through incorrect or defective equipment in use. Eye damage.	Instruction, training, and close supervision by qualified tutors/ technicians. Correct PPE identified, to be worn at all times. Students to be fully informed on all danger regarding these high-risk processes. No flammable or combustible material within immediate vicinity. Hoses and joints in good condition.	3	2	6	Low	Regular checks on gas cylinders and equipment.	Low

Ref No.	What are the hazards?	Who might be harmed and how?	What are you already doing? List the control measures already in place	Likelihood × Severity = Risk (see Risk Table)			Risk Rating H, M, L	What further action, if any, is necessary? If so, what action is to be taken by whom and by when? Use additional sheet if required.	What is the risk rating now – H, M, L?
				L ×	S=	R			
03	Hot flying particles	Staff and students are at risk of burns, potentially causing short-term or permanent disability, blindness, skin damage, life-changing injuries	Instruction, training, and close supervision by qualified tutors/ technicians. PPE to be worn at all times. Students to be fully informed on all danger regarding these high-risk processes. Screening in place to keep non-participating public at least 2 metres from activity.	4	3	12	Medium	Burn dressings in First Aid boxes	Medium
04	Slips, trips and falls	Staff and students fractures, cuts, impact injuries	Cable management/no trailing leads. Good housekeeping policy. Walkways kept clear. All hazards/fabric of the flooring issues reported to organisers. Adequate lighting. Students informed of no-running policy.	2	3	6	Low	No action or low-priority action (continue to review periodically to ensure controls remain effective)	Low
05	Metal splinters	Staff and students Cuts, injection wounds	Student information where appropriate. Suitable and sufficient PPE, e.g. appropriate eyewear, that conforms to British Standards. Suitable and sufficient hand tools for performing task.	2	3	6	Low	No action or low-priority action (continue to review periodically to ensure controls remain effective)	Low
06	Use of hand and electric tools	Staff and students Cuts, puncture wounds, electric shock	Where applicable, all portable appliance equipment should be PAT tested annually. Pre-use inspection of equipment and cables for signs of damage. Student safety information on safe use of equipment. Trained first aiders available.		5	10	Medium	Medium priority (continue to implement additional reasonably practicable controls where possible and monitor regularly)	Med

Ref No.	What are the hazards?	Who might be harmed and how?	What are you already doing? List the control measures already in place	Likelihood × Severity = Risk (see Risk Table) $L \times$	$S=$	R	Risk Rating H, M, L	What further action, if any, is necessary? If so, what action is to be taken by whom and by when? Use additional sheet if required.	What is the risk rating now – H, M, L?
07	Bright light glare	Staff and students Eyestrain	Screening in place to keep non-participating public at least 2 metres from activity. Participating staff and students to wear approved safety glasses.	2	3	6	Low	No action or low-priority action (continue to review periodically to ensure controls remain effective)	Low
08	Aches and strains	Staff and students Back and repetitive strain injury	The equipment is positioned so that it is comfortable to use. This will be checked for each individual in terms of height and reach.	2	3	6	Low	No action or low-priority action (continue to review periodically to ensure controls remain effective)	Low

Tick (√) if any of the identified hazards relate to any of the following specific themes:

Hazardous Substance	Manual Handling	Display Screen Equip	Work Equip/ Machinery	Stress	Individual Person Such as Young Person, New/ Expectant Mother or Student
www.calor.co.uk/business/ lpg-safety-data-sheets	**None**	**None**	None	**None**	**Not appropriate for this group**

Link to data sheet for propane – this is to be stored in secure area when not in use.

Risk Rating
The risk rating is used to prioritise the action required. Deal with those hazards that are high risk first.

Risk Rating	Description	Action Priority
High	Where harm is certain or nearly certain to occur and/or major injury or ill health could result	**Urgent action** (Identify new controls. Activity must not proceed until risks are reduced to a medium or low level.)
Medium	Where harm is possible to occur and/or serious injury could result, e.g. off work for more than 3 days	**Medium priority** (continue to implement additional reasonably practicable controls where possible and monitor regularly)
Low	Where harm is unlikely or seldom to occur and/or minor injury could result, e.g. cuts, bruises, strain, or minor physiological scars	**No action or low-priority action** (continue to review periodically to ensure controls remain effective)

Assessed by: XXXXXXX   **Responsible Person: XXXXXXX**
Date Assessed: 6 June XXXX   **Review Date: June XXXX**

Communication and Review – This risk assessment should be communicated to all employees and relevant persons who may come into contact with the hazards being assessed. The assessment must be reviewed annually or following a significant change, accident or violent incident

Risk Table

L	5	5	10	15	20	25
I K	4	4	8	12	16	20
E L	3	3	6	9	12	15
I H	2	2	4	6	8	10
O O	1	1	2	3	4	5
D		1	2	3	4	5
X	Severity					

Likelihood × Severity = Risk
1 > 6 Low Risk 8 > 12 Medium Risk 15 > 25 High Risk

Likelihood

- Very likely (occurs frequently) = 5
- Likely (recurrent but not frequent) = 4
- Possible (could occur but uncommon) = 3
- Unlikely (occurs rarely) = 2
- Remote (almost never) = 1

Severity

- Fatal (single or multiple) = 5
- Serious (fracture major injury/damage) = 4
- Moderate (3-day injury or more) = 3
- Minor (small cut, abrasion, basic first aid) = 2
- Trivial (discomfort, slight bruising) = 1

Technical Publication

CALOR   Calor Safety Data Sheet - Liquefied Propane Gas

Safety Data Sheet No 2 Revision 11
Replaces Revisions 03/00, 04/03, 08/05, 03/06, 06/09, 02/10, 12/10, 07/11, 12/14, 09/16

This safety data sheet (SDS) has been prepared in accordance with the requirements of EU Regulations 1907/2006 & 453/2010(REACH) and 1272/2008 & 2015/830(CLP).

1. Identification of the Substance/Mixture and of the Company/Undertaking

1.1 Product Identifier	
Substance Name	Liquefied Propane Gas (odorised)
Product Description	Liquefied Propane Gas consisting predominately C_3 Hydrocarbons supplied as a fuel in a closed system meeting the requirements for commercial propane of BS4250 with <0.1% 1,3 Butadiene.
Trade Name(s)	Calor Propane, Calor Autogas, Calor Patio Gas & Calor High Purity Propane
CAS Number	68476-85-7
EC Number	270-704-2
Reach Registration Number	Exempt (Annex V) as Liquefied Petroleum Gas (LPG)
1.2 Relevant identified uses of the substance or mixture and uses advised against	
Identified Use(s)	Multi-purpose product intended for uses including fuels for equipment which has been specifically designed to run on commercial propane, an internal combustion engine and fuel feedstock for the petrochemical industry
Uses Advised Against	Anything other than the above
1.3 Details of the Supplier of the safety data sheet	
Company:	Calor Gas Limited
Address:	Athena House, Athena Drive, Tachbrook Park, Warwick, CV34 6RL
Telephone:	01926 330088
Web Address:	www.calor.co.uk
National Contact:	01926 318497 (Technical Help Desk)
E-mail (competent person)	healthandsafety@calor.co.uk
1.4 Emergency telephone number	
Emergency Number:	0345 7 444 999 (Availability 24/7)
Language spoken	English

2. Hazards Identification

2.1 Classification of the substance or mixture

Classification according to Directive 67/548/EEC or 1999/45/EC
F+; R12 - extremely flammable

Classification according to Regulation (EC) No 1272/2008 as amended.

Physical Hazards	
Flammable gas , category 1	H220: extremely flammable gas
Gases under pressure	H280: contains gas under pressure; may explode if heated

2.2 Label Elements

Label elements according to Regulation (EC) No. 1272/2008

Hazard Pictograms

Signal words: Danger

Hazard statements: H220: extremely flammable gas
                           H280: contains gas under pressure; may explode if heated

Precautionary Statement

Prevention: P102   Keep out of the reach of Children
                           P210: Keep away from heat, hot surfaces, sparks, open
                           flames and other ignition sources. No smoking.
                           P243: Take precautions against static discharge

Response: P377: Leaking gas fire: Do not extinguish, unless leak can
                           be stopped safely.
                           P381: Eliminate all ignition sources if safe to do so.

Storage: P410 + P403: Protect from sunlight. Store in a well-
                           ventilated place.

2.3 Other Hazards   Cold burns (frostbite) will result from skin/eye contact with
                           liquid product.

                           Vapour is heavier than air and may travel to remote
                           sources of ignition (e.g. along drainage systems, into
                           basements etc.).

                           Abuse involving wilful inhalation of very high
                           concentrations of vapour, even for short periods can
                           produce unconsciousness and might prove fatal. Inhalation
                           may cause irritation to the nose and throat, headache,
                           nausea, vomiting, dizziness and drowsiness. In poorly
                           ventilated or confined spaces, unconsciousness or
                           asphyxiation may result.

                           The material does not meet the criteria for PBT or vPvB in
                           accordance with REACH Annex XIII

3. Composition and Information on Ingredients

3.1 Substance   Not Applicable. The material is regulated as a mixture

3.2 Mixtures   Product is to a mixture according to regulation 1907/2006/EC

Identifier Name	Index No.	CAS No.	EC No.	%v/v
Petroleum gases, liquefied	649-202-00-6	68476-85-7	270-704-2	>99%
1,3 Butadiene	601-013-00-X	106-99-0	203-450-8	<0.1%
Methanol (anti-icing agent)	603-001-00-X	67-56-1	200-659-6	<0.125
Ethyl Mercaptan (odorant)	016-022-00-9	75-08-01	200-837-3	<0.005

4. First Aid Measures

4.1 Description of first aid measures

Inhalation: Remove the affected person to fresh air. Keep the patient warm and at rest. If breathing has stopped administer artificial respiration. Give external cardiac massage if necessary. If the person is breathing, but unconscious, place them in the recovery position. Obtain medical assistance immediately.

Skin Contact: Burns should be flushed with tepid water to normalise temperature and until circulation returns. Cover the burns with sterile dressings. Do not apply ointments or powders. Obtain medical assistance immediately.

Eye Contact: Cold burns should be flushed immediately with tepid water to normalise temperature. Hold eyelids apart while flushing to rinse entire surface of the eye and lids with water. Cover the eye with a sterile dressing and obtain medical assistance immediately.

Ingestion: Not applicable

Protection for First-Aiders: No action shall be taken involving any personal risk or without suitable training. It may be dangerous to the person providing aid to give mouth-to-mouth resuscitation.

4.2 Most important symptoms and effects, both acute and delayed

Frostbite (cold burn). The vapour may have narcotic effect. Overexposure may cause adverse effect such as drowsiness, dizziness, disorientation, vertigo. Severity of the symptoms will vary dependent on the concentration and the length of exposure

4.3 Indication of any immediate medical attention and special treatment needed
Notes for the doctor - treat symptomatically

5. Fire Fighting Measures

5.1 Extinguishing Media
Large Fires
- None. Product flow must be stopped and container cooled by water spray. Water fog should be used to assist approach to source of the fire. Large fires should only be fought by the Fire Brigade.
- DO NOT USE WATER JET

Small Fires
- Dry powder
- DO NOT USE WATER OR FOAM

5.2 Special hazards arising from the substance or mixture
These materials are delivered, stored and used at temperatures above their flash point. Avoid all naked flames, sparks, cigarettes, etc.

IN CASE OF FIRE, EVACUATE THE AREA AND IMMEDIATELY ALERT THE FIRE BRIGADE
- Ensure an escape path is always available from any fire.
- If gas has ignited, do not attempt to extinguish but, if safe to do so, stop gas flow and allow to burn out.
- Use water spray to cool heat-exposed containers and to protect surrounding areas and personnel effecting shut-off.

- Beware of vapour accumulating to form explosive concentrations. Explosive vapours may travel, be ignited at remote locations and flash back. A water spray may be used for vapour dispersal.

Pressurised containers are liable to explode violently when subjected to high temperatures

Every precaution must be taken to keep containers cool to avoid the possibility of a boiling liquid expanding vapour explosion (BLEVE).

Fires in confined spaces should be dealt with by trained personnel wearing approved breathing apparatus.

5.3 Advice for the fire-fighters

Fight fire with normal precautions from a reasonable distance. Fight fighters should wear complete protective clothing including self-contained breathing apparatus. Keep containers cool by spraying with water if exposed to fire. Avoid release to the environment.

6. Accident Release Measures

6.1 Personal Precautions, Protective equipment and Emergency Procedures

Immediate Emergency Action:	<ul><li>Clear people away from the area to a safe place</li><li>Do not operate electrical equipment unless flameproof</li><li>Summon aid of emergency services</li><li>Treat or refer casualties if necessary</li></ul>
Further Action – Fire, if safe to do so	<ul><li>Stop product flow</li><li>Use dry powder or carbon dioxide extinguishers</li><li>Cool containers exposed to fire by water fog/spray</li></ul>
Further Action – Spillage, if safe to do so	<ul><li>Extinguish naked lights, e.g. cigarettes – AVOID MAKING SPARKS. Do not use a mobile phone</li><li>Isolate power from sources of ignition and ventilate the area</li><li>Position fire fighting equipment</li><li>Try to stop the flow of liquid product</li><li>Cover drains and sewers. Disperse vapour with water spray</li></ul> **Note:** *Vapour may collect in confined spaces*

6.2 Environmental Precautions
Avoid release to the environment. Contain the spillage. Any large spillage into water courses must be informed to the relevant authorities.

6.3 Methods and material for containment and cleaning up
Only trained and properly protected personnel must be involved in clean-up operations.
Small scale: Contain spillages with sand, earth or any suitable adsorbent material. Allow small spillages to evaporate provided there is adequate ventilation. Transfer to a lidded container for disposal or recovery. Ventilate the area and wash spill site after material pick-up is complete.
Large scale: In case of major fire and large quantities: Evacuate area. Fight fire remotely due to the risk of explosion. Notify police and fire brigade as soon as possible.

6.4 Reference to other sections
See sections 8 & 13

7. Handling and Storage

General

Cylinders containing Calor Liquefied Propane Gas are designed to give liquid or vapour offtake.
- Vapour offtake cylinders must be used in the vertical position with the outlet valve at the top.
- Liquid offtake cylinders must be stored and used in the position indicated on the cylinder.

A face shield or safety goggles and impervious rubber gloves should be worn when transferring this product as a liquid.

7.1 Precautions for safe handling
- No smoking or naked lights
- Switch off mobile phones
- Ensure good ventilation
- Avoid inhalation of vapour
- Avoid contact with liquid and cold storage containers
- When handling cylinders wear protective footwear and suitable gloves.
- Avoid contact with eyes.

7.2 Conditions for safe storage, including any incompatibilities

Calor Liquefied Propane Gas must be stored in purpose designed mild steel cylinder(s) or tank(s) or other systems of suitable pressure rating. These should be segregated from oxidant gases and other oxidants in store. Reference should be made to the relevant Codes of Practice for Safe Storage and Handling of LPG produced by HSE and UKLPG (See Section 16)
- No smoking or naked lights
- Switch off mobile phones
- Store and use only equipment/containers designed for use with this product
- Store and dispense only in well ventilated areas away from heat and sources of ignition.
- Containers must be labelled properly
- Do not remove warning labels from containers
- Check that cylinders are within test date. If overdue for inspection they must be returned to Calor Gas Limited.
- Ensure that Pipework and handling equipment are designed for the purpose, inspected and maintained and are electrically bonded and grounded (earthed) to prevent accumulation of static charge
- Explosive air/vapour mixtures may form at ambient temperature

7.3 Specific End Use(s)

Calor Liquefied Propane Gas is a multi-purpose product intended for uses including;
- Fuel for equipment which has been specifically designed to be powered by Commercial Propane;
- Internal combustion engine fuel;
- Feedstock for petrochemical industry.

Note: Product spilt on clothing may give rise to delayed evaporation and subsequent fire hazard

8. Exposure Controls / Personal Protection

8.1   Control Parameters
The following limits are taken from the Health and Safety Executives Guidance Note EH40 Workplace Exposure Limits.

8.1.1 Occupational Exposure Limit Values
Calor Gas Liquefied Propane Gas* is not subject to a specific OEL. However as a Liquefied Petroleum Gas the following OEL should be applied:

	LTEL (8hr TWA)	STEL (15 min Period)
Liquefied Petroleum Gas	1000 ppm (1750 mg/m^3)	1250 ppm (2180 mg/m^3)

Pure Propane is identified as a simple asphyxiant and EH40 paragraphs 57 & 59 apply.

Recommended Monitoring Procedures Personal or workplace atmosphere monitoring may be required to determine the effectiveness of the ventilation or other control measures and/or the necessity to use respiratory protective equipment. Reference should be made to European Standard EN 689 for methods for the assessment of exposure by inhalation to chemical agents and national guidance documents for methods for the determination of hazardous substances.

8.1.2 Biological limit value: None established

8.2   Exposure Controls

8.2.1 Appropriate Engineering Controls
Provide natural or explosion-proof ventilation that is adequate to ensure flammable gas does not reach its lower explosive limit.

8.2.2 Individual protection measures, such as Personal Protective Equipment (PPE)

Protective clothing
- Wear suitable gloves and overalls to prevent cold burns and frostbite (Neoprene or LPG resistant Gauntlet Glove).
- In filling operations wear protective clothing including impervious gloves, safety goggles or face shields to BS EN 166,167 & 168. (Note: at retail Autogas applications alternative arrangements may be put in place).
- When handling cylinders wear protective footwear to BS EN345

Respiratory protection
- If operations are such that significant exposure to vapour may be anticipated, then suitable approved respiratory equipment should be worn.
- Respiratory Protective Equipment (RPE) may be used and the selection of RPE must be based on actual or anticipated exposure levels and the safe working limits of the selected RPE. Material: Filter AX Guidance: EN14387, *Respiratory protective devices. Gas filter(s) and combined filter(s). Requirements testing and marking.*
- Respirators **must not** be used in oxygen-deficient atmospheres. If there is a risk of oxygen depletion, then suitable breathing apparatus is required.
- The use of respiratory equipment must be strictly in accordance with manufacturers' instructions and any statutory requirements governing its selection and use.
- All wearers of respiratory protection must be trained in its use. The nature of the atmosphere and the working environment will determine the protection required. Equipment must be to the

relevant British and European Standards and this may be determined by reference to BS4275: *Recommendations for the selection, use and maintenance of respiratory protective equipment.*

8.2.3 Environmental Exposure Controls

Not applicable. The substance is a vapour at normal temperature and pressure. In normal use it is not discharged into the atmosphere but used as a fuel.

9. Physical and Chemical Properties

9.1   General Information

Appearances:	Colourless liquefied gas
Odour:	Odorant added to provide a distinctive smell
Odour Threshold:	<20% of Lower Flammable Limit
pH:	Not Applicable
Melting Point:	-159°C
Boiling Point:	-42°C
Flash Point:	-104 °C (PMCC)
Evaporation Rate:	Not Available
Flammability Limits:	2% to 11% in air
Autoflammability:	460-580 °C
Vapour Pressure:	7.5 bar at 15 °C
Specific Gravity of Liquid:	0.512 at 15 °C (Water = 1.0)
Specific Gravity of Vapour:	1.5 at 15 °C (Air = 1.0)
Solubility in Water:	Negligible
Viscosity:	Not established
Explosive properties:	Vapour may create explosive atmosphere:
Oxidising properties:	Not oxidising

9.2   Other Information

No other information is relevant to this product.

10. Stability and Reactivity

10.1 Reactivity: Stable under normal conditions

10.2 Chemical Stability: Calor Liquefied Propane is stable at ambient temperatures. Hazardous polymerization will not occur, however, it can form explosive mixture with air.

10.3 Possibility of Hazardous Reactions: Hazardous polymerisation will not occur

10.4 Conditions to avoid:	• Sources of ignition • Storage at above 50°C.
10.5 Materials to avoid:	Propane reacts violently with strong oxidising agents (e.g. chlorates which may be used in agriculture), peroxide, plastics, chlorine dioxide and concentrated nitric acid.
10.6 Hazardous Decomposition products:	The substance arising from the thermal decomposition of these products will largely depend upon the conditions bringing about decomposition. The following hazardous substances may be expected from normal combustion:

- Carbon Dioxide
- Carbon Monoxide may be produced (if there is insufficient air for complete combustion).

11. Toxicological Information

11.1 Information on the toxicological effects

Eye Contact:	Contact with Calor Liquefied Propane gas will present a risk of serious damage to the eyes.
Skin Contact:	Contact with Calor Liquefied Propane gas will cause cold burns and frostbite to the skin.
Inhalation:	Low vapour concentrations may cause nausea, dizziness, headaches and drowsiness. May have a narcotic effect if high concentrations are inhaled. High vapour concentrations may produce symptoms of oxygen deficiency which, coupled with central nervous system depression may lead to rapid loss of consciousness.
Volatile Substance Abuse:	Under normal conditions of use the product is not hazardous; however, abuse involving deliberate inhalation of very high concentrations of vapour, even for short periods, can produce unconsciousness and/or result in a sudden fatality.
Gem cell mutagenicity:	No known behaviour. Contains <0.1% butadiene
Carcinogenicity:	No known behaviour. Contains <0.1% butadiene
Reproductive Toxicity:	No known behaviour
Aspiration hazard:	Based upon available data, the classification is not met
STOT–single exposure:	Based upon available data, the classification is not met
STOT-repeated exposure:	Based upon available data, the classification is not met

11.2 Other information   None

12. Ecological Information

12.1 Toxicity:	No known ecological damage is caused by this product.
	Air: Calor Liquefied Propane gas is a mixture of volatile components which, when released to air, will react rapidly with hydroxyl radicals and ozone to give carbon dioxide and water.
	Water: If released to water the product will rapidly evaporate.
	Soil: If released to soil the product will rapidly evaporate.
12.2 Persistence and Degradability:	Unlikely to cause long term adverse effects in the environment.
12.3 Bio-accumulative Potential	This material is not expected to bio-accumulate.
12.4 Mobility in soil:	Spillages are unlikely to penetrate the soil.

12.5 Results of PBT and vPvB: Assessment	Not classified as PBT or vPvB
12.6 Other adverse effects:	No known behaviour

13. Disposal Considerations

13.1 Waste treatment methods

Product is suitable for burning in an enclosed controlled burner for fuel value or disposal by supervised incineration at very high temperature to prevent the formation of undesirable combustion products.

13.2 Additional Information

Disposal Considerations:

- Calor Gas Cylinders are the property of Calor Gas Limited and should be returned to the local dealer/ stockist when no longer required.
- Users are recommended to contact their local Calor Gas representative when they wish to dispose of surplus quantities of Calor Propane.
- Emptying of tanks containing Calor Propane is the responsibility of Calor Gas Limited
- Do not discharge product into areas where there is a risk of an explosive mixture with air.
- Empty vessels or cylinders may contain some remaining product.
- Hazard warning labels are a guide to the safe handling of empty packaging and should not be removed.
- Empty containers represent a fire hazard as they may contain flammable product residues and vapour. Never incinerate, crush, weld, solder or braze empty containers.

14. Transport Information

Dangerous for Conveyance

UN Number	1978
UN Proper Shipping name:	Propane
Transport Hazard Class:	Flammable Gas
Packing Group:	Special Containers
ADR/RID Proper Shipping Name:	Propane
Substance Identification Number:	1978
Class:	2
Classification Code:	2F
Label:	2.1
IATA / ICAO Hazard Class:	2.1 (forbidden on passenger aircraft)
IMO Hazard Class:	2.1
Marine Pollutant:	No
Hazard Identification Number:	23
Emergency Action Code:	2YE

15. Regulatory Information

15.1 Safety, Health and Environmental Regulations/Legislation Specific for the Substance or Mixture

This material has been classified according to the requirements of the United Nations "Globally Harmonised System of Classification and Labelling of Chemicals" (GHS), EU Regulation 1271/2008 on the Classification, Labelling and Packaging of Substances and Mixtures (the CLP Regulation) and Article 31 of EU Regulation 1907/2006 (as amended) on the Registration, Evaluation, Authorisation and Restriction of Chemicals (REACH)

Dangerous for Supply

Hazard Statements

H220 Extremely flammable
H280 Contains gas under pressure; may explode if heated

Precautionary Statements

P102 Keep out of the reach of Children
P403 Store in a well-ventilated place
P210 Keep away from heat/sparks/ open flames / hot surfaces – NO SMOKING
P377 Leaking gas fire: Do not extinguish, unless leak can be stopped safely
P381 Eliminate all ignition sources if safe to do so.

Note: *As detailed in Annex I of CLP 2008/1272 "Closed refillable cylinders and non-refillable cylinders within the scope of EN417, containing fuel gases which are only used for combustion are* **only to be labelled with the appropriate pictogram and the hazard and precautionary statements concerning flammability.** *Such cylinders are exempted from carrying the risk and safety phases relating to health effects."*

P410 & P403 may be omitted for gases filled in transportable gas cylinders in accordance with packing instruction P200 of the UN RTDG, Model Regulations

Product label

Danger Extremely Flammable Gas
Contains : Calor Propane, Calor Autogas or Patio Gas
Symbol : Flame

GHS Label

Transport Label

EU Legislation

EU Directive 2012/18/EU (SEVESO III):   Listed Part 2 – Extremely Flammable
EU Regulation 1907/2006 on the Registration, Evaluation, Authorisation and Restriction of Chemicals (REACH);
EU Regulation 1271/2008 on the Classification, Labelling and Packaging of Substances and Mixtures (the CLP Regulation)

Relevant UK Legislation
- Dangerous Substances and Explosive Atmospheres Regulations
- The Regulatory Reform (Fire Safety) Order 2005
- Control of Substances Hazardous to Health Regulation

- Provision and Use of Work Equipment Regulations
- Equipment and Protective Systems Intended for Use in Potentially Explosive Atmospheres Regulations
- Personal Protective Equipment Regulations
- Control of Major Accident Hazards Regulations
- Chemical Hazards Information and Packaging for Supply
- Pressure Systems Safety Regulations
- Carriage of Dangerous Goods and Use of Transportable Pressure Equipment Regulations
- Chemical Hazard Information and Packaging for Supply Regulations
- Dangerous Substances (Notification and Marking of Sites) Regulations

15.2 Chemical Safety Assessment
Not Applicable. Liquefied Petroleum Gas (LPG) is exempt from REACH (Annex V)

16. Other Information

LEGEND

LTEL: Long Term Exposure Limit STEL: Short Term Exposure Limit
DNEL: Derived No Effect Level PNEC: Predicted No Effect Concentration
PBT: Persistent, Bio-accumulative and Toxic STOT: Specific Target Organ Toxicity
vPvB: Very Persistent and very Bio-accumulative

REFRENCES

Various sources of data have been used in the compilation of this SDS, they include but are not exclusive to:
European Chemical Agency: Guidance on the Compilation of Safety Data Sheets
EH40 (as amended): Workplace exposure limits.
BS: 4250: Specification for Commercial Butane and Commercial Propane
UKLPG Code of Practice No. 29: Hazard Information and Packaging Labelling for Commercial LPG Cylinders
Suppliers Safety Data Sheets

Health and Safety Advisory Literature

The UKLPG produce over 30 Industry Codes of Practice which can be obtained from UKLPG. For a comprehensive publication list please access the UKLPG website www.uklpg.org.

Further guidance the legislation listed in section 15 can be obtained from www.hse.gov.uk and publications can be purchased from HSE Books, PO Box 1999, Sudbury, Suffolk, CO10 6FS.Tel: 01787 881165 or www.hsebooks.co.uk

This Safety Data Sheet is solely intended for this product and the information is based upon the present state of knowledge. It is intended to give guidance and believed to be accurate and represent good practice at the time of publication. It does not replace the need to consult other formal documents where further information may be required.

SAFETY DATA SHEET
Simoniz Wash & Wax Shampoo

SECTION 1: IDENTIFICATION OF THE SUBSTANCE/MIXTURE AND OF THE COMPANY/UNDERTAKING

1.1. Product identifier

Product name	Simoniz Wash & Wax Shampoo
Product No.	SAPP0001A, SAPP0002A, SAPP0003A, SAPP0004A
REACH Registration notes	This is a MIXTURE; no registration information contained in this document . Holts are classed as Downstream User.

1.2. Relevant identified uses of the substance or mixture and uses advised against

Identified uses	Car maintenance product. Auto shampoo.

1.3. Details of the supplier of the safety data sheet

Supplier	Holt Lloyd International Ltd
	Barton Dock Road
	Stretford
	Manchester
	M32 0YQ - England, UK
	+44 (0) 161 866 4800
	FAX +44 (0) 161 866 4854
	A Holts Car Care Product
	www.holtsauto.com
Contact Person	Contact Email address: info@holtsauto.com

1.4. Emergency telephone number

UK - 00 44 (0) 161 866 4800 Office hrs = 0900 - 1700 hrs Out of office hours Tel: 020 7358 9167
National Emergency Telephone Number
http://echa.europa.eu/en/web/guest/support/helpdesks/national-helpdesks/list-of-national-helpdesks

SECTION 2: HAZARDS IDENTIFICATION

2.1. Classification of the substance or mixture

Classification (EC 1272/2008)

Physical and Chemical Hazards	Not classified.
Human health	Not classified.
Environment	Not classified.
Classification (1999/45/EEC)	Not classified.

 The Full Text for all R-Phrases and Hazard Statements are Displayed in Section 16.

2.2. Label elements

Detergent Labelling:

5 - < 15%	anionic surfactants
< 5%	amphoteric surfactants
	perfumes
Contains	1,2-BENZISOTHIAZOL-3(2H)-ONE,2-Methylisothiazolin-3-one

Label In Accordance With (EC) No. 1272/2008
No pictogram required.

Precautionary Statements

P102	Keep out of reach of children.
P101	If medical advice is needed, have product container or label at hand.

| P501 | | Dispose of contents/container in accordance with local regulations. |

2.3. Other hazards

SECTION 3: COMPOSITION/INFORMATION ON INGREDIENTS

3.2. Mixtures

Alcohols, C10-16, ethoxylated, sulfates, sodium salts		1-5%
CAS-No.: 68585-34-2	EC No.: 500-223-8	

Classification (EC 1272/2008)	Classification (67/548/EEC)
Not classified.	Xi;R38,R41.

PROPAN-2-OL		< 1%
CAS-No.: 67-63-0	EC No.: 200-661-7	

Classification (EC 1272/2008)	Classification (67/548/EEC)
Flam. Liq. 2 - H225	F;R11
Eye Irrit. 2 - H319	Xi;R36
STOT SE 3 - H336	R67

DIETHANOLAMINE		< 1%
CAS-No.: 111-42-2	EC No.: 203-868-0	

Classification (EC 1272/2008)	Classification (67/548/EEC)
Acute Tox. 4 - H302	Xn;R22,R48/22
Skin Irrit. 2 - H315	Xi;R38,R41
Eye Dam. 1 - H318	
STOT RE 2 - H373	

The Full Text for all R-Phrases and Hazard Statements are Displayed in Section 16.

| REACH Registration notes | This is a MIXTURE; no registration information contained in this document . Holts are classed as Downstream User. |

SECTION 4: FIRST AID MEASURES

4.1. Description of first aid measures

Inhalation

Move the exposed person to fresh air at once. Rinse nose and mouth with water. Get medical attention if any discomfort continues.

Ingestion

Remove victim immediately from source of exposure. Rinse mouth thoroughly. Drink plenty of water. Get medical attention.

Skin contact

Remove affected person from source of contamination. Remove contaminated clothing. Wash the skin immediately with soap and water. Get medical attention if irritation persists after washing.

Eye contact

Make sure to remove any contact lenses from the eyes before rinsing. Promptly wash eyes with plenty of water while lifting the eye lids. Continue to rinse for at least 15 minutes. Get medical attention if any discomfort continues.

4.2. Most important symptoms and effects, both acute and delayed

General information

The severity of the symptoms described will vary dependant of the concentration and the length of exposure.

4.3. Indication of any immediate medical attention and special treatment needed

Treat Symptomatically.

SECTION 5: FIREFIGHTING MEASURES

5.1. Extinguishing media

Extinguishing media
Extinguish with foam, carbon dioxide, dry powder or water fog. Dry chemicals, sand, dolomite etc.

5.2. Special hazards arising from the substance or mixture

Hazardous combustion products
Thermal decomposition or combustion may liberate carbon oxides and other toxic gases or vapours.
Unusual Fire & Explosion Hazards
No unusual fire or explosion hazards noted.

5.3. Advice for firefighters

Special Fire Fighting Procedures
No specific fire fighting procedure given.

SECTION 6: ACCIDENTAL RELEASE MEASURES

6.1. Personal precautions, protective equipment and emergency procedures

For personal protection, see section 8.

6.2. Environmental precautions

Not relevant considering the small amounts used.

6.3. Methods and material for containment and cleaning up

Wear necessary protective equipment. Absorb in vermiculite, dry sand or earth and place into containers. Do not contaminate water sources or sewer. Flush with plenty of water to clean spillage area.

6.4. Reference to other sections

For waste disposal, see section 13.

SECTION 7: HANDLING AND STORAGE

7.1. Precautions for safe handling

Avoid spilling, skin and eye contact.

7.2. Conditions for safe storage, including any incompatibilities

Store in tightly closed original container in a dry, cool and well-ventilated place. Keep in original container.

7.3. Specific end use(s)

The identified uses for this product are detailed in Section 1.2.

SECTION 8: EXPOSURE CONTROLS/PERSONAL PROTECTION

8.1. Control parameters

Name	STD	TWA - 8 Hrs		STEL - 15 Min		Notes
DIETHANOLAMINE	OES	3 ppm	13 mg/m3			
PROPAN-2-OL	WEL	400 ppm	999 mg/m3	500 ppm	1250 mg/m3	

WEL = Workplace Exposure Limit.

NON HAZARDOUS INGREDIENTS

Ingredient Comments
WEL = Workplace Exposure Limits

8.2. Exposure controls

Hand protection
For prolonged or repeated skin contact use suitable protective gloves.

SECTION 9: PHYSICAL AND CHEMICAL PROPERTIES

9.1. Information on basic physical and chemical properties

Appearance	Coloured liquid.
Colour	Blue.
Odour	Lemon. Fruity.
Solubility	Miscible with water
Relative density	1.01 20
pH-Value, Conc. Solution	6.8
Viscosity	168 cP 20
Flash point (°C)	
Not applicable.	

9.2. Other information

No information required.

SECTION 10: STABILITY AND REACTIVITY

10.1. Reactivity

No specific reactivity hazards associated with this product.

10.2. Chemical stability

Stable under normal temperature conditions.

10.3. Possibility of hazardous reactions

Not applicable.

10.4. Conditions to avoid

Avoid contact with acids.

10.5. Incompatible materials

Materials To Avoid
Strong acids.

10.6. Hazardous decomposition products

No specific hazardous decomposition products noted.

SECTION 11: TOXICOLOGICAL INFORMATION

11.1. Information on toxicological effects

Inhalation
May cause irritation to the respiratory system.

Ingestion
May irritate and cause stomach pain, vomiting and diarrhoea.

Skin contact

Prolonged and frequent contact may cause redness and irritation.

Eye contact

Irritating and may cause redness and pain.

SECTION 12: ECOLOGICAL INFORMATION

Ecotoxicity

The product components are not classified as environmentally hazardous. However, this does not exclude the possibility that large or frequent spills can have a harmful or damaging effect on the environment.

12.1. Toxicity

12.2. Persistence and degradability

12.3. Bioaccumulative potential

12.4. Mobility in soil

12.5. Results of PBT and vPvB assessment

12.6. Other adverse effects

SECTION 13: DISPOSAL CONSIDERATIONS

13.1. Waste treatment methods

Dispose of waste and residues in accordance with local authority requirements.

SECTION 14: TRANSPORT INFORMATION

General   The product is not covered by international regulation on the transport of dangerous goods (IMDG, IATA, ADR/RID).

14.1. UN number

Not applicable.

14.2. UN proper shipping name

Not applicable.

14.3. Transport hazard class(es)

Transport Labels

No transport warning sign required.

14.4. Packing group

Not applicable.

14.5. Environmental hazards

Environmentally Hazardous Substance/Marine Pollutant
No.

14.6. Special precautions for user

Not applicable.

14.7. Transport in bulk according to Annex II of MARPOL73/78 and the IBC Code

Not applicable.

SECTION 15: REGULATORY INFORMATION

15.1. Safety, health and environmental regulations/legislation specific for the substance or mixture

Statutory Instruments

The Chemicals (Hazard Information and Packaging for Supply) Regulations 2009 (S.I 2009 No. 716).

EU Legislation

Dangerous Substance Directive 67/548/EEC. Dangerous Preparations Directive 1999/45/EC. Regulation (EC) No 1272/2008 of the European Parliament and of the Council of 16 December 2008 on classification, labelling and packaging of substances and mixtures, amending and repealing Directives 67/548/EEC and 1999/45/EC, and amending Regulation (EC) No 1907/2006 with amendments.

National Regulations

Workplace Exposure Limits 2005 (EH40)

Authorisations (Title VII Regulation 1907/2006)

No specific authorisations are noted for this product.

Restrictions (Title VIII Regulation 1907/2006)

No specific restrictions of use are noted for this product.

Water hazard classification

WGK 1

15.2. Chemical Safety Assessment

SECTION 16: OTHER INFORMATION

Revision Date	11/10/2013
Revision	1
SDS No.	14720
Risk Phrases In Full	
R22	Harmful if swallowed.
R48/22	Harmful: danger of serious damage to health by prolonged exposure if swallowed.
R11	Highly flammable
R36	Irritating to eyes.
R38	Irritating to skin.
NC	Not classified.
R41	Risk of serious damage to eyes.
R67	Vapours may cause drowsiness and dizziness.
Hazard Statements In Full	
H225	Highly flammable liquid and vapour.
H302	Harmful if swallowed.
H315	Causes skin irritation.
H318	Causes serious eye damage.
H319	Causes serious eye irritation.
H336	May cause drowsiness or dizziness.
H373	May cause damage to organs <<Organs>> through prolonged or repeated exposure.

Glossary

This section defines a number of the words and phrases used in vehicle body repair. It comprises the main vocabulary of the vehicle body repairer or restorer.

Abrasive   A substance used for wearing away a surface by rubbing.

Acceleration   Acceleration is the rate of change of speed.

Accelerator   A constituent of synthetic resin mix which hastens a reaction.

Accessories   Optional extras not essential to the running of a vehicle, such as electric windows and satnav. A vehicle that is fitted with all possible accessories is referred to as fully loaded

Acetone   A liquid hydrocarbon capable of dissolving twenty-five times its own volume of acetylene gas at atmospheric pressure.

Acetylene   A combustible gas which is mixed with oxygen and used in oxy-acetylene welding.

Adhesion   The ability to adhere to a surface.

Adhesive   A substance that allows two surfaces to adhere together.

Adjustments   Necessary alterations to improve tolerances in fit.

Air bags   A passive safety restraint system that inflates automatically on vehicle impact to protect the driver. Also called supplementary restraint system (SRS).

Alignment   The operation of bringing into line two or more specified points on a vehicle structure.

All-metal construction   Generally this applies to those body shells of both private cars and light commercial vehicles in which the construction is in the form of steel pressings assembled by welding, thus forming a fabricated unit.

Alloy   A mixture of two or more metals, with or without other metallic or non-metallic elements.

A-post   A structural pillar on which the front door is hung.

Backfire   In gas welding, a momentary return of gases indicated in the blowpipe by a pop or loud bang, the flame immediately recovering and burning normally at the blowpipe.

Backhand welding   Sometimes classified as 'rightward welding'. A technique in which the flame is directed backwards against the completed part of the weld.

Back light   A central window in the rear panel of the driving cab, or the rear window of a saloon body.

BC-post   Central pillar acting as a central roof and side support between the rear and the front of the car.

Bevel angle   The angle of a prepared edge creating a bevel prior to welding.

Billet   An oblong piece of metal having a square section.

Binder   A resin or cementing constituent of a compound.

Blowpipe   A tool used for welding, known as a welding torch.

Body   The structured part of a vehicle which encompasses the passenger, engine and luggage compartments.

Body hardware   Functional accessories of vehicle body, e.g. door handles.

Body lock pillar   A body pillar that incorporates a lock striker plate.

Body mounting   Conventional body mounted on car chassis in composite method of body construction.

Body panels   Pressed metal panels or plastic moulded composite panels which are fastened together to form the skin of a car body.

Body side moulding   The exterior trim moulding fastened to the exterior of the body in a horizontal position.

Body sill   The panel directly below the bottoms of the doors.

Body soldering or body leading   The use of lead or lead substitute for smoothing surfaces.

Body spoon   A body repairer's hand tool.

Body trim   The materials which are used in the interior of the body for lining and upholstery.

Bonnet   The metal cover over the engine compartment.

Boot   A compartment provided in a car body which takes the luggage and often the spare wheel and fuel tank. It may be at the front or rear of the body depending upon the location of the engine.

Boot lid   Door covering luggage compartment.

Bottom side   The frame member of the base of the body extending along the full of the main portion of the body.

Brazing   A non-fusion process in which the filler metal has a lower melting point than the parent metal(s).

Buckles   The resulting distortion of body panels after collision.

Bulkhead   A transverse support in a body structure.

Bumping   Reshaping metal with a hammer and dolly.

Burr   The resulting condition left on a metal edge after cutting or filing.

Bursting disc   A type of pressure-relief device which consists of a disc, usually of metal, which is so held that it confines the pressure of the cylinder under normal conditions. The disc is intended to rupture between limits of over-pressure due to abnormal conditions, particularly when the cylinder is exposed to fire.

Butt joint   A welded joint in which the ends or edges of two pieces of metal directly face each other.

Calibrate   To check irregularities in measuring instruments.

Cant panel   The curved section of the roof top running between the comparatively flat top and the rain drip or gutter.

Cantrail   The longitudinal framing of the roof at the joint.

Carbon dioxide   A heavy colourless and incombustible gas which results from the perfect combustion of carbon.

Carbon fibre   An extremely strong, though expensive, reinforcement which can be used in conjunction with fibreglass. It gives increased rigidity to the laminate.

Carbonizing flame   An oxy-acetylene flame adjustment created by an excess of acetylene over oxygen, resulting in an excess of carbon in the flame.

Case hardening   This is the process of hardening the outer case or shell articles, which is accomplished by inducing additional carbon into the case of the steel by a variety of methods.

Catalyst   A chemical substance which brings about a chemical change to produce a different substance.

Catalyst dispenser   A purpose-designed container for measuring and dispensing liquid catalyst without splashing.

Catalytic converter   Referred to as a cat; helps to clean the exhaust gas.

Centre pillar   The centre vertical support of a four-door saloon.

Chassis   The base frame of a motor vehicle of composite construction to which the body is attached.

Check value   A safety device that controls the passage of gas or air in one direction in order to prevent the reversal of gas flow and a consequent accident.

Chemical reaction   The resulting change when two or more chemical substances are mixed.

Chopped-strand mat   Chopped strands bonded into a mat to produce a popular, economical, general-purpose reinforcement.

Chopped strands   As the name suggests, glass fibre strands chopped into short (about 12 mm) lengths. They can be used as fillers. Useful for bodywork repairs.

Circuit   The path along which electricity flows. When the path is continuous, the circuit is closed and the current flows. When the path is broken, the circuit is open and no current flows.

Cold curing Generic term for materials which harden at room temperature after the addition of catalyst.

Collapsible steering column A safety feature in the form of an energy-absorbing steering column designed to collapse on impact.

Compartment shelf panel The horizontal panel situated between the rear seat back and the back window.

Compression ratio (CR) This is the ratio between the volume of the gas above the piston when it is at BDC compared to that at TDC.

Compressive strength The ability of a material to withstand being crushed. It is found by testing a sample to failure: the load applied, divided by the cross-section of the sample, gives the compressive strength.

Condensation A change of state from a gas to a liquid caused by temperature or pressure changes. It may also be formed by moisture from the air being deposited on a cool surface.

Conductor Any material or substance that allows current or heat to flow through it.

Copper acetylide A spontaneously explosive and inflammable substance which forms when acetylene is passed through a copper tube.

Corrosion The wearing away or gradual destruction of a substance, e.g. rusting of metal.

Curing The change of a binder from soluble fusible state to insoluble infusible state by chemical action.

Curing time The time needed for liquid resin to reach a solid state after the catalyst has been added.

Cutting tip A torch especially adapted for cutting.

Cylinder Steel container used for storage of compressed gases.

Dash panel A panel attached to the front bulkhead assembly which provides a mounting for all instruments necessary to check the performance of the vehicle.

Deposited metal Filler metal from a welding rod or electrode which has been melted by a welding process and applied in the form of a joint or built up.

Detailing An extremely high level of cleaning, polishing and ensuring that a vehicle is clean in every detail. This process can take several days.

Diagnosis The determination, by examination, of the cause of a problem.

Dinging Straightening damaged metal with spoons, hammers or dollies. In the early days the ding-man was the tradesman who worked on completed bodies to remove minor imperfections without injury to the high-gloss lacquer or varnish.

Dinging hammer A special hammer used for dinging or removal of dents.

Direct damage Primary damage which results from an impact on the area in actual contact with the object causing the damage.

Dolly block A hand tool, made from special steel, shaped to suit the contour of various panel assemblies and used in conjunction with a planishing hammer to smooth out damaged panel surfaces.

Door skins Outside door panels.

Door trim The interior lining of a door.

D-post The rear standing pillar providing a shut face for the rear door and forming the rear quarter panel area.

Drip moulding A roof trough to direct water from door openings.

ECU Electronic control unit, This contains an integrated circuit to control the engine management system. In some cases, a separate one is used to control the body functions such as central locking and window operation.

Electrode The usual term for the filler rod which is deposited when using the electric arc welding process.

Electrolyte A substance which dissolves in water to give a solution capable of conducting an electric current.

Epoxy   Based on an epoxy resin which is mixed with a hardener.

Evaporation   A change of state from solid or liquid into vapour.

Expansion   The increase in the dimensions of metals due to heat.

Extrude   To draw into lengths.

Fender   American term for wing.

Filler   Inorganic types used to extend low-pressure resins, usually polyesters.

Filler metal   Metal added to a weld in the form of a rod, electrode or coil.

Fillet weld   A weld in which two surfaces at right angles to one another are welded together.

Firewall   Panel dividing engine compartment from interior of body.

Flange   A reinforcement on the edge of a panel formed at approximately right angles to the panel.

Flashback   Occurs when the flame disappears from the end of the welding tip and the gases burn in the torch.

Flat   A panel is said to be flat when insufficient shaping has caused uneven contours and so flat areas are obvious.

Floor pan   Main floor of the passenger compartment of an underbody assembly.

Flux   A chemical material or gas used to dissolve and prevent the formation of surface oxides when soldering, brazing or welding.

Foams (flexible)   A resin which is often used for cushioning in the automobile industries. These foams are usually urethanes.

Foams (rigid)   A resin with a higher modulus than the flexible foams. These are also normally urethanes and are used in more structural applications such as cores in sandwich constructions.

Force   An external effort which changes a body.

Four-door   Denotes the type of saloon body having four doors.

Frame gauges   Self-centring alignment gauges which are hung from a car's underbody.

Friction   Pressing the brake pads against the discs causes friction, which converts the dynamic energy of the moving vehicle into heat energy; the resistance to motion that a body meets when moving over another.

Fusion welding   A process in which metals are welded together by bringing them to the molten state at the surface to be joined, with or without the addition of filler metal, and without the application of mechanical pressure or blows.

Gas welding   A fusion welding process which uses a gas flame to provide the welding heat.

Gel   Resin takes on a gel-like consistency (gels) usually within 10 to 15 minutes of being catalyzed. At this point it is impossible to spray, paint or pour. Stored resin which has passed its shelf life may gel without being catalyzed.

Gelcoat   A thixotropic resin normally used without reinforcement and applied first to the mould. It forms the smooth, shiny surface of the finished article.

Glass fibre   Glass filaments drawn together into fibres and treated for use as reinforcement.

Hardener   A chemical curing or hardening agent.

Hardening   Heating to a critical temperature followed by a relatively rapid rate of cooling.

Headlining   The cloth or other material used to cover the inner surface of the car roof.

Heat and temperature   Temperature is the hotness or coldness of a body measured in degrees centigrade (C, also called Celsius) or absolute temperature in Kelvin (K). Heat is energy measured in joules (J).

Heelboard   The vertical board or panel under the rear seat which forms the support for the seat cushion.

Hinge pillar   A pillar on which a door is swung.

Hood   American term for bonnet.

Hydraulic pressure   Pressure transmitted by a liquid.

Hydraulics   The use of pressurized liquid to transfer force.

Impregnated   The particles of one substance infused into that of another.

Independent front suspension   Suspension system in which each wheel is independently supported by a spring.

Indirect damage   Secondary damage found in the area surrounding the damage which caused it.

Inertia   Inertia is the resistance of a body to stop doing whatever it is doing; property of an object by which it continues in its existing state of rest or motion in a straight line, unless that state is changed by an external force.

Insulation   A material which is non-conductive of either heat or electricity.

Integral   A necessary part to complete a whole unit.

Interchangeability   The ability to substitute one part for another.

Kerb weight   The weight of an empty vehicle without passengers and luggage.

Kevlar   A synthetic aramid fibre used as a reinforcement for resins. It is noted for its high impact resistance and is used in racing car bodywork.

Kinetic energy   The energy of motion.

Laminate   A material composed of a number of layers.

Lap joint   A form of joint obtained by overlapping the edges of two pieces of metal. The overlapping parts must be in the sample plane.

Latex   A natural rubber used for making flexible moulds. It is a liquid which solidifies in contact with air.

Lay-up   Layers of glass fibres are laid on top of wet resin and then pressed down into the liquid resin.

Leftward welding   This is known as forehand welding.

Levers   Levers are used to increase the force exerted by pivoting about a point.

Mass production   Large-scale, high-speed manufacture.

Metal fatigue   A metal structural failure, resulting from excessive or repeated stress, finally resulting in a crack.

MIG braze   A low-temperature form of materials joining.

MMA   Manual metal arc welding.

Molecule   A minute particle into which a substance can be divided and retain its properties.

Monomer   A simple molecule capable of combining with itself or a compatible similar chemical to form a chain (polymer).

Moulding   The resulting shape of a plastics material when it is removed from its mould.

Mould release   A substance used to coat the mould to prevent sticking of the resin that will be used to make a part. It facilitates the removal of that part from the mould.

Near side   The left-hand side of the vehicle as viewed from the driver's seat.

Neutral flame   A balanced flame, indicating perfect combustion of both oxygen and acetylene gases.

Non-ferrous metals   Metals which do not contain any ferrite or iron.

Normalizing   Heating to a high temperature to produce a refinement of the grain structure of a metal or alloy.

Off side   The right-hand side of a vehicle as viewed from the driver's seat.

Original finish   The paint applied at the factory by the vehicle manufacturer.

Oxidation Chemical reaction between oxygen and some other element resulting in oxides. Oxidizing flame A gas welding flame which has an excess of oxygen when burning. Paddle   A wooden tool shaped for spreading body solder.

Parent metal   The material of a part to be welded.

Pascal's law   Pascal discovered that the pressure in a body is equal in all directions.

Penetration   Depth of fusion or weld penetration.

Pickle   To soak metal in an acid solution in order to free the surface of rust or scale.

Pillar   A vertical support of a body frame.

Pillar face   The front of a pillar visible when the door is opened.

Pinch weld   Two metal flanges butted together and spot welded along the join of the flat surfaces.

Pod   A portable body repair shelter.

Polyurethane   A versatile material used for adhesives, paints, varnishes, resins and foam materials. These are often used in conjunction with polyester-based GRP.

Porosity   The presence of gas pockets or inclusions within a weld.

Power   Power is usually expressed in horsepower (HP), that is, the number of horses which would be needed to do the equivalent amount of work in the same time.

Pressure   The amount of force per unit area.

Prototype   An original model.

Puddle   The small body of molten metal created by the flame of a welding torch.

Quarter light   The window directly above the quarter panel.

Quarter panel   The side panel extending from the door to the rear end of the body (including rear wing).

Reinforcement   Filler material added to plastics (resin) in order to strengthen the finished product.

Relative density (RD)   Relative density is the relationship which other materials have with water.

Repair pods   Portable structures fitted out as small workshops.

Resin   Resins occur in nature as organic compounds, insoluble in water, e.g. amber, shellac. Synthetic resins have similar properties and are normally converted to solids by polymerization.

Return sweep   A reverse curve.

Saloon   An enclosed body not having a partition between the front and rear seats.

Scuttle panel   The panel between the bonnet and windscreen.

Self-tapping screw   A screw that cuts its own threads into a predrilled hole.

Silicone rubbers   Used, amongst other applications, for sealants and flexible mould compounds. They are usually cold curing.

Solvent   A chemical fluid which will dissolve, dilute or liquefy another material.

Specifications   Information provided by the manufacturer on vehicle data in the form of dimensions.

Squab   The rear seat-back construction.

Stream lining   The shaping of a vehicle body to minimize air resistance.

Subframe   Members to which the engine and front-end assembly are attached.

Swage   A raised form of moulding pressed into a piece of metal in order to stiffen it.

Swage line   A design line on a vehicle body, caused by a crease or step in a panel.

Sweating   Uniting two or more metal surfaces by the use of heat and soft solder.

Synthetic   A substance produced artificially.

Temperature   The measurement, in degrees, of the intensity of heat.

Template   A form or pattern made so that other parts can be formed to exactly the same shape.

Tensile strength   The resistance to breaking which metal offers when subject to a pulling stress.

Thermoplastic   Plastic which can be softened by heating and which still retains its properties after it has been cooled and hardened. Typical thermoplastics are polythene and PVC.

Thermosetting   Plastic which hardens by non-reversible chemical reaction, initiated by heat and/or curing agents. Once hardened, it cannot be melted down without being destroyed.

Thixotropic   Generally used to describe substances which have a very high viscosity when stable but low viscosity when stirred or brushed. 'Non-drip' paint is an obvious example; another is gelcoat resin.

Torque   The turning moment about a point expressed in newton-metres (Nm).

Tunnel   A raised floor panel section for driveshaft clearance.

Turret   American term for roof.

Weld bead   One single run of an electrode welding rod in manual metal arc welding.

Weld deposit   Metal which has been added to a joint by one of the welding processes.

Wheel alignment   The adjustment of a vehicle's wheels so that they are positioned to drive correctly.

Wheel arch Panel forming inner housing for rear wheels.
Wheelbase The distance between the centre lines of the front and rear wheels of a vehicle.

Abbreviations and symbols

A	ampere
ABS	acrylonitrile butadiene styrene
ABS	anti-lock braking system
AC	alternating current
AF	across flats (bolt size)
AFFF	aqueous film forming foam (fire-fighting)
bar	1 atmosphere, or 14.7 psi
BS	British Standard
BSI	British Standards Institution
C_d	aerodynamic drag coefficient
C	Celsius (Centigrade)
CAD	computer-aided design
CAE	computer-aided engineering
CAM	computer-aided manufacturing
CIM	computer-integrated manufacturing
cm	centimetre
cm^2	square centimetres
cm^3	cubic centimetres
CO	carbon monoxide
CO_2	carbon dioxide
COSHH	Control of Substances Hazardous to Health (Regulations)
dB(A)	noise level at the ear
DC	direct current dB decibel
deg.	or ° degree (angle or temperature)
dia.	diameter
DTI	dial test indicator
EC	European Community
ECU	electronic control unit
EPA	Environmental Protection Act
F	Fahrenheit
ft	foot
ft/min	feet per minute
g	gram (mass)
gal	gallon (Imperial)
GLS	general lighting service (lamp)
GRP	glass fibre–reinforced plastic
HASAWA	Health and Safety at Work Act
HSE	Health and Safety Executive
HT	high tension (electrical)
HVLP	high-velocity low-pressure (spray guns)
ID	internal diameter
IFS	independent front suspension
IMechE	Institution of Mechanical Engineers
IMI	Institute of the Motor Industry

in	inch
in²	square inches
IR	infrared
ISO	International Organization for Standardization
kg	kilogram (mass)
kW	kilowatt
l	litre
lb	pound (mass)
lbf	pound force
lbf/in²	pound force per square inch
lbft	pound force foot (torque)
$L_{ep,d}$	noise exposure level, personal (daily)
$L_{ep,w}$	noise exposure level, personal (weekly)
LH	left-hand
LHD	left-hand drive
LHThd	left-hand thread
LT	low tension
lumen	light energy radiated per second per unit solid angle by a uniform point source of 1 candela intensity
lux	unit of illumination equal to 1 lumen/m2
m	metre
MAG	metal active gas (welding)
max.	maximum
MIG	metal inert gas (welding)
min.	minimum
mm	millimetre
–	minus (of tolerance)
'	minute (of angle)
–	negative (electrical)
Nm	Newton metre
no.	number
OD	outside diameter
oz	ounce (mass)
ozf	ounce (force)
part	no. part number
%	percentage
PPE	personal protection equipment (regulations)
pt	pint (imperial)
+	plus (tolerance)
+	plus or minus
PVA	polyvinyl acetate
PVC	polyvinyl chloride
+	positive (electrical)
r	radius
ref.	reference
rev/min	revolutions per minute
RH	right-hand
RHD	right-hand drive
"	second (angle)

SAE	Society of Automotive Engineers
std	standard
TIG	tungsten inert gas (welding)
V	volt
VIN	vehicle identification number
VOCs	volatile organic compounds

Index

Page numbers in italic indicate figures. Page numbers in bold indicate tables.